The Best Test Prepa

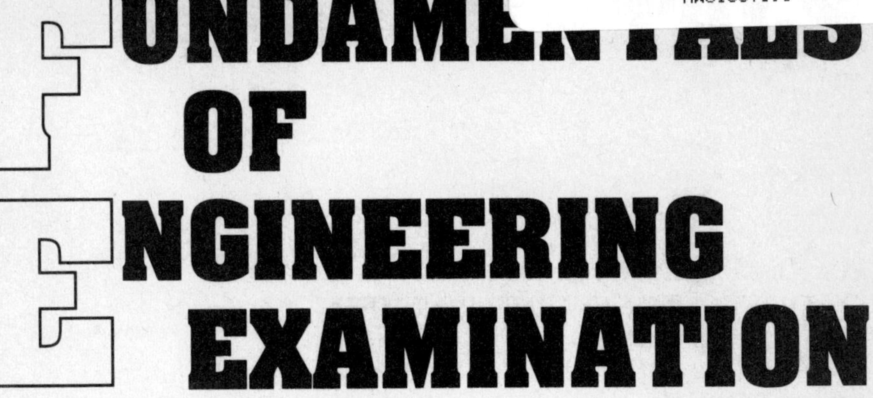

Nesar U. Ahmed, Ph.D.
Assistant Professor of Civil Engineering
Alabama Agricultural & Mechanical University, Normal, AL

Amir Al-Khafaji, Ph.D.
Chairperson & Professor of Civil Engineering
Bradley University, Peoria, IL

S. Balachandran, Ph.D.
Chairperson & Professor of Industrial Engineering
University of Wisconsin - Platteville, Platteville, WI

John M. Cimbala, Ph.D.
Assistant Professor of Mechanical Engineering
Pennsylvania State University, University Park, PA

Leroy Friel, Ph.D., P.E.
Professor of Engineering Science
Montana College of Mineral Science & Technology, Butte, MT

Victor Gerez, Ph.D., P.E.
Chairperson & Professor of Electrical Engineering
Montana State University, Bozeman, MT

Ted Huddleston, Ph.D., P.E.
Chairperson & Professor of Chemical Engineering
University of South Alabama, Mobile, AL

Raouf A. Ibrahim, Ph.D., P.E.
Professor of Mechanical Engineering
Wayne State University, Detroit, MI

Autar K. Kaw, Ph.D.
Assistant Professor of Mechanical Engineering
University of South Florida, Tampa, FL

Siripong Malasri, Ph.D., P.E.
Associate Professor of Civil Engineering
Christian Brothers University, Memphis, TN

Michael R. Muller, Ph.D.
Associate Professor of Mechanical & Aerospace Engineering
Rutgers University, Piscataway, NJ

Enuma Ozokwelu, Ph.D., P.E.
Advanced Research Chemical Engineer
Eastman Kodak Chemicals Company, Kingsport, TN

Yeshant K. Purandare, Ph.D.
Chairperson & Professor of Chemistry
State University of New York-College of Technology, Farmingdale, NY

Gautam Ray, Ph.D.
Associate Dean of the College of Engineering & Design
Florida International University, Miami, FL

Nasser-Eddine Rikli, Ph.D.
Professor of Electrical Engineering
Polytechnic University, Brooklyn, NY

Lt. Colonel Jerry W. Samples, Ph.D., P.E.
Associate Professor of Civil & Mechanical Engineering
United States Military Academy, West Point, NY

Larry Simonson, Ph.D.
Associate Professor of Electrical Engineering
South Dakota School of Mines & Technology, Rapid City, SD

A. Lamont Tyler, Ph.D.
Chairperson & Professor of Chemical Engineering
University of Utah, Salt Lake City, UT

Dev Venugopalan, Ph.D.
Associate Professor of Materials
University of Wisconsin - Milwaukee, Milwaukee, WI

Nikolay G. Zubatov, Ph.D.
Professor of Mechanical Engineering
Purdue University Calumet, Hammond, IN

RESEARCH & EDUCATION ASSOCIATION
61 Ethel Road West • Piscataway, New Jersey 08854

The Best Test Preparation for the
FUNDAMENTALS OF ENGINEERING (FE)
EXAMINATION

REVISED PRINTING, 1992

Printed in the United States of America

Library of Congress Catalog Card Number 90-61940

International Standard Book Number 0-87891-849-3

Research & Education Association
61 Ethel Road West
Piscataway, New Jersey 08854

REA supports the effort to conserve and
protect environmental resources by
printing on recycled papers.

CONTENTS

About Research and Education Association

REA is an organization of educators, scientists, and engineers specializing in various academic fields. REA was founded in 1959 for the purpose of disseminating the most recently developed scientific information to groups in industry, government, high schools and universities. Since then, REA has become a successful and highly respected publisher of study aids, test preparation guides, handbooks and reference works.

REA's Test Preparation series extensively prepares students and professionals for the Medical College Admission Test (MCAT), Graduate Record Examinations (GRE), Graduate Management Admission Test (GMAT), Fundamentals of Engineering Exam, Advanced Placement Exams, and College Board Achievement Tests. **Whereas most Test Preparation books present a limited amount of practice exams and bear little resemblance to the actual exams, REA's series presents numerous exams which accurately depict the actual exams in both degree of difficulty and types of questions. REA's exams are always based on the most recently administered tests and include every type of question that can be expected on the actual tests.**

REA's publications and educational materials are highly regarded for their significant contribution to the quest for excellence that characterizes today's educational goals. We continually receive an unprecedented amount of praise from professionals, instructors, librarians, parents and students for our published books. Our authors are as diverse as the subjects and fields represented in the books we publish. They are well-known in their respective fields and serve on the faculties of prestigious universities throughout the United States.

You Can Succeed on the FE Exam

By reviewing and studying this book, you can succeed on the Fundamentals of Engineering Examination. The FE is an eight-hour exam designed to test knowledge of a wide variety of engineering disciplines. The FE was formerly known as the EIT (Engineer-in-Training) exam. The FE exam format and title have now replaced the EIT exam completely.

The purpose of our book is to prepare you sufficiently for the exam by providing three full-length exams which accurately reflect the FE exam in both degree of difficulty and types of questions. The exams provided are based on the most recently administered FE exams and include every type of question that can be expected on the FE. In addition, a recent format change in the FE exam has made all previous test preparation books obsolete. Our book includes all recent format changes in the FE exam and, therefore, is the most accurate preparation book available for this exam. Following each exam is an answer key complete with detailed explanations and solutions designed to clarify the material to the student. Our objective is not only to provide the correct answers but also to explain the correct manner in which to solve the given problems. A detailed yet general solution is provided for each problem along with other helpful hints and strategies which may help in solving the problem faster. At the same time, use of this book will help you review the material most likely to be encountered on the actual FE exam. By completing all three exams and studying the explanations which follow, you can discover your strengths and weaknesses and thereby concentrate on the sections of the exam which you find more difficult.

About the Test Experts

In order to meet our objectives of providing exams that reflect the FE in both accuracy and degree of difficulty, every exam section was carefully prepared by test experts in the various subject fields in engineering. Our authors have spent quality time examining and researching the mechanics of the actual FE exams to see what types of practice questions will accurately depict the exam and challenge the student. Our experts are highly regarded in the educational community, having studied at the doctoral level and taught in their respective fields at competitive universities and colleges throughout the United States. They have an in-depth knowledge of the questions they have presented in the book and provide accurate questions which appeal to the student's interest. Each question is clearly explained in order to help the student achieve a top score on the FE.

About the Test

The Fundamentals of Engineering Examination (FE) is one part in the four-step process toward becoming a Professional Engineer (P.E.). Graduating from an approved four-year engineering program and passing the FE qualifies you for certification as an "Engineer-in-Training" or an "Engineer Intern." The final two steps toward licensure as a P.E. involve completion of four years of additional engineering experience and passage of the Principles and Practices of Engineering Examination administered by the National Council of Examiners for Engineering and Surveying (NCEES). Registration as a Professional Engineer is deemed both highly rewarding and beneficial in the engineering community.

The FE exam is an eight-hour, multiple-choice, open book test of competency in basic engineering subjects included in accredited engineering programs. The FE is compiled and administered by the NCEES and is offered in the Spring and Fall of every year. Contact your school's engineering department for information on applying to take the test.

Test Format

The FE consists of two distinct sections. One section is given in the morning (AM) while the other is administered in the afternoon (PM). You will have four hours to complete each section. The AM section contains 140 questions covering eleven different engineering subjects. The PM section contains 70 questions arranged in problem set form. A problem set consists of one diagram or general problem statement followed by ten questions referring to that original diagram or statement. The PM section will cover five different engineering subjects. In previous FE exams, you were able to choose five problem sets out of several to answer. This is not the case anymore, as the new format requires you to answer all questions. The subjects and their corresponding numbers of questions for the AM and PM sections are shown below.

AM SECTION

Subject	No. of Problems
Mathematics	20
Electrical Circuits	14
Fluid Mechanics	14
Thermodynamics	14
Dynamics	14
Statics	14
Chemistry	14
Mechanics of Materials	11
Engineering Economics	11
Materials Science and Structure of Matter	14
TOTAL:	140

PM SECTION

Subject	No. of Problems
Engineering Mechanics	20 (2 problem sets)
Applied Mathematics	20 (2 problem sets)
Electrical Circuits	10 (1 problem set)
Engineering Economics	10 (1 problem set)
Thermodynamics/Fluid Mechanics	10 (1 problem set)
TOTAL:	70 (7 problem sets)

It is important to note that the subjects covered in the PM section are not as rigidly specified as they are in the AM section. For example, the Engineering Mechanics part of the PM section may consist of any combination of two problem sets from the following subjects: Statics, Dynamics, Strength of Materials, and Mechanics of Materials. Similarly the Electrical Circuits part may be composed of one problem set dealing with electrical circuits, electronics, or electrical machinery. In addition, there will be one problem set in either Fluid Mechanics or Thermodynamics. The two problem sets in Applied Mathematics and one in Engineering Economics will definitely be on the exam. Remember that all parts of the exam are required; that is, there are no optional subjects or topics, as was the case on the previous EIT exams.

Additional Practice Questions

Following the test sections in this book will be several additional practice problem sets with answer explanations. They are not part of the actual practice tests. These extra problems will serve as study aids to further your practice on the FE. You can find these problems starting on page 489.

Scoring the Exam

Your FE score is based upon the number of correct answers you choose. No points are taken off for incorrect answers. Both the AM and PM sections have an equal weight. A single score from 0 to 100 is given for the whole test. The grade is given on a pass/fail basis. The point between pass and fail varies from state to state, although 70 is a general reference point for passing.

The pass/fail margin is not a percentage of correct answers, nor a percentage of students who scored lower than you. This number fluctuates from year to year and is reestablished with every test administration. It is based on previous exam administrations and relates your score to those of previous FE examinees.

Because this grading system is so variable, there is no real way for you to know exactly what you got on the test. For the purpose of grading the tests in this book, however, REA has provided the following formula to calculate your score on the FE practice tests:

$$\left[\frac{(\text{\# of q's answered correctly in AM}) + (\text{\# of q's answered correctly in PM} \times 2)}{280}\right] \times 100 = \text{your score.}$$

NOTE: Use the total number of correct answers from the AM and two times the PM sections.

Remember that this formula is meant for the computation of your grade for the tests **in this book**. It does **not** compute your grade for the actual FE.

FE Test-taking Strategies

HOW TO BEAT THE CLOCK

Every second counts, and you want to use the available test time for each section in the most efficient manner. Here's how:

1. Bring a watch! There is no guarantee that there will be a clock in the room where you will be taking the test, or, if there is one, that it will work.

2. Memorize the test directions for each section of the test. You do not want to waste valuable time reading directions on the day of the exam. Your time should be spent answering the questions.

3. Pace yourself. Work steadily and quickly. Do not get stuck or spend too much time on any one question. Some questions will be longer or tougher. Remember you can always return to any problems with which you are having difficulties. Try to answer the easiest questions first, and then return to the ones you skipped.

4. Since this is an open book examination, no standardized tables, charts, etc. will be given in the test book. In other words, you will have to find them yourself in the books you bring to the exam. Therefore, it's a good idea to learn where all the major or frequently used tables and charts are located in your engineering textbooks. It may prove extremely rewarding if you indexed any important items in a particular book on the inside cover of that book. Some examples of important items and their subjects are:

   Thermodynamics: steam tables, ideal gas tables

   Fluid Mechanics: Moody Chart

   Engineering Economics: interest factor tables

   Mathematics: integral, derivative, trigonometric tables, Laplace transforms, probability tables

   Statics: fundamental centroids and moments of inertia

   Chemistry: Periodic Table

   General Information: abbreviations, conversion factors

5. Related to the previous suggestion, although you may bring a truckload of books into the test with you, it would probably only hurt your performance on the test. Bring only the books you feel you definitely need. The NCEES allows you to use any textbook, handbook, bound reference materials, and any battery-operated, silent, non-printing calculator or slide rule. But remember, you do not want to bring so much that you confuse yourself and get lost in all that information. You also do not want to waste time looking for a certain table when you should be answering questions on the test. It may be a good idea to invest in a Mechanical Engineering handbook because most of the charts and tables you will need for this exam can be found in this one book. Of course you should still bring other texts since they may show you how to solve an unfamiliar problem. In addition, REA publishes numerous Problem Solver books in several engineering fields which would be excellent study guides as you prepare for the FE.

6. You cannot bring any unbound notes or tables to the exam. Also, check with your state board of engineering to see if your state allows notebooks into the exam room. This rule varies from state to state.

GUESSING STRATEGY

1. When all else fails, guess! The score you achieve depends on the number of correct answers you choose. There is no additional penalty for wrong answers, and it is in your best interest to answer all 210 questions. Therefore, it is favorable to guess on any question that would otherwise remain unanswered.

2. Since the FE exam consists of several different engineering subjects, it is possible you will feel more comfortable with certain parts of the exam than with others. It may be beneficial for you to start with the subject areas you know best. Then go back to the ones you will need more time to contemplate and figure out. Accordingly, this strategy may work within individual subjects. Answer questions you can solve easily and then move on to the more difficult problems.

3. Go about solving each problem in a structured and orderly manner. Use diagrams or drawings whenever possible and be sure to use the correct units throughout the problem. Do not wait to assign units when you finally get an answer. This will allow you to continually return to a problem and know what direction you were heading in the last time you attempted to solve it.

OTHER MUST-DO STRATEGIES

1. As you work on the test, make sure your answers correspond with the numbers and letters on the answer sheet.
2. You will do better if you can eliminate some of the possible answers and reduce your chances of making a wrong choice. It is a good idea to work as much as possible on any unanswered questions and resort to guessing only in the last few remaining minutes of the examination.
3. As mentioned above, you may be able to eliminate some choices immediately. However, resort to this only if you do not know how to solve the problem immediately. Otherwise, go ahead and solve the problem. In some cases, four of the possible answers may be eliminated, leaving the correct answer. Here are some examples of what to look for when eliminating answer choices:

   Thermodynamics: check for signs of heat transfer and work

   Fluid Mechanics: check for signs of pressure readings

   Statics: look for direction of forces and compression/tension units
3. Feel free to write in this book as you are allowed to write in the test book at the actual exam.
4. If you finish early, do not leave the examination or take a nap. Use this extra time to review your test and check your answers. Look for simple mathematical errors or read each question over to make sure you found what the question was asking.

The Test Sections

As discussed earlier, the FE exam is divided into two sections, morning (AM) and afternoon (PM). These two parts are then broken down further into individual engineering topics.

The actual FE exam booklet you receive when you take the test will include a page detailing the breakdown of topics along with their corresponding question numbers. See the **Test Format** section on page viii for a typical breakdown.

The individual topics in the AM section of the exam appear as a distinct set of questions. In other words, a certain number of questions concerning a certain topic will appear one after the other. However, these topics will not be labelled in the test, and they are not broken up into distinct sections. In the PM section of the exam, problem sets of ten questions each dealing with one topic will appear one after the other. Each set will be labelled according to its topic. In some instances, a problem set may consist of only five questions. This is rare, but not impossible.

Overall, there will be eleven different topics on the FE exam (in both the AM and PM sections). The following is a comprehensive summary of the topics included on the exam, along with concepts which may have questions relating to them under each topic. For example, if the topic is Statics then the possible concepts covered by the questions are: vector algebra, 2-D equilibrium, 3-D equilibrium, concurrent force systems, centroid of area, moment of inertia, and friction.

CHEMISTRY

(AM section: 14 questions PM section: none)

The questions in the AM section of the FE exam dealing with chemistry may test your knowledge of the following concepts:

chemical reactions	balancing equations
Law of Definite Proportions	equivalent weight
gram mole	Avogadro's Law
freezing point depression	valences
chemical formulae	nomenclature of compounds

- reaction rate
- heat of reaction
- solubility product
- ionization
- electrolysis
- gas phase equilibrium
- isotopes
- solutions
- acids, metals, bases
- hydrocarbons
- Periodic Table
- atomic structure
- molecular structure
- crystal structure
- bonding
- radioactivity
- gas laws

Note that there will be no questions dealing with chemistry on the PM section of the exam.

DYNAMICS

(AM section: 14 questions PM section: possibly one problem set)

The AM section of the FE exam will definitely contain fourteen questions on the topic of Dynamics. The questions may cover any of the following concepts: kinematics; force, mass, and acceleration; impulse and momentum; work and energy; response of first and second order systems.

The PM section of the exam may or may not contain a problem set dealing with Dynamics. If a problem set on Dynamics is included in the test it could touch on such concepts as: relative or curvilinear motion; inertia and force; particles and solid bodies; impulse and momentum; and work and energy. Keep in mind that a problem set consists of ten questions relating to one diagram or figure. However, several difficult concepts may be tested within the problem set.

ELECTRICAL CIRCUITS

(AM section: 14 questions PM section: one problem set)

The AM section of the FE exam will contain fourteen questions which test your knowledge of Electrical Circuits and related topics. The questions may address the following subjects: AC and DC circuits; three-phase circuits; transients; electrical fields; magnetic fields; and electronics.

The PM section of the exam will definitely have one problem set dealing with electrical circuits, electronics, or electrical machinery. The given diagram or problem statement, along with the corresponding questions, may test your knowledge of: transient and steady state circuits, electronics, power and machines, and electrical and magnetic fields. Note the additional topics included in the PM section which do not appear in the AM section.

ENGINEERING ECONOMICS

(AM section: 11 questions   PM section: one problem set)

The AM section of the FE exam will contain eleven questions covering the topic of Engineering Economics. The questions may test your knowledge of any of the following subjects: time value of money, annual cost, present and future worth, capitalized cost, break-even analysis, valuation, and depreciation.

The PM section of the FE exam will definitely include one problem set concerning Engineering Economics. In addition to the concepts covered in the AM questions, the problem set may also touch upon percent return.

FLUID MECHANICS

(AM section: 14 questions   PM section: possibly one problem set)

Fourteen questions in the AM part of the FE exam will test your knowledge of Fluid Mechanics. The questions may deal with any of the following concepts: fluid properties; fluid statics; impulse and momentum; dimensional analysis; open channel flow; pipe flow; flow measurement; similitude; and compressible flow.

The PM section of the FE exam may or may not contain a problem set covering Fluid Mechanics. If a problem set dealing with Fluid Mechanics is part of the PM section of the exam, the questions may concern: fluid statics; conservation of mass, momentum, and energy; fluid friction; and compressible flow. Keep in mind that if a problem set dealing with Fluid Mechanics appears on the exam, there will **not** be a problem set on Thermodynamics.

MATERIALS SCIENCE/STRUCTURE OF MATTER

(AM section: 14 questions   PM section: possibly one problem set)

The topics of Materials Science and Structure of Matter will be combined on the AM section of the FE exam. In other words, the exam will contain a total of fourteen questions which test a combination of knowledge of Materials Science and Structure of Matter. Concepts which may be included from the subject of Materials Science are: physical properties; chemical properties; atomic bonding; crystallography; phase diagrams; processing tests and properties; diffusion; and corrosion. In addition, some of the questions will test your knowledge of Structure of Matter. Possible subject matter for this topic includes nuclei, atoms, and wave phenomena. It is important to remember that there will be only fourteen questions split almost evenly between the two topics. There will **not** be twenty-eight questions on the two.

The PM section of the exam may or may not include a problem set of ten questions dealing with Materials Science **only.** Questions on Structure of Matter will **not** be included in the afternoon session of the exam. If a Materials Science problem set appears on the exam, it will consist of questions testing any of the following concepts: crystallography, phase diagrams, processing tests, and properties.

MATHEMATICS

(AM section: 20 questions   PM section: two problem sets)

The AM section of the FE exam will include twenty questions which test your knowledge in the following mathematical areas: integral and differential calculus; linear algebra; probability and statistics; differential equations; and analytical geometry. Remember that these are very broad subject areas and any preparation should **not** be limited to only basic concepts.

The PM section of the exam will definitely contain two problem sets, each with ten questions, which may cover any of the topics mentioned above for the AM section.

MECHANICS OF MATERIALS

(AM section: 11 questions PM section: possibly one problem set)

Eleven questions in the AM section of the exam will concentrate on the topic of Mechanics of Materials. The questions may test your knowledge of any of the following concepts: stress and strain; tension and compression; shear; combined stress; beams; columns; and composite sections.

It is possible that a problem set of ten questions relating to Mechanics of Materials will appear in the PM section of the FE. If so, the questions may touch on the following areas: uniaxial loading, torsion, bending, and combined loading.

STATICS

(AM section: 14 questions PM section: possibly one problem set)

Fourteen questions in the AM section of the FE exam will focus on Statics. The questions may deal with any of the following concepts: vector algebra, 2-dimensional equilibrium, concurrent force systems, centroid area, moment of inertia, and friction.

As with Dynamics, Mechanics of Materials, and Materials Science, a problem set involving Statics may appear on the PM section of the exam. Keep in mind that any combination of two of these four topics will be put on the afternoon section of the exam. If statics is one of the two topics, the questions in the problem set may cover the following subjects: resultant force systems, equilibrium of rigid bodies, friction effects, and analysis of internal forces.

THERMODYNAMICS

(AM section: 14 questions PM section: possibly one problem set)

The morning session of the FE exam will contain fourteen questions on Thermodynamics. The questions may test your knowledge of the following topics:

properties
phase change
heat
adiabatic processes
availability
gas flow

work

ideal gases

First Law of Thermodynamics

open flow and open systems

throttling

Second Law of Thermodynamics

simple heat transfer

Carnot cycles

Otto cycles

Diesel cycles

Brayton cycles

Rankine cycles

refrigeration cycles

As mentioned before, either a Fluid Mechanics problem set or a Thermodynamics problem set will appear in the PM section of the exam. If a Thermodynamics problem set is used, it may include:

application of basic laws

chemical reactions

thermal cycles

heat transfer

First and Second Law applications

gas power cycles

vapor power cycles

refrigeration

air conditioning

combustion

gas mixtures

nozzles, turbines, and compressors

Fundamentals of Engineering

A.M. SECTION

Test 1

Fundamentals of Engineering

A.M. SECTION

TEST 1 – ANSWER SHEET

1. (A) (B) (C) (D) (E)	26. (A) (B) (C) (D) (E)	51. (A) (B) (C) (D) (E)
2. (A) (B) (C) (D) (E)	27. (A) (B) (C) (D) (E)	52. (A) (B) (C) (D) (E)
3. (A) (B) (C) (D) (E)	28. (A) (B) (C) (D) (E)	53. (A) (B) (C) (D) (E)
4. (A) (B) (C) (D) (E)	29. (A) (B) (C) (D) (E)	54. (A) (B) (C) (D) (E)
5. (A) (B) (C) (D) (E)	30. (A) (B) (C) (D) (E)	55. (A) (B) (C) (D) (E)
6. (A) (B) (C) (D) (E)	31. (A) (B) (C) (D) (E)	56. (A) (B) (C) (D) (E)
7. (A) (B) (C) (D) (E)	32. (A) (B) (C) (D) (E)	57. (A) (B) (C) (D) (E)
8. (A) (B) (C) (D) (E)	33. (A) (B) (C) (D) (E)	58. (A) (B) (C) (D) (E)
9. (A) (B) (C) (D) (E)	34. (A) (B) (C) (D) (E)	59. (A) (B) (C) (D) (E)
10. (A) (B) (C) (D) (E)	35. (A) (B) (C) (D) (E)	60. (A) (B) (C) (D) (E)
11. (A) (B) (C) (D) (E)	36. (A) (B) (C) (D) (E)	61. (A) (B) (C) (D) (E)
12. (A) (B) (C) (D) (E)	37. (A) (B) (C) (D) (E)	62. (A) (B) (C) (D) (E)
13. (A) (B) (C) (D) (E)	38. (A) (B) (C) (D) (E)	63. (A) (B) (C) (D) (E)
14. (A) (B) (C) (D) (E)	39. (A) (B) (C) (D) (E)	64. (A) (B) (C) (D) (E)
15. (A) (B) (C) (D) (E)	40. (A) (B) (C) (D) (E)	65. (A) (B) (C) (D) (E)
16. (A) (B) (C) (D) (E)	41. (A) (B) (C) (D) (E)	66. (A) (B) (C) (D) (E)
17. (A) (B) (C) (D) (E)	42. (A) (B) (C) (D) (E)	67. (A) (B) (C) (D) (E)
18. (A) (B) (C) (D) (E)	43. (A) (B) (C) (D) (E)	68. (A) (B) (C) (D) (E)
19. (A) (B) (C) (D) (E)	44. (A) (B) (C) (D) (E)	69. (A) (B) (C) (D) (E)
20. (A) (B) (C) (D) (E)	45. (A) (B) (C) (D) (E)	70. (A) (B) (C) (D) (E)
21. (A) (B) (C) (D) (E)	46. (A) (B) (C) (D) (E)	71. (A) (B) (C) (D) (E)
22. (A) (B) (C) (D) (E)	47. (A) (B) (C) (D) (E)	72. (A) (B) (C) (D) (E)
23. (A) (B) (C) (D) (E)	48. (A) (B) (C) (D) (E)	73. (A) (B) (C) (D) (E)
24. (A) (B) (C) (D) (E)	49. (A) (B) (C) (D) (E)	74. (A) (B) (C) (D) (E)
25. (A) (B) (C) (D) (E)	50. (A) (B) (C) (D) (E)	75. (A) (B) (C) (D) (E)

76. Ⓐ Ⓑ Ⓒ Ⓓ Ⓔ
77. Ⓐ Ⓑ Ⓒ Ⓓ Ⓔ
78. Ⓐ Ⓑ Ⓒ Ⓓ Ⓔ
79. Ⓐ Ⓑ Ⓒ Ⓓ Ⓔ
80. Ⓐ Ⓑ Ⓒ Ⓓ Ⓔ
81. Ⓐ Ⓑ Ⓒ Ⓓ Ⓔ
82. Ⓐ Ⓑ Ⓒ Ⓓ Ⓔ
83. Ⓐ Ⓑ Ⓒ Ⓓ Ⓔ
84. Ⓐ Ⓑ Ⓒ Ⓓ Ⓔ
85. Ⓐ Ⓑ Ⓒ Ⓓ Ⓔ
86. Ⓐ Ⓑ Ⓒ Ⓓ Ⓔ
87. Ⓐ Ⓑ Ⓒ Ⓓ Ⓔ
88. Ⓐ Ⓑ Ⓒ Ⓓ Ⓔ
89. Ⓐ Ⓑ Ⓒ Ⓓ Ⓔ
90. Ⓐ Ⓑ Ⓒ Ⓓ Ⓔ
91. Ⓐ Ⓑ Ⓒ Ⓓ Ⓔ
92. Ⓐ Ⓑ Ⓒ Ⓓ Ⓔ
93. Ⓐ Ⓑ Ⓒ Ⓓ Ⓔ
94. Ⓐ Ⓑ Ⓒ Ⓓ Ⓔ
95. Ⓐ Ⓑ Ⓒ Ⓓ Ⓔ
96. Ⓐ Ⓑ Ⓒ Ⓓ Ⓔ
97. Ⓐ Ⓑ Ⓒ Ⓓ Ⓔ
98. Ⓐ Ⓑ Ⓒ Ⓓ Ⓔ
99. Ⓐ Ⓑ Ⓒ Ⓓ Ⓔ
100. Ⓐ Ⓑ Ⓒ Ⓓ Ⓔ
101. Ⓐ Ⓑ Ⓒ Ⓓ Ⓔ
102. Ⓐ Ⓑ Ⓒ Ⓓ Ⓔ
103. Ⓐ Ⓑ Ⓒ Ⓓ Ⓔ
104. Ⓐ Ⓑ Ⓒ Ⓓ Ⓔ
105. Ⓐ Ⓑ Ⓒ Ⓓ Ⓔ
106. Ⓐ Ⓑ Ⓒ Ⓓ Ⓔ
107. Ⓐ Ⓑ Ⓒ Ⓓ Ⓔ
108. Ⓐ Ⓑ Ⓒ Ⓓ Ⓔ
109. Ⓐ Ⓑ Ⓒ Ⓓ Ⓔ
110. Ⓐ Ⓑ Ⓒ Ⓓ Ⓔ
111. Ⓐ Ⓑ Ⓒ Ⓓ Ⓔ
112. Ⓐ Ⓑ Ⓒ Ⓓ Ⓔ
113. Ⓐ Ⓑ Ⓒ Ⓓ Ⓔ
114. Ⓐ Ⓑ Ⓒ Ⓓ Ⓔ
115. Ⓐ Ⓑ Ⓒ Ⓓ Ⓔ
116. Ⓐ Ⓑ Ⓒ Ⓓ Ⓔ
117. Ⓐ Ⓑ Ⓒ Ⓓ Ⓔ
118. Ⓐ Ⓑ Ⓒ Ⓓ Ⓔ
119. Ⓐ Ⓑ Ⓒ Ⓓ Ⓔ
120. Ⓐ Ⓑ Ⓒ Ⓓ Ⓔ
121. Ⓐ Ⓑ Ⓒ Ⓓ Ⓔ
122. Ⓐ Ⓑ Ⓒ Ⓓ Ⓔ
123. Ⓐ Ⓑ Ⓒ Ⓓ Ⓔ
124. Ⓐ Ⓑ Ⓒ Ⓓ Ⓔ
125. Ⓐ Ⓑ Ⓒ Ⓓ Ⓔ
126. Ⓐ Ⓑ Ⓒ Ⓓ Ⓔ
127. Ⓐ Ⓑ Ⓒ Ⓓ Ⓔ
128. Ⓐ Ⓑ Ⓒ Ⓓ Ⓔ
129. Ⓐ Ⓑ Ⓒ Ⓓ Ⓔ
130. Ⓐ Ⓑ Ⓒ Ⓓ Ⓔ
131. Ⓐ Ⓑ Ⓒ Ⓓ Ⓔ
132. Ⓐ Ⓑ Ⓒ Ⓓ Ⓔ
133. Ⓐ Ⓑ Ⓒ Ⓓ Ⓔ
134. Ⓐ Ⓑ Ⓒ Ⓓ Ⓔ
135. Ⓐ Ⓑ Ⓒ Ⓓ Ⓔ
136. Ⓐ Ⓑ Ⓒ Ⓓ Ⓔ
137. Ⓐ Ⓑ Ⓒ Ⓓ Ⓔ
138. Ⓐ Ⓑ Ⓒ Ⓓ Ⓔ
139. Ⓐ Ⓑ Ⓒ Ⓓ Ⓔ
140. Ⓐ Ⓑ Ⓒ Ⓓ Ⓔ

FUNDAMENTALS OF ENGINEERING EXAMINATION

TEST 1

MORNING (AM) SECTION

TIME: 4 Hours
140 Questions

DIRECTIONS: For each of the following questions and incomplete statements, choose the best answer from the five answer choices.

1. Given the following set of simultaneous linear equations

$$4x + 6y = 10$$

$$2x + 3y = 7$$

the solution is

(A) $x = 1, y = 1.$
(B) $x = 2, y = 1.$
(C) number is infinite.
(D) does not exist.
(E) $x = -2, y = 1$

2. The velocity of two bodies is given by:

Body A: $V_A = 2t^2 - 12, t > 0$

Body B: $V_B = -t, t > 0$

The velocity of Body A is twice that of Body B for time $t > 0$ at $t =$

(A) -3
(B) 2
(C) $(-1 + \sqrt{97})/2$
(D) $-1 + \sqrt{97}$
(E) 3

3. The solution to the differential equation

$$dy/dt + 2y = 1, y(0) = 1, \text{ is}$$

(A) $^1/_2 + {}^1/_2\, e^{2t}$

(B) $-{}^1/_2 + {}^1/_2\, e^{2t}$

(C) $^1/_2 + {}^1/_2\, e^{-2t}$

(D) $^1/_2 + e^{-2t}$

(E) $^1/_2 + e^{2t}$

4. The sample standard deviation of the following data 2, 8, 3, 10 is close to

(A) 3.345

(B) 3.862

(C) 5.750

(D) 44.75

(E) 0

5. The inner surface area in square meters of a closed (both ends) cylinder 2 meters in height and 3 meters in diameter is

(A) 21/2 π

(B) 33/4 π

(C) 6 π

(D) 30 π

(E) 21 π

6. Given the following three ordinary differential equations, where x is the dependent variable and t is the independent variable

I. $2\, d^2x/dt^2 + 2x\, dx/dt = 0$

II. $2\, d^2x/dt^2 + 2x\, dx/dt + t^2 = e^{-t}$

III. $2\, d^2x/dt^2 + 2t\;\; dx/dt = 0$

(A) I and II are linear.

(B) I and III are linear.

(C) only III is linear.

(D) II and III are linear.

(E) none of the above.

7. Given the function

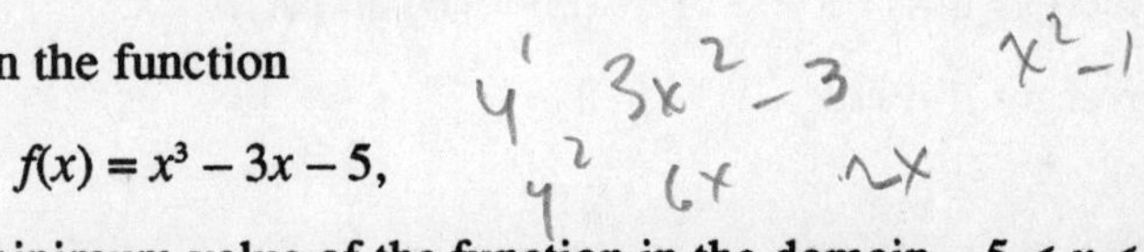

$$f(x) = x^3 - 3x - 5,$$

the minimum value of the function in the domain $-5 < x < 5$ is

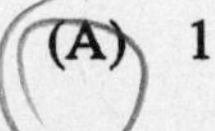

(A) 1

(D) -3

(B) -1 (E) 0

(C) -7

8. In an unbiased coin toss, the probability of getting heads or tails is exactly $^1/_2$. A coin is tossed and one gets heads. If the coin is tossed again, the probability of getting heads is

(A) $^1/_3$ (D) 0

(B) $^1/_4$ (E) $^1/_2$

(C) 1

9. Given two square matrices [A] and [B} of the same order, if [C] = [A] + [B], then

(A) det [C] = det [A] + det [B]

(B) det [C] = det [A] – det [B]

(C) det [C] = det [A] det [B]

(D) det [C] = 0

(E) none of the above

10. The shortest distance between two points (0, 3) and (5, – 6) on an x-y graph is

(A) 34 (D) $\sqrt{34}$

(B) 106 (E) 14

(C) $\sqrt{106}$

11. The series $\sum_{k=1}^{\infty} [(-1)^n]/k$

(A) diverges if $n = 2$ (D) (A) and (B)

(B) converges if $n = 1$ (E) (B) and (C)

(C) converges if $n = k$

12. The inflection point of the function

$f(x) = 3x^{5/3}$ is

(A) 0 (D) -1

(B) nonexistent (E) $^4/_3$

(C) ∞

13. If $f(x)$ is a real function then the equation $f(x) = 0$ has at least one real root in the interval $[a, b]$ if

(A) $f(a)\,f(b) < 0$

(B) $f(a)\,f(b) > 0$

(C) $f(a)\,f(b) < 0$ and $f(x)$ is continuous

(D) $f(a)\,f(b) > 0$ and $f(x)$ is piecewise continuous

(E) $f(a)\,f(b) > 0$ and $f(x)$ is continuous

14. $\lim\limits_{x \to 0} \dfrac{\sin(5x) - 2x}{x}$ is

(A) not defined (D) 1

(B) ∞ (E) 0

(C) 3

15. The acute angle between the two straight lines

$y = 3x + 2$

$y = 4x + 7$

is close to

(A) 0 (D) 28.30°

(B) 90° (E) 5.194°

(C) 4.399°

16. A bag contains 4 red balls, 3 green balls and 5 blue balls. The probability of not getting a red ball in the first draw is

(A) $^1/_3$ (D) 1

(B) $^2/_3$ (E) 2

(C) 0

17. The integral

$$\iint_R xy\,dA,$$

where the region R is $0 \le x \le 1, 1 \le y \le 2$, is given by

(A) $^3/_4$

(B) $^1/_4$

(C) 3

(D) 0

(E) $-^3/_4$

18. The equation of a circle on the x-y plane is given by

$$x^2 - 2x + y^2 + 2y = 0.$$

The center and the radius of the circle are respectively

(A) $(1,-1), \sqrt{2}$

(B) $(-1, 1), 2$

(C) $(1, -1), 2$

(D) $(-1, 1), \sqrt{2}$

(E) $(0,0), \sqrt{2}$

19. The intercept of a straight line on the y-axis is -3. If $(5, 2)$ is a point on the straight line, the slope of the straight line is

(A) -3

(B) 1

(C) 5

(D) 0

(E) 0.2

20. The velocity of a body is given by $v(t) = \sin(\pi t)$, where the velocity is given in meters/seconds and t is given in seconds. The distance covered in meters between $t = {}^1/_4$ and $^1/_2$ seconds is close to

(A) -0.2251

(B) 0.2251

(C) 8.971×10^{-5}

(D) 1.2251

(E) 1

21. Which of the following is true for the electric resistance of a cable:

(A) It is proportional to the diameter of the cable.

(B) It is proportional to the square root of the diameter of the cable.

(C) It is inversely proportional to the diameter of the cable.

(D) It is inversely proportional to the square of the diameter of the cable.

(E) It is proportional to the square of the diameter of the cable.

22. In the circuit shown here, the voltage across the 3 Ω resistance is given by:

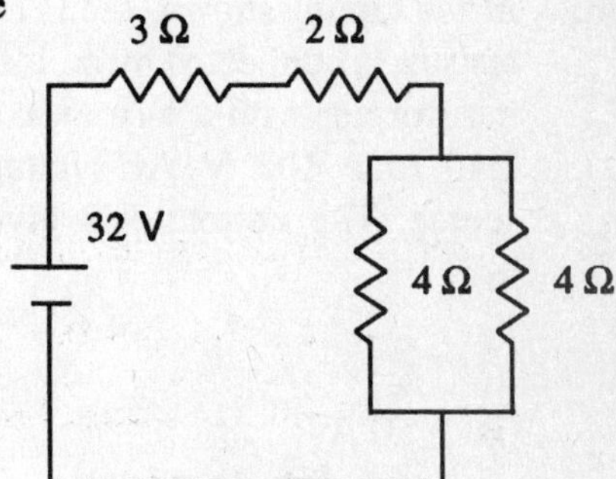

(A) 6.4 V

(B) 13.7 V

(C) 7.4 V

(D) 10.8 V

(E) 5.9 V

23. For the circuit shown here, the resistance between terminals *ab* is equal to:

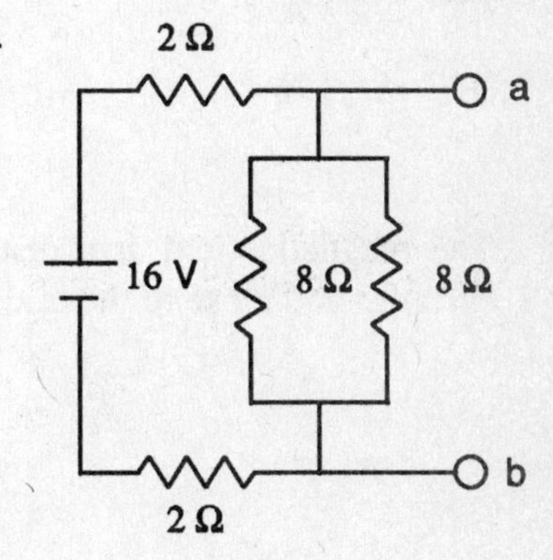

(A) 20 Ω

(B) 8 Ω

(C) 4 Ω

(D) 2 Ω

(E) 10 Ω

24. For the circuit shown, the Thevenin equivalent voltage between terminals *ab* is

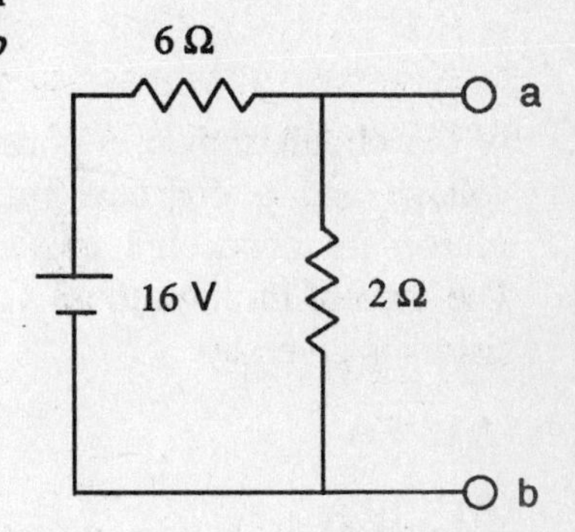

(A) 12 V

(B) 16 V

(C) 4 V

(D) 2 V

(E) 10 V

25. For the circuit shown below, the current through the 2 Ω resistor is:

(A) 3.2 A

(B) 6.4 A

(C) 4.0 A

(D) 8.0 A

(E) 2.0 A

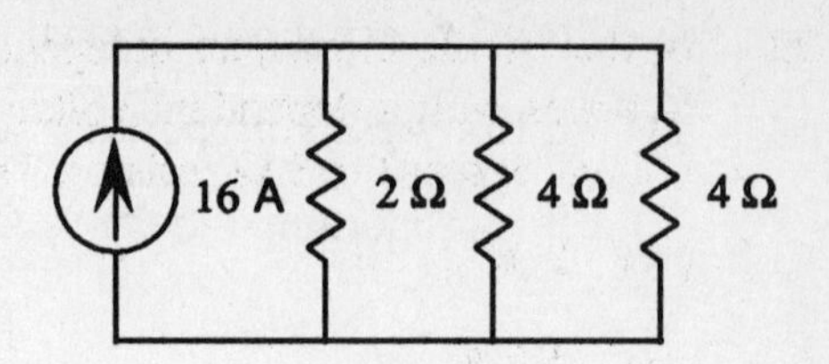

26. In the circuit shown, the 0.1 Ω resistor is connected by an ideal transformer with a turn ratio of 1:10 to a 100 V AC voltage source. The current I is given by

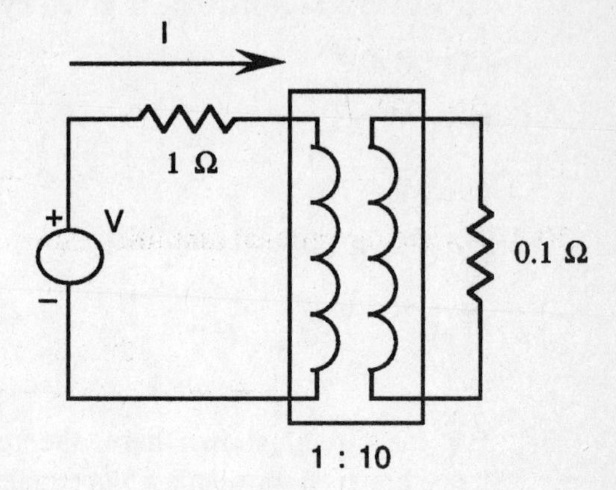

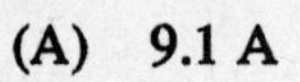

(A) 9.1 A

(B) 50 A

(C) 99.01 A

(D) 9.33 A

(E) 51.3 A

27. The magnetic field at a distance r from a long straight wire carrying a current i and close to the middle of the wire is proportional to:

(A) $i^2 r$

(B) $\frac{i}{r}$

(C) $\frac{i}{r^2}$

(D) $\frac{r}{i}$

(E) $\frac{i}{r}$

28. In the circuit shown, a constant voltage and a constant current source are connected in series. The current through the 3 Ω resistors is given by:

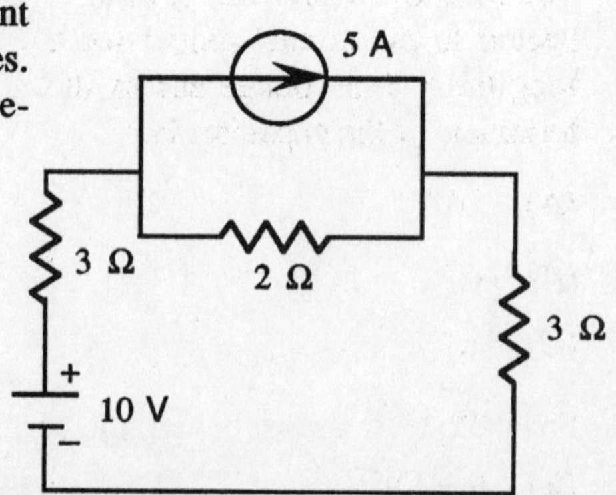

(A) 5 A

(B) 10 A

(C) 1.25 A

(D) 2.5 A

(E) 6.66 A

29. In the circuit shown, an independent voltage source of 24 V is connected in series with a dependent voltage source of value 0.5 V_R, where V_R is the voltage across the 3 Ω resistor. The current i is given by:

(A) 2.29 A

(B) 1.6 A

(C) 1.78 A

(D) 1.45 A

(E) 0.8 A

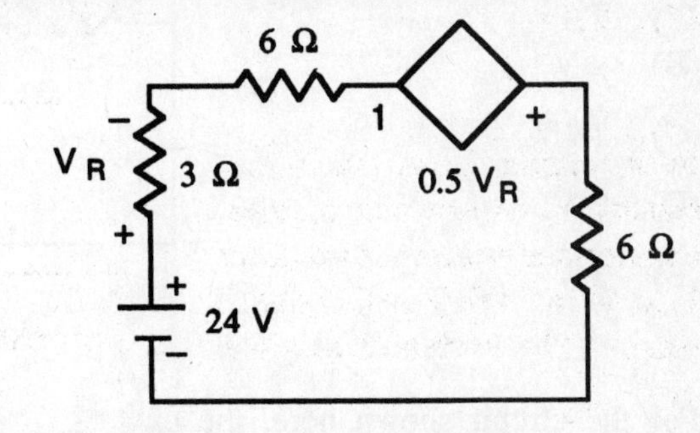

30. For the operational amplifier shown, the output voltage is

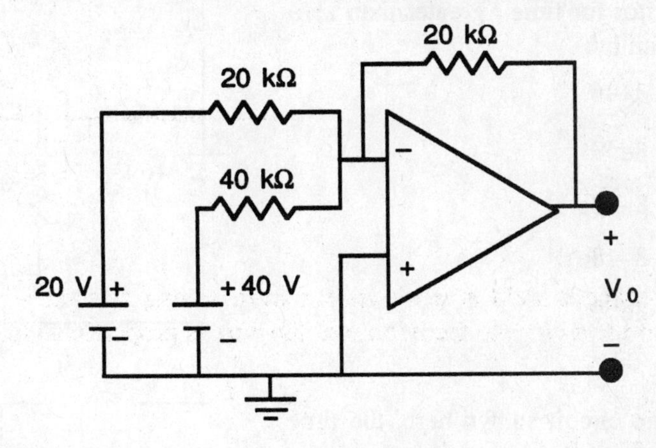

(A) 10 V

(B) – 20 V

(C) 5 V

(D) 40 V

(E) – 40 V

31. The 24 V DC source has been connected to the circuit shown for a long time. The voltage across the terminals of the capacitor is:

(A) 24 V

(B) 0 V

(C) 12 V

(D) 18.18 V

(E) 9.09 V

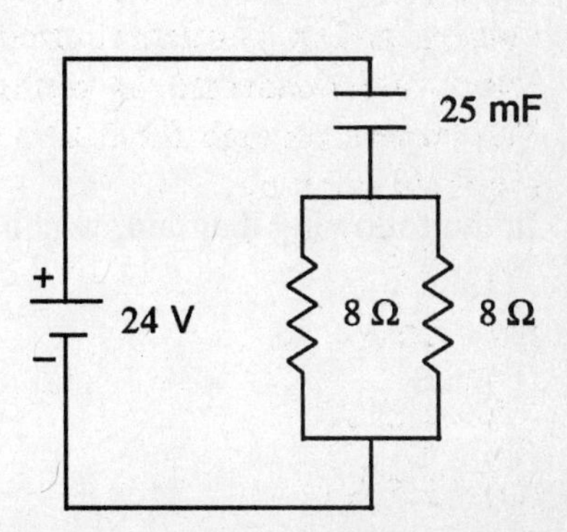

32. In the circuit shown here, the capacitor has been connected to the 36 V source for many hours. The voltage across the capacitor is

(A) 27 V

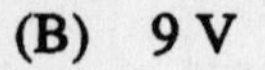
(B) 9 V

(C) 18 V

(D) 0 V

(E) 3 V

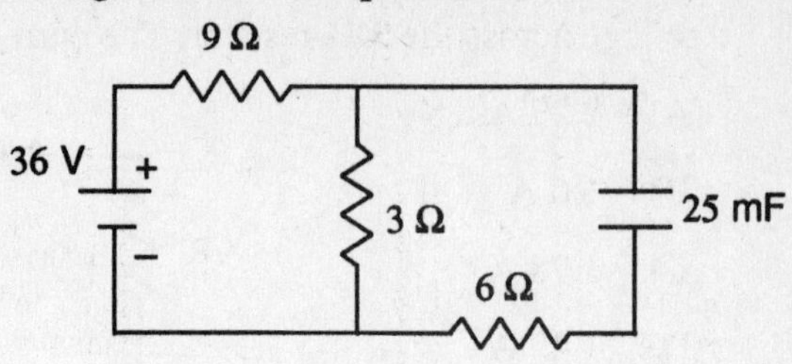

33. For the circuit shown here, the expression for the current in the 50 mF capacitor for time t greater than zero is given by:

(A) $4e^{-10t}$

(B) $8e^{-0.1t}$

(C) $8 - 8e^{-10t}$

(D) $8 - 8e^{-0.1t}$

(E) $8e^{-10t}$

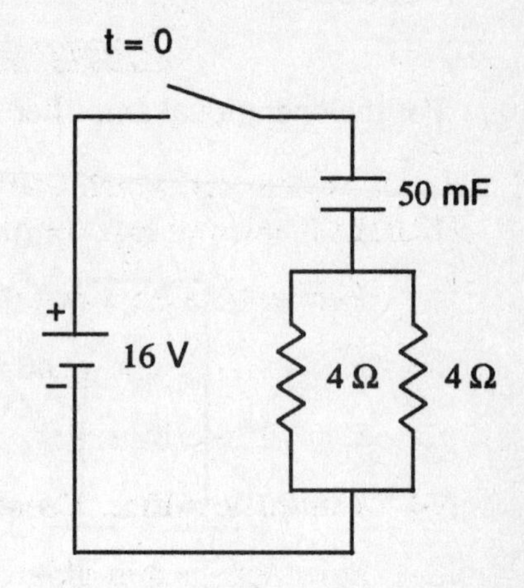

34. For the circuit shown here, the time constant τ is:

(A) 0.4 s

(B) 2.5 s

(C) 0.1 s

(D) 10 s

(E) 2.0 s

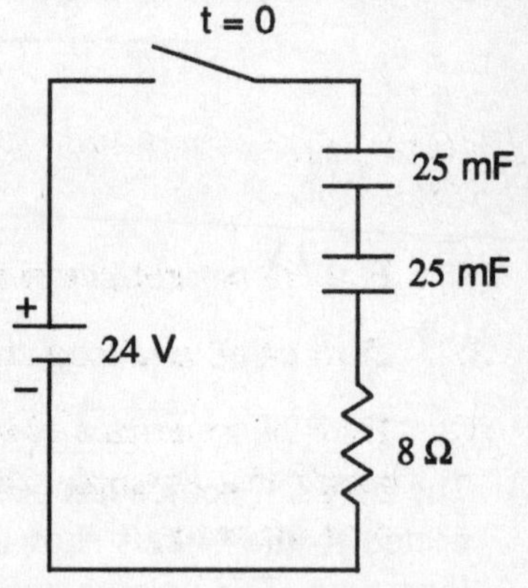

35. In the following diagram, which point has the highest pressure?

(A) A

(B) B

(C) C

(D) D

(E) E

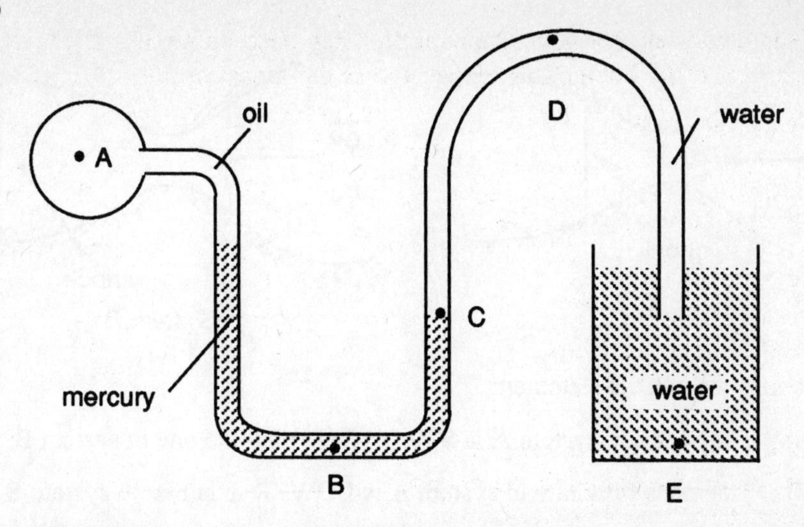

36. Select the answer which correctly ranks fluid meters from LOWEST to HIGHEST pressure loss (i.e., low loss, medium loss, high loss).

    (A) Venturi tube, flow nozzle, thin-plate orifice

    (B) Venturi tube, thin-plate orifice, flow nozzle

    (C) Thin-plate orifice, flow nozzle, Venturi tube

    (D) Thin-plate orifice, Venturi tube, flow nozzle

    (E) Flow nozzle, thin-plate orifice, Venturi tube

37. Identify one of the following problems for which the Moody chart would NOT be of use.

    (A) Flow of natural gas in a long pipe

    (B) Flow of oil in a long channel with square cross section

    (C) Flow of water in a pipe whose inner wall is corroded and extremely rough (for example, the average roughness height being 4% of the pipe diameter)

    (D) Flow in a subsonic diffuser

    (E) Flow of water in a pipe at very high speeds, where the flow is extremely turbulent

Problems 38 and 39 deal with the following two sketches. The geometry and upstream conditions are identical. Neglect friction, and assume adiabatic, one-dimensional flow.

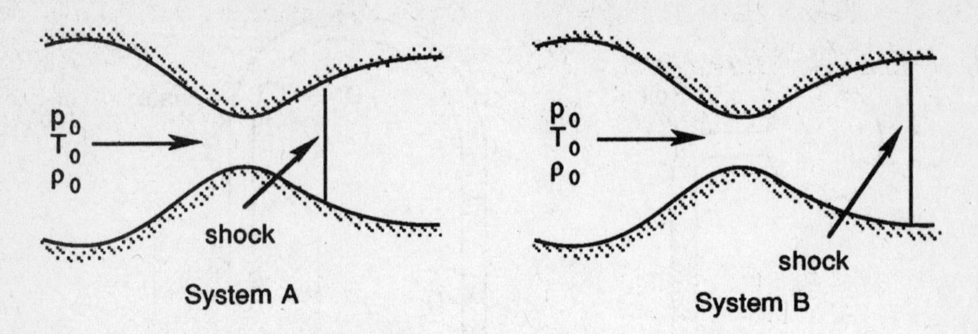

38. Choose the correct statement:

    (A) The shock in system A is STRONGER than the one in system B.

    (B) The mass flow rate in system A is LOWER than that in system B.

    (C) The pressure just UPSTREAM of the shock in system A is HIGHER than the pressure just UPSTREAM of the shock in system B.

    (D) The Mach number just DOWNSTREAM of the shock in system A is HIGHER than the Mach number just DOWNSTREAM of the shock in system B.

    (E) The jump in stagnation temperature, T_o, across the shock in system A is SMALLER than the jump in stagnation temperature, T_0, across the shock in system B.

39. Choose the correct statement:

    (A) The pressure at the exit plane in system A is HIGHER than the pressure at the exit plane in system B.

    (B) The stagnation temperature at the exit plane in system A is LOWER than that at the exit plane in system B.

    (C) The flow downstream of the shock in system A is SUBSONIC, but for system B, if the shock is close to the exit plane, the flow downstream of the shock (but still INSIDE the duct) may be SUPERSONIC.

    (D) If P_o were increased in either system, keeping everything else the same, the shock would remain in the same location since the flow is choked.

    (E) None of the above is correct.

40. Assuming Laminar flow, the volumetric flux Q through a tube is a function of the tube internal radius r, the fluid viscosity μ, and the pressure

gradient (dp/dx). Then $Q = f(r, \mu, dp/dx)$ can be expressed as

(A) $Q = r^3\mu\, dp/dx$

(B) $Q =$ (constant) $r^3\mu\, dp/dx$

(C) $Q =$ (constant) $r^4\, \mu\, dp/dx$

(D) $Q = r^4\mu\, dp/dx$

(E) $Q =$ (constant) $(r^4/\mu)\, dp/dx$

41. For steady, incompressible, inviscid flow, with no shaft work or heat transfer, Bernoulli's equation can be written,

$$\frac{p}{\rho} + \frac{V^2}{2} + gz = \text{constant}$$

This form of the equation is valid

(A) for rotational flow, but ONLY along streamlines

(B) for irrotational flow, but ONLY along streamlines

(C) for rotational flow EVERYWHERE

(D) for rotational flow, but ONLY across streamlines

(E) ONLY for irrotational flow (i.e., it is NEVER valid if the flow is rotational)

Questions 42 – 44 refer to the following: In the figure below, 6 m³/hour of water is pumped from the lake to a reservoir whose surface is 17 m above the lake surface. The total frictional head losses are estimated to be 2.5 m of water.

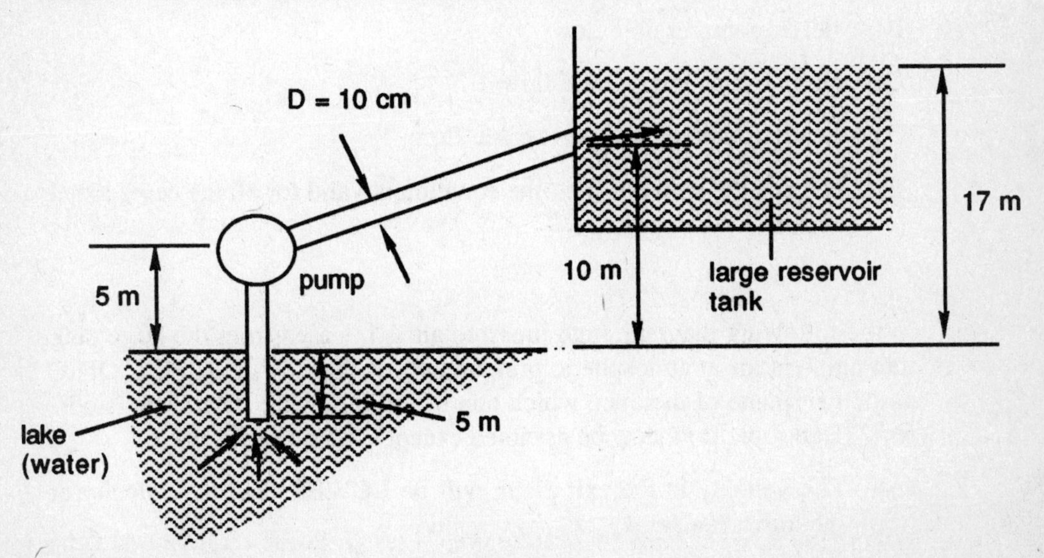

42. The difference in atmospheric pressure between the reservoir surface and the surface of the lake is typically neglected. For air at a constant density of 1.2 kg/m³, this difference in pressure is closest to

(A) 200 Pascals
(B) 350 Pascals
(C) 657 Pascals
(D) 3400 Pascals
(E) 101,000 Pascals

43. The power delivered by the pump to the water is closest to

(A) 128 Watts
(B) 319 Watts
(C) 1048 Watts
(D) 2096 Watts
(E) 5132 Watts

44. If the pump is only 65% efficient, how much electrical power is required to pump 6 m³/hr of water?

(A) 83.2 Watts
(B) 207 Watts
(C) 491 Watts
(D) 681 Watts
(E) 1612 Watts

45. The simplified continuity equation $\nabla \cdot \mathbf{V} = 0$ is NOT valid for:

(A) steady incompressible flow
(B) steady compressible flow
(C) unsteady incompressible flow
(D) steady rotational incompressible flow
(E) none of the above — i.e., the equation is valid for all the cases listed above

46. In the following diagram, high pressure air is released from the large tank into ambient air at atmospheric pressure, P_a. If the flow is SUPERSONIC at the exit plane of the duct, which one of the following statements is correct? (Isentropic flow may be assumed except across a shock wave)

(A) The velocity at the exit plane will be LOWER than the velocity at the throat (i.e., at A_{min}).

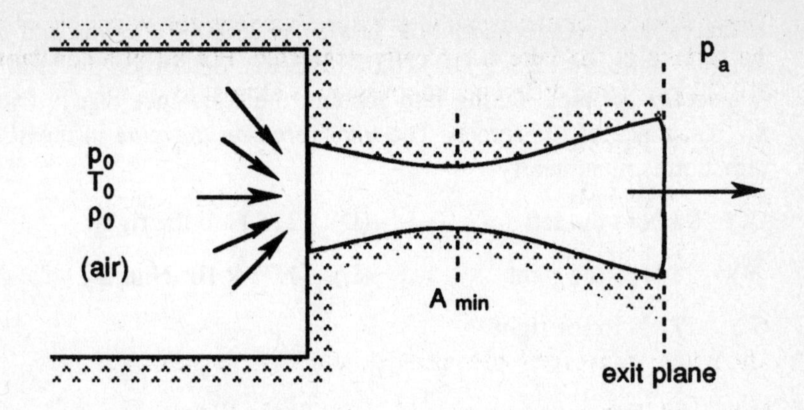

(B) The pressure at the exit plane will be HIGHER than the pressure at the throat (i.e., at A_{min}).

(C) A normal shock wave is likely to exist somewhere inside the DIVERGING section of the duct.

(D) A normal shock wave is likely to exist somewhere inside the CONVERGING section of the duct.

(E) The density at the exit plane will be LOWER than the density at the throat (i.e., at A_{min}).

Questions 47 – 48 refer to the following: Sketched below is incompressible fully developed flow of water ($\rho = 1000$ kg/m³, $\vartheta = 1.0 \times 10^{-6}$ m²/s) in a horizontal round pipe of 5.0 cm inner diameter. Assume the flow field is steady in the mean. At section 2, a high blockage screen is stretched acros the pipe cross section.

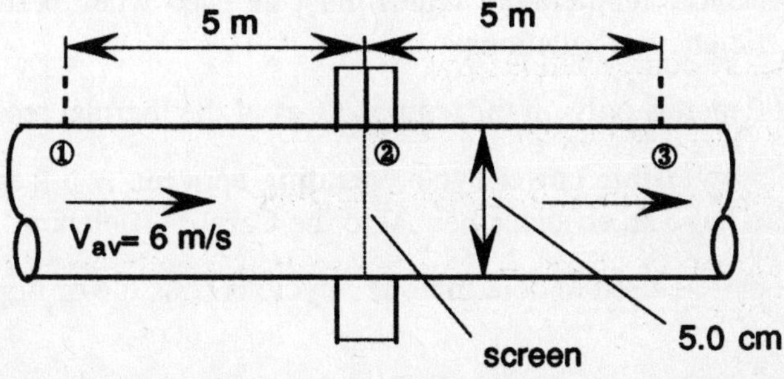

47. Momentum flux, $\dot{M}$ is defined as the momentum per second flowing through a cross section of the pipe. If the average velocity at section 1 is 6 m/s, the *momentum* flux across section 3 is most nearly

(A) $\dot{M}_3 = 12\ \text{kg} \cdot \text{m/s}^2$

(B) $\dot{M}_3 = 24\ \text{kg} \cdot \text{m/s}^2$

(C) $\dot{M}_3 = 36\ \text{kg} \cdot \text{m/s}^2$

(D) $\dot{M}_3 = 72\ \text{kg} \cdot \text{m/s}^2$

(E) $\dot{M}_3 = 7.2 \times 10^5\ \text{kg} \cdot \text{m/s}^2$

48. If the static pressure at section 1 is p_1 = 200,000 Pascals and the static pressure at section 3 is p_3 = 110,000 Pascals, determine the total force on this section of pipe, taking into account the resistance due to friction and the resistance of the screen. The total force *on the pipe* in the streamwise direction is most nearly

    (A) 98 N to the left
    (B) 98 N to the right
    (C) 177 N to the right
    (D) 225 N to the right
    (E) 1.77×10^6 N to the left

49. An adiabatic process is characterized by which of the following?

    (A) The entropy change is zero.
    (B) The heat transfer is zero.
    (C) It is isothermal.
    (D) The work is zero.
    (E) It is reversible.

50. Which of the following statements about the Carnot efficiency is NOT TRUE?

    (A) It is the maximum efficiency any power cycle can obtain while operating between two thermal reservoirs
    (B) Absolute temperature scales must be used when performing Carnot efficiency calculations
    (C) It depends only on the temperatures of the thermal reservoirs
    (D) No reversible power cycle operating between two thermal reservoirs can have an efficiency equal to the Carnot efficiency
    (E) It can be used to determine if a cycle is possible or impossible

51. A rigid container is heated by the sun. There is no shaft work associated with the container. From the First Law of Thermodynamics, you determine the resulting work to be

    (A) equal to the heat transfer.
    (B) equal to the change in internal energy.
    (C) equal to the volume times the change in pressure.

(D) equal to zero.

(E) equal to the change in enthalpy.

52. An inventor claims to have built an engine which will revolutionize the automotive industry. Which of the following would be the best test to determine if the inventory's claims are true?

(A) Conservation of Mass

(B) Zeroth Law of Thermodynamics

(C) First Law of Thermodynamics

(D) Second Law of Thermodynamics

(E) Check to see if the thermal efficiency is less than 100%.

53. For the normal shock shown in the diagram, which statement is inconsistent with the flow direction and the occurrence of the shock, if the flow area is considered constant, the forces acting on the wall are neglected, and the heat transfer and work are zero?

(A) $M_1 > M_2$

(B) $s_2 < s_1$

(C) $P_2 > P_1$

(D) $T_1 > T_2$

(E) $V_1 > V_2$

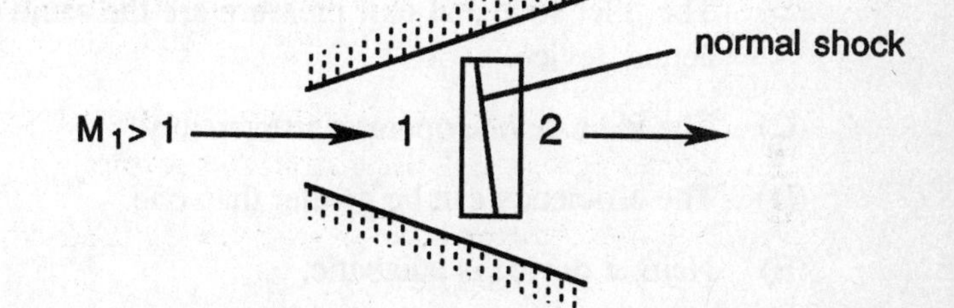

54. Two independent intensive properties are required to fix the state of a pure, simple compressible substance. People often attempt to fix the state of a medium using heat and/or work, which are not properties. Which of the following statements about heat and work is not true?

(A) Heat and work are transient phenomena.

(B) Heat and work are forms of energy.

(C) Heat and work are associated with processes.

(D) Heat and work are boundary phenomena.

(E) Heat and work are point functions.

55. An ideal vapor compression refrigeration cycle requires 2.5 kW to power the compressor. You have found the following data for the cycle:

    – the enthalpy at the condenser entrance is 203 kJ/kg

    – the enthalpy at the condenser exit is 55 kJ/kg

    – the enthalpy at the evaporator entrance is 55 kJ/kg

    – the enthalpy at the evaporator exit is 178 kg/s,

    If the mass flow rate of the refrigerant is 0.10 kg/s, then the coefficient of performance of this refrigerator is most nearly

    (A) 49.2
    (B) 1.203
    (C) 59.2
    (D) 5.92
    (E) 4.92

56. The isentropic (process) efficiency is used to compare actual devices such as turbines, compressors, nozzles, and diffusers to ideal ones. Which statement is true?

    (A) Only the ideal device is considered adiabatic.
    (B) The inlet state and exit pressure are the same for both the ideal and actual device.
    (C) The ideal device operates irreversibly.
    (D) The efficiency can be greater than one.
    (E) Neither device is adiabatic.

57. Referring to the diagrams below for the ideal compression ignition cycle, which of the statements is false?

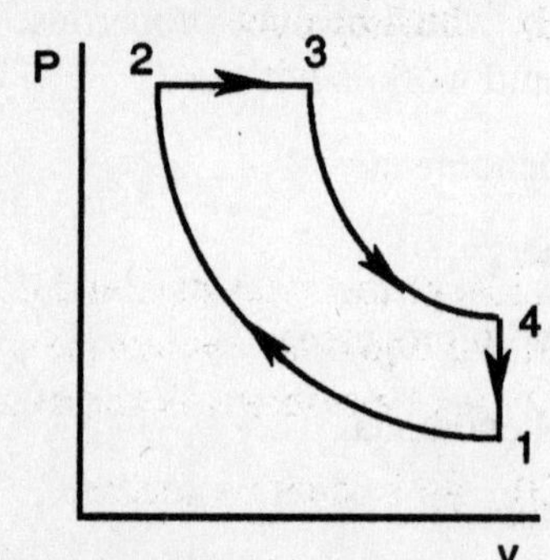

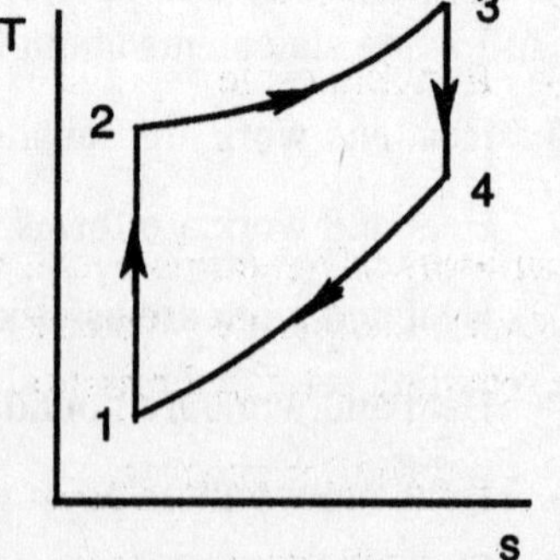

    (A) $W_{cycle} = W_{23} + W_{34} - W_{12}$

(B) $Q_{cycle} = Q_{23} - Q_{41}$

(C) $n_{th} = W_{cycle} / Q_{cycle}$

(D) Compression ratio $(r) = V_1/V_2$

(E) $W_{cycle} = Q_{23} - Q_{41}$

58. Referring to the diagrams below for the ideal spark ignition cycle, which of the following statements is false?

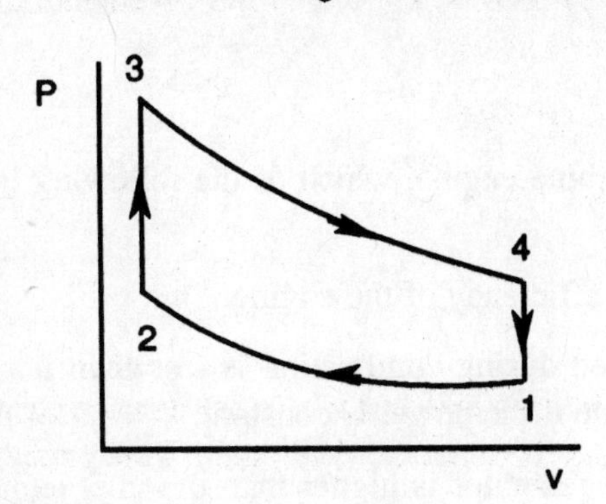

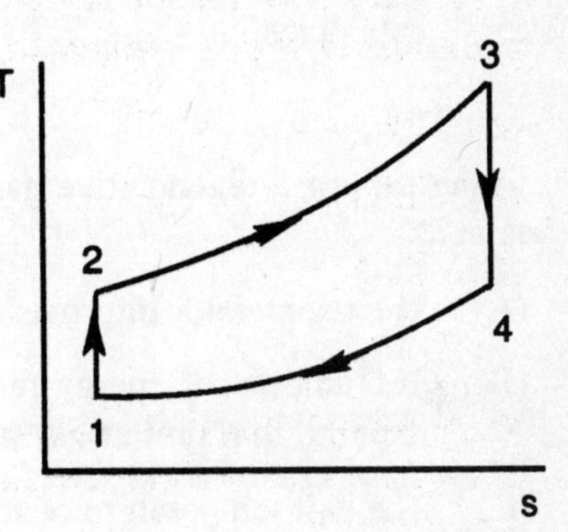

(A) $W_{cycle} = W_{34} - W_{12}$

(B) $W_{cycle} = Q_{23} - Q_{41}$

(C) $n_{th} = 1 - (u_4 - u_1) / (u_3 - u_2)$

(D) Compression ratio $(r) = V_1/V_2$

(E) Compression ratio $(r) = V_3/V_4$

59. A steam power cycle is modeled by the ideal cycle known as the

(A) Otto cycle
(B) Diesel cycle
(C) Rankine cycle
(D) Brayton cycle
(E) Stirling cycle

60. In an ideal refrigeration cycle, liquid leaves the condenser and is expanded in such a manner that the enthalpy of the liquid is equal to the enthalpy of the resulting saturated mixture. This type of expansion is known as

(A) a throttling process.
(B) an isothermal process.
(C) a compression process.
(D) an isochoric process.
(E) a reversible process.

61. For an ideal turbojet, all of the following statements are true EXCEPT:

(A) the process (isentropic) efficiency of the turbine is 100%.

(B) the process (isentropic) efficiency of the compressor is 100%.

(C) the work produced by the turbine is greater than the work required by the compressor.

(D) the process (isentropic) efficiency of the nozzle is 100%.

(E) the exit velocity of the nozzle is a function of the entrance and exit enthalpies.

62. In the case of a regenerative gas turbine engine, which of the following is false?

(A) The regenerator improves the efficiency of the engine.

(B) The amount of energy required during combustion is less than that required in a similar gas turbine if the power is constant.

(C) The exit temperature of the regenerator is higher than the inlet temperature of the compressor.

(D) There is no heat rejected to the atmosphere in a regenerative gas turbine engine.

(E) If the regenerator is reversible, the effectiveness can approach 100%.

63. Given the following acceleration-time curve for a particle, find the distance between points *A* and *B*.

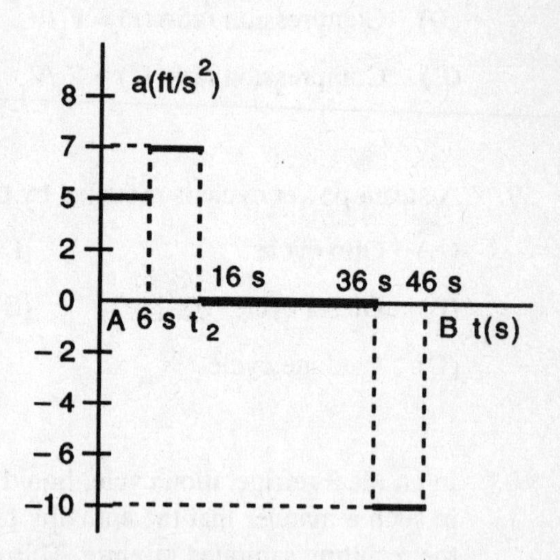

(A) 3240 ft.

(B) 4100 ft.

(C) 2250 ft.

(D) 3100 ft.

(E) 970 ft.

64. A 40 lb. missle moves with a velocity of 150 ft./sec. It is intercepted by a laser beam which causes it to explode into two fragments *A* and *B* which weigh 25 lbs. and 15 lbs. respectively. If the fragments travel as shown immediately after explosion, find the magnitude of velocity of fragment *A*.

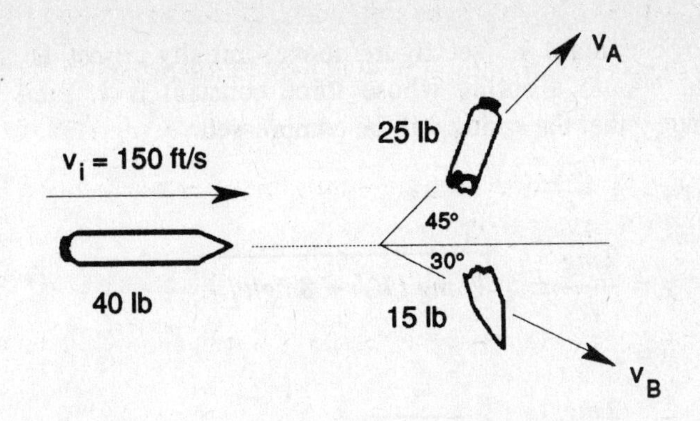

(A) 101.25 ft/s

(B) 73.21 ft/s

(C) 31.06 ft/s

(D) 43.92 ft/s

(E) 146.41 ft/s

65. A 20 kg box is released from rest at point *A*. It travels down a slope of 35°. If the coefficient of friction, μ, is .3, what will be the velocity of the box at point *B*?

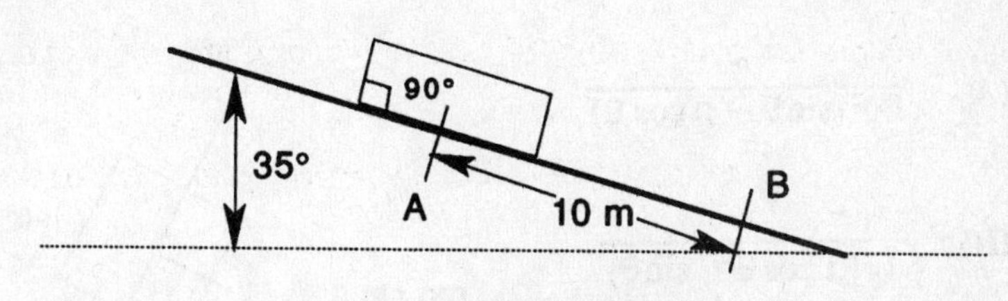

(A) 2.56 m/s

(B) 8.02 m/s

(C) 10.60 m/s

(D) 7.32 m/s

(E) 9.11 m/s

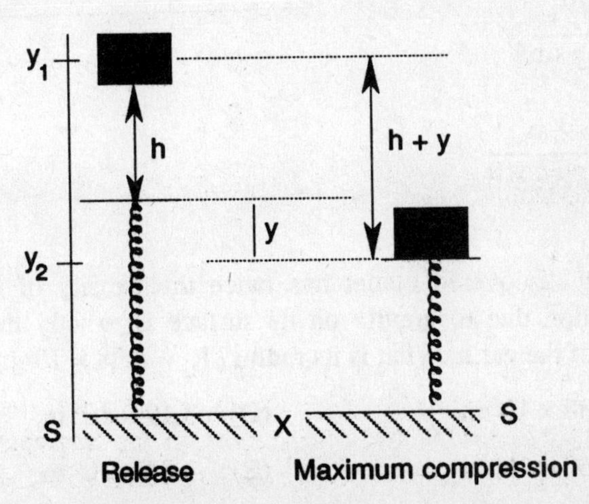

The total fall of the block is h + y.

66. A block of mass *m*, (see figure above) initially at rest, is dropped from a height *h* onto a spring whose force constant is *k*. Find the maximum distance *y* that the spring will be compressed.

(A) $y = \pm \frac{mg}{k}$

(B) $y = \frac{2mg}{k} \pm \sqrt{(2mg/k)^2 + 8mgh/k}$

(C) $y = \pm h$

(D) $y = \frac{2mg}{k} \pm \sqrt{8mgh/k}$

(E) $y = \frac{1}{2}\left[\frac{2mg}{k} \pm \sqrt{(2mg/k)^2 + (8mgh/k)}\right]$

67. A body of mass *m* has an initial velocity v_0 directed up a plane that is at an inclination angle θ to the horizontal. The coeffieicnt of sliding friction between the mass and the plane is μ. What distance *d* will the body slide up the plane before coming to rest?

(A) $\frac{V_0^2}{2g\,(\sin\theta - \mu\cos\theta)}$

(B) $\frac{V_0^2}{2g(\mu\cos\theta + \sin\theta)}$

(C) $\frac{V_0^2}{2g(\mu\sin\theta + \cos\theta)}$

(D) $\frac{V_0^2}{2g\sin\theta}$

(E) $\frac{V_0^2}{2g\mu\cos\theta}$

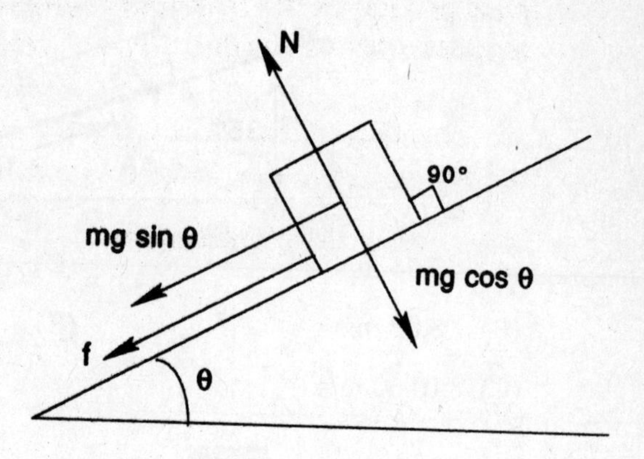

68. A newly discovered planet has twice the density of the earth, but the acceleration due to gravity on its surface is exactly the same as on the surface of the earth. What is its radius? $R_e = 6.38 \times 10^6$ m.

(A) 6.38×10^6 m

(B) 1.60×10^6 m

(C) 1.28×10^7 m

(D) 3.19×10^6 m

(E) 3.84×10^6 m

69. A car, with its door open and free to swing on its hinges is accelerating with a constant acceleration a. Determine the angular acceleration of the door, relative to the car, as a function of the acceleration of the automobile.

(A) 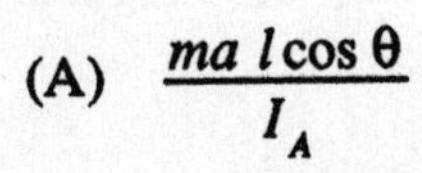

(B) 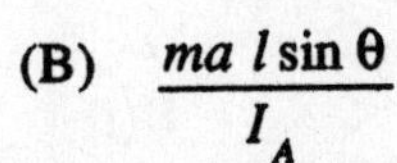

(C) 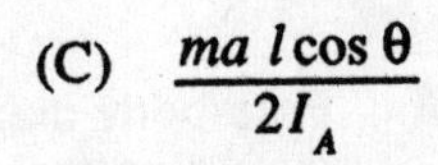

(D) 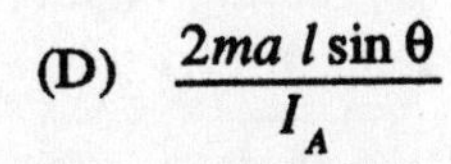

(E) 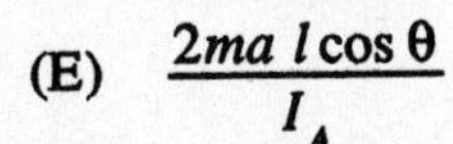

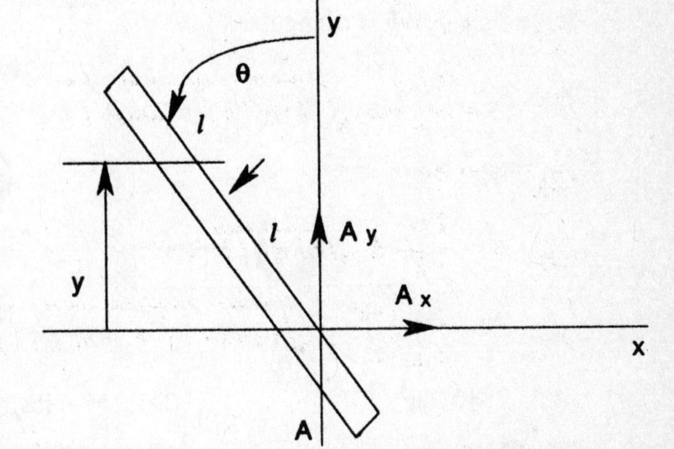

70. Two cars, A and B, are traveling along the same route. Car A is traveling at 4.5 m/s and has a mass of 1150 kg. Car B is traveling at 6.7 m/s and has a mass of 1300 kg. If car B gently bumps into car A and their bumpers lock together, what will be their common velocity?

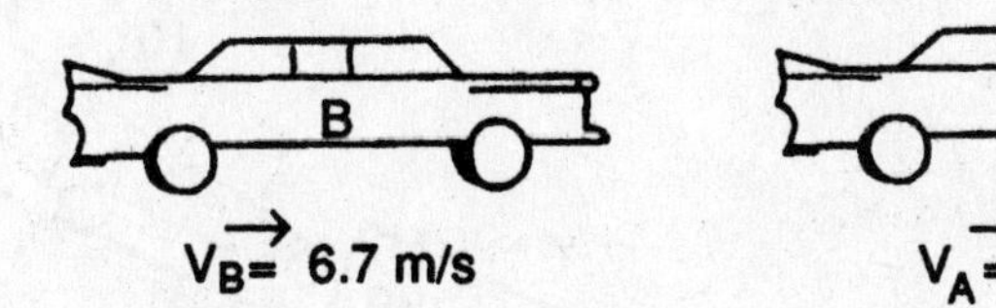

(A) 0

(B) 5.53 m/s

(C) 11.33 m/s

(D) 5.67 m/s

(E) 5.60 m/s

71. A wheel of radius .5 m rolls to the right (without slipping) with its center C having a velocity of 2.5 m/s. Find the velocity of point B.

(A) 2.5i m/s

(B) 1.5i m/s

(C) (3.56i – 1.06j) m/s

(D) (– 1.06i – 3.56j) m/s

(E) (– 2.5i + 1.5i) m/s

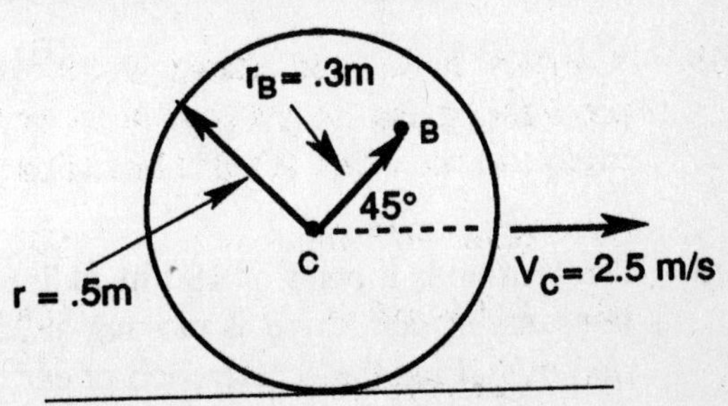

72. A ball weighing 8 lbs. slides along a frictionless surface. The ball strikes a wall with a velocity of 30 ft/s and rebounds with the same velocity. If the time of contact between the ball and the wall is a quarter-second, determine the maximum force of the wall acting on the ball, assuming that the force-time curve is triangular.

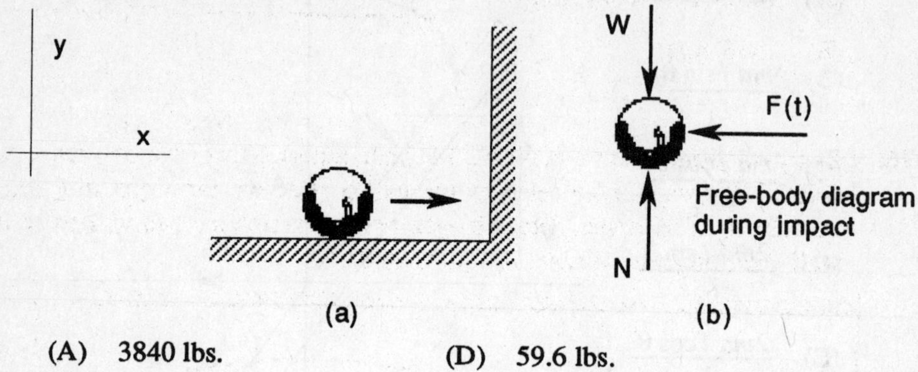

(A) 3840 lbs.
(B) 119 lbs.
(C) 1920 lbs.
(D) 59.6 lbs.
(E) 24.2 lbs.

73. Determine the tension, T, in the cable which will give the 150 kg block a steady acceleration of 4 m/s² up the slope.

(A) 868.6 N

(B) 547.3 N
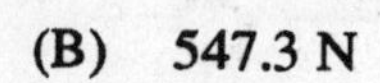
(C) 844.4 N
(D) 810.6 N
(E) 1211.2 N

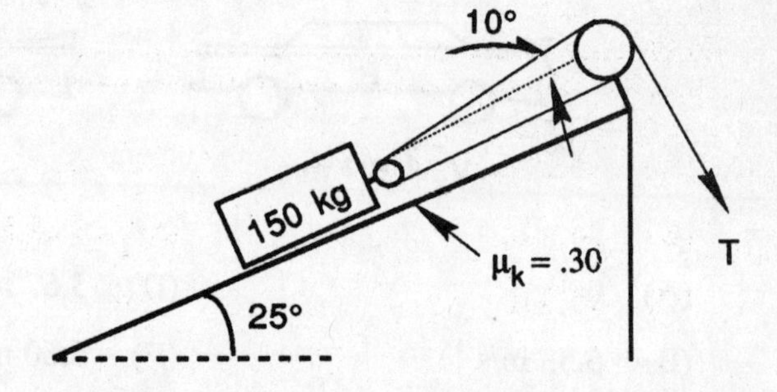

74. The motion of a particle along a straight line is described by the equation $x(t) = t^3 - 3t^2 - 45t = 50$ where x is expressed in feet and t in seconds. Compute the acceleration of the particle at the time in which $V(t) = 0$.

(A) 28 ft/s²
(B) 6 ft/s²
(C) – 112 ft/s²
(D) 1.33 ft/s²
(E) – 65.3 ft/s²

75. Car A rounds a bend of 180 m radius at a constant speed of 40 km/h. At the same instant, car B is moving at 75 km/h but is speeding up at the rate of 3 m/s². Find the acceleration of car A as observed from car B.

(A) $-3.686\mathbf{i}$ m/s²

(B) $2.314\mathbf{i}$ m/s²

(C) $-3{,}000\mathbf{i}$ m/s²

(D) $(-3{,}000\mathbf{i} + .686\mathbf{j})$ m/s²

(E) $.686\mathbf{i}$ m/s²

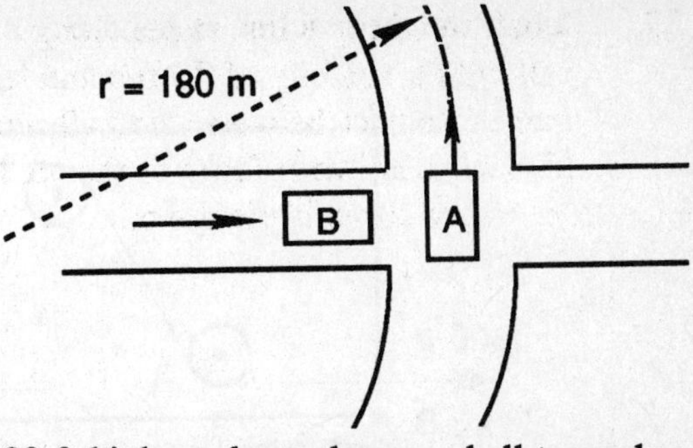

76. In a gymnasium with a ceiling 30 ft high, a player throws a ball towards a wall 80 ft away. If he releases the ball 5 ft above the floor with initial velocity v_0 of 55 ft/sec, determine the point at which the ball will strike the wall if the angle of release, θ, equals 46.8°.

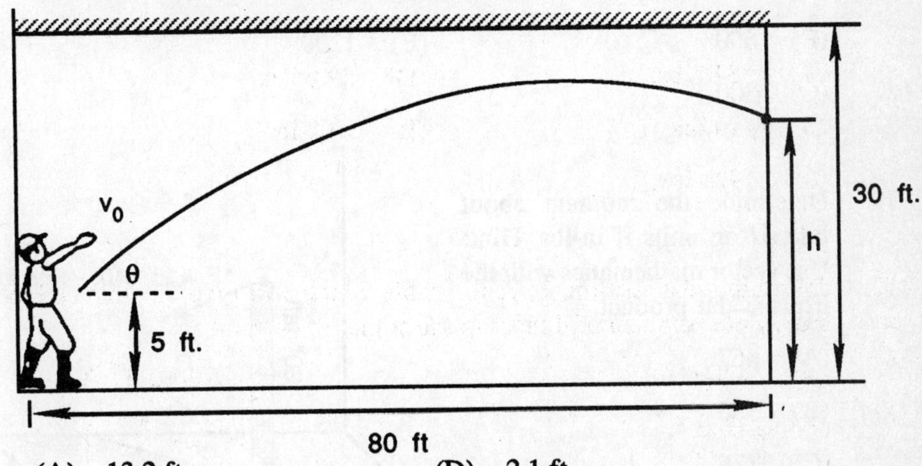

(A) 13.2 ft

(B) 24.5 ft.

(C) 15.9 ft

(D) 2.1 ft

(E) 18.2 ft

77. Determine the rope tension force for the pulley system.

(A) 2000

(B) 1000

(C) 167

(D) 333

(E) 500

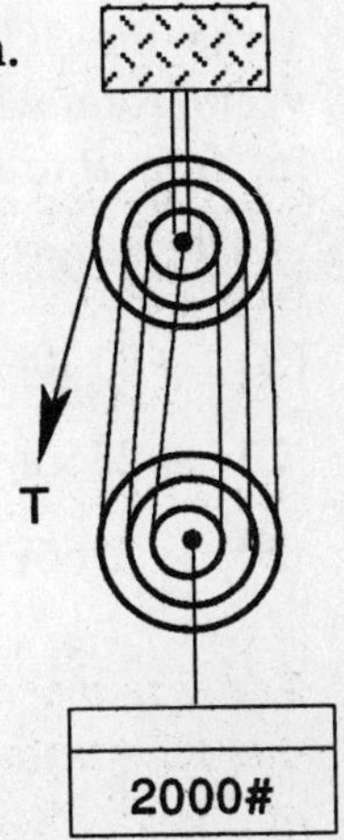

78. Determine the reaction at point *B*.

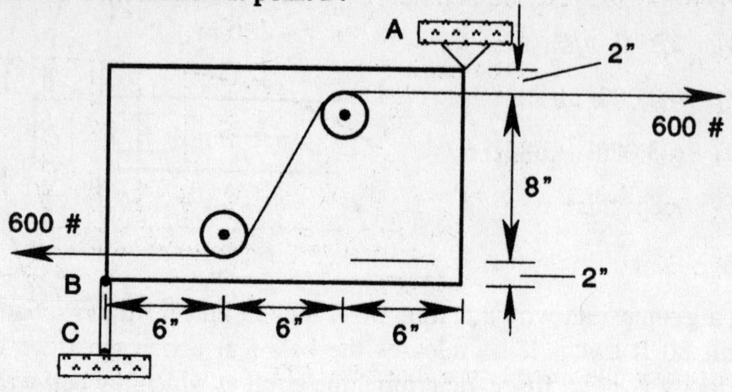

(A) 267
(B) 600
(C) 300
(D) 534
(E) 1200

79. Determine the moment about line *AB* in units if in-lbs. Hint: Use vector mathematics with the triple scalar product.

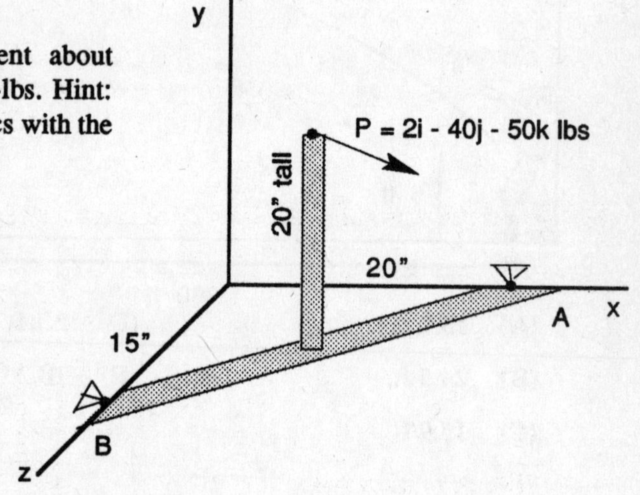

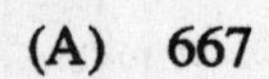

(A) 667
(B) 336
(C) 776
(D) 785
(E) 846

80. Write the 3000 N force T_{BD} in vector form with the vector pointing from *B* to *D*.

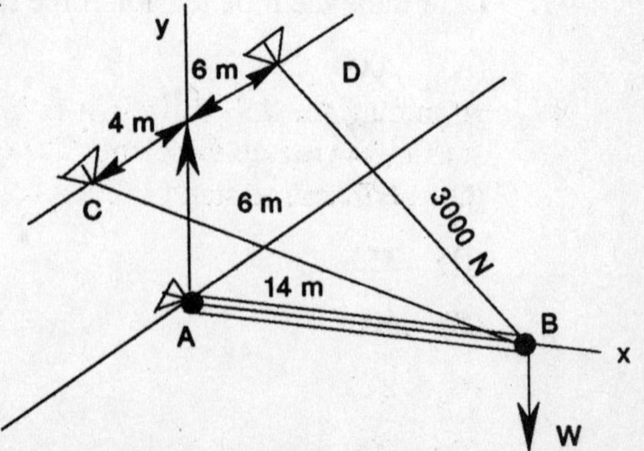

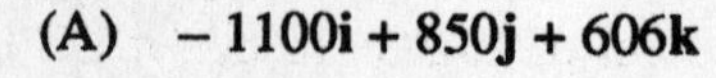

(A) − 1100i + 850j + 606k
(B) + 1100i − 850j − 606k
(C) − 2566i + 1100j + 1100k
(D) − 2566i + 1100j − 1100k
(E) − 408i + 642j − 2850k

81. Determine the reaction at point A.

(A) 48

(B) 96

(C) 38

(D) 58

(E) 32

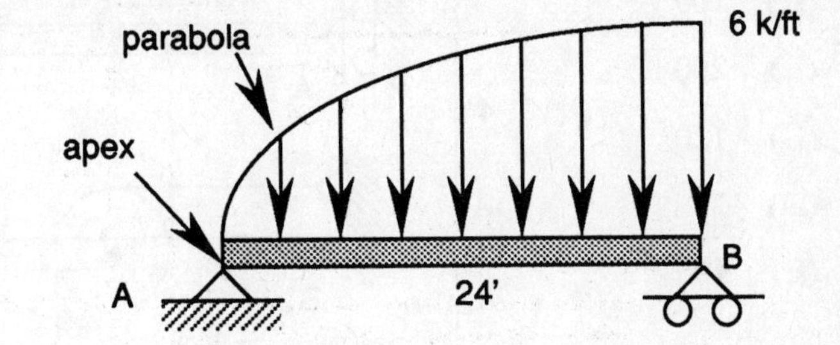

82. Determine the force in truss member CD.

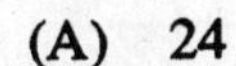

(A) 24

(B) 12

(C) 36

(D) 60

(E) 48

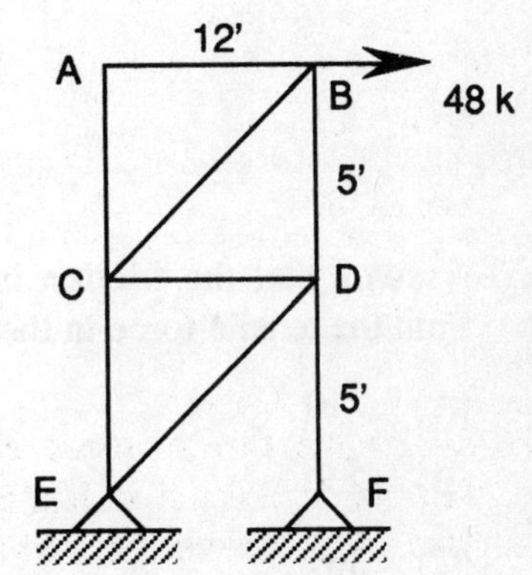

83. Find the centroid of the four 2×2 areas for both the x and y–direction.

(A) $x = 4, y = 2.5$

(B) $x = 4, y = 4$

(C) $x = 2, y = 2.5$

(D) $x = 8, y = 4$

(E) $x = 2.5, y = 4$

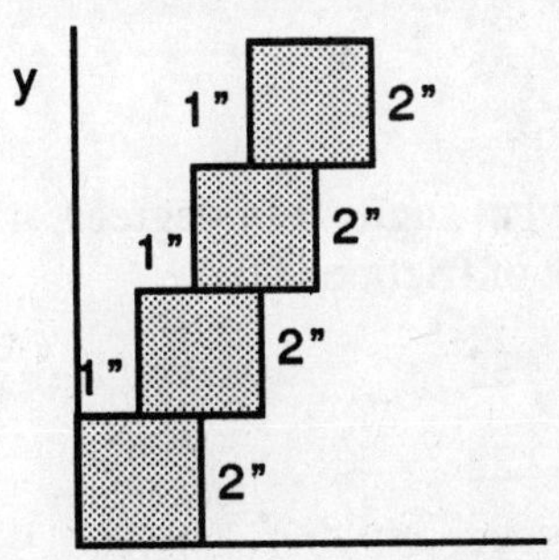

84. Assuming the 250 lb weight is initially at rest, find the force required to start it moving up the plane. The coefficient of kinetic friction is 0.30, and the coefficient of static friction is 0.40.

(A) 75

(B) 50

(C) 212

(D) 190

(E) 125

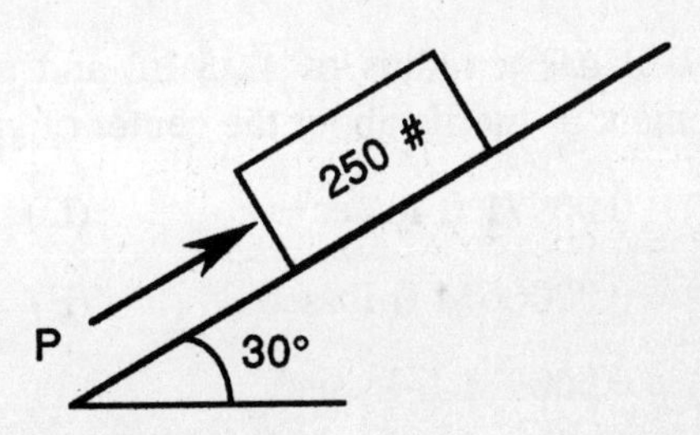

85. Find the vertical reaction at point F.

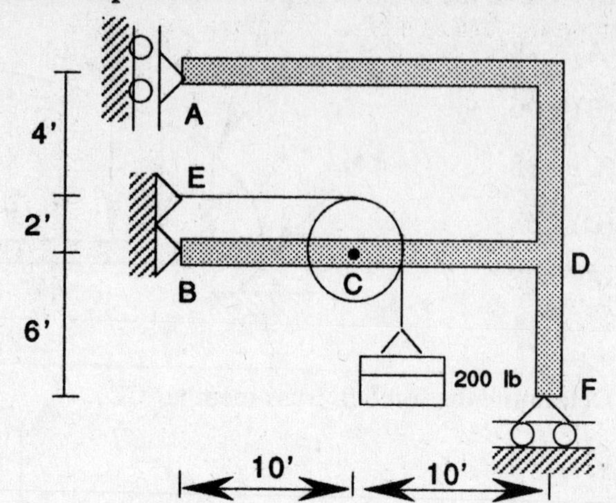

(A) 100

(B) 200

(C) 120

(D) 75

(E) 400

86. Assume that the friction is large enough that the cylinder will not slide. Find the tensile force in the rope.

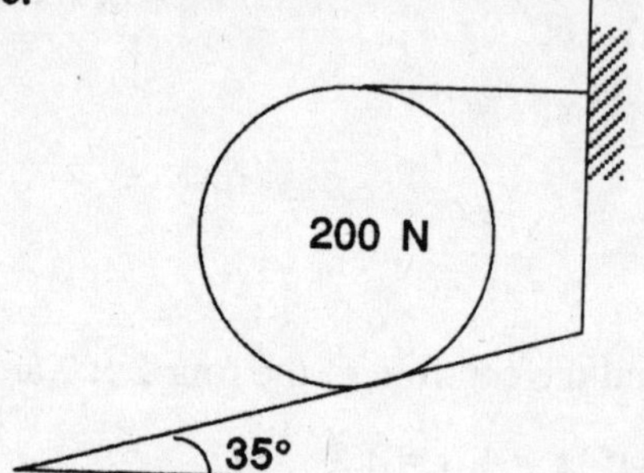

(A) 210

(B) 200

(C) 36

(D) 63

(E) 84

87. At what angle, θ in degrees, will the cylinder start sliding when the coefficient of friction is 0.4.

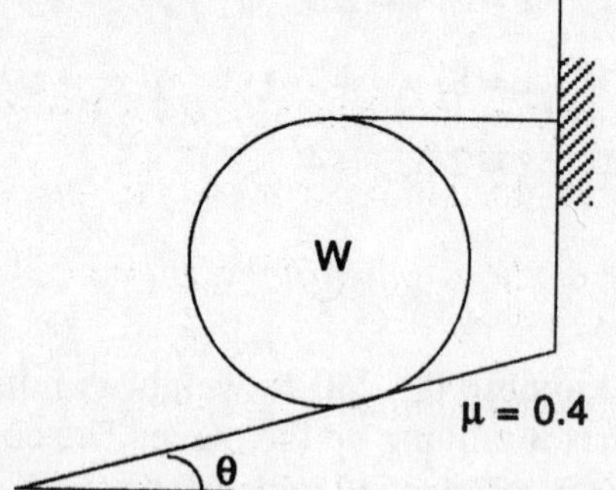

(A) 33

(B) 22

(C) 44

(D) 55

(E) 66

88. A ball has a radius of 1.25 in. and a weight of 0.5 lbs. Find the mass moment of inertia about the center of gravity.

(A) 0.00971 ft-lb-sec^2

(B) 0.0000674 ft-lb-sec^2

(C) 0.000971 ft-lb-sec^2

(D) 1.398 ft-lb-sec^2

(E) 0.313 ft-lb-sec^2

89. A rectangular beam has a cross-sectional area of 4×8 as shown. Determine the moment of inertia about the x-axis.

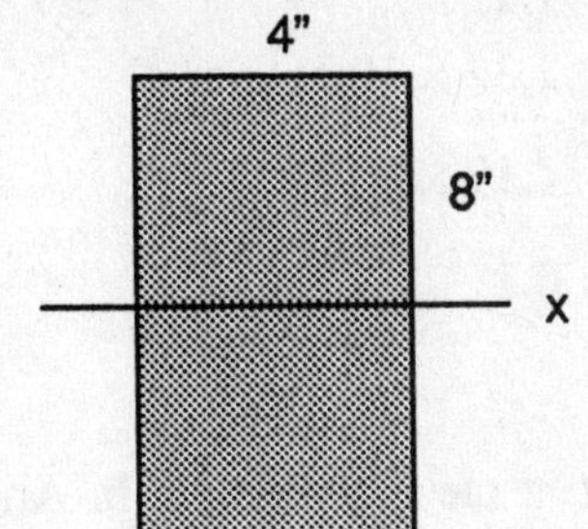

(A) 150

(B) 130

(C) 170

(D) 43

(E) 210

90. A gear system transfers a torque from shaft *AB* to shaft *DE*. Determine the amount of torque in the shaft at *E*.

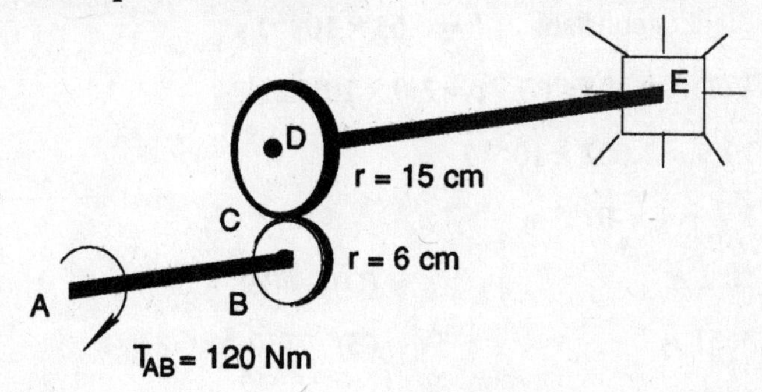

(A) 100

(B) 200

(C) 300

(D) 400

(E) 500

91. Avogadro's Number (6.023×10^{23}) represents:

(A) the number of molecules in 1 gram of any compound.

(B) the number of molecules in 1 liter of any gas at 1 atm pressure and 0° C.

(C) the number of molecules in 1 gram mole of any compound.

(D) the number of hydrogen molecules in 1 gram of hydrogen gas at 1 atm pressure and 0° C.

(E) the number of valence electrons in a mole of any element.

92. The ionization equilibrium constant for the acidic hydrogen of acetic acid ($C_2H_3O_2$–H) in an aqueous solution at 25° C is 1.753×10^{-5}. What is the

hydrogen ion concentraion (in moles/liter) in a 0.1 M solution of acetic acid in water at 25° C?

(A) 1.32×10^{-3}
(B) 4.19×10^{-3}
(C) 0.1
(D) 1.753×10^{-5}
(E) 1.753×10^{-6}

93. What is the wavelength, in Angstrom units, of a photon emitted when an electron in a hydrogen atom falls from the $n = 2$ state ($E_2 = -3.4$ eV) to the ground state ($E_1 = -13.6$ eV)?

The following physical constants are applicable:

Plank's constant $h = 6.63 \times 10^{-34}$ J s

Velocity of light $c = 3.0 \times 10^{8}$ m/s

1 eV = 1.602×10^{-19} J

1 Å = 1×10^{-10} m

(A) 10.2 Å
(B) 3651 Å
(C) 1217 Å
(D) 4868 Å
(E) 730 Å

94. The chemical formula of the most common compound formed from beryllium (a Group IIa element) and iodine (a Group VIIa element) is:

(A) Be_2I_7
(B) BeI
(C) $BeIn_2$
(D) BeI_2
(E) Be_7I_2

95. The chemical formula of the salt formed from the neutralization of potassium hydroxide and sulfuric acid is:

(A) NaCl
(B) P_2SO_4
(C) KSO_4
(D) PSO_4
(E) K_2SO_4

96. Aluminum hydroxide, $Al(OH)_3$, is called an amphoteric hydroxide because:

(A) it contains three hydroxyl groups.

(B) it has limited solubility in water.

(C) either one, two, or all three hydroxyl groups can ionize in water.

(D) it can react as either an acid or a base.

(E) None of the above.

97. Which of the following characteristics of the alkaline earth elements is unique to this group of elements?

(A) The have two "s" electrons in the outer orbit.

(B) They exhibit a valence of +2.

(C) They are in Group IIa of the Periodic Table.

(D) They form hydroxides which are alkaline in aqueous solution.

(E) They are found in soil.

98. The presence of calcium and magnesium ions makes water "hard" because:

(A) they raise the freezing point so that ice crystals form much more easily.

(B) insoluble carbonates precipitate and form a "hard" scale on pipe walls and elsewhere.

(C) they precipitate soap rendering it useless for cleaning purposes.

(D) All of the above

(E) Only answers (B) and (C) above

99. Which of the following reactions is NOT an oxidation-reduction reaction?

(A) $Cu^{++} + SO_4^{--} \rightarrow CuSO_4 \downarrow$

(B) $5H_2O_2 + 2KMnO_4 + 6HCl \rightarrow 2MnCl_2 + 5O_2 \uparrow + 8H_2O + 2KCl$

(C) $H_2S + O_2 \rightarrow 3/2\ SO_2 + H_2O$

(D) $CH_4 + H_2O \rightarrow 3H_2 + CO$

(E) None of the above (i.e., all are oxidation-reduction reactions)

100. The Principle of LaChatelier permits qualitative predictions of:

(A) how equilibrium compositions will shift when the temperature changes.

(B) how equilibrium compositions will shift when pressure changes.

(C) how equilibrium compositions will shift when the concentration of reacting species changes.

(D) how equilibrium compositions will shift when pH changes.

(E) All of the above.

101. From the following Table of Heats of Formation, which statement about the water-gas reaction ($CH_4 + H_2O \rightarrow 3H_2 + CO$) is correct?

Table 1 — Heats of Formation

	$(\Delta H_f)_{298°K}$ (Kcal/mole)
CH_4	–17.899
$H_2O_{(g)}$	–57.798
CO	–26.416

(A) The reaction is exothermic at 298° K.

(B) The reaction is endothermic at 298° K.

(C) The reaction may be either exothermic or endothermic depending on the pressure at which the reaction occurs.

(D) The reaction will not proceed at 298° K.

(E) It is impossible to determine because the value of $(\Delta H)_{f298°}K$ for H_2 is missing from the table.

102. Which of the following statements about the halogens is incorrect?

(A) They are good oxidizing agents.

(B) The react with water to form strong bases.

(C) They ALL exhibit a valence state of – 1.

(D) They are reactive with metals.

(E) They form diatomic molecules.

103. The Group VIII elements, called the noble or rare gases, have the following property:

(A) They are rare.

(B) They do not react with other elements.

(C) They remain gaseous down to just a few degrees above absolute zero.

(D) They are lighter than air.

(E) All of the above.

104. From the following table of atomic weights, determine the amount of iron metal that can be produced, by the indicated reaction, from 10 tons of pure ferric oxide.

$$Fe_2O_3 + 3C \rightarrow 2Fe + 3CO$$

Table of Atomic Weights	
Iron	55.85
Carbon	12
Oxygen	16

(A) 7 tons

(B) 12.3 tons

(C) 2 tons

(D) 7.7 tons

(E) Impossible to determine since atomic weights apply to grams, not tons.

105. Find the normal stress on section ① – ①, if the member has a constant cross-section area of 0.1 sq. in.

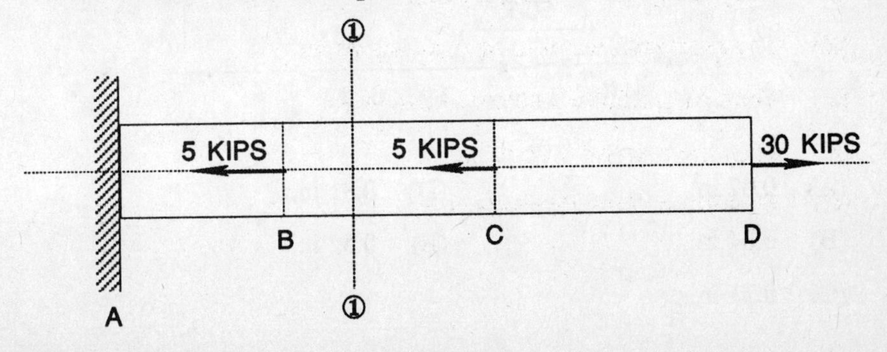

(A) 250 ksi (tension)

(B) 250 ksi (compression)

(C) 50 ksi (tension)

(D) 50 ksi (compression)

(E) 100 ksi (tension)

106. Find the maximum tensile stress of the given beam.

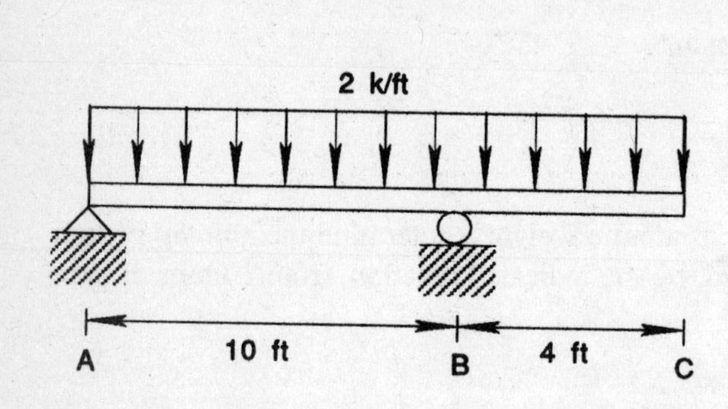

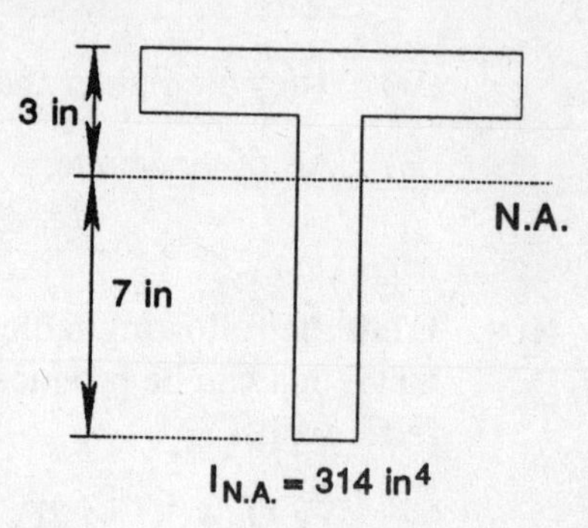

(A) 2 ksi

(B) 5 ksi

(C) 8 ksi

(D) 12 ksi

(E) 15 ksi

107. Determine the elongation of member '*EC*'.

Cross Section area = 0.2 in²
Modulus of Elasticity = 29,000 ksi

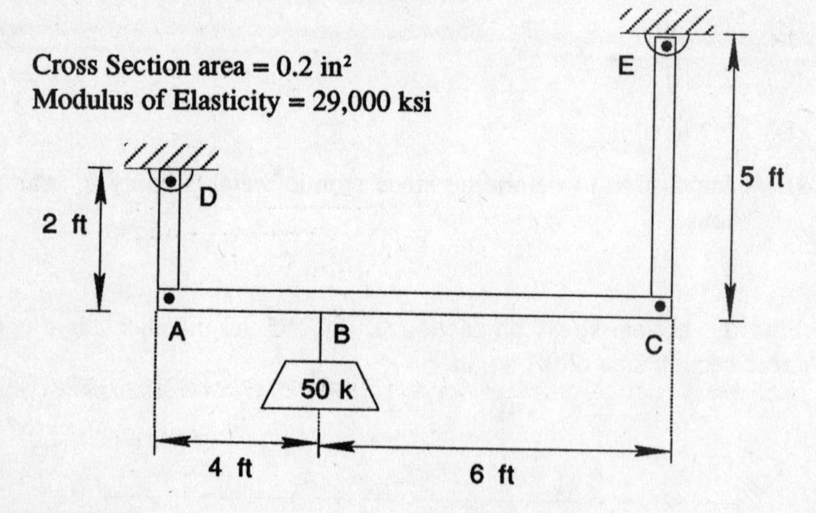

(A) 0.02 in.

(B) 0.04 in.

(C) 0.21 in.

(D) 0.31 in.

(E) 0.52 in.

108. '*AB*' and '*BC*' are circular shafts made from the same material. A twisting moment, or torque, of 200 k-ft is applied at the connection '*B*'. Determine the reaction at '*A*'.

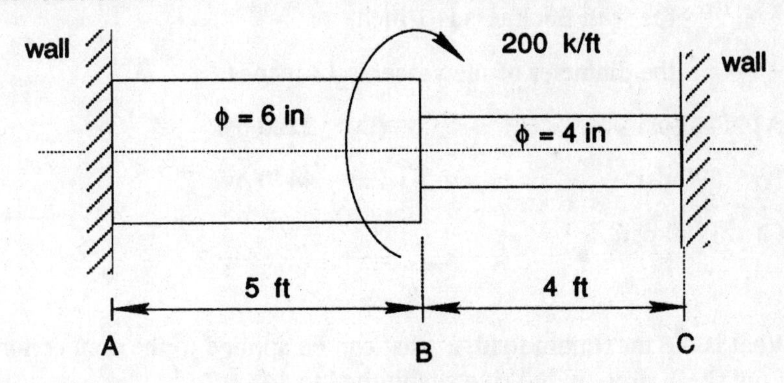

(A) 90 k-ft

(B) 100 k-ft

(C) 110 k-ft

(D) 160 k-ft

(E) 200 k-ft

109. Find the maximum tensile stress.

(A) 100 psi

(B) 108 psi

(C) 112 psi

(D) 206 psi

(E) 224 psi

200 psi

100 psi

50 psi

110. Find the maximum pressure that the cylinderical vessel can withstand.

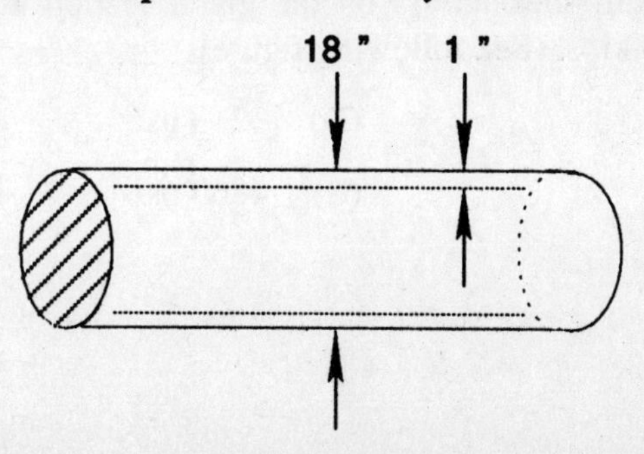

Given: the longitudinal stress cannot exceed 20 ksi,

the circumferential (tangential) stress cannot exceed 8 ksi,

the wall thickness is 1 inch,

the diameter of the vessel is 18 inches.

(A) 445 psi
(B) 890 psi
(C) 1780 psi
(D) 2220 psi
(E) 4440 psi

111. What is the maximum load, P, that can be applied to the connection shown if the shear stress in the rivets is limited to 14 ksi?

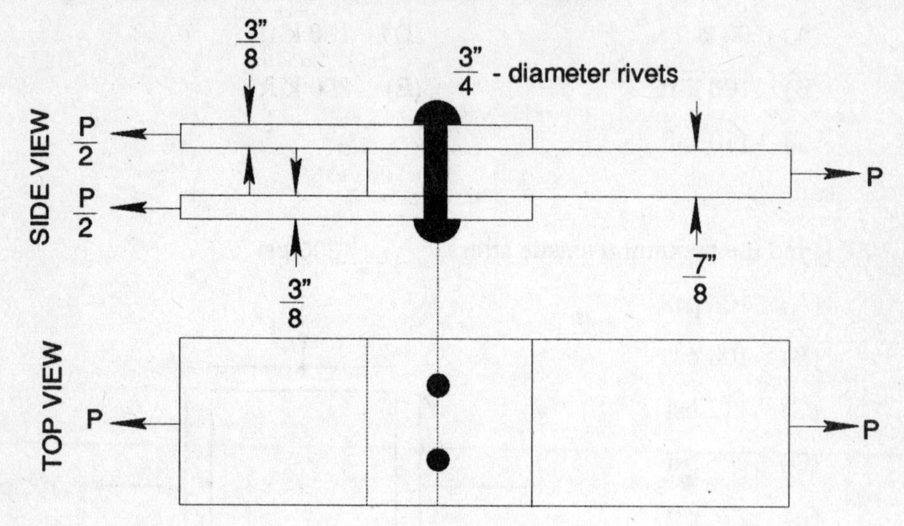

(A) 6 k
(B) 12 k
(C) 25 k
(D) 50 k
(E) 100 k

112. Find the maximum shear stress on the given section due to the vertical shear force of 50 kips. (See following figure)

(A) 0 psi
(B) 1250 psi
(C) 2640 psi
(D) 3210 psi
(E) 4860 psi

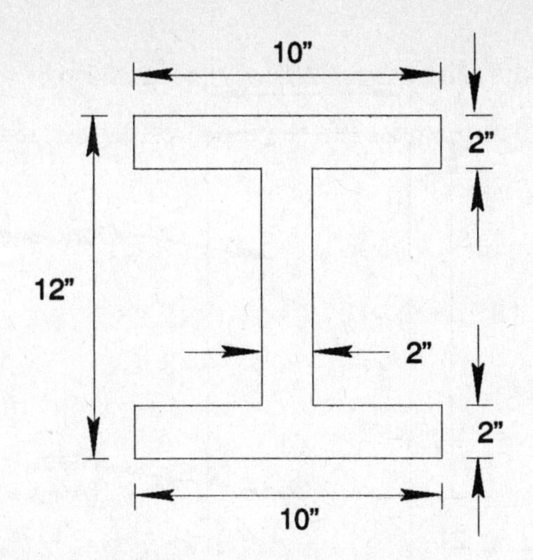

113. Find the maximum shear stress.

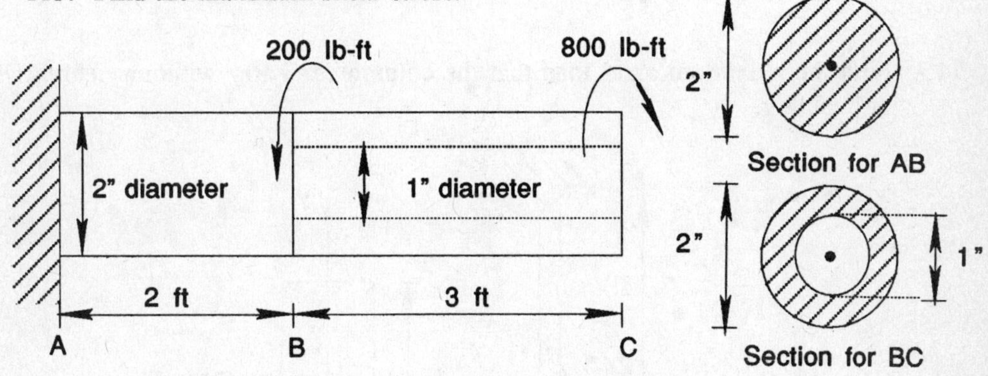

(A) 1940 psi
(B) 4590 psi
(C) 6520 psi
(D) 11100 psi
(E) 40000 psi

114. Find the maximum compressive stress in concrete. Given the moment at the section is 1,000,000 lb-in.

Modulus of Elasticity

Concrete	$E_c = 3 \times 10^6$ psi
Steel	$E_s = 30 \times 10^6$ psi

(A) 750 psi
(B) 1500 psi
(C) 1660 psi
(D) 3220 psi
(E) 30000 psi

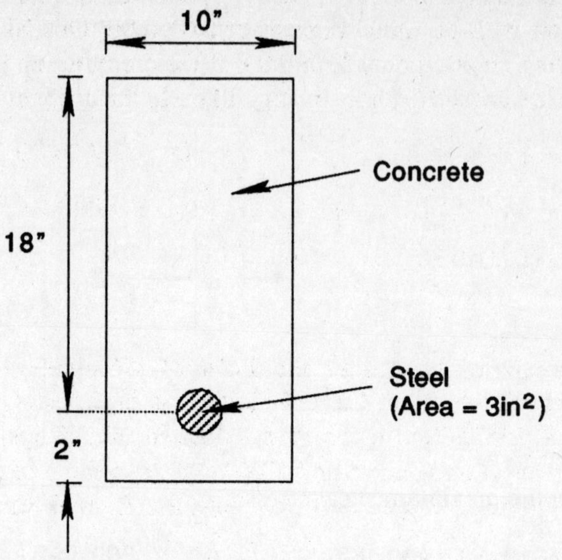

115. Find the maximum axial load that the column can carry without yielding or buckling.

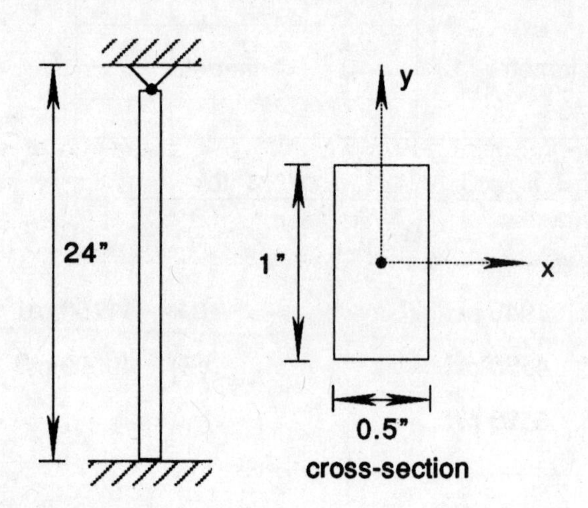

Modulus of Elasticity:

$E = 30 \times 10^3$ ksi

Stress at yield point:

$\sigma_{yp} = 30$ ksi

(A) 4 K

(B) 5 K

(C) 11 K

(D) 15 K

(E) 26 K

116. On December 20, 1989, Dixon opened an account at the Eastman Credit Union with an initial deposit of $1000.00. On February 20, 1990, he deposited an additional $1000.00. If the credit union pays 12% interest compounded monthly, how much will be in the account on March 20, 1990?

(A) $1030.20
(B) $1010.10
(C) $2,040.30
(D) $4,012.70
(E) $3050.50

117. A company requires an initial cost of $75,000.00 to purchase a minicomputer whose useful life is estimated to be 8 years. The computer will sell for $15,000.00 at the end of its useful life. If operating and maintenance costs are estimated to be $10,000.00 per year, and interest rate is 25% per annum, what is the Equivalent Uniform Annual Cost for the investment?

(A) $31,773.91
(B) $50,438.33
(C) $30,046.15
(D) $43,399.61
(E) $61,119.73

118. An engineering firm wants to purchase a device with a useful life of 10 years, at an initial cost of $700,000.00. The uniform annual benefit to be derived is worth $100,000.00, and the salvage value is $180,000.00. At 8% interest rate, what is Net Present Worth of the investment?

(A) – $51.41
(B) – $51,410.00
(C) + $54.38
(D) + $54,386.00
(E) $5,438.00

119. Ozok Systems International is considering buying equipment for $50,000.00. This equipment will have a salvage value of $8,000.00 after a useful life of 14 years. Using sum-of-years-digit (SOYD) method of depreciation, compute the depreciation charge on the equipment for the third year of its useful life.

(A) $16,400.00
(B) $4,800.00
(C) $8,200.00
(D) $2,400.00
(E) $9,600.00

120. Which of the following statements is true?

(A) The capitalized cost is always greater than the present worth of the costs for a project of finite life.

(B) The payback period analysis technique ignores time value of money.

(C) The payback formula: "Payback period = First cost/Annual Benefits" is always valid.

(D) (A) and (B)

(E) (B) and (C)

121. Martha deposited $100.00 per month is an account paying 6% interest compounded monthly for 24 months. Thereafter, she made no deposits or withdrawals for 5 years. How much must have accumulated in Martha's account after the seven-year period?

(A) $8,395.14 (D) $17,616.00

(B) $3,430.78 (E) $4,200.00

(C) $50,820.84

122. A young engineer borrowed $10,000.00 at 12% interest, and paid $2000.00 per annum for the first 4 years. What does he have to pay at the end of the fifth year in order to pay off the loan?

(A) $3,926.00 (D) $6,074.00

(B) $5,674.00 (E) $6,919.28

(C) $3,037.00

123. A 100 million pounds per year DMT Methanalysis plant costs $10,000,000.00 to build. If the DMT sells for 25¢ per pound, what is the return on investment for this plant? Profit after taxes is 10%. Assume that working capital is 25% of fixed capital.

(A) 10% (D) 15%

(B) 25% (E) 30%

(C) 20%

124. A bond pays $500.00 interest per year and has a face value of $5,000.00 at the end of 8 years, when it has to be redeemed. If its current discounted price is $3,900.00, what true interest could be earned on the bond?

(A) 15%
(B) 11%
(C) 13%
(D) 12%
(E) 9%

125. A pump costs $10,000.00. If installed, it will save the company $2,000.00 per year and will have a salvage value of $3,000.00 at the end of 10 years of useful life. The company plans to replace the pump with identical ones every 10 years for 30 years. Estimate the present worth of the entire 30 years of service at 10% interest rate.

(A) $3,445.76
(B) $3000.00
(C) $2,000.00
(D) $10,000.00
(E) $5,287.00

126. OSI company invested $173,000.00 in a fish pond. The useful life of the pond is expected to be six years. If the pond pays a profit of $6.70 per lb of fish sold, how many lbs of fish will OSI sell in order to break even? Assume the company's Minimum Attractive Rate of Return (MARR) is 15%.

(A) 3,412
(B) 4,286
(C) 8,624
(D) 173
(E) 6,824

127. Which of the following statements about the free carrier concentration associated with intrinsic semiconductors at room temperature is valid?

(A) It increases with increasing values of energy gap.
(B) It decreases with increasing values of energy gap.
(C) It increases with increasing carrier mobility.
(D) It decreases with increasing carrier mobility.
(E) It is independent of both energy gap and carrier mobility.

128. At absolute zero temperature (0 K), all the valence electrons in an intrinsic semiconductor:

(A) are in the valence band.

(B) are in the forbidden gap.

(C) are in the conduction band.

(D) are free electrons.

(E) are equally distributed between the valence band and the conduction band.

129. Carrier mobility depends on:

(A) resistivity.

(B) conductivity.

(C) recombination rate.

(D) charge per carrier.

(E) temperature and the regularity of the crystal structure.

130. The movement of charges from an area of high carrier concentration to an area of lower carrier concentration is called:

(A) gradient.

(B) recombination.

(C) diffusion.

(D) lifetime.

(E) mobility.

131. Which term describes a material whose properties depend on the direction of stress?

(A) Anisotropic

(B) Isotropic

(C) Symmetrical

(D) Asymmetrical

(E) Endotropic

132. A heat sink will dissipate heat more rapidly if:

(A) its temperature is the same as the temperature of the surrounding air.

(B) its surface area is increased.

(C) it is covered with a material having a high thermal resistance.

(D) it is kept away from air currents.

(E) its temperature is lower than the temperature of the surrounding air.

133. Materials that emit light in the absence of high heat and continue to emit light after the energy source has been removed are called:

(A) phosphorescent.

(B) fluorescent.

(C) semiconductor-laser diodes.

(D) light-emitting diodes.

(E) translucent diodes.

134. Ions are formed when

(A) electrons are displaced from atoms or molecules by bombardment of high energy particles.

(B) electrons are displaced from atoms or molecules by high energy electromagnetic radiation (for example, x-rays).

(C) a salt forms in an acid-base neutralization reaction.

(D) a salt crystal forms from the reaction of elemental sodium and chlorine.

(E) All of the above.

135. Which of the following statements about the two crystalline forms of carbon, graphite and diamond, is NOT correct?

(A) The properties of the two crystals are, in fact, more similar than dissimilar.

(B) The diamond crystalline lattice permits little or no relative atomic motion while the graphite lattice offers little resistance to relative atomic motion.

(C) The diamond lattice is transparent to visible light, graphite is not.

(D) Diamond is the hardest naturally-occurring substance; graphite is soft.

(E) At high temperatures diamond is a semiconductor; graphite is a conductor.

136. A gamma ray is composed of

(A) high energy protons.

(B) high energy electrons.

(C) high energy neutrons.

(D) high energy neutrinos.

(E) high energy electromagnetic radiation.

137. Which statement about crystal dislocations is NOT correct?

(A) A dislocation is a defect caused by an atomic misalignment in a crystal.

(B) Dislocations are the cause of plastic deformation of crystals under an applied shear stress.

(C) A dislocation is a crystalline lattice located in an unexpected place.

(D) Edge, screw, and mixed dislocations are possible.

(E) Dislocations affect the tensile strength of a crystal.

138. Plank's Constant (6.625 10^{-27} erg sec) represents

(A) the ratio between the magnitude of a quantum of radiated energy and its frequency.

(B) the quantum or smallest amount of energy that can be emitted by radiation.

(C) the smallest number of any physical significance.

(D) the numbers of ergs in a kilogram calorie.

(E) None of the above.

139. From the following table of masses of protons, neutrons and the $_2He^4$ atom and the applicable physical constants, determine the binding energy of the $_2He^4$ nucleus.

Species	Mass (atomic mass units)
proton (including electron)	1.007276
neutron	1.008665
$_2He^4$ atom	4.00387

Applicable Physical Constants

1.0 erg = 6.2422×10^{11} electron volts

Avogadro's Number = 6.023×10^{23} atoms/mole

speed of light = 3×10^{10}cm/sec

(A) 150 eV

(B) 26.1 MeV

(C) 8.5 BeV

(D) 2.5×10^{19} eV

(E) 1.57×10^{31} eV

140. The elements exhibit periodically recurring chemical and physical properties because

(A) they are arranged that way in the Periodic Table.

(B) the properties are largely determined by the quantum numbers of the outer electrons all of which, except for the first which indicates the orbit, recur in each orbit.

(C) probability considerations dictate that properties will recur since there are a limited number of possibilities.

(D) The reason is unknown; it is simply observed experimentally.

(E) None of the above.

TEST 1 (AM)

ANSWER KEY

1.	(D)	26.	(A)	51.	(D)	76.	(E)
2.	(B)	27.	(E)	52.	(D)	77.	(D)
3.	(C)	28.	(D)	53.	(B)	78.	(A)
4.	(B)	29.	(C)	54.	(E)	79.	(C)
5.	(A)	30.	(E)	55.	(E)	80.	(D)
6.	(C)	31.	(A)	56.	(B)	81.	(C)
7.	(C)	32.	(B)	57.	(C)	82.	(E)
8.	(E)	33.	(E)	58.	(E)	83.	(E)
9.	(E)	34.	(C)	59.	(C)	84.	(C)
10.	(C)	35.	(B)	60.	(A)	85.	(A)
11.	(C)	36.	(A)	61.	(C)	86.	(D)
12.	(A)	37.	(D)	62.	(D)	87.	(C)
13.	(C)	38.	(D)	63.	(A)	88.	(B)
14.	(C)	39.	(A)	64.	(D)	89.	(C)
15.	(C)	40.	(E)	65.	(B)	90.	(C)
16.	(B)	41.	(A)	66.	(E)	91.	(C)
17.	(A)	42.	(A)	67.	(B)	92.	(A)
18.	(A)	43.	(B)	68.	(D)	93.	(C)
19.	(B)	44.	(C)	69.	(A)	94.	(D)
20.	(B)	45.	(B)	70.	(D)	95.	(E)
21.	(D)	46.	(E)	71.	(C)	96.	(D)
22.	(B)	47.	(D)	72.	(B)	97.	(C)
23.	(D)	48.	(C)	73.	(C)	98.	(E)
24.	(C)	49.	(B)	74.	(A)	99.	(A)
25.	(D)	50.	(D)	75.	(B)	100.	(E)

101.	(B)	111.	(C)	121.	(B)	131.	(A)
102.	(B)	112.	(C)	122.	(E)	132.	(B)
103.	(B)	113.	(C)	123.	(C)	133.	(A)
104.	(A)	114.	(C)	124.	(A)	134.	(E)
105.	(A)	115.	(C)	125.	(E)	135.	(A)
106.	(B)	116.	(C)	126.	(E)	136.	(E)
107.	(C)	117.	(A)	127.	(B)	137.	(C)
108.	(D)	118.	(D)	128.	(A)	138.	(A)
109.	(B)	119.	(B)	129.	(E)	139.	(B)
110.	(B)	120.	(D)	130.	(C)	140.	(B)

DETAILED EXPLANATIONS OF ANSWERS

TEST 1

MORNING (AM) SECTION

1. **(D)**

A set of equations does not have a unique solution if the determinant of the coefficient matrix is zero. The determinant of the coefficient matrix is

$$\det \begin{vmatrix} 4 & 6 \\ 2 & 3 \end{vmatrix} = 4 \times 3 - 2 \times 6 = 0$$

Since the determinant is zero, (C) and (D) are the possible answers. But since one equation is not an exact multiple of the other, infinite numbers of solutions do not exist. This would have been the case if the right hand side of the second equation had been 5 instead of 7.

2. **(B)**

Since the velocity of Body A is double that of Body B,

$$V_A = 2V_B,$$

$$2t^2 - 12 = -2t$$

$$t^2 + t - 6 = 0$$

$$(t + 3)(t - 2) = 0$$

$$t = -3, t = 2$$

Since the solution sought is for time $t > 0$, the correct answer is $t = 2$. $t = -3$ is not a solution within the domain of required time, $t > 0$.

3. **(C)**

The characteristic equation is

$$s + 2 = 0; \qquad s = -2$$

and the homogeneous part of the solution is

$$y_h = Ke^{-2t}$$

The particular part of the solution is of the form $y_p = A$, which gives $A = {}^1/_2$ by inspection.

The complete solution is

$$y = y_h + y_p = Ke^{-2t} + {}^1/_2$$

Applying the initial condition, $y(0) = 1$, $y = {}^1/_2\, e^{-2t} + {}^1/_2$

Common mistakes: Choice (A) may be chosen if the characteristic equation root is taken as +2. Choice (D) may be chosen by applying the initial condition on only the homogeneous part. Choice (E) may be chosen if mistakes, both in parts (A) and (D), are comitted.

4. **(B)**

The mean of the sample is 5.75

$$S_t = (2 - 5.75)^2 + (8 - 5.75)^2 + (3 - 5.75)^2 + (10 - 5.75)$$

$$= 44.75$$

Standard deviation

$$= \sqrt{S_t/(n-1)}$$

$$= \sqrt{44.75/(4-1)}$$

$$= 3.862$$

Note that the sample standard deviation formula includes division by $(n - 1)$ and not (n). Common mistakes: Choice (A) may be chosen if $(n - 1)$ is replaced by (n). Sample standard deviation is the measure of the spread of the sample about the mean. One reason that division is made by $(n - 1)$, and not (n), is that there is no such thing as the spread of a single data point. For the case of $n = 1$, the formula for sample standard deviation gives a result of infinity.

5. **(A)**

In calculating the inner surface area of a cylinder, the surface area is given by

$$\text{Surface Area} = \pi D^2/4 \times 2 \text{ ends} + \pi DL$$

$$= \pi\, 3^2/4 \times 2 \text{ ends} + \pi \times 3 \times 2$$

$$= 21/2\, \pi$$

Common mistakes: Choice (B) may be chosen as the correct answer if one closed end is not accounted for. Choice (C) may be chosen if both the ends are not accounted for. Choice (D) may be chosen if the formula for the surface is chosen as $\pi D_2 \times 2$ ends + $2\pi DL$, that is, by mistake, substituting the diameter for the radius. Choice (E) may be chosen as the correct answer if the inside and outside surface area are both added.

6. **(C)**

An ordinary differential equation is linear if the coefficients of the dependent variable x and its derivatives are constants or are functions of the independent variable t. Only equation (III) meets this requirement. Common mistakes: The coefficients of the dependent variable terms in a linear differential equation need not be a constant. If all the coefficients are all constants, it is only a special case of linear ordinary differential equations called fixed coefficient linear ordinary differential equations.

7. **(C)**

$$f(x) = x^3 - 3x - 5$$
$$f'(x) = 3x^2 - 3$$

Hence the extremes are at $f'(x) = 0$, that is, $x = \pm 1$. Since $f''(x) = 6x$, the minimum occurs at $x = +1$ ($f''(x) > 0$). The value of the function at this point is

$$f(1) = 1 - 3(1) - 5$$
$$= -7.$$

8. **(E)**

Since each toss is independent and one can get heads or tails with the same probability in the next toss, the probability is still $^1/_2$. Common mistakes: Since heads was got in the earlier try, the probability of getting heads on the next toss may be thought to be zero in the next toss. Also, the probability of the event of getting 2 heads in consecutive tosses is $^1/_4$.

9. **(E)**

det $[C] \neq$ det $[A]$ + det $[B]$. Unlike matrices, determinants of matrices do not follow the addition law. For example

$$[A] = \begin{bmatrix} 2 & 1 \\ 1 & 0 \end{bmatrix}, \ [B] = \begin{bmatrix} -1 & -1 \\ 2 & 3 \end{bmatrix}, \text{ then } [C] = \begin{bmatrix} 1 & 0 \\ 3 & 3 \end{bmatrix}, \text{ then}$$

$$\det[A] = -1, \ \det[B] = -1, \ \det[C] = 3.$$

In mathematics, in order to prove that a statement is true, one needs to prove it for all cases. If the statement is to be proven to be false, one needs to find only one example which shows that it is false.

10. **(C)**

The shortest distance between two points (x_1, y_1) and (x_2, y_2) is given by the straight line distance between two points.

$$\text{Shortest Distance} = \sqrt{(x_1 - x_2)^2 + (y_1 - y_2)^2}$$

$$= \sqrt{(5-0)^2 + (-6-3)^2}$$
$$= \sqrt{106}.$$

11. **(C)**

Although
$$\lim_{k \to \infty} |a_k| = 0,$$
the series may diverge because the condition
$$\lim_{k \to \infty} |a_k| = 0$$
is only a necessary condition for a series to converge. However, if $n = k$, that is, the series is an alternating series, the condition
$$\lim_{k \to \infty} |a_k| = 0$$
becomes a necessary as well as a sufficient condition for convergence of a series.

12. **(A)**

$$f'(x) = 5x^{2/3}, \quad \text{and} \quad f''(x) = {}^{10}/_{3}x^{-1/3}.$$

The second derivative does not exist at $x = 0$ and the second derivative is never zero. So $x = 0$ is the only possibility of an inflection point. $f''(x) > 0$ for $x > 0$ and $f''(x) < 0$ for $x < 0$. This implies f is concave down on $(-\infty, 0]$ and concave up for $[0, \infty)$. Hence $f(x)$ has an inflection point at $x = 0$.

Common mistakes: Choice (B) may be chosen if points where $f''(x) = 0$ are considered to be the only possible points of inflection.

13. **(C)**

For a function $f(x)$, if $f(a)\,f(b) < 0$, there is at least one root in $a < x < b$. However, the function also needs to be continuous in $a < x < b$. For example,

$$f(x) = 1 \quad \text{if } 1 < x \leq 2,$$
$$f(x) = -1 \quad \text{if } 2 < x < 3,$$

satisfies the condition $f(1)\,f(3) < 0$ but does not have a root in the interval $1 < x < 3$ because the function is piecewise continuous.

14. **(C)**

Since the expression is of 0/0 (indeterminate) form at $x = 0$ and the functions on the numerator and the denominator are differentiable, one can apply L'Hôpital's rule to find the limit.

$$\lim_{x \to 0} \frac{\sin(5x) - 2x}{x} = \lim_{x \to 0} \frac{5\cos(5x) - 2}{1} = 3$$

Identical results can be also found by expanding $\sin(5x)$ in terms of a Maclaurin Series and then take the limit.

15. **(C)**

The angle between the two lines can be found by finding the slope of the two lines. The slope of the first line is 3 and the second line 4. Hence the angle between the two lines is

$$|\tan^{-1} 4 - \tan^{-1} 3| = 4.399°.$$

16. **(B)**

The probability of getting a red ball is

$$\begin{aligned} p\{\text{red}\} &= \{\text{no. of red balls}\}/\{\text{total balls}\} \\ &= 4/(4+3+5) \\ &= {}^1/_3 \end{aligned}$$

Hence

$$\begin{aligned} p\{\text{not red}\} &= 1 - p\{\text{red}\} \\ &= 1 - {}^1/_3 \\ &= {}^2/_3 \end{aligned}$$

17. **(A)**

$$\begin{aligned} \iint_R xy\,dA &= \int_1^2 \int_0^1 xy\,dx\,dy \\ &= \int_1^2 y/2\; dy \\ &= 3/4 \end{aligned}$$

The integral is found by integrating over y and then over x. Since the region R is a rectangle, the order of integration can be interchanged to get the same answer.

18. **(A)**

The given equation can be rewritten as follows

$$(x-1)^2 + (y+1)^2 = (\sqrt{2})^2,$$

which shows that the equation is of a circle with radius of $\sqrt{2}$ and center at $(1, -1)$.

19. **(B)**

Since the intercept is on the y–axis at $x = 0$, one of the points on the straight line is (0, –3). The other point (5, 2) is already given. The slope between two points (x_1, y_1) and (x_2, y_2) is given by

$$m = (y_1 - y_2) / (x_1 - x_2)$$
$$= [2 - (-3)] / [5 - 0]$$
$$= 1.$$

20. **(B)**

The distance covered between $t = {}^1/_4$ and ${}^1/_2$ seconds is

$$\int_{1/4}^{1/2} \sin(\pi t)\, dt = 1/\pi\, [-\cos(\pi t)]_{1/4}^{1/2}$$
$$= 1/(\pi\sqrt{2}) = 0.2251$$

Common mistakes are not including the negative sign in the integral of sin (πt). Also be sure that you change the angle mode in the calculator to the radian mode from the default degree mode. If the mode of angle is kept at degrees, you will get an answer of approximately zero, Choice (C).

21. **(D)**

The resistance R of a cable is given by

$$R = \frac{\rho L}{A},$$

where ρ is the resistivity of the cable, L the length of the cable and A its cross section, which is given by

$$A = \frac{D^2 \pi}{4}.$$

The resistance of the cable is therefore inversely proportional to the square of the diameter of the cable.

22. **(B)**

The two 4Ω resistors are connected in parallel and therefore are equivalent to:

$$\frac{1}{R_{eq}} = \frac{1}{R_1} + \frac{1}{R_2}$$
$$= 1/4 + 1/4$$
$$= 1/2$$
$$R_{eq} = 2\Omega.$$

The circuit has been reduced to three resistances in series as shown in the figure below.

The equivalent resistance is therefore equal to

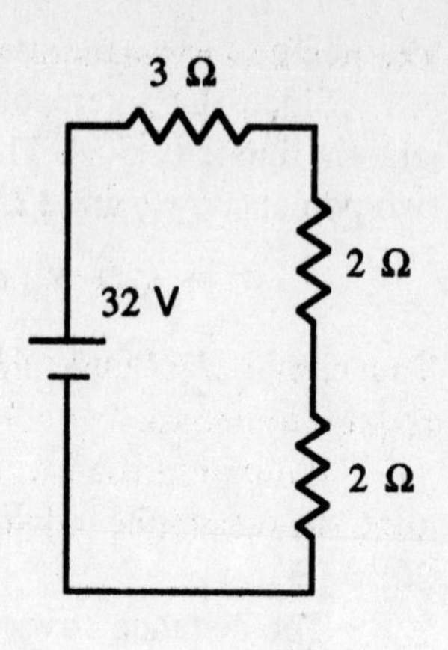

$$R_{eq} = R_1 + R_2 + R_3$$
$$= 2 + 2 + 3$$
$$= 7\ \Omega.$$

By Ohm's law the current in the circuit is given by:

$$i = \frac{V}{R_{eq}}$$
$$= 32/7$$
$$= 4.57\ \text{A}$$

and the voltage across the 3 Ω resistor is given by Ohm's Law

$$V = i R_3$$
$$= 4.57 \cdot 3$$
$$= 13.7\ \text{V}.$$

23. **(D)**

The 16 V voltage source has no resistance. Therefore if one connects an Ohm-meter between terminals *ab*, the device will measure the combinations of R_s shown in Figure (a) below.

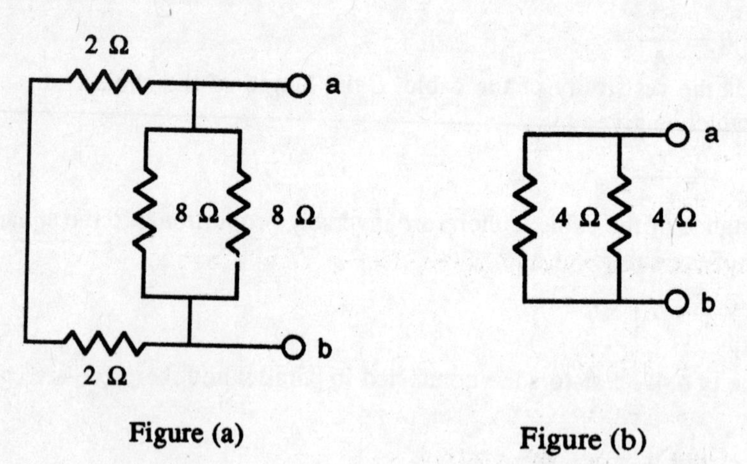

Figure (a) Figure (b)

The two 8 Ω resistances are in parallel and are therefore equivalent to:

$$\frac{1}{R_{eq}} = \frac{1}{R_1} + \frac{1}{R_2}$$
$$= 1/8 + 1/8$$
$$= 1/4$$
$$R_{eq} = 4\Omega.$$

The two 2 Ω resistances are connected in series and therefore are equivalent to:

$$R_{eq} = R_1 + R_2$$
$$= 2\,\Omega + 2\,\Omega$$
$$= 4\,\Omega.$$

The circuit has been simplified to the parallel combinations of R_s shown in Figure (b) above.

Following the procedure shown above, to combine two resistances in parallel, the resistance between terminal *ab* is $R_{ab} = 2\,\Omega$.

24. **(C)**
By the voltage divider rule the voltage between terminals *ab* is given by

$$V_{ab} = V\frac{R_1}{R_1 + R_2}$$
$$= 16\frac{2}{2+6}$$
$$= 4\,V$$

25. **(D)**
The three resistors are connected in parallel; therefore, their equivalent resistance is given by:

$$\frac{1}{R_{eq}} = \frac{1}{R_1} + \frac{1}{R_2} + \frac{1}{R_3}$$
$$\frac{1}{R_{eq}} = \frac{1}{2} + \frac{1}{4} + \frac{1}{4}$$
$$R_{eq} = 1.0\,\Omega.$$

The voltage between nodes *ab* is therefore by Ohm's law:

$$V_{ab} = R_{eq} \cdot i$$
$$= 1.0 \cdot 16$$
$$= 16.0\text{ V}.$$

and by Ohm's law the current i' through the 2 Ω resistor is:

$$i' = \frac{V_{ab}}{R}$$
$$= \frac{16.0}{2}$$
$$= 8.0\text{ A}.$$

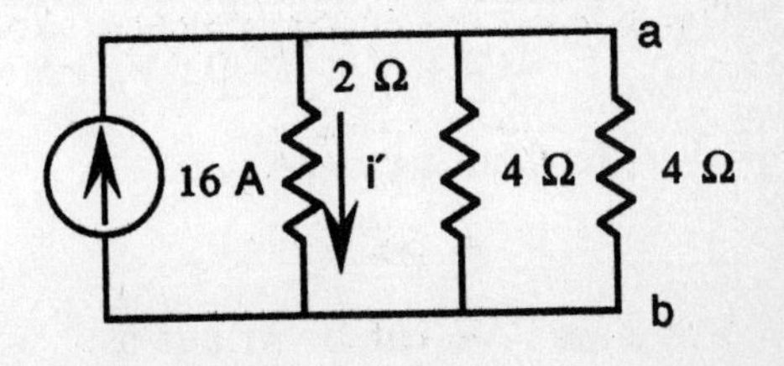

26. (A)

The 0.1 Ω resistor is connected to the secondary of an ideal transformer with a turn ratio of 1:10. It therefore is equivalent to a $0.1 \cdot 10^2 = 10$ Ω resistor connected to the primary of the transformer. The total resistance seen by the 100 V AC source is therefore 1 + 10 = 11 Ω and the primary current I is by Ohm's law

$$\frac{100}{11} = 9.1 \text{ A}$$

27. (E)

Because we are examining the magnetic field in the middle of a long wire, the wire can be considered infinite. Therefore by Biot-Savart's Law the magnetic field B at a distance r of an infinite straight long wire carrying a current i is given by

$$B(r) = \frac{\mu_0 i}{2\pi r},$$

where μ_0 is the permeability constant.

The magnetic field is therefore proportional to i / r.

28. (D)

A 5 A current source in parallel with a 2 Ω resistor can be transformed into a voltage source with $V = RI$ in series with an R resistor. The circuit can therefore be transformed into the circuit shown here.

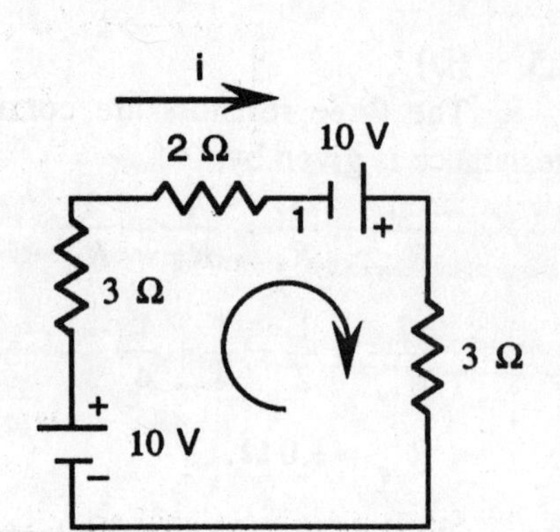

Kirchoff's voltage law around the loop establishes that:

$$-10 + 3i + 2i - 10 + 3i = 0.$$

Therefore:

$$i = 20/8$$
$$= 2.5 \text{ A}.$$

29. (C)

Kirchoff's voltage law (KVL) around the loop establishes that:

$$-24 + 3i + 6i - 0.5\, V_R + 6i = 0,$$

where by Ohm's Law:

$$V_R = 3i.$$

Therefore KVL can be written as:

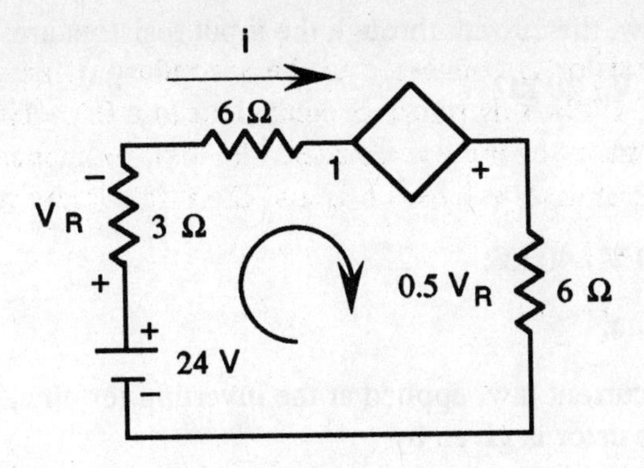

$$-24 + 3i + 6i - (0.5 \cdot 3i) + 6i = 0, \text{ or } 24 = 13.5i.$$

The current is therefore 1.78 A.

30. **(E)**

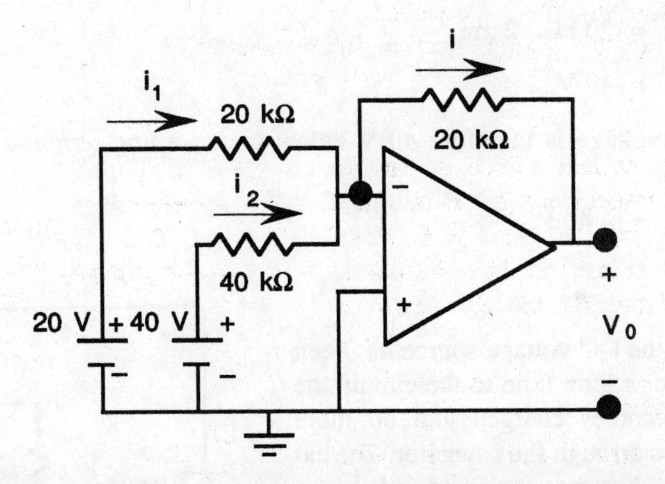

Since the operational amplifier has a very large input resistance, no current flows into either the inverting or the non-inverting terminals, and therefore the sum of the current i_1 supplied by the 20 V, and the current i_2 supplied by 40 V sources flows through the 20 kΩ feedback resistor to the output.

The output voltage of an operational amplifier is given by $v_0 = G\,(v_+ - v_-)$, where G is the gain

$$\frac{v_0}{G} = (v_+ - v_-).$$

As operational amplifiers have a very large gain, G tends to infinity and therefore

$$\frac{v_0}{G} = 0 \quad \text{and} \quad v_+ = v_-.$$

As the non-inverting terminal is grounded, the inverting terminal is also at ground potential (virtual ground), and $v_- = 0\ V$.

By Ohm's law, the current through the input resistors are:

$$i_1 = 20\text{ V} / 20\text{ k}\Omega$$

$$= 1\text{ ma}$$

and

$$i_2 = 40\text{ V} / 40\text{ k}\Omega$$

$$= 1\text{ ma},$$

and by Kirchoff's current law, applied at the inverting terminal node, the current through the 2 kΩ resistor is given by

$$i = i_1 + i_2$$

$$= 1 + 1$$

$$= 2\text{ ma}.$$

By Ohm's law the voltage drop across the 20 kΩ feedback resistor will be:

$$V = 20\text{ k}\Omega \cdot 2\text{ ma}$$

$$= 40\text{ V}.$$

The output voltage is therefore 40 V below the inverting terminal which is at 0 V; therefore

$$V_0 = -40\ V.$$

31. **(A)**

After the DC voltage source has been connected for a long time to the circuit, the capacitor becomes charged and no more current flows through the capacitor. The capacitor can therefore be modeled as an open circuit as shown here.

The voltage across the terminals of the capacitor is therefore 24 V.

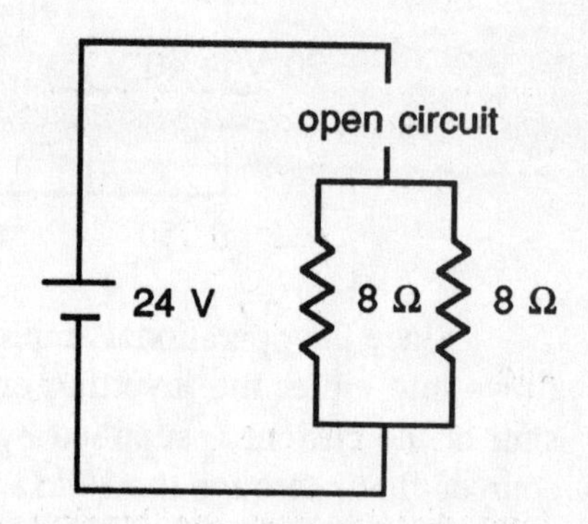

32. **(B)**

Since the capacitor has been connected to the circuit for many hours, it has been charged and no more current flows through it. The circuit has an open branch as shown below:

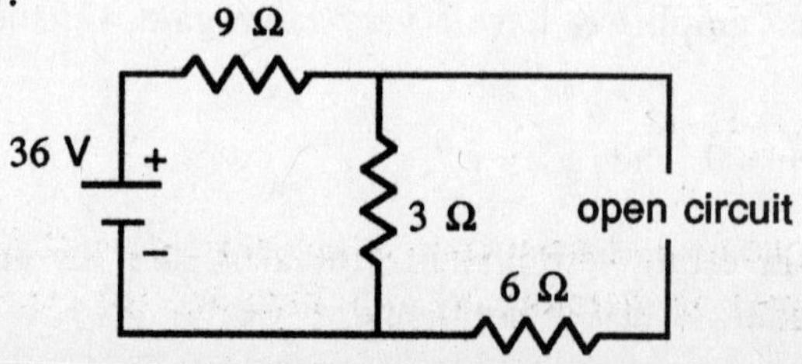

and

$$R_{eq} = R_1 + R_2$$
$$= 9 + 3$$
$$= 12 \text{ ohm.}$$

By Ohm's law the current is given by

$$i = \frac{V}{R_{eq}}$$
$$= \frac{36}{12}$$
$$= 3 \text{ A,}$$

and by Ohm's law

$$V_{ac} = R\, i$$
$$= 3 \cdot 3$$
$$= 9 \text{ V.}$$

Since no current flows through the 6 ohm resistor, $V_b = V_c$, and therefore $V_{ab} = V_{ac}$.

$$V_{ad} = 9 \text{ V.}$$

33. **(E)**

The two 4 Ω resistors are connected in parallel and can therefore be substituted by an equivalent resistance of 2 Ω. The time constant τ is given by

$$\tau = RC$$
$$= 2 \cdot 0.050$$
$$= 0.1 \text{ s}$$

and

$$1/\tau = 10.$$

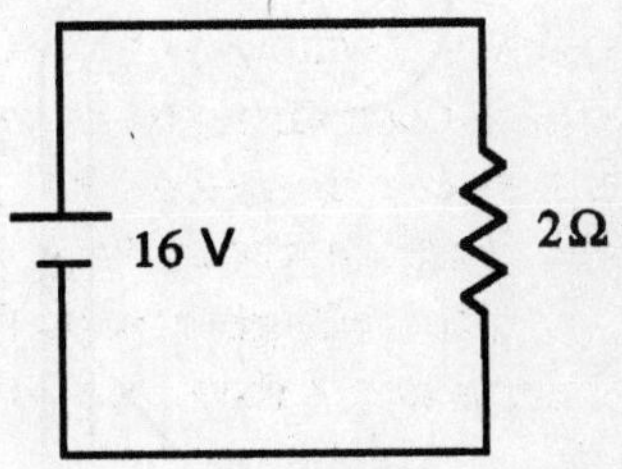

16 V

2Ω

Immediately after the breaker is closed the capacitor is not charged and therefore can be modeled by a short circuit as shown here:

By Ohm's law the initial current is given by:

$$i = \frac{V}{R_{eq}}$$
$$= 16/2$$
$$= 8 \text{ A.}$$

After the capacitor has been charged no more current flows through the circuit. The final current is therefore equal to zero and the expression for the current in the capacitor is given by

$$i(t) = 8e^{-10t},$$

where the response of a discharging RC circuit is given by

$$i(t) = I_o e^{-t/RC}.$$

34. **(C)**

The two capacitors are connected in series, and their equivalent capacitance is therefore given by:

$$\frac{1}{C_{eq}} = \frac{1}{C_1} + \frac{1}{C_2}$$

$$= 1/25 + 1/25$$

$$C_{eq} = 12.5 \text{ mF}.$$

The time constant τ is given by

$$\tau = RC_{eq}$$

$$= 8 \cdot 0.0125$$

$$= 0.1 \text{ s}.$$

35. **(B)**

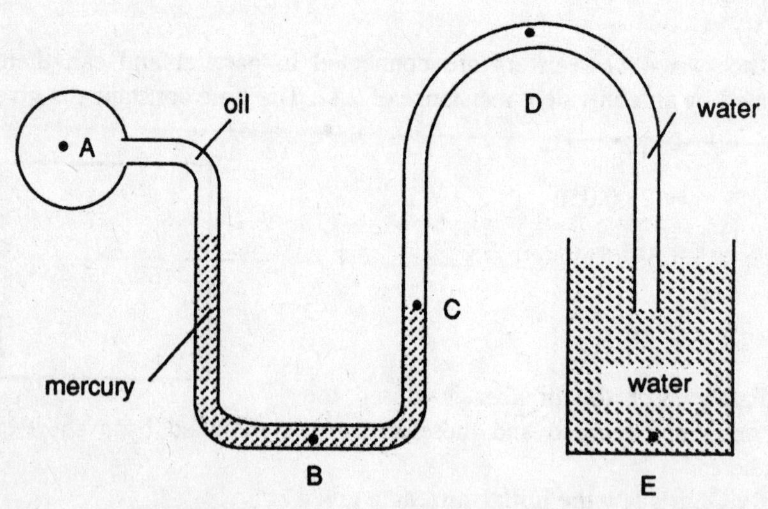

In the horizontal plane cutting through point C, the pressure through any of the three sections must be constant. Then, since mercury is denser than water, the pressure at B will be greater than that at E. All other pressures are lower.

36. **(A)**

The three fluid metering devices are sketched below.

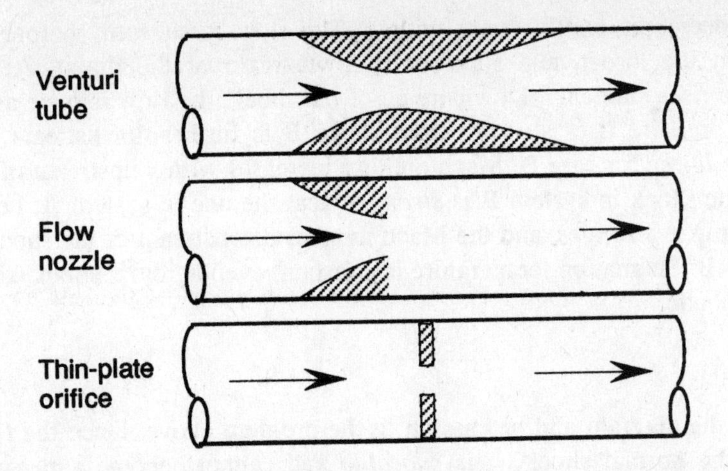

All three devices serve the same purpose: to measure volumetric flow rate in a pipe. Because of its smooth design, the Venturi tube has the lowest pressure loss. The thin-plate orifice, which is not smooth on either side, has the highest pressure loss. The flow nozzle is in between.

37. **(D)**

The Moody chart is designed for use with long straight sections of pipe. The flow in the pipe may be laminar or turbulent, and the walls may be smooth or rough. The only restriction is that the pipe must have constant cross-sectional shape. Even if the cross section is not circular, the Moody chart is still fairly accurate when hydraulic diameter is substituted for pipe diameter.

A subsonic diffuser is an expanding pipe, which does not have constant cross section; hence the Moody chart is of no use.

38. **(D)**

For a converging diverging nozzle without friction or heat addition, the pressure ratio p/p_0 is plotted against downstream distance x:

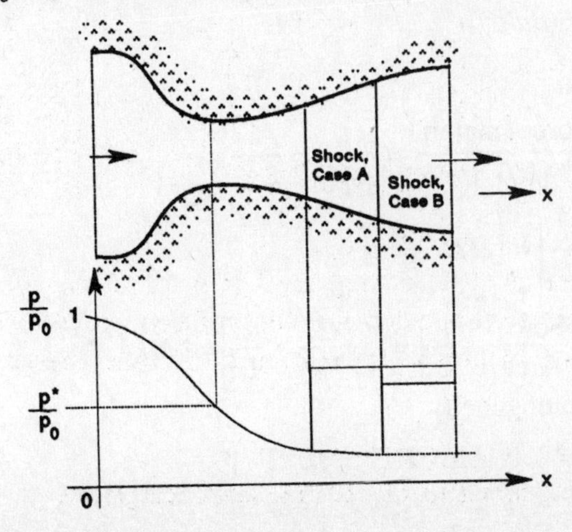

The pressure decreases continuously with x. The flow is subsonic before the throat, sonic at the throat, and supersonic downstream of the throat. At the shock, pressure rises suddenly. Downstream of the shock, the flow is once again *subsonic*, and pressure rises slowly. Since shock B is further downstream, the exit pressure is *lower* for case B. Mach number increases with x upstream of the shock. Thus, the shock in system B is *stronger* than the one in system A; hence the pressure jump is *stronger*, and the Mach number downstream of the shock is *lower* for case B. Stagnation temperature is constant, even across a shock, since the flow is adiabatic.

39. **(A)**

Refer to the diagram and discussion in the problem above. Once the flow goes through the normal shock, it is *subsonic* and cannot become supersonic again unless another throat would be added downstream. Choice (D) is incorrect because even though the flow is *choked*, changing stagnation pressure will still have an effect on the flow. Remember that it is the ratio of exit pressure to stagnation pressure that determines the shock location.

40. **(E)**

We use dimensional analysis to find the function relating Q and the variables r, M and dp / dx.

We can write the following

$$Q = (\text{constant})\,(r)^a\,(\mu)^b\,(dp/dx)^c$$

Our aim is to determine the powers a, b, c.

Express the variables in dimensional form:

$$Q = [L^3 T^{-1}]$$

$$r = [L]$$

The viscosity coefficient μ can be shear stress per unit velocity gradient, i.e.,

$$\mu = \tau / (du/dy)$$

$\therefore$ dimensions of μ:

$$\mu = \frac{[\text{Force/meter}]}{[L/T]} = \frac{ML}{LT^2} \cdot \frac{T}{L}$$

$$\mu = \left[\frac{M}{TL}\right] = \left[ML^{-1}T^{-1}\right]$$

$$\frac{dp}{dx} = \left[\frac{\text{Force}}{\text{Area}} \cdot \frac{1}{L}\right] = \left[\frac{ML}{L^2T^2} \cdot \frac{1}{L}\right] = \left[ML^{-2}T^{-2}\right]$$

For dimensional homogeneity

$$[L^3 T^{-1}] = [L]^a\,[ML^{-1}T^{-1}]^b\,[ML^{-2}T^{-2}]^c$$

$$[L^3 T^{-1}] = [L]^{a-b-2c}\,[M]^{b+c}\,[T]^{-b-2c}$$

Equating respective exponents:

Length: $3 = a - b - 2c$ (1)

Mass: $0 = b + c$ (2)

Time: $-1 = -b - 2c$ (3)

The above is a set of simultaneous equations

from (2) $b = -c$

from (3) $-1 = +c - 2c = -c$

$\therefore c = 1$ and $b = -1$

from (1) $a = 3 + b + 2c = 3 - 1 + 2 = 4$

$\therefore$ $Q = (\text{constant})\ r^4\mu^{-1}\frac{dp}{dx}$

$Q = (\text{constant})\ \frac{r^4}{\mu}\frac{dp}{dx}$

41. **(A)**

This form of the steady incompressible Bernoulli equation is valid everywhere for *irrotational* flow; but if the flow is *rotational*, it is valid only along streamlines of the flow. In other words, the constant in the right-hand side of the equation may be different for each streamline in a rotational flow field.

42. **(A)**

For static conditions, a close approximation for variation of atmospheric pressure over short height differences is

$$\Delta p = \rho g\,\Delta z$$

where density has been assumed to be constant.

Here, $\Delta z = 17m$, $\rho = 1.2\text{ kg/m}^3$, $g = 9.8\text{ m/s}^2$.

Thus

$$\Delta p = (1.2\text{ kg/m}^3)\ (9.8\text{ m/s}^2)(17\text{ m})\left(\frac{N\text{ s}^2}{\text{kg/m}}\right) = 200\text{ N/m}^2$$

43. **(B)**

Use the one-dimensional energy equation (in terms of head) from point ① to point ②, which are on the surface of the lake and the surface of the reservoir tank, respectively.

$$\frac{p_1}{\rho g} + \alpha_1\frac{V_1^2}{2g} + z_1 = \frac{p_2}{\rho g} + \alpha_2\frac{V_2^2}{2g} + z_2 + h_f + h_s - h_q \quad (1)$$

Here h_s = shaft work head done by the fluid (to be solved for)

h_q = heat transfer head (neglect this term)

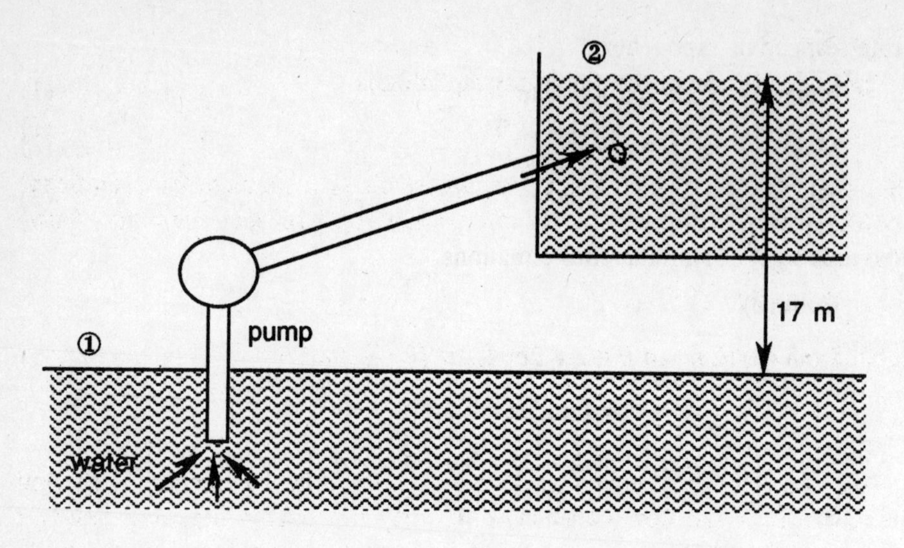

h_f = frictional head loss (given at 2.5 m of water)

$V_1 = V_2$ = average velocities (≈ 0 on the surfaces)

$p_1 \approx p_2 = p_a$ = atmospheric pressure on the surfaces. (Note that from above, $p_2 - p_a$ is negligible)

(1) becomes $h_z = z_1 - z_2 - h_f + -17$ m $- 2.5$ m $= -19.5$ m (h_s is negative because work is being done *on* the fluid).

Finally,

$$h_s = \frac{\dot{W}_s}{\dot{m}g}$$

where $\dot{W}_s$ = shaft power done by the water and $\dot{m}$ = mass flow rate = ρQ where Q = volume flow rate. Hence,

$$W_s = \rho Q g h_s$$

$$= (100 \text{ kg/m}^3)(6 \text{ m}^3/\text{hr})\left(\frac{1 \text{ hr}}{60 \text{ min}}\right)\left(\frac{1 \text{ min}}{60 \text{ s}}\right)(9.8 \text{ m/s}^2)(-19.5 \text{ m})$$

$$= -(319 \text{ kg m}^2/\text{s}^3)\left(\frac{\text{N s}^2}{\text{kg m}}\right)\left(\frac{\text{W s}}{\text{N}\cdot\text{m}}\right)$$

$$= -319 \text{ Watts}$$

Finally $\dot{W}_s$ is negative because power is supplied *to* the water, rather than *by* the water. The power supplied by the pump *to* the water is $-\dot{W}_s$ or 319 Watts.

44. **(C)**

For a pump with efficiency $\eta = 0.65$,

$$P_{\text{electrical}} = \frac{-\dot{W}_s}{\eta} = \frac{319 \text{ Watts}}{0.65} = 491 \text{ Watts}$$

45. **(B)**

The general form of the continuity equation is

$$\frac{\partial \rho}{\partial t} + \nabla \cdot (\rho \mathbf{V}) = 0 \qquad (1)$$

which is valid for steady or unsteady, compressible or incompressible, rotational or irrotational flow. Only when ρ = constant (i.e., only for *incompressible* flow) does (1) reduce to

$$\nabla \cdot \mathbf{V} = 0.$$

Thus, choices (A), (C), and (D) are valid, but (B) is not.

46. **(E)**

This is a standard converging-diverging nozzle. In the converging section (upstream of A_{min}), the flow is *subsonic*. Hence the velocity increases, pressure decreases, and density decreases in the flow direction. Downstream of the throat (i.e., in the diverging section of the duct), the flow must be everywhere supersonic. Note that if a normal shock were to appear in the duct, the flow would change suddenly from supersonic to subsonic. We know this is not the case since the flow was given as *supersonic* at the exit plane. In the converging section then, the velocity increases, pressure decreases, and density decreases. Hence, the density at the exit plane must be *less than* the density at A_{min}.

47. **(D)**

For one-dimensional flow,

$$\dot{M} = \iint_{\text{cross section}} \rho u^2 dA$$

For a uniform flow, $\dot{M} = \rho V^2 A$ where A is the cross-sectional area of the pipe.

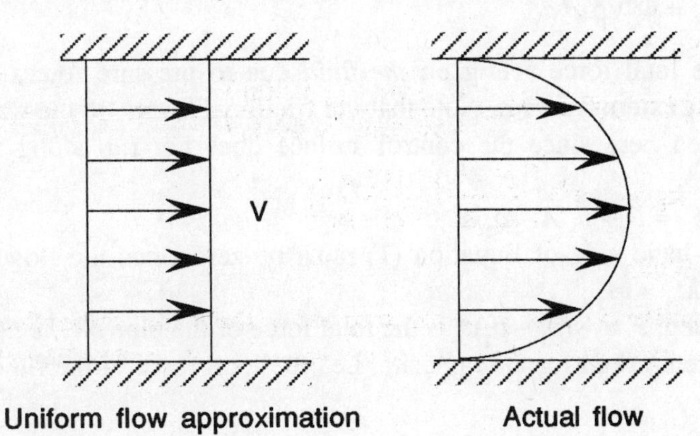

If the flow were uniform,

$$\dot{M}_1 = (1000 \text{ kg/m}^3)(6 \text{ m/s})^2 (\pi / 4)(0.05 \text{ m})^2 = 70.7 \text{ kg}\cdot\text{m/s}^2$$

For fully developed flow, the velocity is *not* uniform; hence the momentum flux will be somewhat greater than this value. Here, $Re = V_{av} D / \upsilon$ = (6 m/s) (0.05 m) / 1.0×10^{-6} m²/s = 300,000. Thus, the flow is *turbulent* and the correction factor for momentum flux will only be a few percent.

Thus

$$\dot{M}_1 \text{ actual} = (70.7 \text{ kg m/s}^2)(1.02) \approx 72.0 \text{ kg m/s}^2$$

Finally, $\dot{M}_1 = \dot{M}_3$ since the flow is fully developed. (In other words, the velocity profile shape does not change with downstream distance.)

Hence $\dot{M}_3 \approx 72.0$ kg m/s²

48. **(C)**

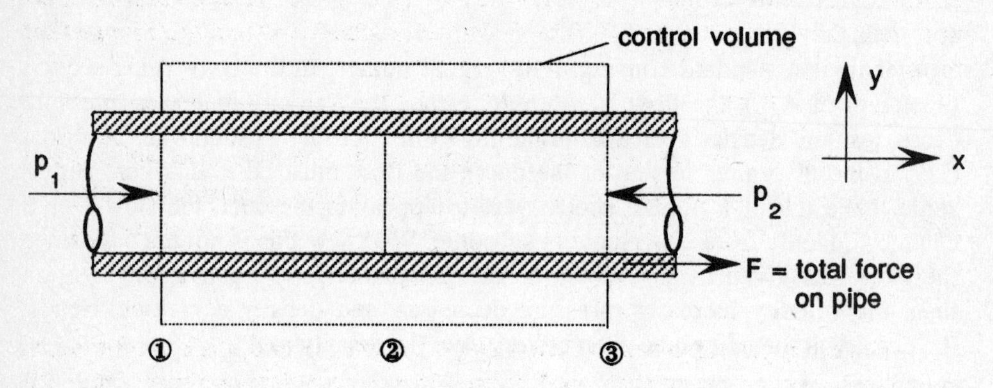

By conservation of momentum, the total force **F** acting on the pipe can be found by summing forces on the fluid and calculating momentum fluxes and pressure forces on the control volume sketched above. Note that the control volume cuts through the pipe walls at sections 1 and 3.

Conservation of momentum in the x direction is

$$\Sigma F_x = \oint\!\!\oint_{cs} \rho u^2 dA \qquad (1)$$

Here ΣF_x is the total force acting *on the fluid* due to pressure forces, viscous forces, and other external forces. Note that the frictional forces on the walls need not be calculated here since the control surface does not run along the wall surface.

Hence, $\Sigma F_x = F + p_1 A - p_2 A$

The right-hand side of Equation (1) must be zero since the flow is fully developed, and $\dot{M}_1 = \dot{M}_3$.

Finally, then, $F = -(p_1 - p_2)A$ is the total force of the pipe *on the fluid*.

The problem asked for the opposite, i.e., the total force of the fluid on the pipe,

$$F = (p_1 - p_2)A$$

$$= (200{,}000 \text{ N/m}^2 - 110{,}000 \text{ N/m}^2)(\pi / 4)(0.05\text{m})^2$$

$$F = 177 \text{ N}$$

49. **(B)**

An adiabatic process has no associated heat transfer; the heat transfer is zero. An adiabatic process can be isothermal, and an isothermal process can be adiabatic, however, neither is the precondition for the other. A word about entropy; if the process occurs, the entropy must change, unless the process is reversible and is then adiabatic and reversible. The last case is referred to as an isentropic process.

50. **(D)**

The Carnot efficiency is the standard against which we measure the cycle efficiency of other cycles. If the efficiency of other cycles exceeds that of Carnot, then they are not possible. The Carnot efficiency is expressed in equation form as

$$n_{th,rev} = 1 - T_c / T_h \quad \text{or} \quad n_{th,rev} = (T_h - T_c) / T_h$$

where T_c and T_h are the absolute temperatures of the cold and hot thermal reservoirs, respectively. Only another reversible cycle operating between the same thermal reservoirs can have an efficiency EQUAL to the Carnot efficiency.

51. **(D)**

For the rigid container, a closed system, there are two work modes; work associated with a rotating shaft, and expansion and compression work. Since the container is rigid the latter is zero. Thus the work associated with the heating process is zero.

52. **(D)**

The Second Law of Thermodynamics provides the mechanism to test the applicability of solutions provided by the other laws. The direction of processes, the theoretical performances of cycles, relationships of properties, and the definition of the absolute temperature scale are important aspects of the Second Law. Often the First Law will be satisfied, but if the Second Law is not satisfied, the process/cycle fails the critical test.

53. **(B)**

The Mach number decreases across a normal shock. It follows that the associated velocity will also decrease. From the energy equation and continuity, the rise in pressure and temperature are shown to be inversely proportional to the local Mach number. Pressure and temperature increase as the Mach number decreases. Since there is no heat transfer, the entropy (ies) must be either constant (reversible) or must increase (irreversible). A shock is one of the best examples of an irreversible process, thus the entropy must increase.

54. **(E)**

Properties are point functions while heat and work are path functions, depending on both the end points and the path followed. Only properties can be used to fix the state of the medium. Heat and work can be used to evaluate properties.

55. **(E)**

The coefficient of performance of a refrigerator is defined as the cooling capacity divided by the work required to make the refrigerator operate. The cooling effect is associated with the evaporator. The work is associated with the compressor. In this problem units must be converted to either kW or kJ/kg for each device. The mass flow rate is provided for this reason. Answer (A) is obtained if the units are not changed. (C) and (D) would be obtained if the device was analyzed as a heat pump, with (D) being correct in this case.

56. **(B)**

The isentropic (process) efficiency is a comparison of the adiabatic actual device with the adiabatic reversible (ideal) device, operating between the same inlet state point and exit pressure. The efficiency must be less than one unless the actual device is adiabatic and reversible.

57. **(C)**

The efficiency of the compression ignition cycle is calculated by dividing the work of the cycle by the heat added. The heat is added during the constant pressure process and is labeled Q_{23}. For a cycle the work equals the heat transfer, thus the efficiency in answer (C) would be 100%. This is totally impossible to obtain.

58. **(E)**

The compression ratio for the spark ignition engine results in a value of the order of 8-10. The equation given in answer (E) would result in a fractional compression ratio. Since V_4 equals V_1 and V_3 equals V_2, the ratio should be V_4/V_3 to be correct.

59. **(C)**

The Rankine cycle is the ideal cycle used to model the steam power cycle. The others are: Otto-spark ignition cycle, Diesel-compression ignition cycle, Brayton-gas turbine cycle, and Stirling (like the Carnot cycle with the two isentropic processes replaced by two constant-volume regeneration processes).

60. **(A)**

The expansion process in an ideal refrigeration cycle is modeled as an adiabatic process with no work, and no change in potential or kinetic energy.

Although the enthalpy is probably not constant throughout the process, the end state enthalpies are the same. This can be easily verified by applying the first law for a control volume. The strongly irreversible process which takes place is known as throttling.

61. **(C)**

In the ideal turbojet engine, the sole purpose of the turbine is to produce power to run the compressor. The magnitude of the turbine work is exactly equal to the magnitude of the compressor work. The process efficiencies are equal to 100%, and since the nozzle is considered adiabatic, the nozzle exit velocity is a function of the inlet and exit temperatures.

62. **(D)**

Even with the regenerator installed, there must be energy "wasted" in every power cycle. The case described in (D) violates the Kelvin-Planck statement of the second law because it implies a cyclic device which produces power and exchanges heat with a single reservoir.

63. **(A)**

First, find the velocity-time curve for the particle. Since the acceleration is either constant or zero, and velocity is the integral of acceleration, the velocity curve consists of straight lines connecting points whose abscissa are the times when the acceleration changes ($t = 0, 6, 16, 36, 46$ s) and whose ordinates are the total area under the a - t curve up to that value of t.

$t = 0 \quad v = 0$ (0, 0)

$t = 6, \quad v = (5)(6)$ (6, 30)

$t = 16, v = (5)(6) + (7)(10)$ (16, 100)

$t = 36, v = (5)(6) + (7)(10)$ (36, 100)

$t = 46, v = (5)(6) + (7)(10) + (-10)(10)$ (46, 0)

This curve is shown in Figure 1.

Now find the position-time curve for the particle. Position is the integral of velocity, and the points to be connected on the x-t curve will be determined by the same method as those on the v-t curve. However, now the velocity is constant only between $t = 16s$ and $t = 36s$. The only straight line segment will be present over this interval of the x-t curve. The other sections will have parabolic curves, turning upward for $t < 16s$, and downward for $t > 36s$. The defining points are:

$t = 0 \quad x = 0$ (0, 0)

$t = 6 \quad x = \frac{1}{2}(6)(30)$ (6, 90)

$t = 16,\ x = {}^1/_2(6)\ (30) + {}^1/_2(10)\ (30 + 100)$ (16, 740)

$t = 36, x = 740 + (20)\ (100)$ (36, 2740)

$t = 46, x = 3740 + {}^1/_2(10)\ (100)$ (46, 3240)

This curve is shown in Figure 2. Also, the result of $x(t = 46s) = 3240$ ft yields the distance between points *A* and *B*.

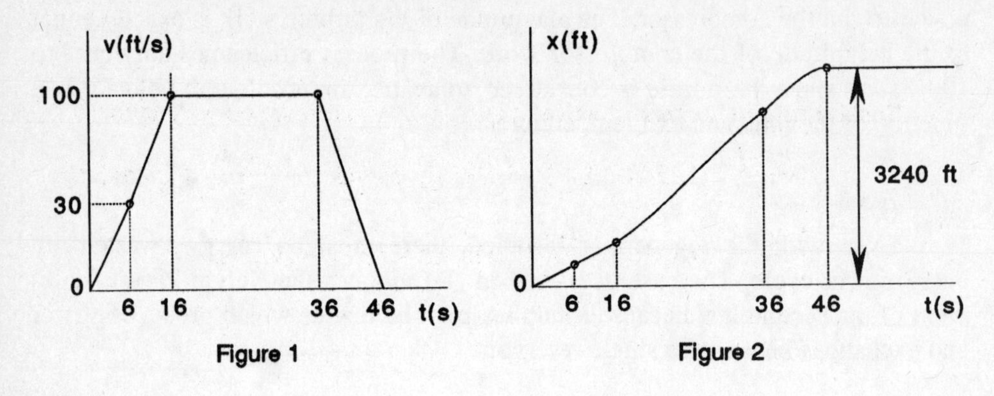

Figure 1 Figure 2

64. **(D)**

There are no external forces in this system since the explosion is caused by an internal force. Therefore linear momentum of the system is conserved.

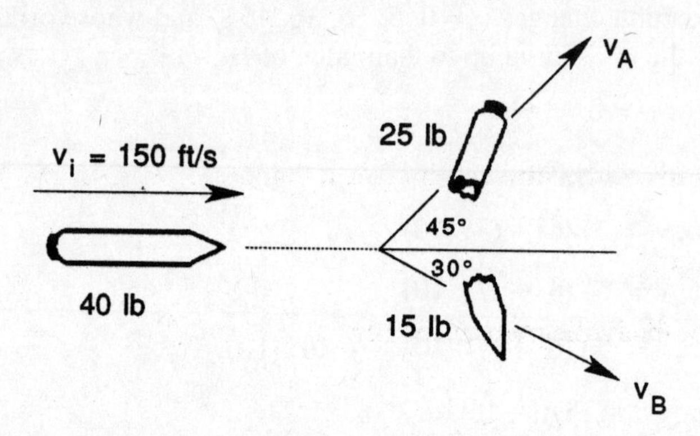

Momentum before = Momentum after

$$(m_A + m_B)\ v_0 = m_A\ v_A + m_B\ v_B$$

This equation can be broken into components. In the *x*–direction

$$(m_A + m_B)\ v_0 = m_A\ v_A \cos 45° + m_B\ v_B \cos 30°$$

and in the *y*–direction

$$(m_A + m_B)\ (0)\ = m_A\ v_A \sin 45° - m_B\ v_B \sin 30°$$

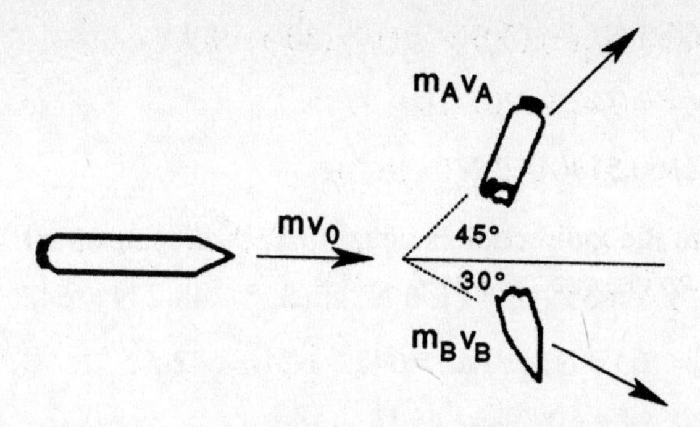

Substituting in numerical values yields a set of simultaneous equations.

$$\left[\frac{40\text{ lb}}{g}\right](150\text{ ft/s}) = \left(\frac{25\text{ lb}}{g}\right)V_A \cos 45° + \left(\frac{15\text{ lb}}{g}\right)V_B \cos 30°$$

$$\left[\frac{40\text{ lb}}{g}\right](0) = \left(\frac{25\text{ lb}}{g}\right)V_A \sin 45° - \left(\frac{15\text{ lb}}{g}\right)V_B \sin 30°$$

Cancelling out g, and then solving for v_A and v_B yields

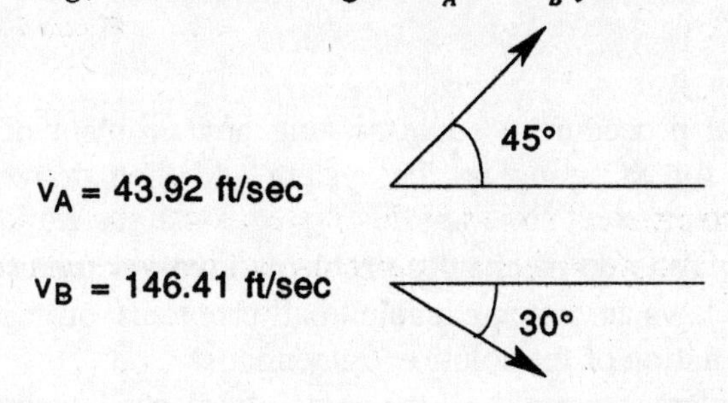

65. **(B)**

The velocity of the box can be found using the work-kinetic energy relation:

$$W = \Delta KE$$

First, draw a free-body diagram of the box:

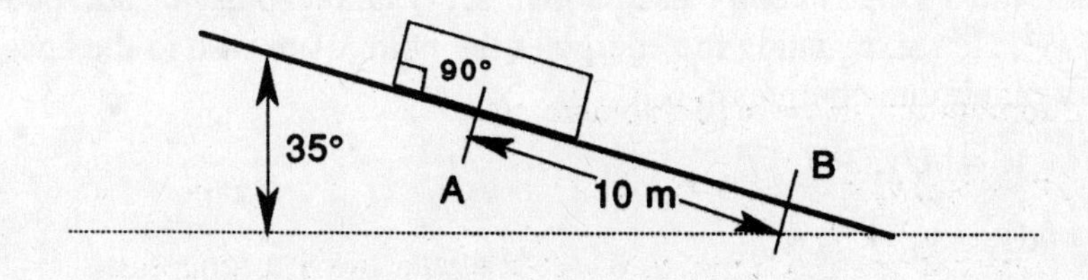

$$W = (20\text{ kg})(9.8\text{ m/s}^2) = 196\text{ N}$$

Using the equation of equilibrium in the y-direction,

$$\Sigma F_y = 0$$

$$(196\text{ N})\cos 35° + N = 0$$

$$N = 166.5 \text{ N}$$

$$f = \mu N$$

$$f = .3\ (160.5) = 48.2 \text{ N}$$

The work done in the x-direction is equal to $\Sigma F_x \cdot$ displacement.

$$\Sigma F_x = W \sin 35° - f = (196 \text{ N}) \sin 35° - 48.2 \text{ N} = 64.3 \text{ N}$$

$$\text{Work} = (64.3 \text{ N})\ (10 \text{ m}) = 642.5 \text{ Nm} = 642.5 \text{ J}$$

$$\text{Work} = \Delta KE = {}^1/_2\, m\, V_b^{\,2} - {}^1/_2\, m\, V_a^{\,2}$$

$$V_a = 0$$

$$m = 20 \text{ kg}$$

$$647.5 \text{ N} \cdot m = {}^1/_2\ (20 \text{ kg})\ (V_a)^2$$

$$V_b = 8.02 \text{ m/s.}$$

66. **(E)**

The general procedure used in solving any problem in mechanics is to calculate all the forces acting on the system and then derive the equation of motion of the system.

An easier way to do mechanics problems involves the use of conservation principles. These laws are not applicable to all problems, but when they are, they simplify the calculation of the solution tremendously.

In this problem, we may use the principle of conservation of energy. We relate the energy of the block before it was released to the block's energy at the point of maximum compression (see figure). At the moment of release, the kinetic energy is zero. At the moment when maximum compression occurs, there is also no kinetic energy.

As shown in the figure, the reference level for gravitational potential energy is the surface S. The initial gravitational potential energy of m is mgy_1. At the point of maximum compression, the potential energy of m is mgy_2. However, at this point, the spring is compressed a distance y and also has elastic potential energy ${}^1/_2\, ky^2$. Hence, equating the energy at the point of release to the energy at the point of maximum compression,

$$mgy_1 = mgy_2 + {}^1/_2\, ky^2$$

$$mg(y_1 - y_2) = {}^1/_2\, ky^2$$

But $y_1 - y_2 = h + y$ and

$$mg(h + y) = {}^1/_2\, ky^2$$

$$y^2 = \frac{2\, mg}{k}(h + y)$$

$$y^2 - \left(\frac{2\,mg}{k}\right)y - \frac{2\,mgh}{k} = 0$$

Therefore, using the quadratic formula to solve for y,

$$y = \frac{1}{2}\left(\frac{2mg}{k} \pm \sqrt{(2mg/k)^2 + (8mgh/k)}\right).$$

67. **(B)**

The forces on the body, resolved in the plane and perpendicular to the plane are shown in following figure.

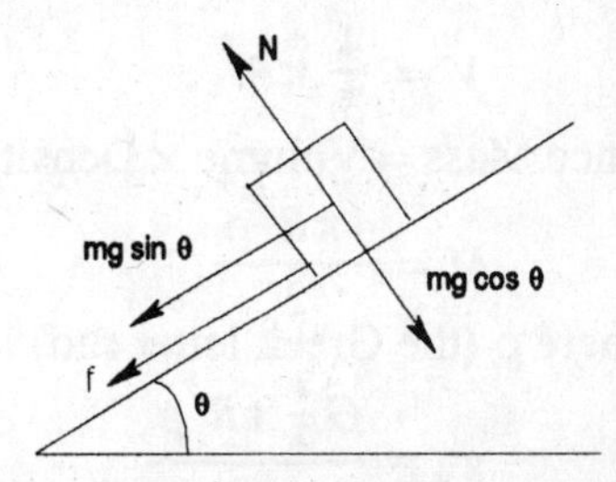

The motion is perpendicular to the normal force N and the $mg\cos\theta$ component of gravity: they do not work on the block. The other two forces $mg \sin \theta$ and $f = \mu N = \mu\, mg \cos \theta$, are along the path of motion and do work. The amount of which is equal to their magnitudes, which are constant, times the distance d the body travels;

$$W = -(mg \sin \theta)\, d - (\mu\, mg \cos \theta)\, d.$$

This quantity of work is equal to the energy loss, from the body's initial kinetic energy,

$$\Delta KE = -\tfrac{1}{2} m v_0^2$$

$$-\tfrac{1}{2} m v_0^2 = -(mg \sin \theta)\, d - \mu\, (mg \cos \theta)\, d \qquad (1)$$

$$d = \frac{v_0^2}{2g(\mu \cos \theta + \sin \theta)}.$$

The purist may say this analysis misleadingly puts the non-conservative force, friction, on equal footing with the conservative force, gravity. The results, however, are identical. Using the most general energy conservation law,

$$W_{nc} = \Delta E + \Delta V$$

yields equation (1) again. $W_{nc} = -\mu mgd\cos\theta$, $\Delta E = -\tfrac{1}{2} m v_0^2$ and $\Delta V = mg\Delta h = mgd\sin\theta$ so

$$-\mu mgd\cos\theta = -\tfrac{1}{2} m v_0^2 + mgd \sin \theta$$

$$\theta = \frac{v_0^2}{2g(\mu \cos\theta + \sin\theta)}.$$

68. **(D)**

This problem must be approached carefully. We must express the acceleration due to gravity in terms of the density and the radius of the planet. If the radius is R and the mass of the planet M, then the acceleration due to gravity on its surface is found from Netwon's Second Law, $F = ma$. Consider an object of mass m on the surface of the planet. Then the only force on m is the gravitational force F, and

$$F = \frac{GMm}{R^2}$$

But a is the acceleration of m due to the planet's gravitational field, or g_p. Then

$$g_v = \frac{GM}{R^2}$$

Assuming the planet is spherical, its volume is the volume of a sphere of radius R:

$$V = \frac{4}{3}\pi R^3$$

Since Mass = Volume × Density.

$$M = \frac{4\pi R^3 \rho}{3}$$

where ρ (the Greek letter rho) is the density of the planet. Therefore

$$g_v = \frac{G\frac{4}{3}\pi R^3 \rho}{R^2}$$

$$= \frac{4\pi}{3} GR\rho$$

Similarly, the acceleration due to gravity on the surface of the earth is

$$g = \tfrac{4}{3}\pi GR_{e}\rho_e$$

where ρ_e is the density of the earth, and R_e is its radius. If

$$g_v = g$$

Then

$$\tfrac{4}{3}\pi GR\rho = \tfrac{4}{3}\pi GR_{e}\rho_e$$

Canceling $^4/_3\ \pi\ G$ on both sides

$$R\rho = R_{e}\rho_e.$$

If the density of the planet is twice that of the earth,

$$\rho = 2\rho_e$$

So $\quad R2\rho_e = R_{e}\rho_e$

Whence

$$R = {}^1/_2 R_e$$

$$= {}^1/_2 \times 6.38 \times 10^6 \text{ m}$$

$$= 3.19 \times 10^6 \text{ m}$$

The radius of the planet is one half of the radius of the earth, or 3.19×10^6 meters.

69. **(A)**

Shown is a free-body diagram of the door, of mass m and length $2l$. Not shown is the fictitious force *ma* acting at the center of mass of the door. In the car's frame of reference we consider this force to be real and use it in developing a solution.

The forces A_x and A_y are unknown. When one takes moments about *A*, these two unknown forces passing through point *A* would not appear in the moment equation. The moment equation then, is

$$ma_y = I_A \alpha$$

$$(ma)\ l \cos\theta = I_A \alpha \text{ (since } y = l\cos\theta)$$

Solving for the angular acceleration of the door,

$$\alpha = \frac{ma\ l\cos\theta}{I_A}.$$

70. **(D)**

This is a conservation of linear momentum problem. The common velocity can be found by equating the momentum of the system (cars $A + B$) before they locked together to the momentum of the system *after* they locked together.

With

M_A : mass of car $A = 1150$ kg

M_B : mass of car $B = 1300$ kg

V_A : speed of car $A = 4.5$ m/s

V_B : speed of car $B = 6.7$ m/s

V_C : common speed

$$\underbrace{M_A V_A + M_B V_B}_{\text{momentum before collision}} = \underbrace{(M_A + M_B)V_C}_{\text{momentum after collision}}$$

$$(1150 \text{ kg})(4.5 \text{ m/s}) + (1300 \text{ kg})(6.7 \text{ m/s}) = (1150 \text{ kg} + 1300 \text{ kg})V_c$$

$$V_C = \frac{13.885 \text{ kgm/s}}{2450 \text{ kg}}$$

$$V_C = 5.67 \text{ m/s}.$$

71. **(C)**

To find the velocity of point *B* we will use the following equation:

$$\mathbf{V}_B = \mathbf{V}_C + \mathbf{V}_{B/C}$$

where $\mathbf{V}_B$ = velocity of point *B*

$\mathbf{V}_C$ = velocity of point *C*

$\mathbf{V}_{B/C}$ = velocity of point *B* relative to point *C*.

In addition, we know that

$$\mathbf{V}_{B/c} = \omega \times \mathbf{r}_B$$

or $\quad \mathbf{V}_B = \mathbf{V}_C + (\omega \times \mathbf{r}_B)$

where $\quad \mathbf{w} = (-V_c / r)\,\mathbf{k} = (-2.5 \text{ m/s} / .5m)\,\mathbf{k} = -5k\mathbf{k}\text{rad/s}$

Note: The vector ω is directed into the paper by the right-hand rule, hence the negative sign.

$$\mathbf{r}_B = .3\,m\,(\cos 45°\mathbf{i} + \sin 45\,°\mathbf{j})\,m = (.212\,\mathbf{i} + .212\,\mathbf{j})\,m$$

$$\mathbf{V}_C = 2.5\,\mathbf{i} \text{ m/s}$$

Now, by solving the vector equation,we find $\mathbf{V}_B$:

$$\mathbf{V}_B = 2.5\mathbf{i} \text{ m/s} + [-5\mathbf{k} \text{ rad/s} \times (.212\mathbf{i} + .212\mathbf{j})\text{m}]$$

$$\mathbf{V}_B = 2.5\mathbf{i} \text{ m/s} + [-1.06\mathbf{j} \text{ m/s} + 1.06\mathbf{i} \text{ m/s}]$$

$$\mathbf{V}_B = (3.56\mathbf{i} - 1.06\mathbf{j}) \text{ m/s}$$

72. **(B)**

The impulse-momentum equation may be employed since the change in velocity of the particle is known and since it is desired to determine the force-time characteristic of the force of the wall acting on the ball during the impact:

$$\int_A^B F\,dt = \frac{8\text{ lb}}{32.2}[(-30) - (+30)]\text{ ft/s}$$

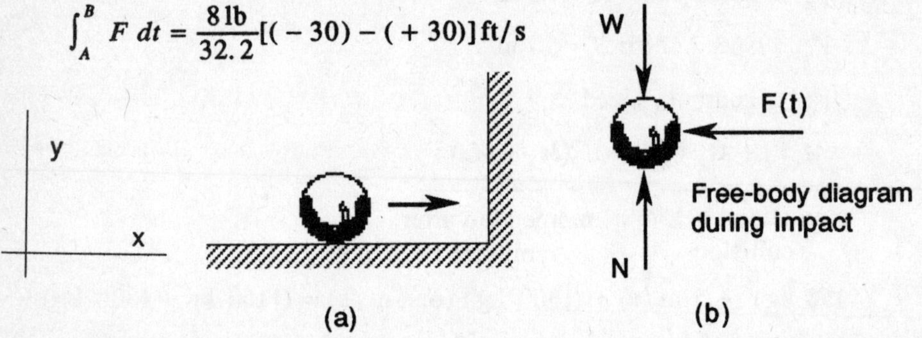

where A is the time immediately preceding impact and B is the time immediately after impact.

Since the velocity before and after impact is known, the impulse may be obtained by evaluating the change in momentum. First, the velocity before and after impact is

$$v_A = 30\mathbf{i} \text{ ft/s}$$

$$v_B = -30\mathbf{i} \text{ ft/s}$$

Dropping the vector notation since motion occurs in one dimension only, the impulse-momentum equation becomes

$$\int_A^B F\,dt = \frac{8\text{ lb}}{32.2}[(-30) - (+30)]\text{ ft/s} = -14.9\text{ lbs}$$

The negative sign indicates $F(t)$ is to the left.

The free-body diagram of the particle during impact is shown in the above Figure (b). Since the net force in the y direction is equal to zero, momentum will be conserved in the y direction. In this case, the velocity is zero and will remain zero during the motion. The force $F(t)$ is the force that the wall exerts on the ball; thus, it acts during contact only. Experiments have shown that the time distribution of that force during impact is roughly of the form shown in Figure (a) following. In this case it is assumed for analytical purposes that the force-time distribution is triangular, as shown in the following Figure (b).

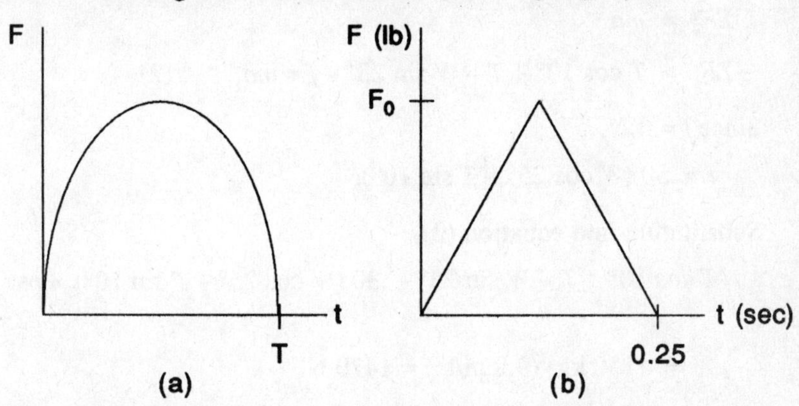

But the impulse is just the area under the force-time curve, so that

$$\text{Impulse} = -\tfrac{1}{2}(0.25)(F_0)$$
$$= -0.125\,F_0$$

The minus sign must be chosen because $F(t)$ acts in the negative x direction. However, we already know what the impulse must be for the ball to rebound as it does. Thus,

$$(-0.125)\,s\;F_0 = -14.9 \text{ lb–s}$$

and, solving for F_0,

$$F_0 = 119 \text{ lb.}$$

73. **(C)**

First, draw a free-body diagram of the block.

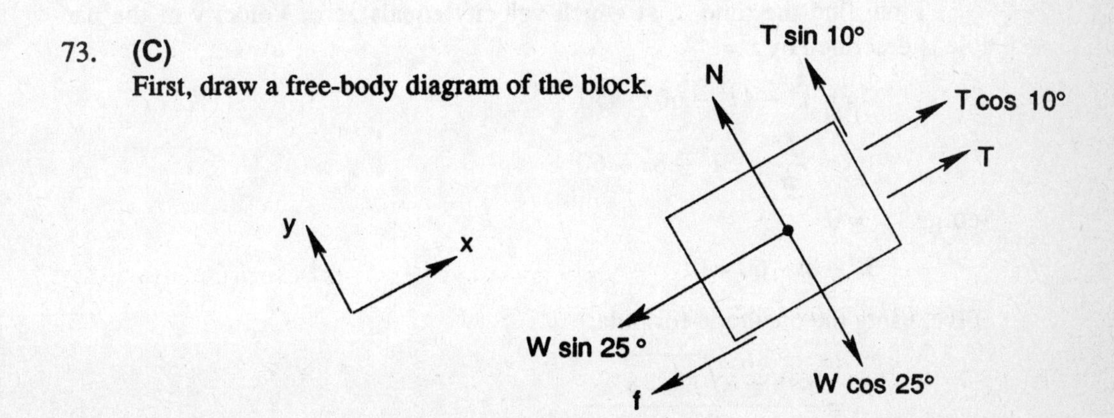

Using Netwon's law,

$$\left.\begin{aligned} \Sigma F_y &= ma_y \\ \Sigma F_y &= 0 \end{aligned}\right\} \text{ there is no acceleration in the } y \text{ direction.}$$

$$\Sigma F_y = N + T\sin 10° - W\cos 25° = 0$$

$$N = W\cos 25° - T\sin 10° \qquad (1)$$

From Newton's Law

$$\Sigma F_x = ma$$

$$\Sigma F_x = T\cos 10° + T - W\sin 25° - f = ma_x \qquad (2)$$

Since $f = \mu_k N$,

$$f = .30\,(W\cos 25° - T\sin 10°).$$

Substituting into equation (2),

$$T\cos 10° + T - W\sin 25° - .30\,(W\cos 25° - T\sin 10°) = ma_x$$

with

$$W = (150\text{ kg})\,(9.8\text{ m/s}^2) = 1470\text{ N}$$

$$a_x = 4\text{ m/s}^2$$

$$m = 150\text{ kg}$$

$$T\,(1 + \cos 10°) - 1470\text{ N}\sin 25° - .30\,(1470\text{ N}\cos 25°$$

$$- T\sin 10°) = (150\text{ kg})\,(4\text{ m/s}^2)$$

$$T\,(1.866) - 621.2\text{ N} - 400\text{ N} + .052\,T = 600\text{ N}$$

$$192\,T = 1621.2\text{ N}$$

$$T = 844.4\text{ N}.$$

74. **(A)**

First, find the time, t, at which velocity equals zero. Velocity of the particle is described by $\dot{x} = \frac{dx}{dt}$.

$$x = t^3 - 4t^2 - 60t + 50$$

$$\dot{x} = \frac{dx}{dt} = 3t^2 - 8t - 60$$

Setting $\frac{dx}{dt} = 0$,

$$3t^2 - 8t - 60 = 0.$$

Solve, using the quadratic formula:

$$t = \frac{-b \pm \sqrt{b^2 - 4ac}}{2a}$$

$$t = \frac{8 \pm \sqrt{64 - 4(3)(60)}}{2(3)}$$

$$t = \frac{8 \pm \sqrt{784}}{6}$$

$$t = 1.33 + 4.66 \quad \text{or} \quad t = 1.33 - 4.66$$

$$t = 6.0\,s \quad \text{or} \quad t = -3.33s$$

Since $t > 0$, choose 6.0 s.

Check your answer:

$$V(6) = 3(36) - 8(6) - 60 = 0$$

$$t = 6\,s \checkmark$$

Now find

$$a(t) = \ddot{x} = \frac{d^2x}{dt^2} = \frac{dv}{dt}$$

$$\dot{x}(t) = v(t) = 3t^2 - 8t - 60 = 0$$

$$a(t) = \ddot{x}(t) = \frac{dv}{dt} = 6t - 8$$

$$a(6\text{ s}) = 6(6) - 8 = 28 \text{ ft/s}^2$$

75. **(B)**

To find the acceleration of car A with respect to B ($\mathbf{a}_{A/B}$), use the relative acceleration equation:

$$\mathbf{a}_A = \mathbf{a}_B + \mathbf{a}_{A/B}$$

The acceleration of B is given as

$$\mathbf{a}_B = -3\text{ m/s}^2\mathbf{i}$$

The acceleration of A is normal to the curve in the **i** direction and has magnitude

$$a_A = \frac{V_A^2}{r} = \frac{[(40\text{ km/h})(1000\text{ m/km})(h/3600\text{ s})]^2}{180\text{ m}} = .686\text{ m/s}^2$$

Therefore $\mathbf{a}_A = -.686\,\mathbf{i}\text{ m/s}^2$.

Solving for $\mathbf{a}_{A/B}$,

$$\mathbf{a}_{A/B} = \mathbf{a}_A - \mathbf{a}_B = -.686\,\mathbf{i}\text{ m/s}^2 - (-3\,\mathbf{i}\text{ m/s}^2)$$

$$\mathbf{a}_{A/B} = 2.314\,\mathbf{i}\text{ m/s}^2.$$

76. (E)

To find the height where the ball strikes we first find the time required to reach the wall. This requires v_{ox}, the horizontal component of velocity, which is found as

$$v_{ox} = v_0 \cos 46.8^\circ = 37.6 \text{ ft/s}$$

Since there is no horizontal acceleration, the time required to hit the wall at this velocity is found from

$$s_h = v_{ox} t = 80 \text{ ft.}$$

Therefore

$$t = \frac{80 \text{ ft}}{37.6 \text{ ft/s}} = 2.1 \text{ sec.}$$

Now, the vertical height at which the ball strikes after this time is found from

$$h = v_{oy} t - {}^1/_2\, g t^2 + s_0$$

where s_0 is the height of the release point. Substitution then gives

$$h = 40.1 \text{ ft/s } (2.1 \text{ s}) - {}^1/_2\, (32.2 \text{ ft/s}^2)(2.1 \text{ s})^2 + 5 \text{ ft}$$

$$h = 18.2 \text{ ft.}$$

77. (D)

Six ropes are connecting the lower pulley to the upper pulley; therefore, the force in the rope is

Tension = 2000 / 6 = 333 lbs.

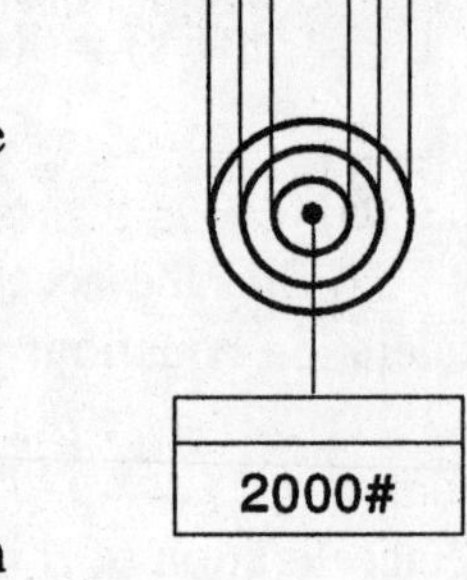

78. (A)

A reactive couple results at reactions A and B from the applied couple of 600 × 8 in-lbs.

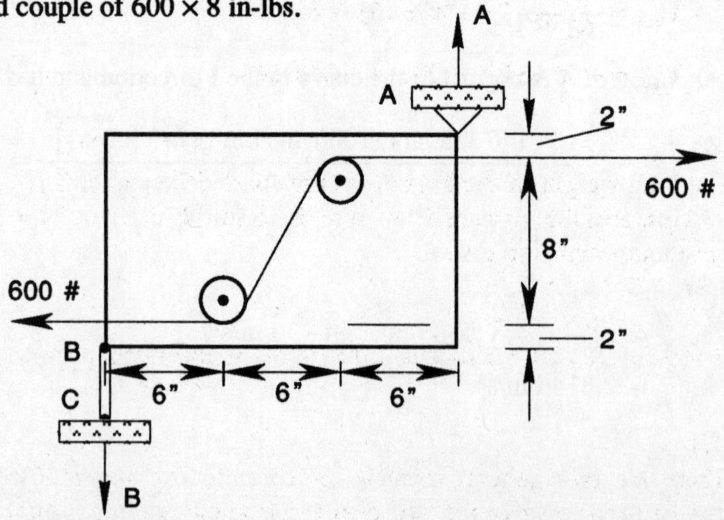

$(A \text{ or } B) \times 18 = 600 \times 8$ in-lbs.

$A = B = 600 \times 8 / 18 = 267$ lbs.

A second approach would be to sum moments about point *A*.

$\Sigma M_a = 0$

$600 \times 8 - 18 \times B = 0$

$B = 600 \times 8 / 18 = 267$ lbs.

79. **(C)**
The moment about the line *AB* results in a vector triple scalar product.

$$\lambda_{AB} = \frac{-20\mathbf{i} + 15\mathbf{k}}{25}$$

$$M_{AB} = \lambda_{AB} \cdot \mathbf{r} \times \mathbf{P}$$

$$M_{AB} = \begin{vmatrix} -0.8 & 0.0 & 0.6 \\ 0 & 20.0 & 0 \\ 2 & -40 & -50 \end{vmatrix} = 0.8 \times 20 \times 50 - 0.6 \times 20 \times 2$$

$$M_{AB} = 776 \text{ in-lbs}$$

80. **(D)**
The general form of the equation is

$$T = \frac{|\mathbf{T}|}{d}(d_x\mathbf{i} + d_y\mathbf{j} + d_z\mathbf{k})$$

where $d = \sqrt{14^2 + 6^2 + 6^2} = 16.37$ m

$$\mathbf{T}_{BD} = \frac{3000}{16.37}(-14\mathbf{i} + 16\mathbf{j} - 6\mathbf{k})$$

$$\mathbf{T}_{BD} = -2566\mathbf{i} + 1100\mathbf{j} - 1100\mathbf{k} \text{ N}$$

81. **(C)**
The total weight of the load is the area under the parabola $^2/_3$ bh = $(^2/_3)$ 24 $\times$ 6 = 96 kips, and the distance from *B* to the centroid is 0.4 $\times$ 24 = 9.60 ft. Sum moments about point *B* to give

$\Sigma M_B = 0$

$24 \times R_A - 96 \times 9.6 = 0$

$R_A = 9.6 \times 96 / 24 = 38.4$ kips (See figure on following page)

82. **(E)**
There are two general methods of solution for trusses. The method of joints can be used to solve for two member unknowns at each joint in sequence

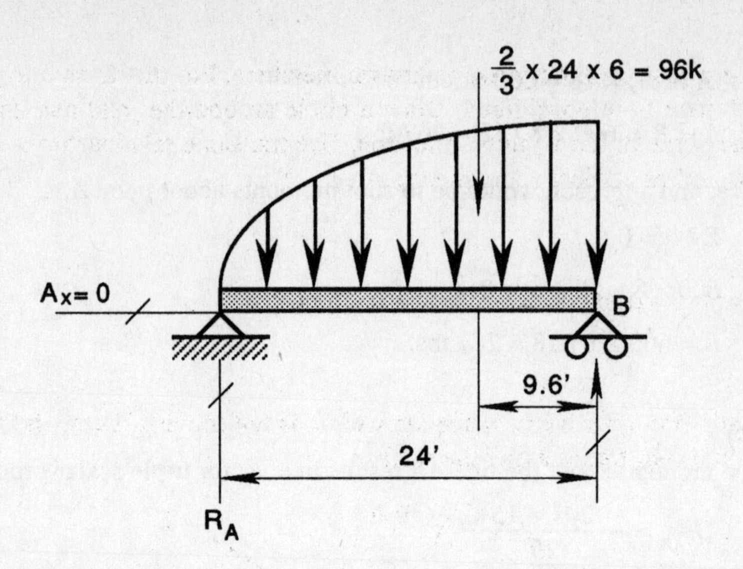

until all members' forces are found or the desired member force is found. The method of sections is used to find the force of a single member unknown.

Method of Sections:

Draw the free-body diagram and sum forces $F_x = 0$

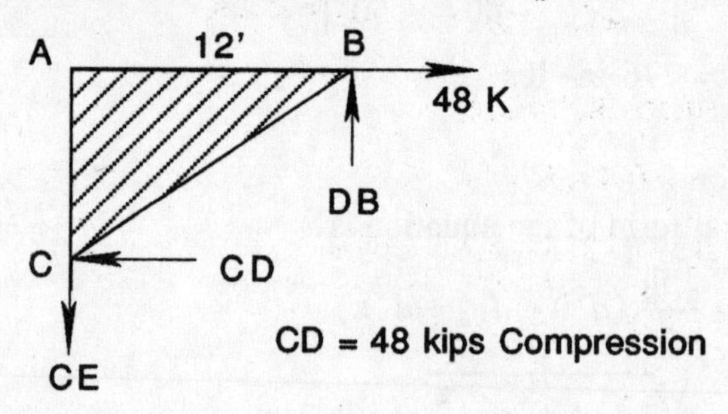

Method of Joints:

Make a larger drawing of the truss. Sum forces at each joint in sequence

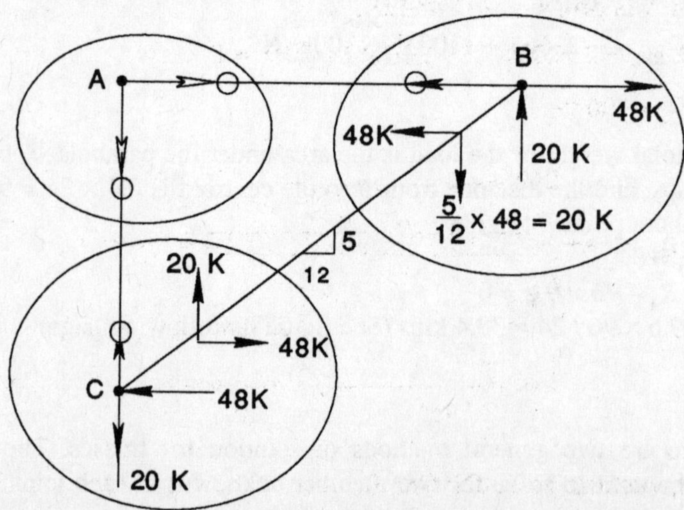

starting at a joint with only two unknown members. For this truss only joints *A*, *B*, and *C* have been considered. Draw a circle around the joint and sum forces within the circle in the *x*- and *y*-direction. Use the slope relationship to help find the horizontal or vertical components.

83. **(E)**
By observation $\overline{x} = 2.5$ in. and $\overline{y} = 4.0$ in.

84. **(C)**
Static friction governs since the weight is not moving. Draw the free-body diagram and sum forces along the surface.

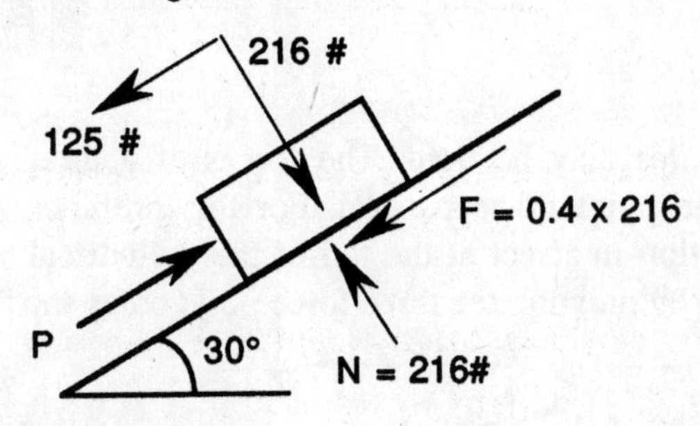

$N = 250 \cos 30° = 216$

$\text{Friction} = 0.4 \times N$

$\Sigma F = 0 : P - 125 - 0.4 \times 216 = 0$

$P = 212$ lbs.

85. **(A)**
This problem must be worked in stages using at least two free-body diagrams. First, working with *BD* find D_y.

$\Sigma M_B = 0 : 200 \times 12 - 200 \times 2 - 20D_y = 0$

$D_y = 100$ lbs.

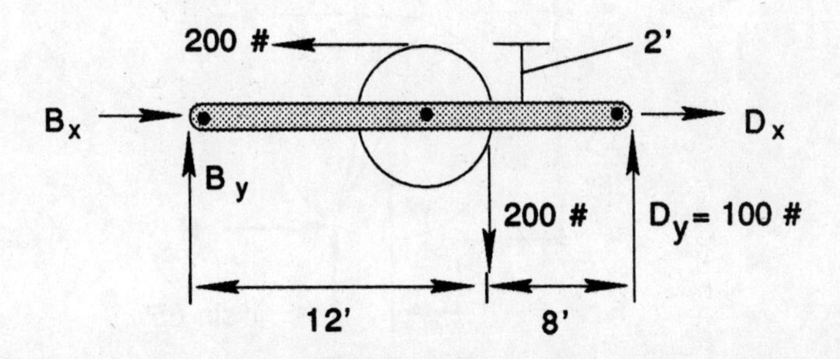

Next, working with *ADF* find F_y.

$$\Sigma F_y = 0 : F_y - \quad 100 = 0$$

$$F_y = 100 \text{ lbs}$$

A x

Dy = 100 #

Dx

Fy

86. **(D)**

Two procedures may be used. The forces of a three force system must either intersect at a point or be parallel. For this problem, the weight, tensile force and the reaction intersect at the top of the cylindrical weight. The tensile force can be found by drawing the three force polygon as shown.

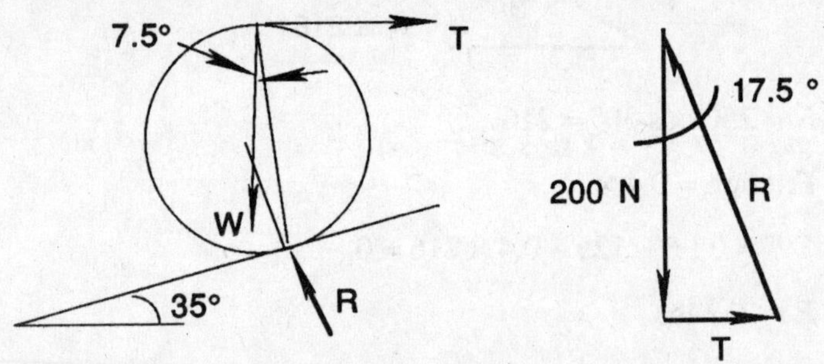

A second procedure is to sum moments about the point of contact with the inclined surface.

$$\Sigma M = 0$$

$$T_r(1 + \cos 35°) - 200\, r \sin 35° = 0$$

$$T = 63 \text{ N}$$

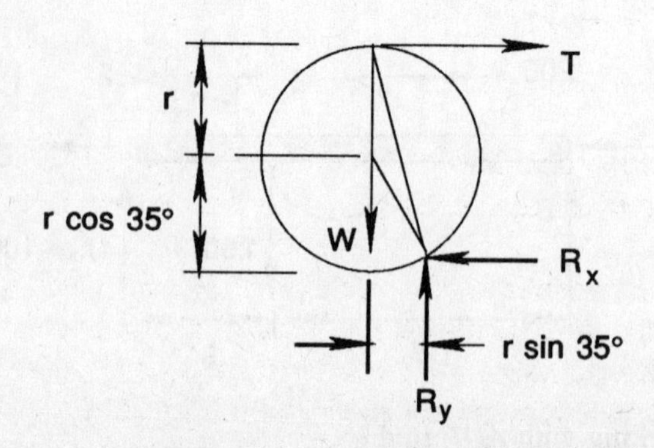

Please refer to the figure in the solution to Problem 87 (below) to see the derivation of the angle 17.5°.

87. **(C)**

Normally, a weight will slide when the slope angle is equal to the angle of friction. The cylinder of this problem, however, is restrained by the rope. Because of this the slope angle is larger than the friction angle of 22 degrees ($\tan\theta = 0.4$). The three forces *T*, *W*, and *R* intersect at the top of the cylinder. The force *R* can be divided into the normal force (*N*) and the friction force (*F*). The resulting slope angle in this case is $2\theta = 44$ degrees.

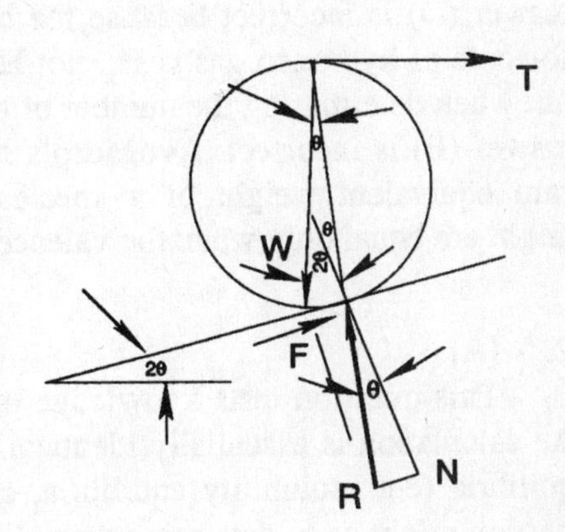

88. **(B)**

The mass moment of inertia of a sphere is $(2/5)\ mr^2$.

$$I = (2/5) \times (0.5/32.2) \times (1.25/12)^2$$

$$I = 0.0000674 \text{ ft-lb-sec}^2$$

89. **(C)**

The moment of inertia of the rectangular area is $bh^3/12$

$$I_x = 4 \times 8^3/12 = 171 \text{ in}^4.$$

90. **(C)**

The torque in shaft *AB* is transferred through gear *B* to the gear teeth at point *C* resulting in a force of T_{AB}/r_B. Then the force is transferred through gear *D* to shaft *DE*. The resulting torque in shaft *DE* is

$$T_{DE} = T_{AB}(r_D/r_B) = 120(15/6) = 300 \text{ Nm}.$$

91. **(C)**

This question tests knowledge of the concept of a gram mole—the key to understanding the gas law, Avogadro's Law, equivalent weight, and many other concepts.

The correct answer is (C). The number of molecules in a gram mole (the mass equal to the molecular weight in grams) of a compound is the same for all compounds. That number (6.023×10^{23}) is called Avogadro's number in honor of the Italian Renaissance physicist, Amedeo Avogadro.

Answer (A) is incorrect because the number of molecules in a gram depends on the mass (or molecular weight) of the molecule, and it is different for each species. Answer (B) is incorrect. The volume of 1 mole (which contains 6.023×10^{23} molecules) of any gas at standard conditions is 22.4 liters. One liter of a gas at standard conditions will contain only $(6.023 \times 10^{23})/22.4$ molecules. Answer (D) is incorrect because the molecular weight of hydrogen gas is 2. A molecule of hydrogen gas is H_2, not H. The temperature and pressure are irrelevant when determining the number of molecules in a specified mass of a species. Answer (E) is incorrect. Avogadro's number of valence electrons exists in one gram equivalent weight of a species. The equivalent weight and molecular weight are equal only when the valence is ±1.

92. **(A)**

This question tests knowledge of equilibrium calculations and ionization. The calculation is essentially identical to that used in calculating other kinds of equilibria (e.g., solubility equilibria, chemical equilibria, etc.). {An alternative question might be to determine the pH of the solution. [$pH = -\log_{10}(C_{H+})$]}

The correct answer is (A).

If the acetate radical $(C_2H_3O_2)^{-1}$ is abbreviated as Ac-, then the ionization of acetic acid can be expressed by the following chemical equation:

$$HAc \leftrightarrow H^+ + Ac^-$$

The ionization equilibrium constant is defined as

$$K = [a_H+][a_{Ac}-]/[a_{HAc}]$$

where (a_i) represents the chemical activity of species i. In dilute solutions, the activity is equal to the concentration. For each mole of H+ formed, one mole of Ac^- is also formed and one mole of acetic acid is consumed. The concentrations of H^+ and Ac^- are, therefore, equal (let that concentration = x), and the concentration of HAc, diminished by the amount of H^+ or Ac^- formed, is $(0.1 - x)$.

$$K = x^2/(0.1 - x)$$

If $x << 0.1$, the equation becomes $0.1K = x^2$ or $x = 1.32 \times 10^{-3}$.[Answer (A)]

(Note that the assumption that $x<<0.1$ is justified.)

Answer (B) is obtained if the effect of the initial concentration of HAc is neglected. Answer (C) is obtained if complete ionization is assumed. Answer (D) is obtained if both the error leading to Answer (B) and the error leading to Answer (E) are made. Answer (E) is obtained if the equation, $x = 0.1$ K, is used rather than $x^2 = 0.1$ K.

93. **(C)**

This equation tests knowledge of the structure of electronic energy states in matter and the quantum theory of matter and energy which predicts the frequency or wavelength associated with quanta of radiated energy.

$$E = h\nu = hc/\lambda$$

where E is the energy, h is Plank's constant, ν is the frequency of the radiation,

c is the velocity of light, and λ is the wavelength.

$$E = E_2 - E_1 = (-3.4) - (-13.6) = 10.2 \text{ eV}$$

$$E = 10.2 \text{ eV} = (6.63 \times 10^{-34} \text{ J s})(3 \times 10^8 \text{ m/s})/\lambda \times 1.602 \times 10^{-19} \text{ J/eV})$$

$$\lambda = 1.2172 \times 10^{-7} \text{ m} = 1217 \text{ Å}$$

Therefore, the correct answer is (C).

Answers (A), (B), (D), and (E) might be obtained by making one of a number of somewhat obvious errors. Answer (A) is the energy of the photon in eV. Answer (B) is three times the correct answer and may be obtained if the correct value of C is not used. Answer (D) is four times the correct answer. Answer (E) is obtained if the energy of the photon is mistakenly calculated as 17 eV (3.4 eV + 13.6 eV).

94. **(D)**

This question tests knowledge of chemical formulas, valence, and the Periodic Table. It also tests knowledge of the chemical symbol for iodine.

The correct answer is (D). Beryllium has, as do all Group IIa elements, a valence of +2 meaning that it can donate two electrons in a reaction. The only negative valence of iodine (and all other Group VIIa elements) is –1, meaning it can accept 1 electron in a reaction. Since the algebraic sum of the valences of all elements in a compound will be zero (i.e., the electrons donated by one element must be accepted by some other element), two iodine atoms are required to react with one beryllium atom.

Answer (A) is incorrect. It may be selected if the examinee confuses the numbers assigned to groups in the Periodic Table with number of valence electrons of elements in the group. Answer (B) is incorrect because all Group IIa elements, including beryllium, donate two electrons in a reaction and a single iodine atom can accept only 1.

Answer (C) is incorrect because the chemical symbol of iodine is "I". "In" is the chemical symbol of indium. Answer (E) is incorrect. It, like Answer (A), may be selected if the examinee confuses the numbers assigned to groups in the periodic table with the number of valence electrons of elements in the group.

95. **(E)**

This question tests knowledge that sulfate, $(SO_4)^{-2}$, is the anion in sulfuric acid (the most common mineral acid), knowledge of the valence of the sulfate radical, $(SO_4)^{-2}$, and of potassium, the chemical symbol for potassium, and that any product of acid-based neutralization is called a salt.

The correct answer is (E). Two molecules of potassium hydroxide (KOH) are required to neutralize one molecule of sulfuric acid (H_2SO_4).

Answer (A) is incorrect. The chemical formula for common table salt is NaCl, but the product of any acid-base neutralization reaction is also called a salt. NaCl is the product of the neutralization of sodium hydroxide (NaOH) and hydrochloric acid (HCl). Answer (B) is incorrect because the chemical symbol for potassium is "K". "P" is the chemical symbol for phosphorus. Answer (C) is incorrect because two atoms of potassium, which has a valence of +1, are required to react with one sulfate radical, which has a valence of –2. Answer (D) is incorrect for each of the reasons that Answers (B) and (C) are incorrect.

96. **(D)**

This question tests knowledge of amphoteric hydroxides which can provide either an H^+ ion or an OH^- ion and react as either an acid or a base.

The correct answer is (D). Aluminum hydroxide, like many other metal hydroxides, can provide either an H^+ ion or an OH^- ion in reaction. The circumstances of the reaction determine whether it behaves as an acid or base.

Answer (A) is incorrect. Although an amphoteric hydroxide may contain three hydroxyl groups, that is not what makes it amphoteric. Answer (B) is incorrect. The solubility is immaterial in determining whether or not the material is amphoteric. Answer (C) is incorrect. The number of hydroxyl groups which ionize is immaterial in determining whether or not the material is amphoteric.

97. **(C)**

This question tests knowledge of the Periodic Table and the names of the groups in the Periodic Table.

The correct answer is (C). The alkaline earth metals are those in Group IIa.

Answer (A) is incorrect. Only the Group Ia elements do not have two "s" electrons in the outer orbit. Answer (B) is incorrect. Many other elements also exhibit a valence of +2. Answer (D) is incorrect. Most metal hydroxides form alkaline solutions in water. Answer (E) is incorrect. The soil contains, somewhere on Earth, essentially all of the elements in the Periodic Table.

98. **(E)**

This question tests knowledge of what makes water "hard" and the chemistry associated with hard water.

The correct answer is (E).

Answer (A) is incorrect. Solutes tend to lower, not raise, the freezing point. Although the high school (or perhaps junior high school) riddle, "How do you spell 'hard water' with three letters?" (ice) is still oft repeated, it is not ice crystals that make water hard. Answers (B) and (C) are both correct statements, hence the correct answer is (E).

99. **(A)**

This question tests knowledge of the class of reactions known as oxidation-reduction reactions. These reactions result in a change in the oxidation or valence states of at least two of the elements in the reaction. The total of the oxidations states of all elements must add to zero for each neutral species in the reaction.

The correct answer is (A). There is no change in the oxidation state of any of the elements in this reaction. It is simply the precipitation of copper sulfate from copper and sulfate ions. Answer (B) is incorrect. Mn is reduced from a valence of +7 to +2. Oxygen (in H_2O_2) is oxidized from a valence of –1 to 0. Answer (C) is incorrect. Sulfur is oxidized from –2 to +4. Oxygen is reduced from a valence state of 0 to –2. Answer (D) is incorrect. Carbon is oxidized from a valence of –4 to +2. Hydrogen is reduced from a valence of +1 to 0.

100. **(E)**

This question tests knowledge of the Principle of LaChatelier which states that a system in equilibrium will shift to offset any imposed change or stress. Therefore, the system will tend to shift to offset changes in temperature, pressure, or concentration (of which pH is simply a special case).

The correct answer is, therefore, (E).

101. **(B)**

This question tests knowledge of heats of formation and how to calculate the heat of reaction from data on heats of formation.

The heat of formation of a species is defined as the heat of reaction to make the species from the elements. Since the elements are conserved in a chemical reaction, the First Law of Thermodynamics dictates that the heat of reaction is the sum of the heats of formation of the products minus the sum of the heats of formation of the reactants. Note that from the definition, the heat of formation of an element is zero.

For this case (at 298°K):

$$(\Delta H)_r = 3(\Delta H)_{fH_2} + (\Delta H)_{fCO} - (\Delta H)_{fCH_4} - (\Delta H)_{fH_2O}$$

Since $(\Delta H)_{fH_2}$ is zero by definition, $(\Delta H)_r = 49.281$ Kcal/mole.

Since $(\Delta H)_r > 0$, the reaction is endothermic. The correct answer is (B).

Answer (A) is incorrect. If $(\Delta H)_r$ were negative, the reaction would be exothermic. Answer (C) is incorrect. The pressure does not affect $(\Delta H)_r$ for ideal gases. Even at extreme pressures where departure from ideal behavior is significant, the effect is insufficient to change this reaction from being endothermic to exothermic. Answer (D) is incorrect. All reactions proceed to some extent at equilibrium. Answer (E) is incorrect. $(\Delta H)_f$ for the elements is zero by definition and is frequently not tabulated in tables.

102. **(B)**

This question tests knowledge of the characteristics of the halogens (the elements of Group VIIa).

The correct answer is (B). The halogen gases form strong acids and do not react appreciably with water.

Each of the other answers, (A), (C), (D), and (E), is a true statement about the halogens.

103. **(B)**

This question tests knowledge about the Group VIII elements which are called the noble gases, rare gases, or the inert gases.

The correct answer is (B). The s and p electron orbitals of the Group VIII elements are full. They have no driving force to enter reactions to fill these orbitals.

Answer (A) is incorrect. While some Group VIII elements may be rare, argon forms 1% of the atmosphere — hardly rare by any reasonable definition. Answer (C) is incorrect. Only helium of the Group VIII elements remains gaseous to within a few degrees of absolute zero. Answer (D) is incorrect. Only helium (at. wt. = 4) and neon (at. wt. = 20) are lighter than air (molecular wt. ≅ 29). All of the other noble gases have atomic weights greater than 29.

Argon	39.9
Krypton	83.8
Xenon	131.3
Radon	222

104. **(A)**

This question tests knowledge of atomic weights and how to use them to calculate the material balance for a reaction.

The elements are conserved in a chemical reaction. The mass of iron present in the initial ferric oxide is calculated by taking the ratio of the sum of the atomic weights of the two iron atoms to the molecular weight of ferric oxide.

$$\text{M.W. ferric oxide} = 2(55.85) + 3(16) = 159.7$$

$$\Sigma\,(\text{at. wt. of iron}) = 2(55.85) = 111.7$$

The ratio of iron/ferric oxide equals 111.7/159.7 = 0.7

Therefore, in 10 tons of ferric oxide, there are 7 tons of iron

Answer (A) is the correct answer.

Answer (B) is incorrect. It is the sum of the weight of ferric oxide plus the stoichiometric amount of carbon required. Answer (C) is incorrect. It is 20% of the initial ferric oxide present. Since ferric oxide contains five atoms, selection of this answer might be possible if the examinee misunderstands how the calculation is made. Answer (D) is incorrect. It is the difference between the initial ferric oxide present and the stochiometric carbon required for reaction. Answer (E) is incorrect. From the atomic weights it is possible to calculate the ratio of wt. iron to wt. ferric oxide. The units are immaterial.

105. **(A)**

Step 1. Cut the section at section ① – ① and draw a free-body diagram. There are two possible free-body diagrams, one on the left and one on the right. If the left part is chosen, the reaction at *A* must be determined first. It is easier in this problem to use the right part as a free-body diagram. The free-body diagram is shown below with the unknown internal reaction at the section.

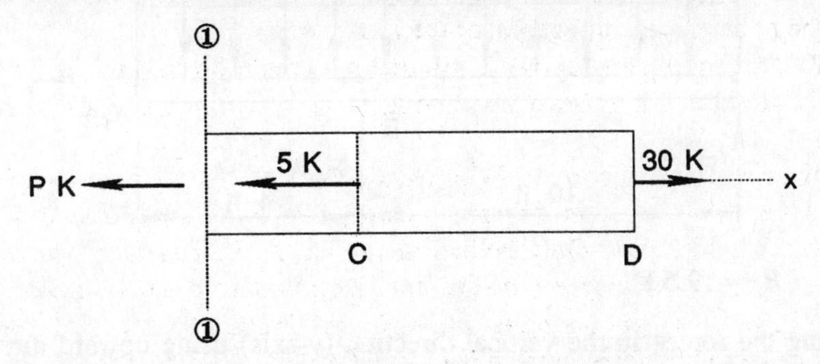

Step 2. To determine the unknown internal reaction P, a static equilibrium equation is used. To maintain the equilibrium in the horizontal direction (x-axis), the sum of forces along the x-axis must be zero. To be consistent among forces in the equation, forces pointing to the right are considered as positive forces.

$\Sigma F_x = 0$ (to the right is positive)

$-P^K - 5^K + 30^K = 0$

Therefore, $P = +25$ kips

The positive value of the result indicates that the assumed direction of 'P' is correct. Since the force 'P' is acting away from the section (pulling the section), the internal reaction is in tension.

Step 3. Assuming that all forces act through the centroidal axis of the member, the axial stress is distributed uniformly across the cross section. The normal (axial) stress is then computed using the following equation:

$\sigma = P/A$

where σ = normal stress (kips per square inch or ksi)

ρ = normal force (kips or k)

A = area (square inches)

Thus, σ = (25 kips) / (0.1 square inches) = 250 ksi (tension).

106. **(B)**

Step 1. Support reactions are determined using equilibrium equations. To maintain the equilibrium, the sum of moments about any point must be zero, and the sum of forces in any direction must also be zero.

Summing the moments about point 'A' using the clockwise direction as positive direction.

$\Sigma M_A = 0$

$(2\ \text{k/ft})(14\ \text{ft})(7\ \text{ft}) - (R_B\text{k})(10\ \text{ft}) = 0$

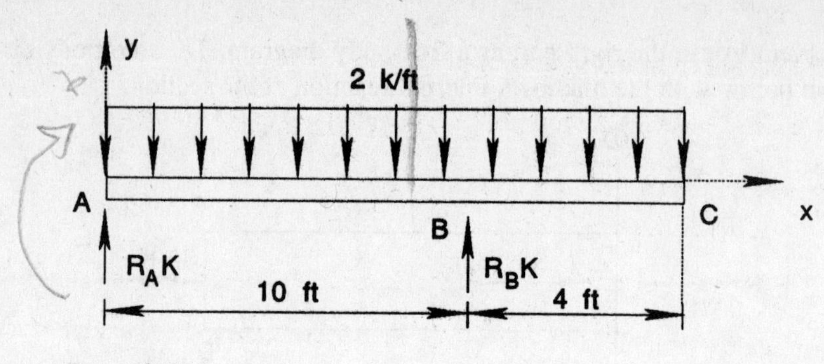

$R_B = 19.6$ k

Summing the forces in the vertical direction (y-axis) using upward direction as positive direction.

$$\Sigma F_y = 0$$

$$(R_A k) + (19.6\text{ k}) - (2\text{ k/ft})(14\text{ ft}) = 0$$

$$R_A = 8.4\text{ k}$$

Step 2. Sketch the shear and moment diagrams to determine critical sections for bending moment.

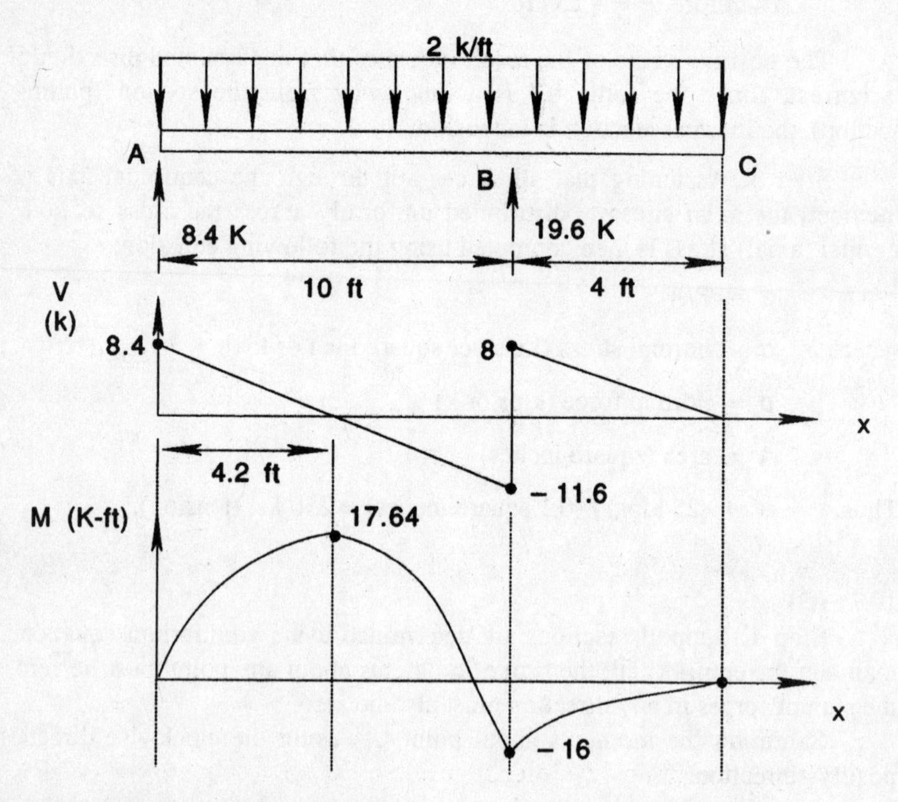

Maximum Bending Moments:

Positive Moment = 17.64 k-ft

Negative Moment = 16 k-ft

Step 3. Bending stresses can be computed using the following equation:

$\sigma = My/I$

where σ = bending stress (ksi)

y = distance from the neutral axis (N.A.) to the point of interest. (in)

I = moment of interia (I) with respect to the neutral axis. (in^4)

M = bending moment (k-in)

The maximum tensile stress at the maximum positive moment section occurs on the bottom fiber of the section.

$$\sigma = \frac{(17.64 \text{ k-ft} \times 12 \text{ in/ft})(7 \text{ in})}{(314 \text{ in}^4)}$$

$$= 4.72 \text{ ksi}$$

The maximum tensile stress at the maximum negative moment section occurs on the top fiber of the section.

$$\sigma = \frac{(16 \text{ k-ft} \times 12 \text{ in/ft})(3 \text{ in})}{(314 \text{ in}^4)}$$

$$= 1.83 \text{ ksi}$$

The maximum tensile stress of the entire beam is then 4.72 ksi and occurs at the maximum positive moment section.

107. **(C)**

Step 1. Draw the free-body diagram of member *ABC*.

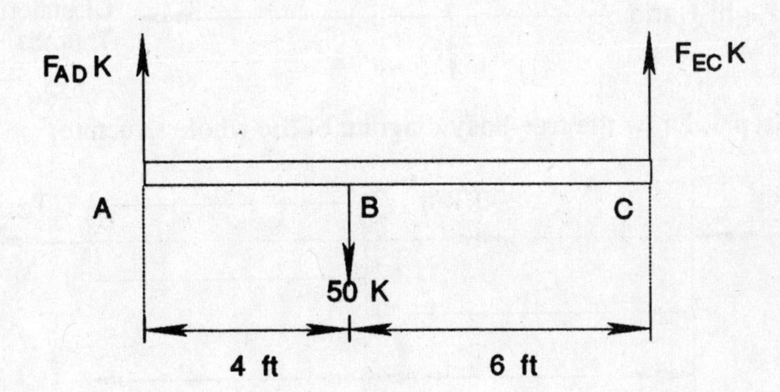

Step 2. Determine the force in member EC using an equilibrium equation.

Summing the moments about point *'A'* using the closewise direction as positive direction:

$$\Sigma M_A = 0$$

$$(50 \text{ k})(4 \text{ ft}) - (F_{EC} \text{ k})(10 \text{ ft}) = 0$$

$$F_{EC} = 20 \text{ k}$$

Step 3. Calculate the elongation of member *'EC'* using the following equation:

$$\delta = \frac{PL}{AE}$$

where δ = elongation (in)

P = force (k)

L = member length (in)

A = cross section area (in²)

E = modulus of elasticity (ksi)

Thus

$$\delta_{EC} = \frac{(20 \text{ k})\ (5 \text{ ft} \times 12 \text{ in/ft})}{(0.2 \text{ in}^2)\ (29{,}000 \text{ k/in}^2)}$$

$$= 0.21 \text{ in.}$$

108. **(D)**

In the following solution, double-headed arrows are used to represent the twisting moments or torques. The direction of a double-headed arrow is determined using the right-hand rule, where the fingers represent the direction of the twisting moment and the thumb represent the direction of the double-headed arrow.

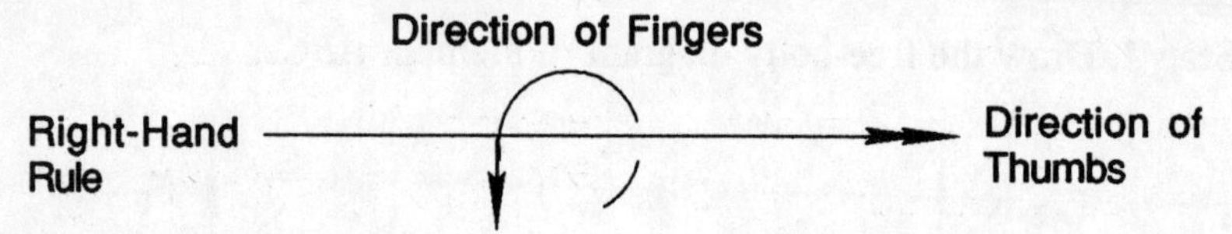

Step 1. Draw the free-body diagram of the whole structure.

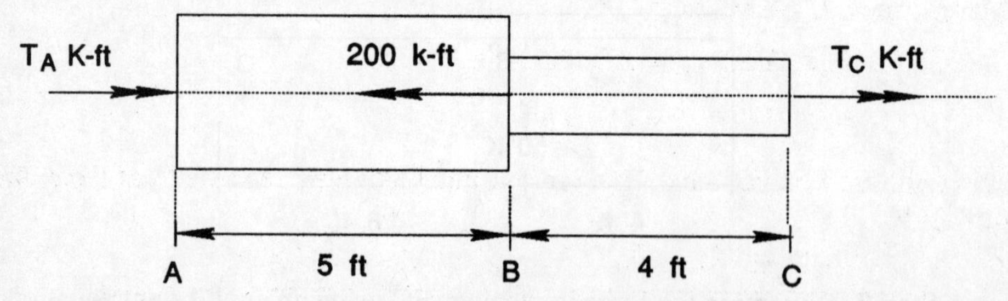

Step 2. Write an equilibrium equation by summing the twisting moments using the double-headed arrow to the right as positive direction.

$$\Sigma T = 0$$

$$(T_A \text{ k-ft}) + (T_C \text{ k-ft}) - (200 \text{ k-ft}) = 0$$

which can be simplified to:

$$T_A + T_C = 200 \text{ k-ft} \qquad (1)$$

Step 3. Write a compatibility equation using the condition that the total angle of twist from A to C is zero.

$$\theta_{AC} = 0$$

$$\theta_{AB} + \theta_{BC} = 0$$

The angle of twist in a member can be determined from:

$$\theta = \frac{TL}{JG}$$

where θ= angle of twist (radian)

T= twisting moment (k-ft)

L= length (ft)

G= shear modulus (k-ft)

J= polar moment of inertia (ft^4) = $(\pi/2)\, r^4$ for circular section. (r = radius)

Therefore,

$$\theta_{AB} + \theta_{BC} = 0 \Rightarrow \frac{T_{AB}L_{AB}}{J_{AB}G} + \frac{T_{BC}L_{BC}}{J_{BC}G} = 0$$

Free-body diagrams showing internal twisting moments of AB and BC are:

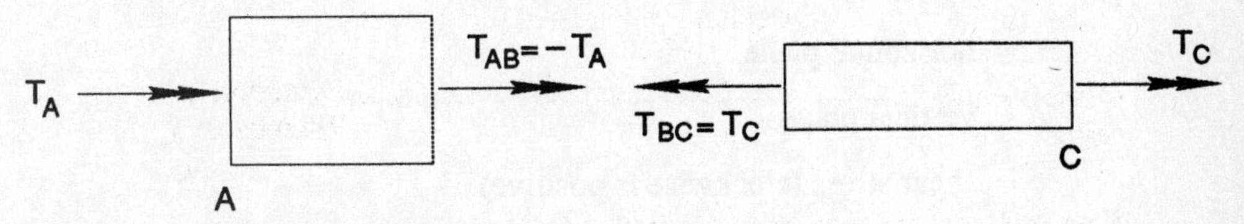

Polar moment of inertias are:

$$J_{AB} = \pi\,(3 \text{ in})^4 / 2 = 127.23 \text{ in}^4$$

$$J_{BC} = \pi\,(2 \text{ in})^4 / 2 = 25.13 \text{ in}^4$$

Therefore, the compatibility equation becomes (G was cancelled out from the equation):

$$\frac{(-T_A \text{ k-ft})\ (5 \text{ ft})}{(127.23 \text{ in}^4 \times 1 \text{ ft}^4/12^4 \text{ in}^4)} + \frac{(T_C \text{ k-ft})\ (4 \text{ ft})}{(25.13 \text{ in}^4 \times 1 \text{ ft}^4/12^4 \text{ in}^4)} = 0$$

which can be simplified to:

$$T_C = 0.25\ T_A \qquad (2)$$

Step 4. Solve the unknown (T_A) by substiting T_C from equation (2) into equation (1).

$$T_A + 0.25\ T_A = 200 \text{ k-ft}$$
$$T_A = 160 \text{ k-ft.}$$

109. **(B)**

Problem 109 may be found by sketching Mohr's circle or by using the principal stress equations.

Solution 1 (Mohr's Circle)

Step 1. Sketch the Mohr's Circle

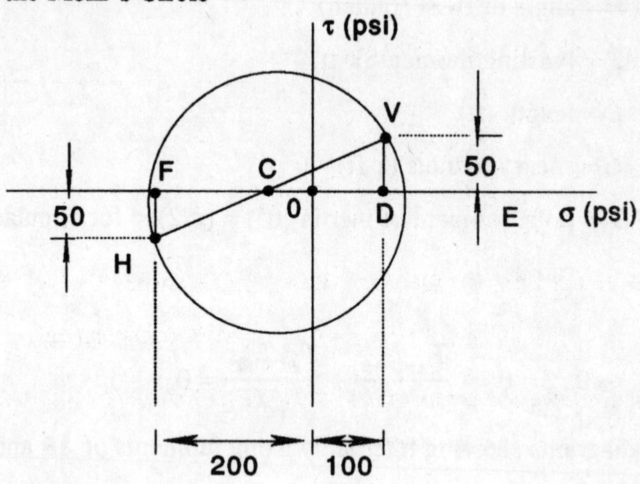

H = horizontal plane

V = vertical plane

τ = shear stress (clockwise is positive)

σ = normal stress (tension is positive)

C = center of the Mohr's Circle

Step 2. Calculation based on geometry.

$$CD = DF/2 = (OD + OF)/2 = (100 \text{ psi} + 200 \text{ psi})/2 = 150 \text{ psi}$$

Radius of the Mohr's Circle = R

$$R = CV = \sqrt{CD^2 + DV^2} = \sqrt{150^2 + 50^2} = 158 \text{ psi}$$
$$CO = CD - OD = 150 \text{ psi} - 100 \text{ psi} = 50 \text{ psi}$$

Maximum Tensile Stress = $OE = R - CO = 158 \text{ psi} - 50 \text{ psi} = 108 \text{ psi.}$

Solution 2 (Principal Stress Equations)

The maximum tensile stress or the positive principal stress may be found with the following equation

$$\sigma_{1,2} = \frac{\sigma_x + \sigma_y}{2} \pm \sqrt{\left(\frac{\sigma_x - \sigma_y}{2}\right)^2 + \tau_{xy}^{\ 2}}$$

From the problem statement

σ_x = 100 psi (tension)

σ_y = – 200 psi (compression)

τ_{xy} = 50 psi

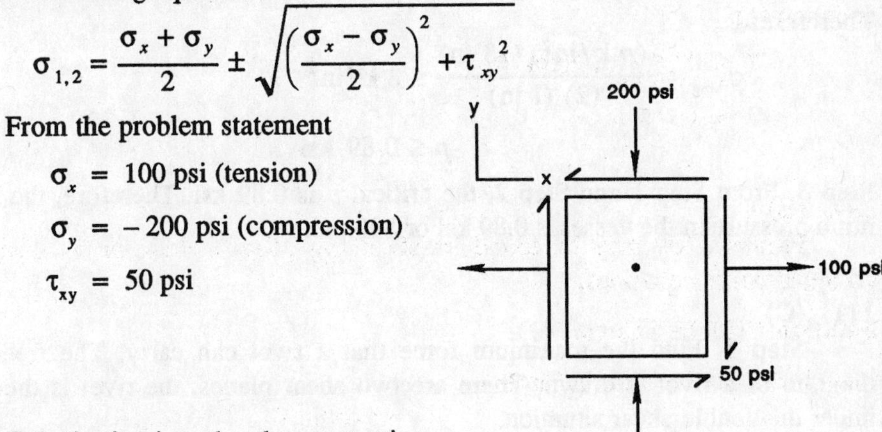

Substituting into the above equation

$$\sigma_{1,2} = \frac{100 \text{ psi} - 200 \text{ psi}}{2} \pm \sqrt{\left(\frac{100 \text{ psi} + 200 \text{ psi}}{2}\right)^2 + (50 \text{ psi})^2}$$

$$\sigma_{1,2} = -\ 50 \text{ psi} \pm \sqrt{25000 \text{ psi}^2}$$

$$\sigma_1 = -\ 50 \text{ psi} + 158 \text{ psi} = 108 \text{ psi (Tension)}$$

$$\sigma_2 = -\ 50 \text{ psi} - 158 \text{ psi} = -\ 208 \text{ psi (Compression)}$$

Since we're looking for maximum *tensile* stress we choose σ_1 or 108 psi as our answer.

110. **(B)**

Step 1. The longitudinal stress in the cylindrical vessel can be computed from:

$$\sigma_{long.} = \frac{pD}{4t}$$

where
p = pressure in the vessel (ksi)
D = diameter of the vessel (in)
t = thickness of the vessel (in)
$\sigma_{long.}$ = longitudinal stress (ksi)

Therefore;

$$\sigma_{long.} = \frac{(p \text{ k/in}^2)\ (18 \text{ in})}{(4)\ (1 \text{ in})} \le 20 \text{ k/in}^2$$

$$p \le 4.44 \text{ ksi}$$

Step 2. The circumferential (tangential) stress can be computed from:

$$\sigma_{tang.} = \frac{pD}{2t}$$

Therefore;

$$\sigma_{tang.} = \frac{(p\ \text{k/in}^2)\ (18\ \text{in})}{(2)\ (1\ \text{in})} \leq 8\ \text{k/in}^2$$

$$p \leq 0.89\ \text{ksi}$$

Step 3. From Step 1 and Step 2, the critical p is 0.89 ksi. Therefore, the maximum pressure in the vessel is 0.89 ksi or 890 psi.

111. **(C)**

Step 1. Find the maximum force that a rivet can carry. The free-body diagram of a rivet is drawn. There are two shear planes, the rivet is therefore under the double-shear situation.

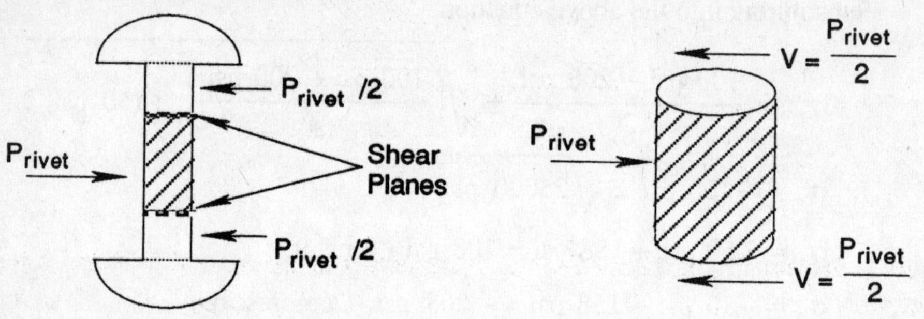

The area of the rivet =

$$A = \frac{\pi}{4} d^2 = \frac{\pi}{4} (\frac{3}{4}\ \text{in})^2 = 0.44\ \text{in}^2$$

The shear stress can be computed from:

$$\tau = V / A$$

where τ = shear stress (ksi)

V = shear force (k)

A = area (in²)

Therefore,

$$\tau = \frac{V}{A} = \frac{(P_{rivet}/2\ \text{k})}{(0.44\ \text{in}^2)} \leq 14\ \text{k/in}^2$$

$$P_{rivet} \leq 12.32\ \text{k}$$

Step 2. Since we have two rivets at the connection, the applied P is then the capacity of two rivets.

$$P = (2\ \text{rivets})\ (12.32\ \text{k/rivet}) = 24.64\ k \approx 25\ k.$$

112. **(C)**

Step 1. The centroid is at half-depth of the section due to symmetry. The moment of inertia about the neutral axis can be computed by:

$$I\text{n.a.} = I\text{n.a.}_{ABCD} - 2[I\text{n.a.}_{EFGH}]$$

$$= \frac{1}{12}(10\text{ in})(12\text{ in})^3 - 2[\frac{1}{12}(4\text{ in})(8\text{ in})^3]$$

$$= 1099\text{ in}^4$$

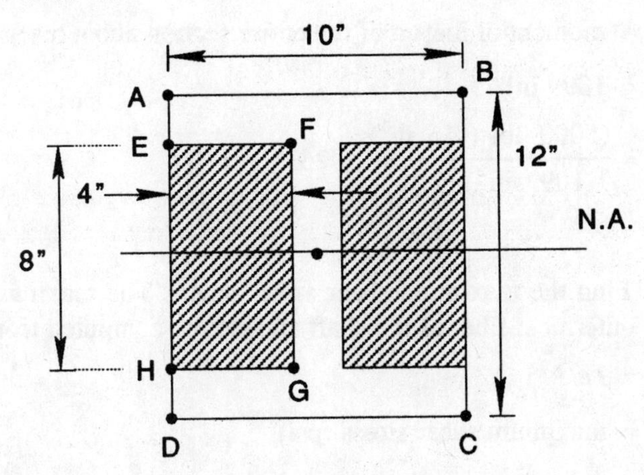

Step 2. The maximum stress for *I*-sections occurs at the neutral axis and can be computed from:

$$\tau = \frac{VQ}{It}$$

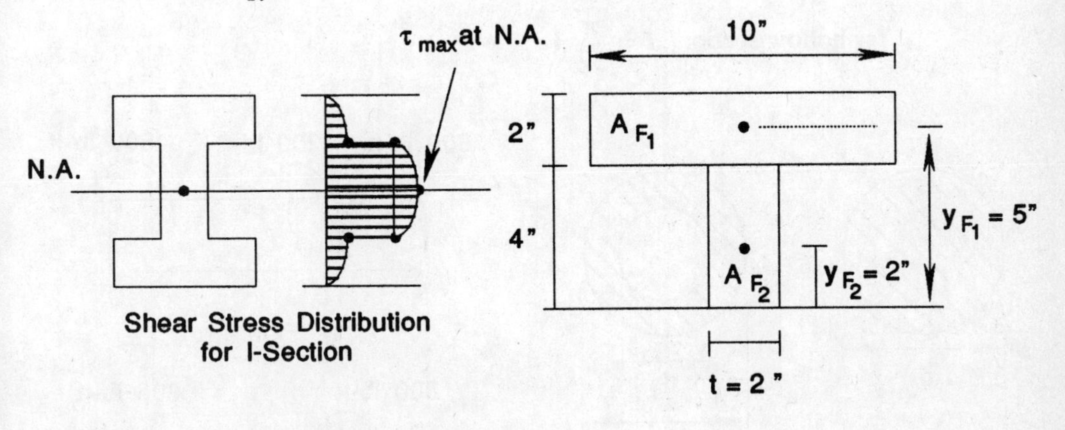

Shear Stress Distribution for I-Section

where τ = shear stress (psi)

V = vertical shear force (lb)

= 50,000 lbs

Q = moment of area (in^3)

$$= A_{F_1} y_{F_1} + A_{F_2} y_{F_2}$$

$$= (10 \times 2 \text{ in}^2)(5 \text{ in}) + (4 \times 2 \text{ in}^2)(2 \text{ in})$$

$$= 100 \text{ in}^3 + 16 \text{ in}^3 = 116 \text{ in}^3$$

t = width of the section where the shear stress is considered (in)

= 2 in

I = moment of inertia of the entire section about n.a.

= 1099 in^4

Thus $$\tau = \frac{(5000 \text{ lb})(116 \text{ in}^3)}{(1099 \text{ in}^4)(2 \text{ in})} = 2639 \text{ psi}$$

113. **(C)**

Step 1. Find the maximum shear stress in *AB*. The maximum shear stress occurs on the outermost fiber of the shaft and can be computed from:

$$\tau_{max} = Tc/J$$

where $\tau\sigma_{max}$ = maximum shear stress (psi)

T = twisting moment (lb-in)

C = radius of the shaft (in)

J = polar moment of inertia (in^4)

for solid section $J = \frac{\pi}{32} d^4$

for hollow section $J = \frac{\pi}{32}(d_0^4 - d_i^4)$

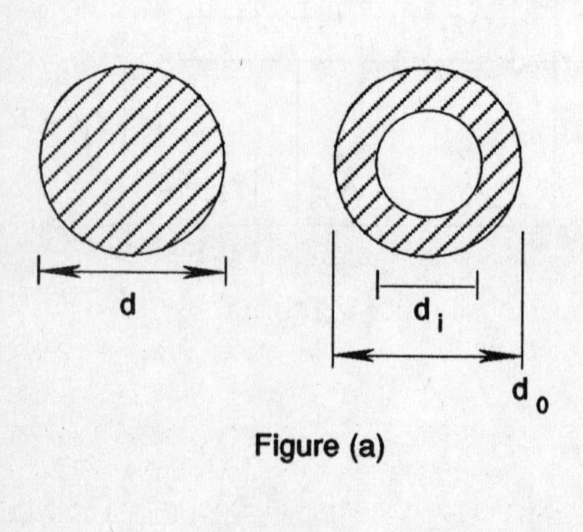

Figure (a)

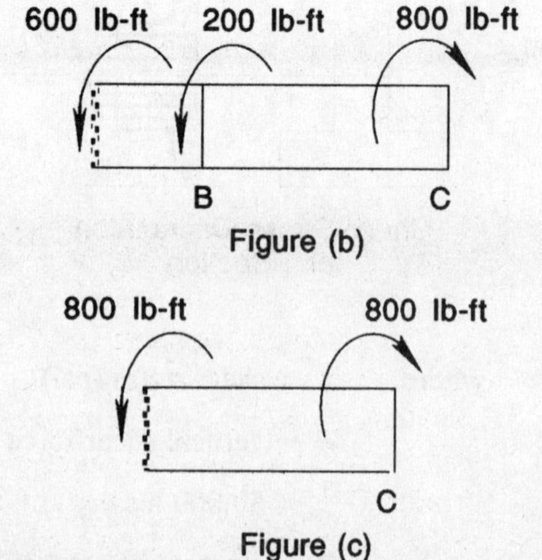

Figure (b)

Figure (c)

For *AB*, T = 800 lb-ft – 200 lb-ft = 600 lb-ft

$$J = \frac{\pi}{32}(2 \text{ in})^4 = 1.57 \text{ in}^4$$

Thus,

$$\tau_{max} = \frac{(600 \text{ lb-ft} \times 12 \text{ in/ft})(1 \text{ in})}{(1.57 \text{ in}^4)} = 4586 \text{ psi}$$

Step 2. Find the maximum shear stress in *BC*. (See Figure (c))

$$T = 800 \text{ lb-ft}$$

$$J = \frac{\pi}{32}[(2 \text{ in})^4 - (1 \text{ in})] = 1.472 \text{ in}^4$$

Thus,

$$\tau_{max} = \frac{(800 \text{ lb-ft} \times 12 \text{ in/ft})(1 \text{ in})}{(1.472 \text{ in}^4)} = 6522 \text{ psi}$$

Step 3. Compare the maximum stresses from Step 1 and Step 2. The maximum shear stress = 6522 psi.

114. **(C)**

Step 1. When there is more than one material, the transformed section method is normally utilized. In this method, the section is transformed into an equivalent section of only one material. All materials are transformed into one material. In this case, the steel area can be transformed into an equivalent area of concrete using the following equations:

$$Acs = n\,As \quad \text{and} \quad n = Es/Ec$$

where Acs = equivalent concrete area for steel (in^2)

As = steel area (in^2)

Es = modulus of elasticity of steel (psi)

Ec = modulus of elasticity of concrete (psi)

Therefore;

$$n = (30 \times 10^6 \text{ psi}) / (3 \times 10^6 \text{ psi}) = 10$$

$$Acs = (10)(3 \text{ in}^2) = 30 \text{ in}^2$$

The transformed section becomes as shown in the following figure.

In reinforced concrete bending members, the concrete on the tension side is cracked at a very low stress (200–400 psi), while the concrete in the compression side can withstand much higher (about 10 times of the tensile strength). The steel can even withstand higher stress (about 30,000 psi to 60,000 psi). It is therefore very common to neglect the concrete in the tension side, since once cracked it cannot transfer the stress. The transformed section is then changed to the concrete on the compression side, and the steel which has been transformed

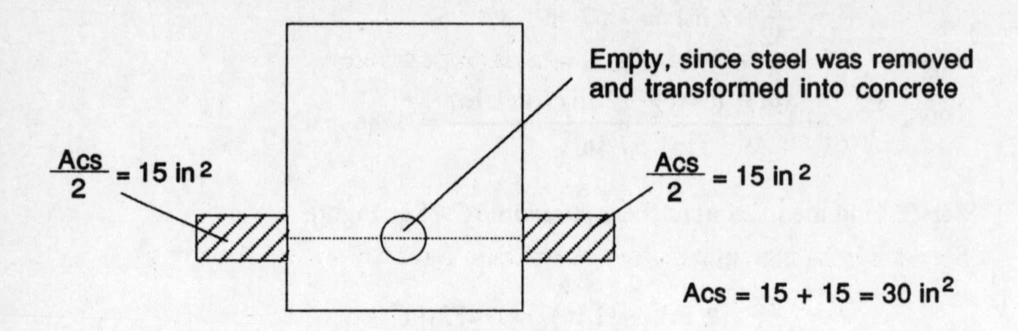

to an equivalent concrete area. The section becomes:

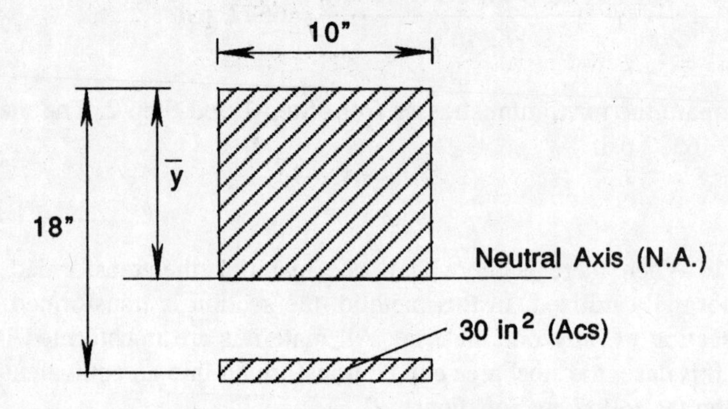

Step 2. Find the neutral axis of the transformed section after taking out the unusable concrete in the tension side. The neutral axis passes through the centroid of the section. The summation of the moments of the areas about the centroidal axis must be zero.

Thus,

$(10x\bar{y} \text{ in}^2)\ (\bar{y}/2 \text{ in}) - (30 \text{ in}^2)\ (18 - \bar{y} \text{ in}) = 0$

$\bar{y}^2 + 6\bar{y} - \quad 108 = 0$

$\bar{y} = 7.82$ in (Ignore the negative value of $\bar{y}$)

Step 3. Find the moment of inertia of the transformed section about the neutral axis.

$I_{NA} = {}^1/_3\ (10 \text{ in})\ (7.82 \text{ in})^3 + (30 \text{ in}^2)\ (18 \text{ in} - 7.82\text{in})^2 = 4703 \text{ in}^4$

Step 4. Find the stress on the top of the beam (maximum compressive stress in concrete).

$\sigma = Mc/I = (1000000 \text{ lb-in})\ (7.82 \text{ in})\ /\ (4703 \text{ in}^4) = 1663$ psi.

115. **(C)**

Step 1. Find the maximum load the column can carry without yielding.

$$\sigma = P/A \leq \sigma_{yp}$$

where P = axial load, and A = cross sectional area.

Thus, $$\frac{(P \text{ k})}{(0.5 \text{ in} \times 1 \text{ in})} \leq 30 \text{ ksi}$$

$$P \leq 15 \text{ k}$$

Step 2. Find the maximum load the column can carry without buckling.

$$P = EI\,\pi^2/(Le)^2$$

where E = modulus of elasticity = 30×10^3 ksi

I = moment of inertia = $I_y = {}^1/_{12}(1)\,(0.5)^3 = 0.0104 \text{ in}^4$

Le = effective length = 0.7 (length of column)

= 0.7 (24 in) = 16.8 in

Note: effective length is dependent on column's end conditions, i.e., fixed, pinned, free, etc.

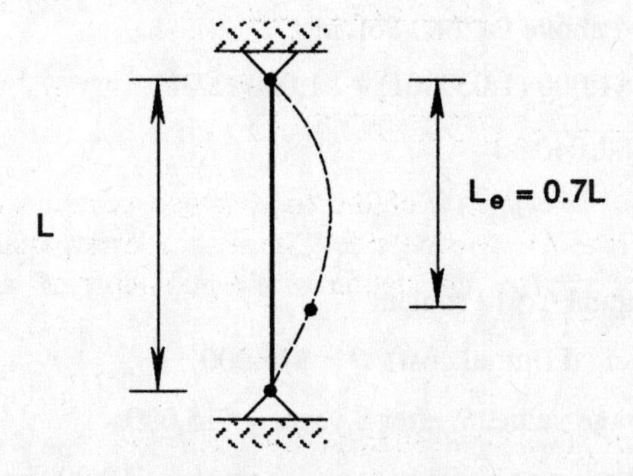

Thus,

$$P = \frac{(30 \times 10^3 \text{ k/in}^2)\,(0.0104 \text{ in}^4)\,\pi^2}{(16.8 \text{ in})^2} = 10.9 \text{ k}$$

Step 3. Compare P from Step 1 and Step 2.

P = 10.9 k.

116. **(C)**

Given: Effective Interest / Future Worth / Equivalence Problem.

(i) Two deposits of \$1,000. One at zero time, P_0 and one at the end of one second month, P_2.

(ii) Interest rate is 12% compounded monthly, i.e., Interest rate per interest period of one month

= 12% nominal rate / 12 interest periods

= 1% per month.

Required To Find: Future amount (accrual) at the end of the 3rd month.

Cash Flow Diagram:

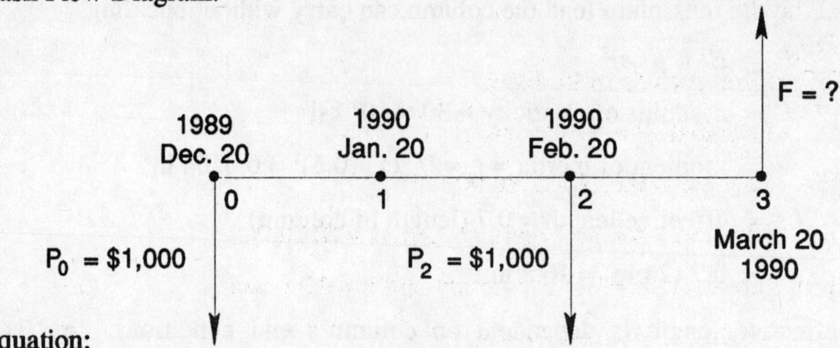

Equation:

$$F = \$P_0 (F/P, 1\%, 3) + \$P_2 (F/P, 1\%, 1)$$

Use tables to find above factors. Solving,

$$= \$1{,}000\ (1.030301) + \$1{,}000\ (1.01)$$

$$= \$2{,}040.30$$

117. **(A)**

Given: Annual Cost Problem

(i) Principal (initial cost), P = \$75,000

(ii) Salvage value, S, after 8 years = \$15,000

(iii) Operating and Maintenance costs, A = \$10,000 per year

(iv) Interest rate = 25%

Required to Find: Equivalent Uniform Annual Cost, EUAC

Cash Flow Diagram:

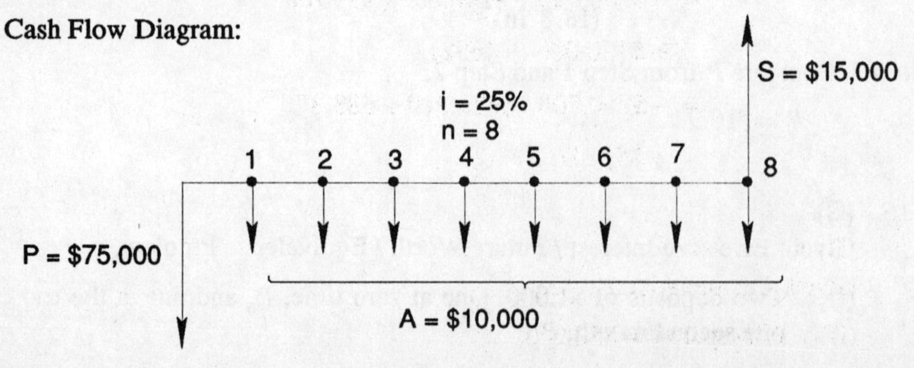

Equation:

$$EUAC = P(A/P, 25\%, 8) + A - S(A/F, 25\%, 8)$$

Use tables to find above factors. Solving,

$$= \$75{,}000\ (0.3004) + \$10{,}000 - \$15{,}000\ (.0504)$$

$$= \$31{,}773.91$$

118. **(D)**

Given: Present Worth Problem

(i) Principal, P = \$700,000

(ii) Salvage value, S = \$180,000

(iii) Uniform Annual Benefit (UAB), A = \$100,000

(iv) Interest rate = 8% per annum and n = 10 years

Required To Find: Net Present Worth, *NPW*

Cash Flow Diagram:

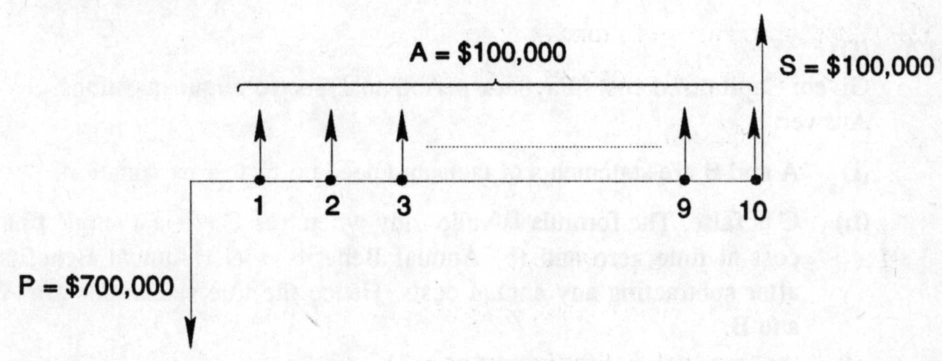

Equation:

$$NPW\ (8\%) = -P + A\ (P/A, 8\%, 10) + S(P/F, 8\%, 10)$$

Use tables to find the above factors. Solving,

$$= -\$700{,}000 + \$100{,}000\ (6.7101)$$

$$+ \$180{,}000\ (0.4632)$$

$$= -\$700{,}000 + \$671{,}010 + \$83{,}376$$

$$= +\$54{,}386$$

119. **(B)**

Given: Depreciation Problem

(i) Principal, P = \$50,000

(ii) Salvage, $S = \$8{,}000$

(iii) Useful life, $n = 14$ years

Required To Compute: The depreciation during the 3rd year, using Sum of Years Digit (*SOYD*) method.

Solution: Sum of Years Digit,

$$SOYD = \frac{n}{2}(n+1)$$

$$= \frac{14}{2}(14+1) = 105$$

Now *SOYD* depreciation for k^{th} year

$$= \frac{\begin{pmatrix}\text{Remaining Life} \\ \text{beginning of } k^{th} \text{ year}\end{pmatrix}}{(SOYD)}(P - S)$$

Substituting, $= \left(\frac{14-3+1}{105}\right)(\$50{,}000 - \$8{,}000)$

$= \$4{,}800$

120. **(D)**

Given: Capitalized cost / Payback period analysis technique questions.
Answer:

(i) A and B are statements of facts and need no further explanation.

(ii) C is false. The formula is valid only when: (a) There is a single first cost at time zero and (b) Annual Benefits = Net Annual Benefits after subtracting any annual costs. Hence the true statements are A and B.

121. **(B)**

Given: Future worth / Effective Interest / Multiple Compounding Problem.

(i) Monthly deposits, $A = \$100$, for 24 months.

(ii) Interest rate

$= 6\%$ compounded monthly

$= {}^{6}/_{12}$ or ${}^{1}/_{2}\%$ per month.

(iii) Total Period = 7 years ≡ (60 + 24) months

Required To Find: Future Worth, *FW*, accrued.

Cash Flow Diagram:

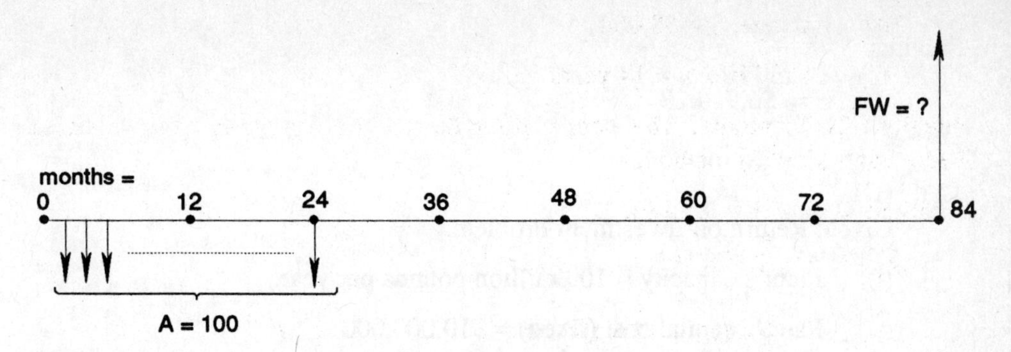

Equation:

$$FW(^1/_2\%) = A(F/A, {}^1/_2\%, 24)\ (F/P, {}^1/_2\%, 60)$$

Use tables to find the above factors. Solving,

$$= \$100\ (25.432)\ (1.349)$$

$$= \$3{,}430.78.$$

122. **(E)**

Given: Cash Flow Equivalence Problem.

(i) Principal, P = \$10,000

(ii) Interest rate, I = 12% per year and n = 5 years

(iii) Annual Payments, A = \$2,000 for first 4 years.

Required To Calculate: Payment in year 5, required to pay off the loan. In other words, equate present worth of cash flow to amount of loan.

Cash Flow Diagram:

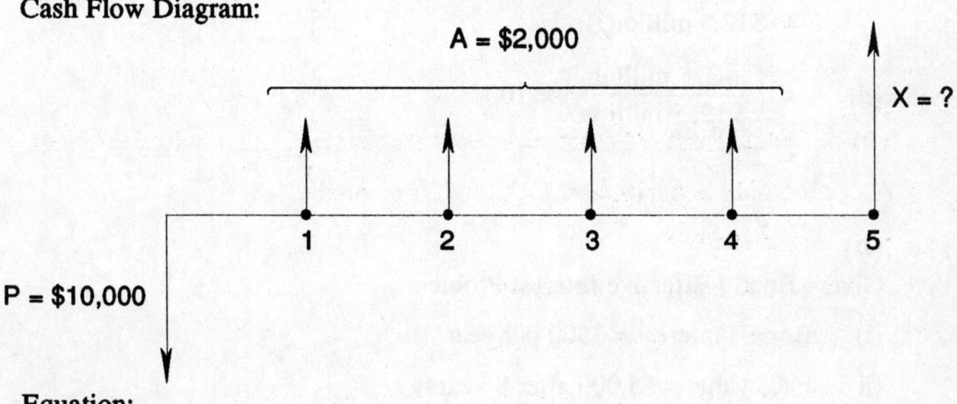

Equation:

$$P = A(P/A, 12\%, 4) + X(P/F, 12\%, 5)$$

Using tables,

$$10{,}000 = 2{,}000\ (3.037) + X(0.5674)$$

Solving,

$$X = \$6{,}919.28$$

123. **(C)**

Given: Return on Investment Problem

(i) Plant's capacity = 100 million pounds per year

(ii) Plant's capital cost (fixed) = $10,000,000

(iii) Plant's working (operating) capital = 25% of fixed.

(iv) Selling price = 25¢ per pound

(v) Profit after taxes = 10% of sales

Required To Calculate: The Return on Investment (R.O.I.) for the Plant.

Solution:

$$\text{R.O.I.} \equiv \left(\frac{\text{Profit after Taxes}}{\text{Total Capital (Investment)}}\right) 100\%$$

Now Profit after taxes

$$= 0.10\,(100{,}000{,}000 \text{ lb/yr} \times \$0.25/\text{lb})$$

$$= \$2.5 \text{ million per year}$$

Total capital

$$= (\text{Fixed} + \text{Working}) \text{ capital}$$

$$= \$10{,}000{,}000\,(1 + 0.25)$$

$$= \$12.5 \text{ million.}$$

$$\therefore \text{R.O.I.} = \left(\frac{\$2.5 \text{ million}}{\$12.5 \text{ million}}\right) \times 100$$

$$= 20\%$$

124. **(A)**

Given: Bond / Effective Interest Problem

(i) Bond's interest = $500 per year

(ii) Face value = $5,000 after 8 years

(iii) Current price = $3,900

Required To Calculate: Bond's true interest rate

Cash Flow Diagram:

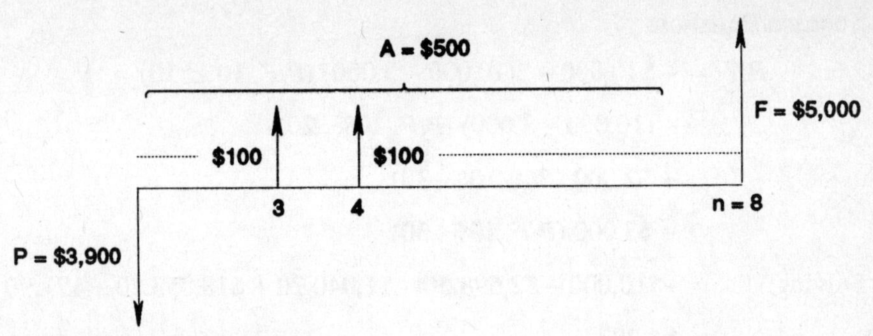

Solution: Let the real interest be i

Equation:

$$P = A(P/A, i\%, 8) + F(P/F, i\%, 8)$$

Solving,

$$3{,}900 = 500(P/A, i\%, 8) + 5{,}000\,(P/F, i\%, 8)$$

$$^{39}/_{5} = (P/A, i\%, 8) + 10(P/F, i\%, 8)$$

Through a trial and error process using the tables,

$$i \approx 14.9\% \approx 15\%.$$

125. **(E)**

Given: Replacement Analysis / Present Worth Problem

(i) Annual Savings, A = $2,000

(ii) Salvage value, S(F) = $3,000

(iii) Analysis Period = 30 years

(iv) Equipment Life = 10 years

(v) Interest rate = 10%

Required To Compute: Present worth of entire 30 years of service.

Assumption: Cash flow for the equipment remains constant for the 30-year period.

Cash Flow Diagram:

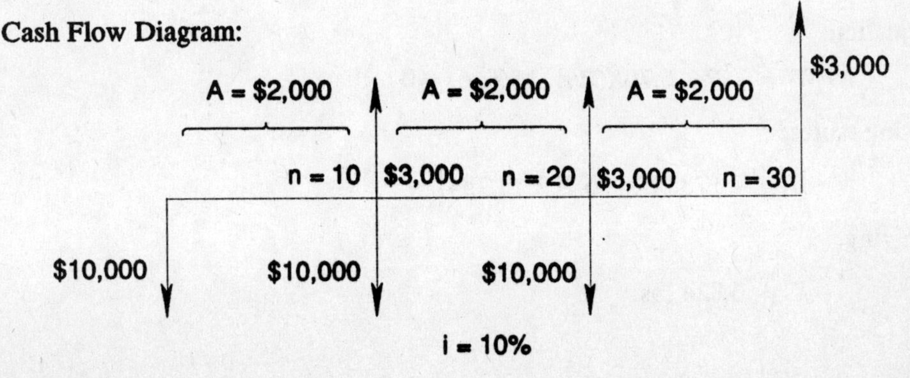

Solution/Equation:

$$PW = -\$10{,}000 - \$(10{,}000 - 3{,}000)\ (P/F, 10\%, 10)$$
$$- (10{,}000 - 3{,}000)\ (P/F, 10\%, 20)$$
$$+ \$2{,}000\ (P/A, 10\%, 30)$$
$$+ \$3{,}000\ (P/F, 10\%, 30)$$

Solving, $= -\$10{,}000 - \$2{,}698.50 - \$1{,}040.20 + \$18{,}853.80 + 171.90$

$= \$5{,}287.$

126. **(E)**

Given: Break-even Analysis Problem

(i) Capital Investment (Principal) = \$173,000

(ii) Profit = \$6.70 per lb of fish sold

(iii) Project Life, $n = 6$ years

(iv) Minimum Attractive Rate of Return, MARR = 15%

Required To Calculate: Pounds of fish to sell to break even.

Solution: Break-even point is given by Net Present Worth, NPW = 0.
Assume X lbs of fish sold required to break even.
Annual Profit = \$6.70X

Cash Flow Diagram:

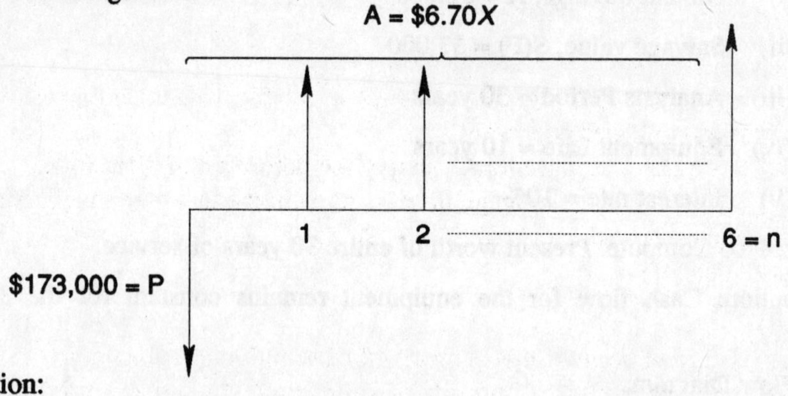

Equation:

$$NPW = -P + 6.70X(P/A, 15\%, 6) = 0$$

Using tables,

$$0 = -173{,}000 + 6.70X\ (3.784)$$

Solving,

$$X = 6{,}824 \text{ lbs.}$$

127. **(B)**

(A) is incorrect because it is the exact opposite of the answer.

(B) is correct because for an intrinsic semiconductor the concentration of free electrons (n) must equal the concentration of free holes (p) to satisfy charge neutrality.

$$n = p = n_i \alpha e^{-E_G/2kT}$$

As E_G increase, n_i decreases.

(C) and (D) are both incorrect because free carrier concentration has little to do with mobility. (E) is incorrect because it is not independent of the energy gap.

128. **(A)**

No free electrons exist at 0 K temperature. They are located at their lowest energy levels — the valence band.

129. **(E)**

(A) is wrong because resistance of a crystal will not impede the movement of a carrier, only free electrons. (B) is wrong for the same reason; conductivity is the opposite of resistance. (C) recombination rate will not affect the movement of a carrier in a crystal. (D) The charge per carrier does not affect the rate, since this remains constant.

(E) is correct because mobility is defined as the ease with which carriers can be made to move through the crystal. Consequently, it only depends on temperature and the regularity of the crystal structure.

130. **(C)**

(A) A gradient is what the charges move through in this question. (B) Recombination can destroy charges.

(C) is correct because this is simply the definition of diffusion.

(D) Lifetime is the average lifteime of a charge in a system. (E) Mobility is the movement through a crystal.

131. **(A)**

(A) is correct because this is simply the definition of aniostropic.

(B) Isotropic is just the opposite of anisotropic; the properties of the material are the same in all directions. (C) and (D) refer to geometric structure, not physical properties.

132. **(B)**

(A) This will reduce the dissipation of heat from it into the air.

(B) is correct because the purpose of a heat sink is to increase the flow of heat away from the source of heat. Thermal resistance can be likened to electrical resistance ($R = \rho l/A$). The thermal resistance can be reduced with a heat sink having a larger surface area to allow more rapid heat dissipation.

(C) This will keep heat from radiating away from it, thus lowering its efficiency. (D) Lack of air currents will also make it less capable of taking heat away.

133. **(A)**

(A) For this to be correct the question would have to state that these materials continue to emit light for up to several seconds after the energy source has been removed; this is simply the definition of phosphorescent.

(B) is correct if the light is emitted only for a few milliseconds after the energy source is removed; .

(C) and (D) only emit light when an energy source is connected to them, not after it is removed.

134. **(E)**

This question tests knowledge of the nature of ions and how they can be formed. The correct answer is (E). Ions are charged atoms or molecular fragments. Each of the four phenomena in Answers (A) through (D) represent a method of forming ions.

Answer (A) represents a method of forming ions that is used in many analytical instruments — e.g., mass spectrometers, ESCA spectrometers, etc. Answer (B) also represents a method of forming ions; in fact, high energy radiation, such as x-rays, is frequently referred to in safety regulations as "ionizing" radiation. Answer (C) represents a third method of forming ions. For example the neutralization of acetic acid (largely un-ionized in aqueous solution) with a base such as sodium hydroxide results in the formation of sodium acetate which dissociates into sodium and acetate ions.

Answer (D) is correct. Crystalline sodium chloride consists of a lattice of sodium ions and chloride ions.

135. **(A)**

This question tests knowledge of the crystalline structure of diamond and graphite – two very important crystalline forms of carbon. The correct answer (i.e., the statement that is *not* correct) is (A). The two crystalline forms of carbon are very different. Each of the other statements is correct. Answer (B) is correct. The tetrahedral structure of diamond permits little relative atomic motion. The hexagonal structure of graphite, on the other hand, permits slipping along the crystalline planes; hence graphite is an excellent lubricant. Answer (C) is also correct. Diamonds are transparent to visible light with a large angle of internal refraction which causes gems to sparkle. Graphite, on the other hand, is opaque and has a characteristic gray color. Answer (D) is also correct. Diamond is the hardest of all naturally occurring substances and is widely used for high performance cutting and grinding tools. Graphite, on the other hand, is so soft that it can be scratched with a fingernail. Answer (E) is also correct. Diamond is an excellent high-temperature semiconductor because of the large band gap or Fermi level. Graphite is a conductor with the unusual property that the resistivity decreases with increasing temperatures.

136. **(E)**

This question tests knowledge of the nature of gamma rays and subatomic particles. The correct answer is (E).

Answers (A), (B), (C), and (D) are all incorrect because they are particles. Gamma rays are electromagnetic waves of extremely high frequency.

137. **(C)**

This question tests knowledge of the structure of crystals and the importance of dislocations on the bulk properties of crystals. The correct answer (i.e., the one that is *not* a correct statement) is (C). Each of the other statements are correct statements about dislocations in crystals.

138. **(A)**

This question tests knowledge of the significance of Plank's constant in quantum descriptions of the structure of matter and energy. The correct answer is (A). The energy of a quantum of radiated energy is given by the equation

$$E = h\nu$$

where h is Plank's constant and ν is the frequency of the radiation.

Answer (B) is incorrect. The size of the quantum of energy is a function of the frequency or wavelength of the radiation. Answer (C) is incorrect. The size of a number is not a consideration in determining whether or not it is of physical significance. It is, of course, quite possible to construct smaller numbers of physical significance. Answer (D) is incorrect. The number of ergs in a kilogram calorie is 4.186×10^{10}. Answer (E) is incorrect since answer (A) is correct.

139. **(B)**

This question tests knowledge of the relationship between energy and matter, the structure of nuclides including which subatomic particles form nuclides, and the ability to make the appropriate units conversion in the calculation.

The binding energy of the $_2He^4$ nucleus is the energy equivalent (calculated as $e = mc^2$) of the difference in the mass of the subatomic particles from which the atom is "built" (2 protons [incl. electrons] plus 2 neutrons) and the $_2He^4$ atom.

$$\Delta m = (1.007276) + 2(1.008665) - 4.00387$$

$$= 0.02801 \text{ amu/atom} = 0.02801 \text{ g/mole}$$

The binding energy is calculated as follows:

$$e(\text{ergs}) = m(g)\,(c(\text{cm/sec}))^2 = .02801 \text{g/mole } (3.0 \times 10^{10} \text{ cm/sec})^2$$

$$= 2.5209 \times 10^{19} \text{ ergs/mole}$$

$$\frac{2.5209 \times 10^{19} \text{ ergs/mole} \times 6.2422 \times 10^{11} \text{ eV/erg}}{6.023 \times 10^{23} \text{ atoms/mole}}$$

$$= 2.61 \times 10^{7} \text{ eV} = 26.1 \text{ MeV}$$

Therefore, the correct answer is (B).

Answer (A) is incorrect. "MeV" stands for million electron volts; "eV" for electron volts. Answer (C) is incorrect. "BeV" stands for billion electron volts. Typical values of the binding energies of nuclei range from 1 to 50 million electron volts. Answer (D) is incorrect. It is, numerically, the binding energy is ergs/mole. Answer (E) is incorrect. It is the binding energy per mole. Binding energies are expressed per atom or per nucleus.

140. (B)

This question tests knowledge of the Periodic Table and the recurring nature of the properties of the elements. The correct answer is (B). The reactive electrons are those in the outer orbits and the electronic structure recurs periodically.

Answer (A) is incorrect. The Periodic Table is a reflection of observed phenomena, not its cause. Answer (C) is incorrect. The possible properties, as shown by the wide range of properties in newly developed synthetic materials, seems limitless. Answer (D) is incorrect. The reasons for periodically recurring properties of the elements are well understood.

Fundamentals of Engineering

P.M. SECTION

Test 1

Fundamentals of Engineering

P.M. SECTION

TEST 1 – ANSWER SHEET

1. (A) (B) (C) (D) (E)
2. (A) (B) (C) (D) (E)
3. (A) (B) (C) (D) (E)
4. (A) (B) (C) (D) (E)
5. (A) (B) (C) (D) (E)
6. (A) (B) (C) (D) (E)
7. (A) (B) (C) (D) (E)
8. (A) (B) (C) (D) (E)
9. (A) (B) (C) (D) (E)
10. (A) (B) (C) (D) (E)
11. (A) (B) (C) (D) (E)
12. (A) (B) (C) (D) (E)
13. (A) (B) (C) (D) (E)
14. (A) (B) (C) (D) (E)
15. (A) (B) (C) (D) (E)
16. (A) (B) (C) (D) (E)
17. (A) (B) (C) (D) (E)
18. (A) (B) (C) (D) (E)
19. (A) (B) (C) (D) (E)
20. (A) (B) (C) (D) (E)
21. (A) (B) (C) (D) (E)
22. (A) (B) (C) (D) (E)
23. (A) (B) (C) (D) (E)
24. (A) (B) (C) (D) (E)
25. (A) (B) (C) (D) (E)
26. (A) (B) (C) (D) (E)
27. (A) (B) (C) (D) (E)
28. (A) (B) (C) (D) (E)
29. (A) (B) (C) (D) (E)
30. (A) (B) (C) (D) (E)
31. (A) (B) (C) (D) (E)
32. (A) (B) (C) (D) (E)
33. (A) (B) (C) (D) (E)
34. (A) (B) (C) (D) (E)
35. (A) (B) (C) (D) (E)
36. (A) (B) (C) (D) (E)
37. (A) (B) (C) (D) (E)
38. (A) (B) (C) (D) (E)
39. (A) (B) (C) (D) (E)
40. (A) (B) (C) (D) (E)
41. (A) (B) (C) (D) (E)
42. (A) (B) (C) (D) (E)
43. (A) (B) (C) (D) (E)
44. (A) (B) (C) (D) (E)
45. (A) (B) (C) (D) (E)
46. (A) (B) (C) (D) (E)
47. (A) (B) (C) (D) (E)
48. (A) (B) (C) (D) (E)
49. (A) (B) (C) (D) (E)
50. (A) (B) (C) (D) (E)
51. (A) (B) (C) (D) (E)
52. (A) (B) (C) (D) (E)
53. (A) (B) (C) (D) (E)
54. (A) (B) (C) (D) (E)
55. (A) (B) (C) (D) (E)
56. (A) (B) (C) (D) (E)
57. (A) (B) (C) (D) (E)
58. (A) (B) (C) (D) (E)
59. (A) (B) (C) (D) (E)
60. (A) (B) (C) (D) (E)
61. (A) (B) (C) (D) (E)
62. (A) (B) (C) (D) (E)
63. (A) (B) (C) (D) (E)
64. (A) (B) (C) (D) (E)
65. (A) (B) (C) (D) (E)
66. (A) (B) (C) (D) (E)
67. (A) (B) (C) (D) (E)
68. (A) (B) (C) (D) (E)
69. (A) (B) (C) (D) (E)
70. (A) (B) (C) (D) (E)

FUNDAMENTALS OF ENGINEERING EXAMINATION

TEST 1

AFTERNOON (PM) SECTION

TIME: 4 Hours
70 Questions

DIRECTIONS: For each of the following questions and incomplete statements, choose the best answer from the five answer choices. You must answer all questions.

QUESTIONS 1–10 refer to the following problem.

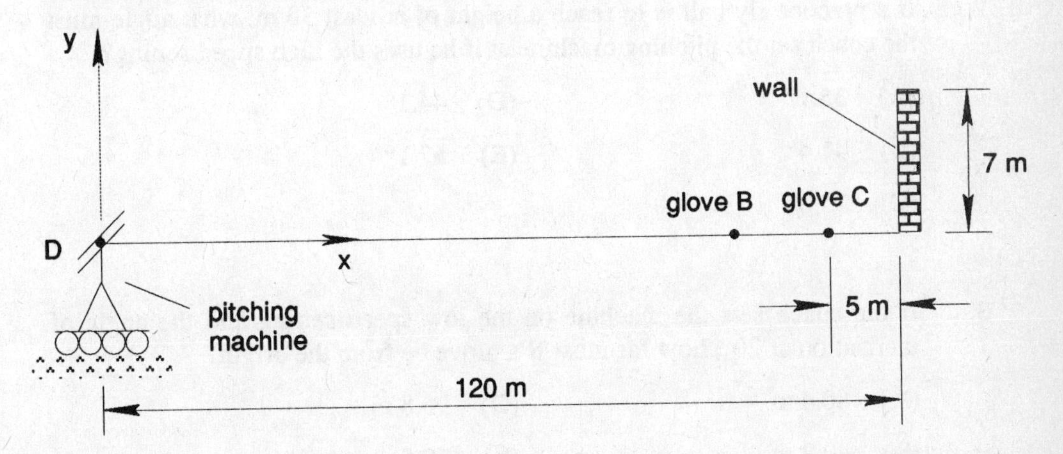

An automatic pitching machine is used by a baseball coach to give his outfielders and catchers practice fielding fly balls. The machine is set up so that the angle of inclination may be adjusted to any angle between 0° and 90°. The pitching machine has only two speed settings. The high speed setting will give the baseball an initial velocity of 45 m/s while the low speed setting corresponds to a velocity of 35 m/s. The machine is placed on rollers and has a mass of 35 kg. The mass of the baseballs which the machine uses is .16 kg.

Assumptions and Simplifications

- neglect the effect of air resistance
- neglect any effects of wind or air movement
- neglect friction between rollers and the ground
- assume the starting point for the baseball is at the origin
- assume the outfielders' gloves all lie on the x-axis
- assume the machine will be moved back to its original position after every fly ball

1. If the catcher, *D*, stands at homeplate (the origin), what is the maximum height a practice fly ball for him can attain?

(A) 206.6 m (D) 45.0 m

(B) 103.3 m (E) 110.0 m

(C) 77.5 m

2. If a practice fly ball is to reach a height of at least 50 m, what angle must the coach set the pitching machine at if he uses the high speed setting?

(A) 35.1° (D) 44.1°

(B) 41.4° (E) 52.1°

(C) 75.0°

3. If the coach sets the machine on the low speed setting and the angle of inclination at 20°, how far must *B*'s glove be from the origin?

(A) 80.4 m (D) 95.8 m

(B) 49.2 m (E) 60.5 m

(C) 158.3 m

4. What is the initial recoil velocity of the pitching machine for the fly ball to outfielder *B* in question 3?

(A) 6934.4 m/s (D) .1500 m/s

(B) .0676 m/s (E) 3.510 m/s

(C) .1860 m/s

5. What is the minimum time it will take for a fly ball to reach outfielder C's glove if the machine is on low setting?

(A) 7.65 s
(B) 6.05 s
(C) 2.67 s
(D) 4.30 s
(E) 3.92 s

6. What is the maximum horizontal range one can attain with the pitching machine?

(A) 180.7 m
(B) 206.6 m
(C) 125.0 m
(D) 301.2 m
(E) 70.3 m

7. By how much did the fly ball in question 6 clear the wall vertically?

(A) 32.1 m
(B) 50.2 m
(C) 43.2 m
(D) 39.1 m
(E) It didn't clear the wall.

8. A fly ball hits the wall with an initial horizontal velocity of 13.84 m/s. If the ball leaves the wall with the same speed in the opposite direction and the time of contact between the wall and ball was 1 ms, what was the average horizontal force acting on the ball?

(A) 4.4 kn left
(B) 0 kn
(C) 4.4 kn right
(D) 2.1 kn left
(E) 2.1 kn right

9. How high on the wall did the ball in question 8 hit? Assume the machine was on high speed setting.

(A) 4.35 m
(B) 6.55 m
(C) 3.20 m
(D) 2.75 m
(E) 5.37 m

10. A ball is shot into the air vertically ($\theta = 90°$) at the high speed setting. The ball narrowly misses a seagull flying at a height of 50 m and at a speed of 10 m/s in the positive x-direction. What is the ball's velocity as seen by the bird?

    (A) (10 **i** + 32.4 **j**) m/s
    (B) (32.4 **i** + 10 **j**) m/s
    (C) (–10 **i** + 45 **j**) m/s
    (D) (– 10 **i** + 32.4 **i**) m/s
    (E) (10 **i** + 45 **j**) m/s

QUESTIONS 11-20 refer to the following diagram.

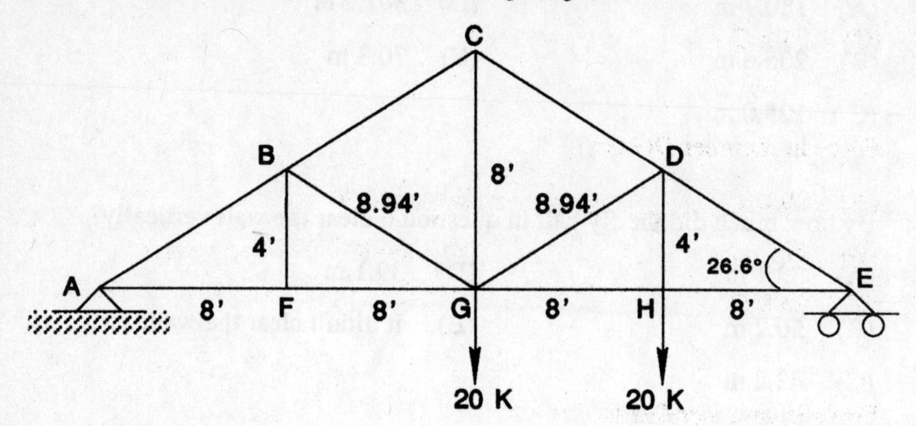

11. Reaction at A is

    (A) 0 k
    (B) 10 k
    (C) 15 k
    (D) 25 k
    (E) 20 k

12. Force in member AB is

    (A) 0 k
    (B) 15 k C
    (C) 30 k C
    (D) 34 k C
    (E) 50 k T

13. Force in member AF is

    (A) 0 k
    (B) 15 k C
    (C) 30 k T
    (D) 34 k T
    (E) 50 k C

14. Force in member *BF* is

(A) 0 k
(B) 15 k *C*
(C) 30 k *T*
(D) 34 k *T*
(E) 50 k *C*

15. Force in member *DE* is

(A) 20 k *T*
(B) 22 k *T*
(C) 25 k *C*
(D) 34 k *C*
(E) 56 k *C*

16. Force in member *DG* is

(A) 20 k *T*
(B) 22 k *C*
(C) 25 k *C*
(D) 34 k *T*
(E) 56 k *C*

17. Force in member *DH* is

(A) 20 k *T*
(B) 22 k *C*
(C) 25 k *C*
(D) 34 k *T*
(E) 50 k *C*

18. Force in member *GH* is

(A) 25 k *T*
(B) 25 k *C*
(C) 50 k *T*
(D) 50 k *C*
(E) 30 k *T*

19. Force in member *CG* is

(A) 0 k
(B) 30 k *T*
(C) 20 k *T*
(D) 56 k *T*
(E) 15 k *C*

20. Force in member *BG* is

(A) 0 k (D) 56 k *C*

(B) 40 k *T* (E) 15 k *C*

(C) 20 k *T*

QUESTIONS 21-30 refer to the following problem.

The following functions are given:

$$g(x) = 2x + 3, \quad 1 < x < 3,$$
$$= 3x^2 \quad 3 < x < 5,$$
$$= 0 \quad \text{elsewhere.}$$
$$p(x) = x^3$$
$$h(x) = x$$
$$q(x) = x^{2/3}$$

21. If $g(x) = f'(x)$, $f(2) = 10$ and given that $f(x)$ is a continuous function in the interval $1 < x < 5$, then $f(x)$ is given by the following equations in the interval $1 < x < 5$,

(A) $x^2 + 3x, \quad 1 < x \leq 3$
$x^3 - 9, \quad 3 \leq x < 5$

(B) $x^3 - 3x, \quad 1 < x < 5$

(C) $x^3 - 9, \quad 1 < x \leq 5$

(D) cannot be found

(E) $x^2 + 3x, \quad 1 < x \leq 3$
$x^3, \quad 3 \leq x < 5$

22. The value of the integral is $\int_2^5 g(x)\, dx$ is

(A) 113 (D) 15

(B) 30 (E) 106

(C) 117

23. The minimum value of function $g(x)$ in the interval $1 < x < 5$ is

(A) 75 (B) 9

(C) 5 (D) 3

(E) 1

24. The function $g(x)$ in the interval $1 < x < 5$ is

(A) continuous (D) undefined

(B) not defined at $x = 3$ (E) monotonically decreasing

(C) constant

25. The function $h(x)$ intersects the function $g(x)$ at

(A) $x = -3$ (D) no points

(B) $x = 1/3$ (E) $x = 0$ and $x = 1/3$

(C) $x = 0$

26. The area enclosed by the two functions $p(x)$ and $h(x)$ in the first quadrant is

(A) 1/2 (D) 1

(B) 3/4 (E) 0

(C) 1/4

27. The volume of the solid of revolution generated by revolving about the y–axis the area enclosed by the graphs of the two functions $p(x)$ and $h(x)$ in the first quadrant is

(A) $\pi/2$ (D) $\pi/4$

(B) $3\pi/4$ (E) $4\pi/21$

(C) π

28. The length of the arc of the function $q(x)$ from $x = 0$ to $x = 8$ is

(A) $(80\sqrt{10} - 8)/27$ (D) 96/5

(B) $\sqrt{80}$ (E) 32

(C) $(80\sqrt{10} + 8)/27$

29. The mean value of the function $p(x)$ between $x = 1$ and $x = 4$ is

(A) 255
(B) 65/2
(C) 255/12
(D) 65
(E) 255/3

30. The curvature of the function $q(x)$ at $x = 8$ is

(A) $-3/(80\sqrt{10})$
(B) $3/(80\sqrt{10})$
(C) 1
(D) 4
(E) 1/4

QUESTIONS 31–40 refer to the following problem.

Given the following differential equation

$$d^2y/dx^2 + b\,dy/dx + cy = 6e^{-2x} + x^2,$$

$$y(0) = 3,\ dy/dx(0) = 2$$

31. If

$$z_1 = (-b + \sqrt{b^2 - 4c})\,/2,$$

$$z_2 = (-b - \sqrt{b^2 - 4c})\,/2,$$

the homogeneous part of the solution of the differential equation decays to zero as $x \rightarrow \infty$ if

(A) $Re(z_1) > 0$ and $Re(z_2) > 0$
(B) $Re(z_1) > 0$ and $Re(z_2) < 0$
(C) $Re(z_1) < 0$ and $Re(z_2) < 0$
(D) $Re(z_1) < 0$ and $Re(z_2) > 0$
(E) $Re(z_1) = 0$ and $Re(z_2) > 0$

32. If $b = 6$ and $c = 9$, the homogeneous part of the solution is of the form (A and B are real constants)

(A) $Ae^{-3x} + Be^{3x}$
(B) Ae^{-3x}

(C) $Ae^{-3x} + Bxe^{-3x}$ (D) $Ae^{-3x} + Bxe^{3x}$

(E) $A \sin(3x) + B \cos(3x)$

33. If $b = 4$ and $c = 25/4$, the homogeneous part of the solution is of the form (A and B are real constants)

(A) $e^{-2x}[A \sin(3x) + B \cos(3x)]$

(B) $A \sin(3x/2) + B \cos(3x/2)$

(C) $Ae^{-7/2x} + Be^{-1/2x}$

(D) $Ae^{-2x} \sin(3x/2 + B)$

(E) $e^{2x}[A \sin(3x/2) + B \cos(3x/2)]$

34. If $b = 0$ and $c = 4$, the particular part of the solution is

(A) $e^{-2x} + 1/4x^2 - 1/8$ (D) $3/4e^{-2x}$

(B) $3/4e^{-2x} + 1/4x^2 - 1/8$ (E) $1/4x^2 - 1/16$

(C) $3/4e^{-2x} + 1/4x^2$

35. The order of the differential equations is

(A) one (D) four

(B) zero (E) two

(C) three

36. The Laplace transforms $L[(f(x))] = F(s)$ of the right hand side function, $6e^{-2x} + x^2$, in the differential equation is

(A) $6/(s + 2) + 2/s^3$ (D) $6/(s - 2) + 1/s^3$

(B) $6/(s - 2) + 2/s^3$ (E) $6/(s + 2) + 2/s^2$

(C) $6/(s + 2) + 1/s^3$

37. The value of d^2y/dx^2 at $x = 0+$ is

(A) $6 - 2b - 3c$ (D) 6

(B) $6 + 2b + 3c$ (E) 0

(C) $-2b - 3c$

38. The definition of dy/dx in the differential equation is

(A) $\lim_{x \to 0} [y(x+h) + y(x)] / h$

(B) $\lim_{h \to 0} [y(x-h) - y(x)] / h$

(C) $\lim_{x \to 0} [y(x-h) + y(x)] / h$

(D) $\lim_{h \to 0} [y(x+h) - y(x)] / h$

(E) $\lim_{h \to 0} [y(x+h) - y(x)] / (2h)$

39. If $b = 8$ and $c = 4$, the undamped natural frequency in radians per second is

(A) 2
(B) 4
(C) 0
(D) 16
(E) $\sqrt{2}$

40. If $b = 8$ and $c = 4$, the differential equation governs a system which is

(A) underdamped
(B) overdamped
(C) critically damped
(D) undamped
(E) none of the above

QUESTIONS 41 – 50 refer to the following circuit.

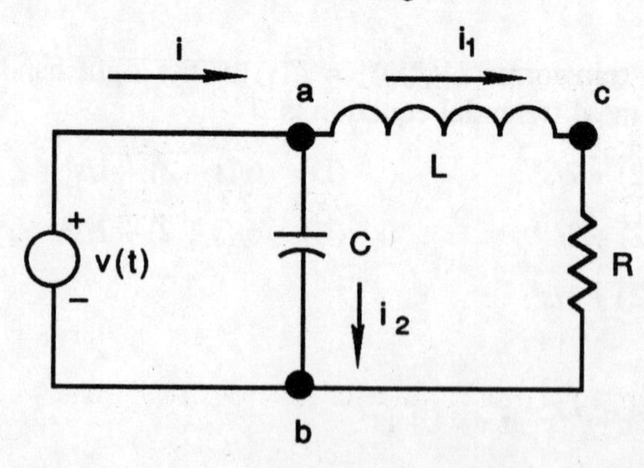

The applied voltage is given by the time function $v(t) = 141.4 \sin 314t$ V. Consider that the phasor that represents the periodic function $1.41 \cos t$ is $1\underline{|0°}$. Remember that the magnitude of a phasor is equal to the root mean square value of the amplitude of the periodic function represented by the phasor. Bold letters are used for phasors and plain letters for their magnitude. **V** would indicate a phasor while *V* its magnitude.

The parameters of the circuit are: $R = 4\Omega$, $L = 0.00955$ H and $C = 382\mu F$. Questions 41 – 49 refer to the steady state behavior of the circuit.

41. The phasor voltage $\mathbf{V}_{ab}$ is given by:

(A) $141.4 \angle 0° V$
(B) $200.0 \angle 90° V$
(C) $100.0 \angle -90° V$
(D) $141.4 \angle 180° V$
(E) $141.4 \angle -90° V$

42. The root mean square value of the current $\mathbf{I}_2$ is:

(A) 12 A
(B) 24 A
(C) 833 A
(D) 16.9 A
(E) 33.9 A

43. The phase displacement of the $\mathbf{I}_1$ with respect to the voltage **V** is given by:

(A) – 37°
(B) + 37°
(C) + 53°
(D) 0°
(E) – 53°

44. The current phasor **I** is given by:

(A) $16.0 \angle -30°$
(B) $16.0 \angle 0°$
(C) $22.6 \angle -90°$
(D) $16.0 \angle -90°$
(E) $22.6 \angle +90°$

45. The root mean square magnitude of the voltage $\mathbf{V}_{ac}$ is:

(A) 33.3 V
(B) 36.0 V
(C) 80.0 V
(D) 18.0 V
(E) 60.0 V

46. The reactive power delivered to the capacitor is:

    (A) 600 VARs
    (B) 1200 VARs
    (C) 17.28 VARs
    (D) 1967 VARs
    (E) 848 VARs

47. The power factor of the current *I* is:

    (A) 0.8
    (B) 0.6
    (C) 1.0
    (D) 0.707
    (E) 0.0

48. The complex power in the *RL* branch is given by:

    (A) $1600 + j1200$ VA
    (B) $1200 - j1600$ VA
    (C) $1600 - j1967$ VA
    (D) $1600 - j1200$ VA
    (E) $1600 - j848$ VA

49. Input impedance is given by:

    (A) $8.84 \angle -90° \ \Omega$
    (B) $6.25 \angle 0° \ \Omega$
    (C) $4.42 \angle 0° \ \Omega$
    (D) $6.25 \angle -30° \ \Omega$
    (E) $4.42 \angle -30° \ \Omega$

50. When the voltage source is connected to or disconnected from the circuit, the current $i(t)$ will have a transient component. This transient response will be

    (A) over-damped with one time constant of 4.77 ms
    (B) over-damped with one time constant of 7.28 ms
    (C) under-damped with a time constant of 4.77 ms
    (D) under-damped with a time constant of 7.28 ms
    (E) critically damped with a time constant of 7.28 ms

QUESTIONS 51–60 refer to the following problem.

Amichi is a growing city in east Tennessee. The city council is embarking on a number of development projects. The city borrows money at an interest rate of 8% and uses the straight line depreciation method for tax purposes. The city is taxed at a rate of 50%.

51. The construction of a motor raceway sits atop the list of the city's development projects. $500,000.00 has been earmarked for the project. This project, which will have an infinite life, will cost $400,000.00. Annual expenses are estimated at $50,000.00. In addition, the price per ticket will be based on an average of 1,000 sales per month. As a public project, the project should be analyzed using the Benefit-Cost Ratio analysis technique. The price of each ticket should be:

    (A) $1.50
    (B) $2.00
    (C) $3.00
    (D) $3.50
    (E) $4.00

52. If the city has a pricing policy of $5.00 per adult and $2.50 per child for the raceway tickets, estimate the annual worth of the investment if the whole $500,000.00 is committed. Ticket sales now average 1,200 per month and the number of adults is twice that of children. Use a Minimum Attractive Rate of Return of 10%.

    (A) – $40,000.00
    (B) + $40,000.00
    (C) – $60,000.00
    (D) + $60,000.00
    (E) $0

53. The city also bought a street sweeper for $700,000.00 This vehicle has a salvage value of $100,000.00 at the end of 5 years of useful life. What will be the book value of this equipment at the end of year 2?

    (A) $240,000.00
    (B) $100,000.00
    (C) $700,000.00
    (D) $200,000.00
    (E) $460,000.00

54. If the city had adopted a SOYD depreciation method, how much tax would have been saved in the first year on the street sweeper?

(A) $80,000.00 (D) $40,000.00

(B) $200,000.00 (E) $20,000.00

(C) $120,000.00

55. If the bank were to change its compounding from annually to monthly, what would be the true or effective interest rate the city would pay?

(A) 8.1% (D) 8.4%

(B) 8.2% (E) 8.5%

(C) 8.3%

56. The city recently hired a chemical engineer to evaluate the city's industrial projects. The chemical engineer wishes to buy a house for $120,000.00 with a $20,000.00 down payment. The city bank will finance the house at 15% interest over 25 years. How much would the engineer pay per month?

(A) $1,207.74 (D) $1,432.00

(B) $1,171.00 (E) $178.70

(C) $1,285.60

57. The city wishes to set up a retirement fund. At 10% interest rate compounded continuously, how large a fund in millions of dollars is required to guarantee an average annual retirement income of $20,000.00? 3,000 retirees are expected to participate.

(A) $600M (D) $570M

(B) $590M (E) $560M

(C) $580M

58. The chemical engineer has the following financial data on two competing public projects

	Project A	Project B
First Cost	$50,000	$60,000
Annual Maintenance	$ 5,000	$ 4,000
Useful Life	7 years	7 years
Salvage	$ 5,000	$ 5,000
Annual Earnings	$20,000	$21,000
Interest Rate	7%	7%

As is customary with analysis of public projects, he used the modified Benefit-Cost Ratio analysis technique. He first computed the Benefit-Cost Ratio for Project A. His answer was

(A) 1.68
(B) 1.70
(C) 1.72
(D) 1.74
(E) 1.76

59. Refer to Problem 8. Having found Project A acceptable, he proceeded to calculate

$$\left(\frac{\Delta B}{\Delta C}\right)_{B \rightarrow A}$$

to see whether Project B was better than A. His answer was

(A) 1.00
(B) 1.02
(C) 1.04
(D) 1.06
(E) 1.08

60. In Problems 58 and 59

(A) Inflation will change the decisions already made.
(B) Inflation will not affect the decisions.
(C) Lowering the interest rate will favor the choice of Project A.
(D) Lowering the interest rate will favor the choice of Project B.
(E) (B) and (D)

QUESTIONS 61–70 refer to the following problem.

This problem set concerns a turbojet engine with several simplifying assumptions. Assume that only air (ideal gas) flows through the engine and that specific heats are constant:

$C_p = 1.005$ kJ/kg-K; $k = 1.4$;

$C_v = 0.718$ kJ/kg-K; $R = 0.287$ kJ/kg-K

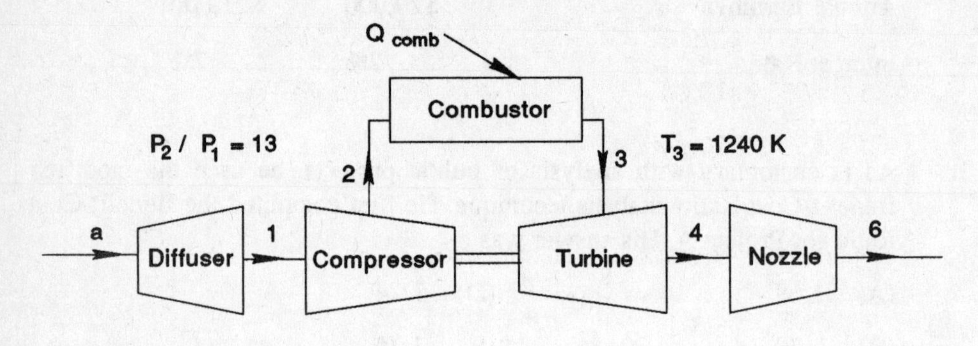

$P_a = 22$ kPa $P_6 = 22$ kPa

$T_a = 220$ K $V_6 = 954.8$ m/s

$T_1 = 259$ K

QUESTIONS 61–66 concern an ideal turbojet without afterburner, with the turbine work equal to 755.15 kW, and only air flowing through the engine.

61. The aircraft speed is most nearly

(A) 8.854 m/s
(B) 198.0 m/s
(C) 279.3 m/s
(D) 280.0 m/s
(E) 310.0 m/s

62. The pressure at state 1 is most nearly

(A) 38.9 kPa
(B) 27.6 kPa
(C) 23.5 kPa
(D) 23.1 kPa
(E) 21.7 kPa

63. The mass flow rate of the air is most nearly

(A) 1.629 kg/s
(B) 1.933 kg/s
(C) 2.684 kg/s
(D) 2.693 kg/s
(E) 3.715 kg/s

64. The entropy change across the combustor is most nearly

(A) 0.8374 kJ/kg-K
(B) 0.8324 kJ/kg-K
(C) 0.7319 kJ/kg-K
(D) 0.6354 kJ/kg-K
(E) 0.5977 kJ/kg-K

65. The temperature of state 6 is most nearly

(A) 494.9 K
(B) 497.2 K
(C) 503.7 K
(D) 506.4 K
(E) 511.0 K

66. If the process from state 1 to state 2 is made irreversible, which of the following statements is true?

(A) The entropy would remain constant.
(B) The work required by the compressor would increase.
(C) The net cycle work would increase.
(D) The temperature at state three would decrease.
(E) The process efficiency would increase.

QUESTIONS 67–70 concern an actual turbojet with an afterburner. The operating conditions are the same, however, a compressor efficiency of 80 percent, a turbine efficiency of 85 percent, an air mass flow rate of 2.8 kg/s, and an exit velocity (V_6) of 1050 m/s are to be assumed. (The work is no longer the same).

$C_p = 1.005$ kJ/kg-K; $k = 1.4$;

$C_v = 0.718$ kJ/kg-K; $R = 0.287$ kJ/kg-K

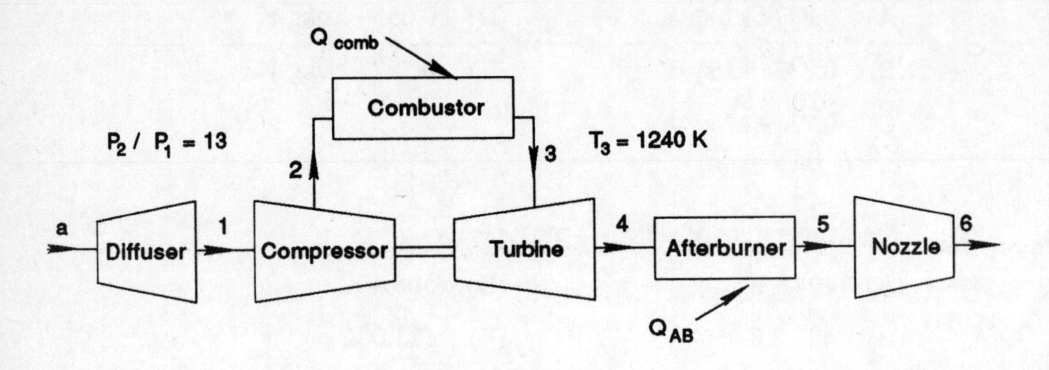

$P_a = 22$ kPa $P_6 = 22$ kPa

$T_a = 220$ K $V_6 = 954.8$ m/s

$T_1 = 259$ K

67. The temperature at state 2 is most nearly

(A) 608.9 K
(B) 588.8 K
(C) 539.4 K
(D) 497.3 K
(E) 483.3 K

68. If the lower heating value of the fuel entering the combustor is 44000 kJ/kg, the mass flow rate of the fuel is most nearly

(A) 0.04327 kg/s
(B) 0.04036 kg/s
(C) 0.04009 kg/s
(D) 0.02880 kg/s
(E) 0.02001 kg/s

69. The temperature at state 4 is most nearly

(A) 749.45 K
(B) 791.21 K
(C) 865.32 K
(D) 890.10 K
(E) 942.11 K

70. If the afterburner receives heat a at a rate of 1127 kW, the temperature at state 6 is most nearly

(A) 742.09 K
(B) 936.71 K
(C) 1037.54 K
(D) 1468.03 K
(E) 1838.54 K

TEST 1 (PM)

ANSWER KEY

1.	(B)	21.	(A)	41.	(C)	61.	(D)
2.	(D)	22.	(E)	42.	(A)	62.	(A)
3.	(A)	23.	(C)	43.	(A)	63.	(C)
4.	(D)	24.	(B)	44.	(D)	64.	(A)
5.	(E)	25.	(C)	45.	(E)	65.	(D)
6.	(B)	26.	(C)	46.	(B)	66.	(B)
7.	(C)	27.	(D)	47.	(C)	67.	(A)
8.	(A)	28.	(A)	48.	(A)	68.	(B)
9.	(D)	29.	(C)	49.	(B)	69.	(D)
10.	(D)	30.	(B)	50.	(C)	70.	(A)
11.	(C)	31.	(C)	51.	(D)		
12.	(D)	32.	(C)	52.	(E)		
13.	(C)	33.	(D)	53.	(A)		
14.	(A)	34.	(B)	54.	(D)		
15.	(E)	35.	(E)	55.	(C)		
16.	(B)	36.	(A)	56.	(A)		
17.	(A)	37.	(A)	57.	(D)		
18.	(C)	38.	(D)	58.	(C)		
19.	(B)	39.	(A)	59.	(E)		
20.	(A)	40.	(B)	60.	(E)		

DETAILED EXPLANATIONS OF ANSWERS

TEST 1

AFTERNOON (PM) SECTION

1. **(B)**

The maximum height attainable would correspond to the high speed setting and an angle of inclination of 90°. Therefore,

$$V_o = 45 \text{ m/s}$$

$$V_{0x} = (45 \cos 90°) \text{ m/s} = 0$$

$$V_{0y} = (45 \sin 90°) \text{ m/s} = 45 \text{ m/s}$$

Using the equation

$$V_y = V_{0y} + a_y t$$

with $a_y = -g = -9.8 \text{ m/s}^2$

$$V_{0y} = 45 \text{ m/s}$$

$$V_y = 0 \text{ (at maximum height)}$$

$$t = \frac{0 - 45 \text{ m/s}}{-9.8 \text{ m/s}^2} = 4.59 \text{ s}$$

Using the equation

$$y - y_0 = V_{0y} t + {}^1/_2\, a_y t^2$$

with y = final height

$$y_0 = 0$$

$$V_{0y} = 45 \text{ m/s}$$

$$t = 4.59 \text{ s}$$

$$y = (45 \text{ m/s})(4.59 \text{ s}) + {}^1/_2\,(-9.8 \text{ m/s}^2)(4.59 \text{ s})^2 = 103.3 \text{ m}$$

2. **(D)**

Given:

i) $x_0 = 0$

ii) $y_0 = 0_n$ $y_{max} = 50$ m

iii) $V_0 = 45$ m/s

Find: θ (angle of inclination)

Solution: First, find the time it takes for the ball to reach its maximum height as a function of θ.

Using the equation:

$$V_y = V_{0y} + a_y t$$

with $V_y = 0$ (at maximum height)

$V_{0y} = (45 \sin \theta)$ m/s

$a_y = -9.8$ m/s

$$t = \frac{-(45 \sin\theta) \text{ m/s}}{-9.8 \text{ m/s}^2} = (4.59 \sin\theta) \text{ s}$$

Using the equation:

$$y - y_0 = V_{0y} t + {}^1/_2 a_y t^2$$

with $y = 50$ m

$y_0 = 0$

$V_{0y} = (45 \sin \theta)$ m/s

$t = (4.59 \sin \theta)$ s

$a_y = -g = -9.8$ m/s²

$$50 \text{ m} = (45 \sin \theta)(4.59 \sin \theta) \text{ m} + {}^1/_2 (-9.8)(4.59 \sin \theta)^2 \text{ m}$$

$$50 \text{ m} = (206.6 \sin^2 \theta) \text{ m} - (103.3 \sin^2 \theta) \text{ m}$$

$$50 \text{ m} = (103 \sin^2 \theta) \text{ m}$$

$$\sin^2 \theta = .484$$

$$\sin \theta = .696$$

$$\theta = 44.1°$$

3. **(A)**

Given:

i) $V_0 = 35$ m/s

ii) $\theta = 20°$

Find: X (Range)

Solution: First, find the time it takes for the ball to reach its maximum height using the equation:

$$V_y = V_{0y} + a_y t$$

with V_{0y} = (35 sin 20°) m/s = 11.97 m/s

V_y = 0 (at maximum height)

$a_y = -g = -9.8\ \text{m/s}^2$

$0 = 11.97\ \text{m/s} - (9.8\ \text{m/s}^2)\,t$

$$t = \frac{-11.97\ \text{m/s}}{-9.8\ \text{m/s}^2} = 1.22\ \text{s}$$

Therefore, the total time the ball is in the air equals:

$$t_T = 2(1.22\ \text{s}) = 2.44\ \text{s}$$

Now, find the distance the ball travels in the x-direction in t_T, using the equation:

$$x - x_0 = V_{0x} t + {}^1/_2\, a_x t^2$$

with $x_0 = 0$

V_{0x} = (35 cos 20°) m/s = 32.9 m/s

t_T = 2.44 s

$a_x = 0$

x = (32.9 m/s) (2.44 s) = 80.4 m

4. **(D)**

Given:

i) V_{0B} = 35 m/s

ii) θ = 20°

iii) m_B = .16 kg

iv) m_p = 35 kg

Find: Recoil velocity of pitching machine.

Solution: Use the concept of the conservation of momentum in the x-direction since there are no external forces acting in that direction.

The conservation of momentum states:

$$m_p V_{px} = m_B V_{0Bx}$$

with V_{0Bx} = (35 cos 20°) m/s = 32.9 m/s

m_B = .16 kg

m_p = 35 kg

$$(35\text{ kg})\,V_{px} = (.16\text{ kg})(32.9\text{ m/s})$$

$$V_{px} = .150\text{ m/s}$$

5. **(E)**

Given:

i) $V_0 = 35$ m/s

ii) $x_{max} = 115$ m

Find: T, time for ball to travel 115 m in the x-direction.

Solution: First, find the angle the machine must be set at for the ball to travel 115 m in the x-direction with $V_0 = 35$ m/s.

We can find the time the ball takes to reach its maximum height as a function of θ using the equation:

$$V_y = V_{0y} + a_y t$$

with

$$V_y = 0 \text{ (at maximum height)}$$

$$V_{0y} = (35 \sin \theta)\text{ m/s}$$

$$a_y = -g = -9.8\text{ m/s}^2$$

$$0 = (35 \sin \theta)\text{ m/s} - (9.8\text{ m/s}^2)\,t$$

$$t = \frac{(-35 \sin \theta)\text{ m/s}}{-9.8\text{ m/s}^2} = (3.6 \sin \theta)\text{ s}$$

Therefore, the total tine in the air, T, equals 2(3.6 sin θ) s = (7.2 sin θ) s.

Now, by using the equation:

$$x - x_0 = V_{0x} t + {}^1/_2\, a_x t^2$$

with

$$x = 115\text{ m}$$

$$V_{0x} = (35 \cos \theta)\text{ m/s}$$

$$t = (7.2 \sin \theta)\text{ s}$$

$$a_x = 0$$

$$115\text{ m} = (35 \cos \theta)(7.2 \sin \theta)\text{ m} + 0$$

$$115\text{ m} = (252 \cos \theta \sin \theta)\text{ m}$$

Using the trigonometric relation

$$\sin 2\theta = 2 \cos \theta \sin \theta$$

$$115\text{ m} = 126\text{ m}\,(2 \cos \theta \sin \theta)$$

$$115\text{ m} = 126\text{ m}\,(\sin 2\,\theta)$$

$$\sin 2\theta = .91$$

$$2\theta = 65.9° \quad \text{or} \quad 2\theta = 114.2°$$

$$\theta = 32.9° \quad \text{or} \quad \theta = 57.1°$$

$$T = (7.2 \sin 32.9°)\text{ s} \quad \text{or} \quad T = (7.2 \sin 57.1°)\text{ s}$$

$$T = 3.92\text{ s} \quad \text{or} \quad T = 6.05\text{ s}$$

To minimize T choose $\theta = 32.9$ or $T = 3.92$ s.

6. **(B)**
Given:

i) $V_0 = 45$ m/s

Find: Maximum range attainable, x_{max}

Solution: It can be shown that the maximum range corresponds to an angle of inclination of 45°. Given a velocity, V_0, the time of flight, T, is equal to twice the amount of time it takes for the projectile to reach its maximum height, or:

$$V_y = V_{0y} + a_y t$$

with $V = 0$ (at maximum height)

$$V_{0y} = (V_0 \sin\theta)\text{ m/s}$$

$$a_y = -g = -9.8\text{ m/s}^2$$

$$0 = (V_0 \sin\theta)\text{ m/s}^2 - (g\text{ m/s}^2)\, t$$

$$t = \frac{-(V_0 \sin\theta)\text{ m/s}}{-g\text{ m/s}^2} = \left(\frac{V_0 \sin\theta}{g}\right) s$$

$$T = 2t = \left(\frac{2V_0 \sin\theta}{g}\right)\text{s}$$

The range can then be found using:

$$x - x_0 = V_{0x} t + {}^1\!/_2\, a_x t^2$$

with $x = \text{Range } (R)$

$$V_{0x} = (V_0 \cos\theta)\text{ m/s}$$

$$t = T = \left(\frac{2V_0 \sin\theta}{g}\right)\text{s}$$

$$a_x = 0$$

$$R = \left[\left(\frac{2V_0 \sin\theta}{g}\right)(V_0 \cos\theta)\right]\text{m}$$

$$R = \frac{2V_0^2 \sin\theta}{g} \cos\theta$$

Given $\sin 2\theta = 2 \sin \theta \cos \theta$

$$R = \frac{V_0^2 \sin 2\theta}{g}$$

Since $\sin 2\theta$ is maximum at $2\theta = 90°$ or $\theta = 45°$, R is maximum at $\theta = 45°$.

So, for the given velocity, $V_0 = 45$ m/s,

$$R_{max} = (45 \text{ m/s})^2 / 9.8 \text{ m/s}^2 = 206.6 \text{ m}.$$

7. **(C)**

Given:

i) $V_0 = 45$ m/s

ii) $\theta = 45°$

iii) $x_w = 120$ m

iv) $y_w = 7$ m

Find: The difference between the height of ball (at $x = 120$ m) and the height of the wall.

Solution: First, find the time at which the ball passes over the wall using the equation:

$$x - x_0 = V_{0x}\, t + {}^1/_2\, a_x t^2$$

with

$$x = 120 \text{ m}$$

$$V_{0x} = (45 \cos 45) \text{ m/s} = 31.8 \text{ m/s}$$

$$a_x = 0$$

$$120 \text{ m} = (31.8 \text{ m/s})\, t$$

$$t = 3.77 \text{ s}$$

Now, find the height of the ball at time t using the equation:

$$y - y_0 = V_{0y}\, t + {}^1/_2\, a_y t^2$$

with

$$y = \text{height of ball}$$

$$v_{0y} = (45 \sin 45°) \text{ m/s} = 31.8 \text{ m/s}$$

$$t = 3.77 \text{ s}$$

$$a_y = -9.8 \text{ m/s}^2$$

$$y = (31.8 \text{ m/s})(3.77 \text{ s}) + {}^1/_2(-9.8 \text{ m/s}^2)(3.77 \text{ s})^2$$

$$y = 50.2 \text{ m}$$

This is the height of the ball above the x-axis. For the height of the ball above the wall we subtract the height of wall above the x-axis.

$$d = 50.2\text{ m} - 7\text{ m} = 43.2\text{ m}$$

8. **(A)**
Given:

i) $V_{x_1} = 13.84$ m/s right

ii) $V_{x_2} = 13.84$ m/s left

iii) $\Delta t = 1 \times 10^{-3}$ s

iv) $m_B = .16$ kg

Find: F_x, the constant average force acting on the ball when in contact with the wall.

Solution: Using the equation

$$\Delta p_x = \mathbf{F}_x \Delta t$$

$$(mV_{x_2} - mV_{x_1}) = \mathbf{F}_x \Delta t$$

$$.16\text{ kg }(-13.84\text{ m/s} - 13.84\text{ m/s}) = \mathbf{F}(1 \times 10^{-3}\text{ s}).$$

$$\mathbf{F} = -4.4 \times 10^{3}\text{ N}$$

$$\mathbf{F} = 4.4\text{ kN acting to the left.}$$

9. **(D)**
Given:

i) $V_{0x} = 13.84$ m/s

ii) $V_0 = 45$ m/s

iii) $x = 120$ m

Find: y, height of the ball at $x = 120$ m.

Solution: First, find the time it took for the ball to reach the wall.

$$x = V_{0x} t$$

$$120\text{ m} = (13.84\text{ m/s})\, t$$

$$t = 8.67\text{ s}$$

Now, find θ using the equation:

$$V_0 x = V_0 \cos\theta$$

$$13.84\text{ m/s} = (45 \cos\theta)\text{ m/s}$$

$$\cos\theta = .308$$

$$\theta = 72.1°$$

Now, find y using the equation:

$$y - y_0 = V_{0y}t + {}^1/_2\, a_y t^2$$

with $V_{0y} = (45 \sin 72.1°)$ m/s = 42.8 m/s

$$t = 8.67 \text{ s}$$

$$a_y = -9.8 \text{ m/s}^2$$

$$y = (42.8 \text{ m/s})(8.67 \text{ s}) - (4.9 \text{ m/s}^2)(8.67 \text{ s})^2$$

$$y = 2.75 \text{ m}$$

10. **(D)**

Given:

i) $V_{BO} = 45$ m/s

ii) $\theta = 90°$

iii) $\mathbf{V}_{SE} = (10\ \mathbf{i})$ m/s (velocity of seagull relative to earth)

iv) $y = 50$ m

Find: $\mathbf{V}_{BS}$, relative velocity of the ball to the seagull.

Solution: First, find the velocity of the ball relative to the earth at $y = 50$ m using the equation:

$$y - y_0 = V_{0y}t + {}^1/_2\, a_y t^2$$

with $y = 50$ m

$$V_{0y} = (45 \sin 90°) \text{ m/s} = 45 \text{ m/s}$$

$$ay = -9.8 \text{ m/s}^2$$

$$50 \text{ m} = (45 \text{ m/s})\,t - (4.9 \text{ m/s}^2)\,t^2$$

$$(4.9 \text{ m/s}^2)\,t^2 - (45 \text{ m/s}\ t + 50 \text{ m} = 0$$

Solve for t using the quadratic formula

$$t = \frac{45 \text{ m/s} \pm \sqrt{(45 \text{ m/s})^2 - 4(50 \text{ m})(4.9 \text{ m/s}^2)}}{2(4.9 \text{ m/s}^2)}$$

$$t = 1.29 \text{ s} \quad \text{or} \quad t = 7.90 \text{ s}$$

Choose the lower value of t to correspond to the upward flight of the ball. (The ball will be at $y = 50$ m at $t = 7.90$ s also, but this will be on the way down.)

$$V_y = (45 \text{ m/s}) - (9.8 \text{ m/s}^2)(1.29 \text{ s}) = 32.4 \text{ m/s}$$

The velocity of the ball relative to earth is

$$\mathbf{V}_{BE} = V_y \mathbf{j} = (32.4 \mathbf{j}) \text{ m/s}$$

To find the relative velocity of the ball to the seagull, $\mathbf{V}_{BS}$, use the equation:

$$\mathbf{V}_{BS} = \mathbf{V}_{BE} + \mathbf{V}_{ES} = \mathbf{V}_{BE} - \mathbf{V}_{SE}$$

$$\mathbf{V}_{BS} = (32.4 \mathbf{j} - 10 \mathbf{i}) \text{ m/s}$$

$$\mathbf{V}_{BS} = (-10 \mathbf{i} + 32.4 \mathbf{j}) \text{ m/s}$$

11. **(C)**

Determine the reactions using the basic equations of equilibrium:

$$\Sigma F_x = 0 \quad \Sigma F_y = 0 \quad \Sigma M = 0$$

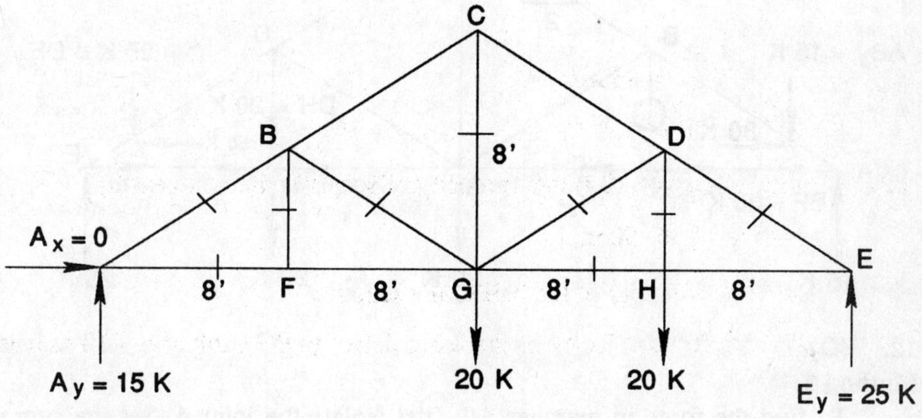

$$\Sigma F_x = 0 \quad A_x = 0$$

$$\Sigma M_A = 0$$

$$16 \text{ ft} \times 20 \text{ k} + 24 \text{ ft} \times 20 \text{ k} - 32 \text{ ft } E_y = 0$$

$$E_y = 25 \text{ k}$$

$$\Sigma F_y = 0$$

$$A_y + 25 \text{ k} - 20 \text{ k} - 20 \text{ k} = 0$$

$$A_y = 15 \text{ k}$$

Solution of Truss Forces

Truss member forces can be solved by either the method of joints or the method of sections.

The procedure for the method of joints is to draw a large sketch of the truss, circle each joint and sum forces, $F_x = 0$ and $F_y = 0$, within each circle. Solution must start at a joint where there are only two member unknowns and proceed through the truss solving each joint in sequence. Most of the members for this truss can be solved easily by the method of joints. At joint D, however, members *CD* and *DG* must be found by using two simultaneous equations. For this reason the method of sections is used here to avoid the simultaneous equations.

The procedure for the method sections is to cut the truss including the member force to be found. Either the left or right hand portion of the section can be used. There are internal forces in the members that have not been cut. These internal forces, counteract each other and drop out. To avoid using the extra forces assume the section is an odd shaped sheet of plywood with forces acting on it. Three equations, $\Sigma F_x = 0$, $\Sigma F_y = 0$, $\Sigma M = 0$, are available by the method of sections. The force equations may be replaced by additional moment equations if the three moment points do not form a straight line.

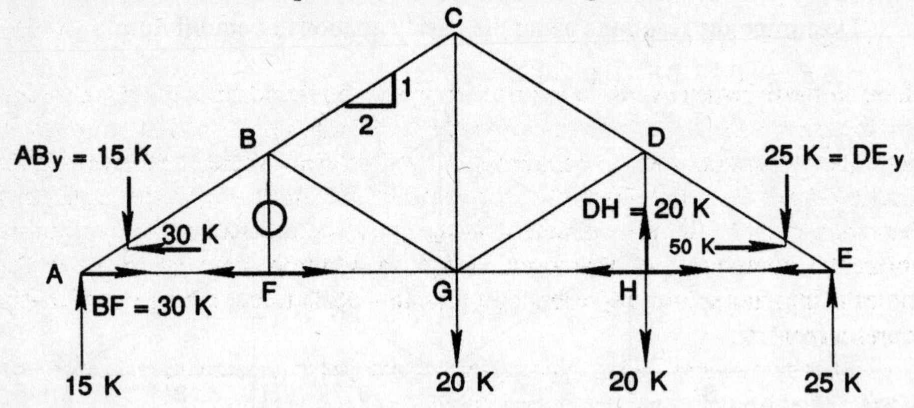

12. **(D)**

Method 1.

To find the force in member *AB*, first isolate the joint *A*. Set the sum of vertical forces, at joint *A*, equal to zero.

$$\Sigma F_y = 0 \quad AB_y = 0$$

Method 1

Method 2

Use the slope relationship to find the horizontal component of AB

$$AB_x = 2AB_y = 30\text{ k}$$

$$AB = \sqrt{(15\text{ k})^2 + (30\text{ k})^2} = 33.5\text{ k}\,C$$

Method 2.

The following is an alternative method of finding the force in member AB using the angles between members rather than resolving each force into its x and y components. Note: The rest of the solutions concerning joints will follow Method 1, however, Method 2 can also be used.

To find the force in member AB, isolate joint A as shown in the previous figure.

From geometry,

$$\tan\theta = 4/8 \Rightarrow \theta = 26.6°$$

Note: in most instances, direction of forces may be found through visualization. For instance, at joint A, it can be seen that the force in AB must act downward (and, therefore, into joint A) because it is the only force capable of opposing the reaction force R_{Ay}. Subsequently, the force in BF must act to the right (and, therefore, out of joint A) because it is the only force capable of opposing the horizontal component of the force in AB. In addition, convention states that forces acting into a joint represent compression while forces acting out of a joint represent tension.

However, when using equilibrium equations it is convenient to use the usual convention of positive forces acting up and to the right.

Back to the calculations:

To find the force in AB, set the sum of the vertical forces equal to zero

$$\Sigma F_y = R_{Ay} - AB\sin 26.6° = 0$$

$$15\text{ k} - AB\sin 26.6° = 0$$

$$AB = 33.5\text{ k (compression)}$$

13. **(C)**

To find the force in member AF, sum the horizontal forces at joint A and set equal to zero.

$$\Sigma F_x = 0, \quad AF = 30\text{ k}\,T$$

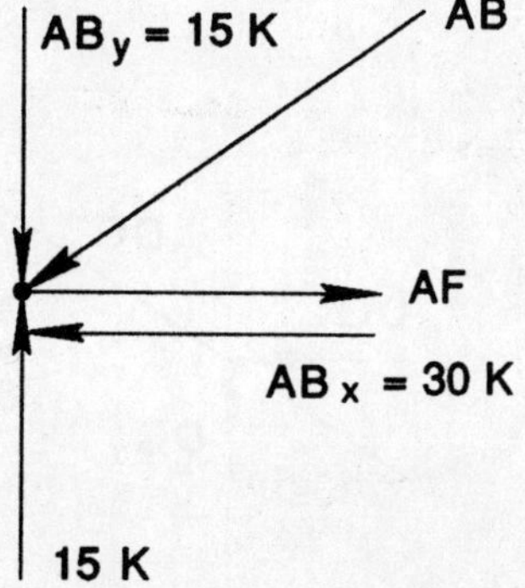

14. **(A)**

By summing the vertical forces at joint F and setting them equal to zero, we find that BF is a zero force member. It follows from equilibrium that BG is also a zero force member.

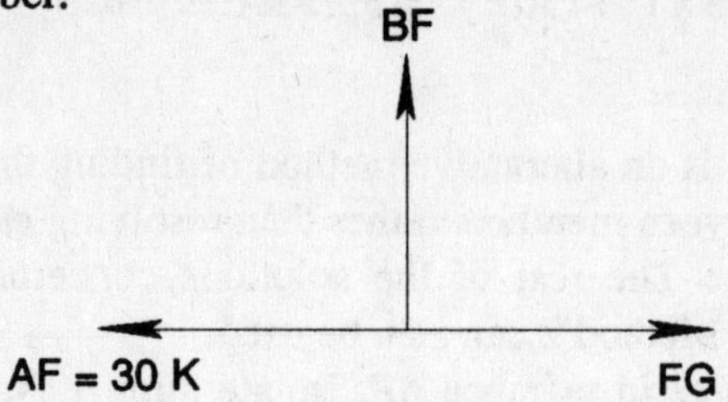

15. **(E)**

To find the force in member DE sum vertical forces at joint E and set equal to zero. Use the slope relationship to find the horizontal component of DE.

$$\Sigma F_y = 0, \quad DE_y = 25 \text{ k}$$

$$DE_x = 2DE_y = 50 \text{ k}$$

$$DE = \sqrt{(25 \text{ k})^2 + (50 \text{ k})^2} = 56 \text{ k } C$$

DE
DE y
1
2
DE x
25 K

16. **(B)**

Force DG can be found most easily by the method of sections. Draw a section as shown and sum moments about point E.

$$\Sigma M_E = 0, 16 \text{ ft } DG_y - 8 \text{ ft} \times 20 \text{ k} = 0$$

$$DG_y = 10 \text{ k}, DG_x = 2DGy = 20 \text{ k}$$

$$DG = \sqrt{(10 \text{ k})^2 + (20 \text{ k})^2} = 22.4 \text{k } C$$

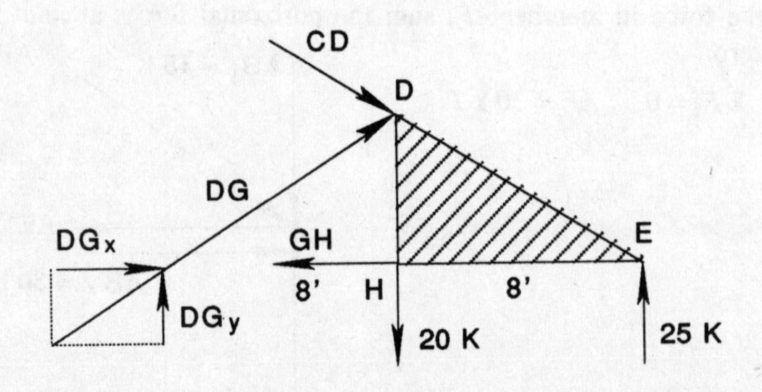

17. **(A)**

The force in member DH can be found by summing the vertical forces at joint H, and setting equal to zero.

$$\Sigma F_y = 0, \quad DH = 20\text{ k } T$$

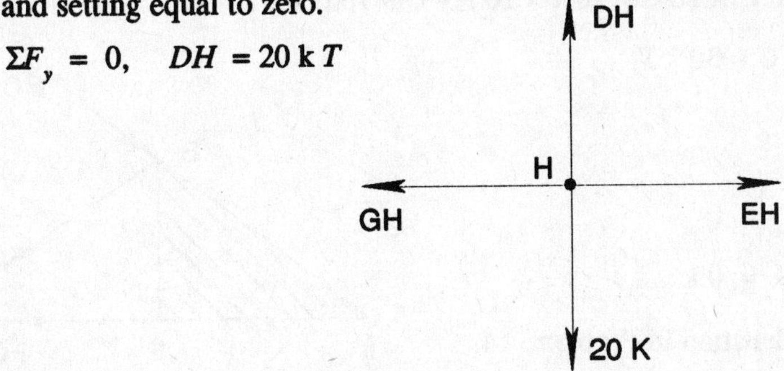

18. **(C)**

To find the force in member GH follow through the calculations with the method of joints from joint E to H.

$$GH = 50\text{ k T}$$

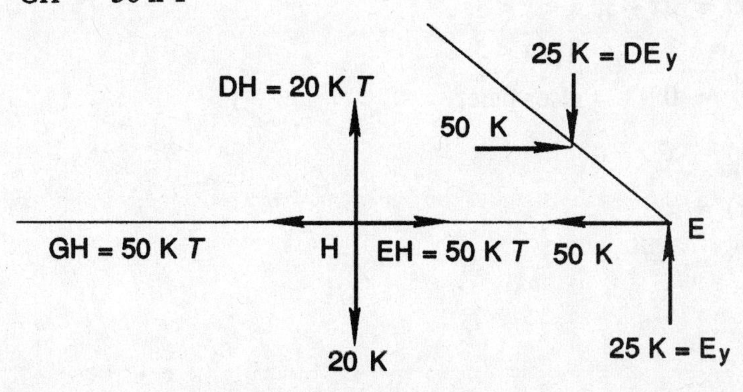

19. **(B)**

There are several approaches to finding the force CG, but none are easy. The following solution uses the vertical component of DG found earlier in Problem 16 and the followig section.

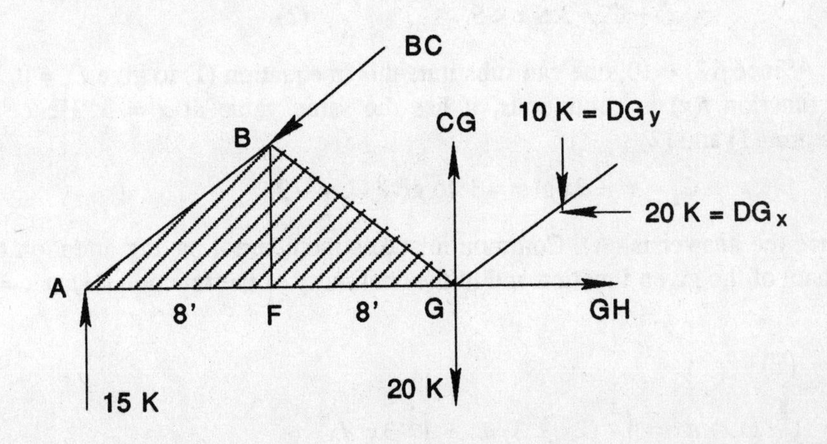

$\Sigma M_A = 0$

$20\text{ k} \times 16\text{ ft} + 10\text{ k} \times 16\text{ ft} - CG\ 16\text{ ft} = 0$

$CG = 30\text{ k}\ T$

20. **(A)**

$\Sigma M_A = 0$

$BG = 0\text{ k}$

See the explanation in Problem 14.

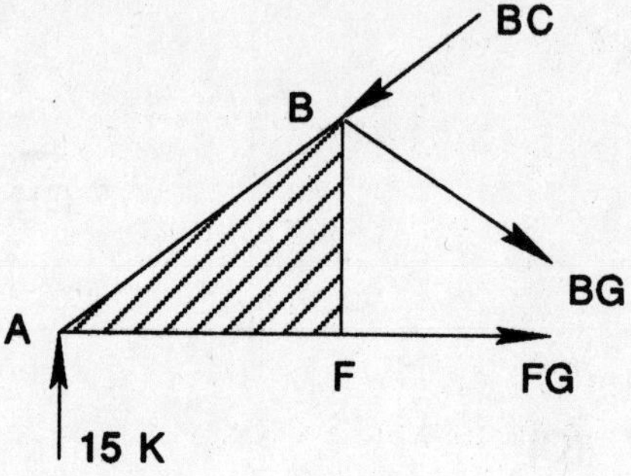

ANSWERS 21 – 30 refer to the following functions.

$$g(x) = 2x + 3, \quad 1 < x < 3,$$
$$= 3x^2 \quad 3 < x < 5,$$
$$= 0 \quad \text{elsewhere.}$$

$$p(x) = x^3$$

$$h(x) = x$$

$$q(x) = x^{2/3}$$

21. **(A)**

$$g(x) = 2x + 3, \quad 1 < x < 3,$$
$$= 3x^2, \quad 3 < x < 5.$$
$$f(x) = x^2 + 3x + C_1, \quad 1 < x \leq 3, \qquad (1)$$
$$= x^3 + C_2, \quad 3 \leq x < 5. \qquad (2)$$

Since $f(2) = 10$, one can substitute this in equation (1) to give $C_1 = 0$. Since the function $f(x)$ is continuous, it has the same value at $x = 3$. Hence from equation (1) and (2)

$$x^3 + C_2 = x^2 + 3x \text{ at } x = 3 \text{ to give } C_2 = -9.$$

Hence the answer is (A). Common mistakes include not paying attention to the domain of the given function and to the continuity of the function $f(x)$ at $x = 3$.

22. **(E)**

$$\int_2^5 g(x)\,dx = \int_2^5 (2x + 3)\,dx + \int_2^5 3x^2 dx$$

$$= (x^2 + 3x)\Big|_2^3 + (x^3)\Big|_3^5$$

$$= 106$$

One can also save time by using the results of Problem 21 since $f(x)$ is the integral of $g(x)$.

$$\int_2^5 g(x)\,dx = f(5) - f(2)$$

$$= 116 - 10$$

$$= 106$$

23. **(C)**

In the interval $1 < x < 3$

$$g'(x) = 2$$

$$g''(x) = 0$$

Since the second derivative is zero, the minimum can occur only at the end points $x = 1$ or $x = 3$.

$$g(1) = 5$$

$$g(3) = 9$$

In the interval $3 < x < 5$

$$g'(x) = 6x$$

$$g''(x) = 6$$

$g'(x) = 0$ at $x = 0$ which is not a point in the interval $3 < x < 5$. Hence the minimum may occur only at the end point $x = 3$ or $x = 5$.

$$g(3) = 27$$

$$g(5) = 75$$

Hence the minimum occurs at $x = 1$ and is $g(1) = 5$.

24. **(B)**

The function $g(x)$ is given as a function of x in the interval $1 < x < 5$. The given expressions on the sub intervals are differentiable and hence continuous on the sub intervals. One needs then only to check the value of the function at the break points. $x = 3$ is a break point in this case.

Since $g(3-) = 9$ and $g(3+) = 27$, the function is not defined at $x = 3$.

25. **(C)**

Points were $g(x) = h(x)$ are the points where the two functions intersect.

In the interval $1 < x < 3$, $g(x) = 2x + 3 = h(x) = x$ gives $x = -3$. But this point is outside the interval $1 < x < 3$.

In the interval $3 < x < 5$, $g(x) = 3x^2 = h(x) = x$ gives $x = {}^1/_3$, $x = 0$. But again these points are outside the interval $3 < x < 5$.

In the interval $x < 1$ and $x > 5$, $g(x) = 0 = h(x) = x$ gives $x = 0$. This point is in the interval $x < 1$ and $x > 5$. Hence $x = 0$ is the only point of intersection.

26. **(C)**

The functions $p(x)$ and $h(x)$ intersect at

$$p(x) = h(x)$$

$$x^3 = x$$

$$x(x^2 - 1) = 0$$

$$x = 0, 1, -1.$$

So the area enclosed by the two curves in the first quadrant is between $x = 0$ and $x = 1$. The area under the curve $p(x)$ between $x = 0$ and $x = 1$ is

$$\int_0^1 x^3\, dx = 1/4$$

The area under the curve $h(x)$ between $x = 0$ and $x = 1$ is

$$\int_0^1 x\, dx = 1/2$$

Hence the area enclosed is the difference of the areas under the two curves between $x = 0$ and $x = 1$ and is given by

$$\text{Area} = {}^1/_2 - {}^1/_4 = {}^1/_4.$$

27. **(D)**

The area lies between ($x = 0$, $y = 0$) and ($x = 1$, $y = 1$). The volume generated by the enclosed graph by revolving about the y-axis is

$$V = \int_0^1 \pi\, [y^{1/3} - y]\, dy$$

$$= \pi\, [3/4 y^{4/3} - y^2/2] \Big|_0^1$$

$$= \pi(3/4 - 1/2)$$

$$= \pi/4$$

Common mistakes include finding the volume generated by the enclosed graph by revolving about the x-axis instead of the y-axis. In that case, since the function $h(x) \geq p(x) \geq 0$ in the interval $0 \leq x \leq 1$, the volume generated by the enclosed graph around the x-axis is

$$V = \int_0^1 \pi\left[\{h(x)\}^2 - \{p(x)\}^2\right] dx$$

$$= \int_0^1 \pi\,[x^2 - x^6]\,dx$$

$$= \pi(1/3 - 1/7)$$

$$= 4\pi/21.$$

28. **(A)**

Since the function $q(x)$ has a continuous derivative in the interval [0, 8], the length of the arc of $q(x)$ is given by

$$s = \int_0^8 \sqrt{1 + \{q'(x)\}^2}\,dx$$

Substituting

$$q(x) = x^{2/3}$$

$$q'(x) = 2/3x^{-1/3}$$

$$s = \int_0^8 \sqrt{1 + 4/(9x^{2/3})}\,dx$$

$$= 1/3\int_0^8 \sqrt{4 + 9x^{2/3}}\,x^{-1/3}\,dx$$

If $u = 4 + 9x^{2/3}$, $du = 6x^{-1/3}\,dx$, $u = 4$ at $x = 0$ and $u = 40$ at $x = 8$,

$$s = \int_4^{40} \sqrt{u}\,du/18 = (80\sqrt{10} - 8)/27.$$

Common mistakes include finding the integral of the function instead in which case the answer is choice (D). Other mistakes include finding the straight line distance between the two extreme points (0, 0) and (8, 4) on the curve $q(x)$ in which case the answer is choice (B).

29. **(C)**

The mean value of a function $p(x)$ in the interval $[a, b]$ is given by

$$p = 1/(b-a)\int_a^b p(x)\,dx$$

$$= 1/(4-1)\int_1^4 x^3 dx = 255/12$$

Common mistakes include using the average of the function values at the end points ($x = 1$ and $x = 4$) in which case the value of the answer is choice (B). This is true only for a function which is either constant or linear.

30. **(B)**

The curvature κ of a function $q(x)$ at any point is given by

$$\kappa = |q''(x)|/[1 + \{q'(x)\}^2]^{3/2},$$

$$q(x) = x^{2/3},$$

$$q'(x) = 2/3x^{-1/3}$$

$$q''(x) = -2/9x^{-4/3}.$$

At $\quad x = 8$

$$q'(8) = 1/3$$

$$q''(8) = -1/72.$$

Hence $\quad \kappa = |-1/72| / (1 + 1/9)^{3/2}$

$$= 3/(80\sqrt{10}).$$

Note that curvature of a curve is always a positive number.

ANSWERS 31–40 refer to the following equations

$$d^2y/dx^2 + b\, dy/dx + cy = 6e^{-2x} + x^2,$$

$$y(0) = 3,\ dy/dx(0) = 2$$

31. **(C)**

The characteristic equation is

$$s^2 + bs + c = 0.$$

Note that the coefficients of this quadratic equation are those on the left hand side of the differential equation. The roots of this equation are

$$z_1 = (-b + \sqrt{b^2 - 4c})/2,$$

$$z_2 = (-b - \sqrt{b^2 - 4c})/2.$$

The homogeneous part of the solution then is

$$y_H = A e^{z_1 x} + B e^{z_2 x}.$$

For the special case of $z_1 = z_2$, then

$$y_H = A e^{z_1 x} + B x e^{z_1 x}.$$

In both case, y_H will decay will decay to zero $x \to \infty$ only if the real part of all the roots is strictly less than zero.

32. **(C)**

If $b = 4$ and $c = 9$, the characteristic equation is

$$m^2 + 6m + 9 = 0$$

$$(m + 3)^2 = 0$$

which gives $z_1 = z_2 = -3$ as the repeated roots of the equation. Since the roots are identical, the form of the homogeneous part should contain two independent terms to represent the solution as

$$y_H = A e^{-3x} + B x e^{-3x}$$

33. **(D)**

If $b = 4$, $c = 25/4$, the characteristic equation is

$$m^2 + 4m + 25/4 = 0$$

The roots of the equation are

$$z_1 = -2 + 3/2 j,$$

$$z_2 = -2 - 3/2 j,$$

where $j = \sqrt{-1}$.

The homogeneous part of the solution then is

$$y_H = E e^{(-2 + j\,3/2)x} + F e^{(-2 - j\,3/2)x}.$$

Using the identity

$$e^{jx} = \cos x + j \sin x$$

This solution can be rewritten as

$$y_H = e^{-2x}[C \cos(3/2x) + D \sin(3/2x)],$$

where $C = E + F$

and $D = j(E - F)$.

The equation can also be rewritten as

$$y_H = A e^{-2x} \sin(3/2x + B),$$

where A and B are real constants.

34. **(B)**

Before finding the particular part of the solution, one should compute the homogeneous part of the solution. For $b = 0$ and $c = 4$, the homogeneous part of the solution is

$$y_H = E \cos(2t) + F \sin(2t).$$

Looking at the right hand side (also called the forcing function) of the differential equation, the form of the particular part of the solution is given by the forcing function and its derivatives.

$$y_p = A e^{-2x} + Bx^2 + Cx + D$$

Since the assumed form of y_p of the differential equation does not contain any terms which are identical to the homogeneous terms, one can proceed to find the value of the constants in y_p. Substituting y_p in the differential equation

$$8A\, e^{-2x} + 4Bx^2 + 4Cx + (2B + 4D) = 6e^{-2x} + x^2.$$

Equating the corresponding coefficients, one gets

$$A = 3/4, \quad B = 1/4, \quad C = 0, \quad D = -1/8.$$

Hence the particular part of the solution is

$$y_p = 3/4e^{-2x} + 1/4x^2 - 1/8.$$

35. **(E)**

The order of an ordinary differential equation is the highest non-zero derivative term in the differential equation. In this case it is two.

36. **(A)**

The Laplace transform follows the superposition rule, that is

$$L[f(x) + g(x)] = L[f(x)] + L[g(x)].$$

Since

$$L(e^{-2x}) = 1\,/\,(s + 2) \text{ and}$$

$$L(x^2) = 2\,/\,s^3, \text{ then}$$

$$L(6e^{-2x} + x^2) = 6/(s + 2) + 2/s^3.$$

37. **(A)**

One does not need to find the solution to the differential equation to find the value of d^2y/dx^2 at $x = 0+$. Since the differential equation is valid for any time and all the forcing functions are differentiable for $x > 0$, one can write it at $x = 0+$ as

$$d^2y/dx^2(0+) + b\, dy/dx\,(0+) + cy(0+) = 6e^{-2(0+)} + (0+)^2$$

$$d^2y/dx^2\,(0+) = 6 - 2b - 3c$$

38. **(D)**

The definition of the derivative of a function is

$$dy/dx = \lim_{h \to 0} \frac{y(x + h) - y(x)}{h}.$$

39. **(A)**

The differential equation for $b = 8$ and $c = 4$ can be written as

$$d^2y/dx^2 + 8\,dy/dx + 4y = 6e^{-2x} + x^2$$

Since the order of the differential equation is second order, it can be written as

$$d^2y/dx^2 + 2\xi\omega_n\,dy/dx + \omega_n^2 y = 6e^{-2x} + x^2,$$

where ξ is the damping ratio and is dimensionless, and ω_n is the undamped natural frequency and has the units of radians per second. Hence,

$$\omega_n^2 = 4,$$

which gives $\omega_n = 2$ radians/second.

40. **(B)**

Depending on the value of the damping ratio, a second order system can be classified as an overdamped, underdamped, critically damped or undamped system. Since

$$\omega_n^2 = 4 \text{ and } 2\xi\,\omega_n^2 = 8, \quad \text{then } \xi = 2.$$

Since $\xi > 1$, the system is overdamped. If $0 < \xi < 1$, then the system is underdamped. If $\xi = 1$, then the system is critically damped, while when $\xi = 0$, the system is undamped.

41. **(C)**

In the time function $v(t) = 141 \sin 314t$, 141.4 is the amplitude of the signal and 314 the angular frequency ω. For AC signals the ratio between amplitude and mean square value is $\sqrt{2}$,therefore mean square value of the applied voltage is:

$$\frac{141.4}{\sqrt{2}} = 100$$

The periodic function $\sin \omega t$ lags 90° behind the reference periodic function $\cos \omega\tau$, the phasor that represents the time function $v(t)$ is therefore:

$$V = 100 \angle{-90^\circ}\text{ V}$$

42. **(A)**

The impedance of a capacitor is given by:

$$Z_c = -j\frac{1}{\omega C}$$

$$= -j\frac{1}{314 \cdot 382 \cdot 10^{-6}}$$

$$= -j\,8.33\ \Omega$$

or in polar form:

$$Z_c = 8.33 \angle{-90^\circ}\ \Omega$$

and the current phasor I_2 will be:

$$I_2 = V / Z_c$$

$$= \frac{100 \angle -90°}{8.33 \angle -90°}$$

$$= 12 \angle 0° \, A.$$

As the phasor diagram shows, the current I_2 through the capacitor leads the voltage V by 90°.

43. **(A)**

The impedance Z_{RL} of the RL branch is given by:

$$Z_{RL} = R + j\omega L$$

$$= 4 + j314 \cdot 0.00955$$

$$= 4 + j3$$

or in polar form:

$$Z_{RL} = \sqrt{4^2 + 3^2} \angle \text{atan } 3/4$$

$$= 5 \angle 37° \, \Omega$$

and the current I_1 is given by

$$I_1 = V / Z_{RL}$$

$$= 100 \angle -90° / 5 \angle 37°$$

$$= 20 \angle -127° \text{ A}$$

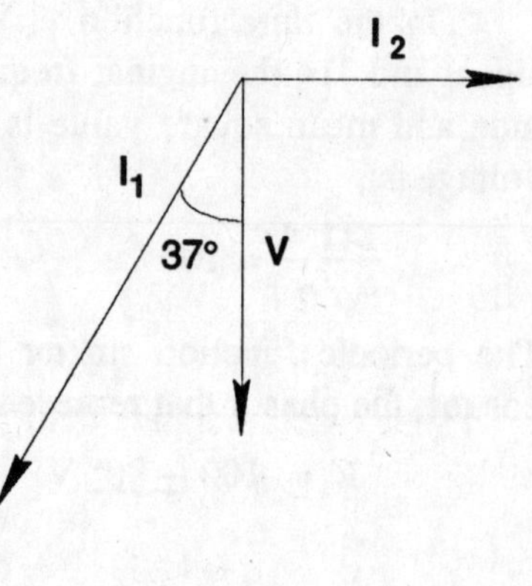

The phasor diagram illustrates the phase relation between I_1, the current in the RL branch and the applied voltage V. The diagram shows that the current I_1 lags 37° behind the voltage, because the branch has an inductance.

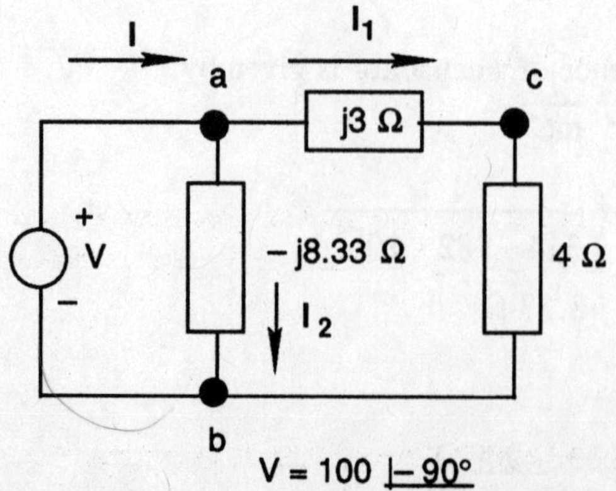

As a reference the circuit diagram in the frequency domain is included above, with the circuit elements substituted by the value of their impedance, and time functions by their phasor representation.

44. **(D)**

Applying Kirchoff's current law to node *a*:

$$I = I_1 + I_2$$

$$= 20 \angle -127° + 12 \angle 0$$

transforming into rectangular form:

$$= 20 \cos(-127°) + j20 \sin(-127°) + 12 \cos(0°) + j12 \sin(0°)$$

$$= -12 + j16 + 12$$

$$= j16$$

$$= 16 \angle -90° \text{ V}$$

As reference the graphic addition of the two current phasors has been included:

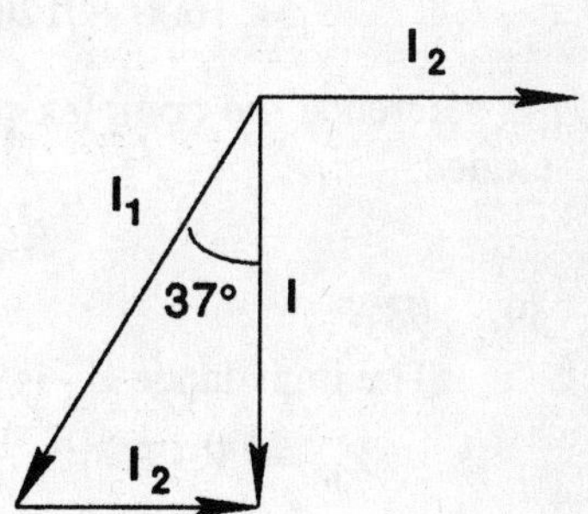

45. **(E)**

The voltage V_{ac} is given by:

$$V_{ac} = I_1 \cdot Z_L$$

$$= 20 \angle -127° \cdot 3 \angle 90°$$

$$= 60 \angle -37° \text{ V}$$

46. **(B)**

The reactive power Q in a capacitor is given by:

$$Q = V I_2$$

$$= 100 \cdot 12$$

$$= 1200 \text{ VAR}$$

47. **(C)**

The diagram in Problem 44 shows that since the current phasor I and the voltage phasor V have the same phase, they are in phase, therefore the power factor is 1.

48. **(A)**

The complex power S in branch *RL* is given by:

$$S = P + jQ$$

where Q is the real power in the resistance given by:

$$P = I_1^2 \cdot R$$
$$= 20^2 \cdot 4$$
$$= 1600 \text{ W}$$
$$Q = I_1^2 \cdot X_L$$
$$= 20^2 \cdot 3$$
$$= 1200 \text{ VAR}$$
$$S = 1600 + j1200 \text{ VA}$$

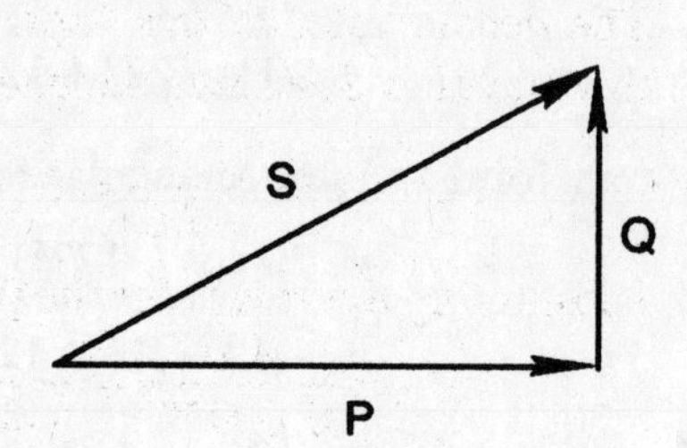

As reference the complex power diagram is included:

49. **(B)**

The impedance Z_{in} is given by the following relation:

$$Z_{in} = V / I$$
$$= 100 \angle -90° \ / \ 16 \angle -90°$$
$$= 6.25 \angle 0° \ \Omega$$

50. **(C)**

To find the transient behavior of the circuit find the input impedance $Z(s)$. The circuit diagram in the "s" domain is shown. In this diagram the elements have been substituted by their impedances in the "s" domain.

$$Z_R = R$$
$$Z_L = sL$$
$$Z_C = \frac{1}{sC}$$

Notice that in this diagram Z_R is in series with Z_L, and this series combination is in parallel with Z_C. Therefore the impedance seen by $V(s)$ is:

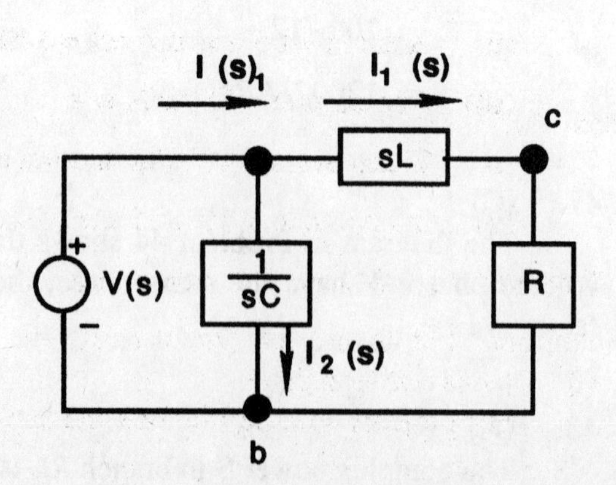

$$Z(s) = \frac{\frac{1}{sC} \cdot (R + sL)}{\frac{1}{sC} + (R + sL)}$$

$$= \frac{R + sL}{LC \cdot [s^2 + s\frac{R}{L} + \frac{1}{LC}]}$$

The denominator of this expression is the characteristic equation. The roots of this equation determine the transient behavior of the system, therefore:

$$s^2 + s\frac{R}{L} + \frac{1}{LC} = 0$$

substituting the values for the parameters the characteristic equation is:

$$s^2 + s\frac{4.0}{0.00955} + \frac{1}{0.00955 \cdot 382 \cdot 10^{-6}} = 0$$

$$s^2 + 418.85s + 274115.3s^2 = 0$$

The roots of this equation are given by:

$$s_{1,2} = \frac{-418.85 \pm \sqrt{418.85^2 - 4 \cdot 274115.3}}{2}$$

$$s_{1,2} = -209.42 \pm j479.85$$

The roots of the characteristic equation are complex conjugates. The system is therefore under-damped and the time constant τ is:

$$\tau = 1/209.42$$

$$= 4.77 \text{ ms}$$

51. **(D)**

Given: Analysis of public projects / capitalized cost or perpetual life problem.

(i) Budget for the project = \$500,000

(ii) Actual cost of the project = \$400,000

(iii) Annual cost = \$50,000 $\equiv A$

(iv) Ticket sales = 1000/month = \$12,000/year

(v) Interest rate = 8% and analysis period is perpetual

Required To Calculate: Price per ticket.

Solution: Public projects are usually analyzed using Benefit – Cost – Ratio (B / C) analysis technique.

For projects with infinite life, the capitalized cost, P is given by

$$P = \frac{A}{i} = \frac{400,000}{0.08}$$

Total cost

$$= \$400,000 + \$50,000/0.08 \equiv C$$

Suppose annual ticket sales = $T, capitalized cost of these perpetual annual sales

$$= T / i = T / 0.08$$

Hence total benefits

$$= \$500,000 + \$T / 0.08 \equiv B$$

realizing that, for the project to be effective, the benefits must at least equal the costs or the benefit-cost ratio must be greater than or equal to one. Taking the minimum requirement

$$B / C = 1$$

$$\frac{\$500,000 + \frac{\$T}{0.08}}{\$400,000 + \frac{\$50,000}{0.08}} = 1$$

Solving, T = $42,000 for a total of 12,000 tickets/year.
Price per ticket

$$= \$43,000 / 12,000$$

$$= \$3.50 / \text{ticket}.$$

52. **(E)**
Given: Depreciation problem

(i) Initial Investment, P = $700,000

(ii) Salvage value, S = $100,000

(iii) Useful life, n = 5 years.

(iv) Depreciation type is straight line

Required To Compute: Book value at the end of year 2.

Solution: S.L. Depreciation/ year

$$= \frac{P - S}{n}$$

$$D = \frac{\$(700,000 - 100,000)}{5} = \$120,000/\text{year}$$

Book value, B_t at end of year t

$$= P - \sum_{i=1}^{t} D$$

where D = Depreciation/year

Hence Book value, B_2

= \$700,000 – 2(\$120,000)

= \$460,000

53. **(A)**

Given: Annual Worth Problem

(i) Ticket price: \$5.00 per adult and \$2.50 per kid.

(ii) Number of kids = half number of adults

(iii) Initial Investment (capital) = \$500,000

(iv) Ticket sales = 1,200 per month

(v) Interest rate (MARR) = 10%

Required To Calculate: Annual Worth of the Investment.

Solution: Annual ticket sales

$$= \$\{\overbrace{(1,200)(\tfrac{2}{3})(5)}^{\text{Adults}} + \overbrace{(1,200)(\tfrac{1}{3})(2.5)}^{\text{Kids}}\}12$$

$$= \$(4,000 + 1,000)12 = \$60,000$$

Annual expenses

= \$50,000 given.

Annual capital expenses

= $P \cdot i$ (remember $P = A / i$)

= \$(500,000) (0.10) = \$50,000

Total expenses

= \$50,000 + \$50,000 = \$100,000

∴ Annual Worth of Investment

= \$60,000 – \$100,000

= – \$40,000.

54. **(D)**

Given: Depreciation / Taxes Problem

(i) First cost, P = \$700,000

(ii) Salvage value, S = \$100,000

(iii) Useful life, n = 5 years

Required To Compute: difference between depreciation found through Sum of Years Digits method and Straight line method.

Solution: First, using Sum of Years Digits (SOYD) method

$$\text{SOYD} = \frac{n}{2}(n+1) = \frac{5}{2}(5+1) = 15,$$

Depreciation (m^{th} year)

$$= \frac{n-m+1}{\text{SOYD}}(P-S)$$

$$\text{D}(1^{st}\text{ year}) = \frac{5-1+1}{15}(\$700,000 - \$100,000)$$

$$\text{D}(1^{st}\text{ year}) = \$200,000$$

Now, using Straight Line (SL) method
SL Depreciation (each year)

$$= (P-S)/N$$

$$= (\$700 - \$100,000)/5$$

D(each year) =

$$\text{D}(1^{st}\text{ year}) = \$120,000$$

Tax Savings = Tax rate (difference in calculated depreciation)

$$= .5\,(\$200,000 - \$120,000)$$

$$= \$40,000$$

55. **(C)**
Given: Effective Interest / Multiple Compounding Problem

$$i_{\text{effective}} \equiv \{(1 + \text{Interest rate/period})^{\#\text{ of periods / year}} - 1\}$$

∴ 8% compounded monthly 8/12 % per month

$$\therefore \quad i_{\text{effective}} = \left(1 + \frac{\frac{8}{12}}{100}\right)^{12} - 1$$

$$= 0.083$$

$$= 8.3\%$$

56. **(A)**

Given: Multiple Compounding / Effective Rate Problem

(i) Amount being borrowed = \$100,000

(ii) $i = 15\%$

Required: Monthly payment for 25 years

Solution:

$$i_{\text{effective}} \equiv \{(1 + \text{Interest rate/period})^{\#\text{ of periods / year}} - 1\}$$

$$0.15 = \left(1 + \frac{i}{12}\right)^{12} - 1$$

$\therefore \quad i = 0.1406, i / 12 = 0.01171$

Number of months

$= 25 \times 12 = 300$ months.

Cash Flow Diagram:

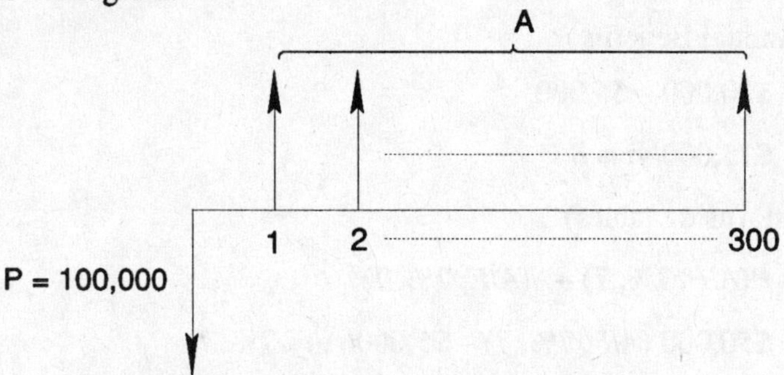

$\therefore \quad A = P(A/P, i/12, 300)$

$= \$100,000\ (A/P, 1.171\%, 200)$

$$= \$\,100,000\left[\frac{0.01171\,(1 + 0.01171)^{300}}{(1 + 0.01171)^{300} - 1}\right]$$

Solving $= \$1207.74$ / month.

57. **(D)**

Given: Perpetual Life / Continuous Compounding Problem.

(i) Annual payments = \$20,000

(ii) Life = ∞,

(iii) Interest rate = 10% compounded continuously

(iv) No of retirees = 300

Required: Initial Investment

Solution: For continuous compounding,

$$i_{\text{eff}} = e^{i} - 1$$
$$= e^{0.10} - 1$$
$$= 0.1052$$

Now for capitalized cost problems involving perpetual life projects,

$$P = A / i_{\text{eff}}$$
$$= \frac{\$\,20{,}000 \times 3{,}000}{0.1052}$$
$$= \$570 \text{ million}$$

58. **(C)**

Given: Benefit / Cost Ratio Analysis Problem

Required To Compute: B/C ratio for Project A

Solution: Net Annual Benefits

$$= \$20{,}000 - \$5{,}000$$
$$= \$15{,}000/\text{yr} \equiv B$$

Net Annual Cost (using tables)

$$= P(A/P, 7\%, 7) - S(A/F, 7\%, 7)$$
$$= \$50{,}000\ (A/P, 7\%, 7) - \$5{,}000(A/F, 7\%, 7)$$
$$= \$50{,}000\ (0.1856) - \$5{,}000\ (0.1156)$$
$$= \$9{,}280 - \$578 = \$8{,}702 \equiv C$$

$\therefore\ (B/C)_A = 15{,}000/8{,}702 = 1.72$

$\therefore\ (B/C)_A > 1$ means A is acceptable.

59. **(E)**

Given: Extension of Problem 8 / Incremental Analysis

$$\left(\frac{\Delta B}{\Delta C}\right)_{B-A} = \frac{(\$21{,}000 - \$20{,}000) - (\$4{,}000 - \$5{,}000)}{(\$60{,}000 - \$50{,}000)\ (A/P, 7\%, 7) - (\$5{,}000 - \$5{,}000)(A/F, 7\%, 7)}$$

$$= \frac{\$2{,}000}{\$10{,}000 (A/P, 7\%, 7)}$$

$$= 1.08$$

$\therefore\ (\Delta B / \Delta C)_{B \to A} > 1$ means B is better than A.

60. **(E)**

Given: Practical questions on effect of inflation and interest rate changes on investment analysis using *B / C* ratio technique.

Solution: Generally, since all cash flows are responsive to inflation, decisions based on cash flow analysis, using *B / C* ratio technique are not affected by inflation because any effects of inflation will divide out in the ratio.

Lowering interest rate usually favors higher cost alternatives.

Hence, correct answer is (E).

ANSWERS 61-66 refer to the following figure.

$C_p = 1.005$ kJ/kg-K; $k = 1.4$;

$C_v = 0.718$ kJ/kg-K; $R = 0.287$ kJ/kg-K

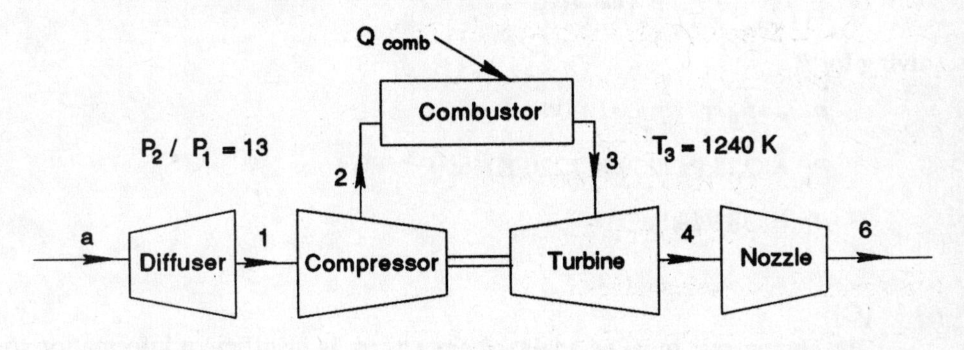

$P_a = 22$ kPa $P_6 = 22$ kPa

$T_a = 220$ K $V_6 = 954.8$ m/s

$T_1 = 259$ K

61. **(D)**

To determine the aircraft velocity the diffuser should be analyzed. For the ideal turbojet, the diffuser is also ideal which implies adiabatic and reversible or isentropic. The first law is

$$\dot{Q}cv + \dot{m}_a (h_a + V_a^2/2 + gz_a) = \dot{m}_1(h_1 + V_1^2/2 + gz_1) + \dot{W}_{cv}$$

The change in potential energy is zero, V_1 = zero,

$$\dot{Q}cv = \dot{W}_{cv} = 0.$$

The mass flow rate divides out leaving

$$h_a + V_a^2/2 = h_1$$

rearranging and solving for V_a

$$V_a^2 / 2 = 2(h_1 - h_a)$$

$$h_1 - h_a = C_p(T_1 - T_a)$$

$$V_a^2 = 2(C_p(T_1 - T_a))$$

$$V_a^2 = 2(1.005 \text{ kJ/kg K}(259 - 220)\text{K})$$

$$V_a^2 = 78.39 \text{ kJ/kg } (1000 \text{ m}^2/\text{s}^2) / (\text{kJ/kg}) = 78{,}390 \text{ m}^2/\text{s}^2$$

$$V_a = 279.98 \text{ m/s}$$

62. **(A)**

Using the isentropic relationships around the diffuser,

$$T_1 / T_a = (P_1 / P_a)^{((k-0/k)}$$

solving for P_1

$$P_1 = Pa(T_1 / T_a)^{(k/(k-1))}$$

$$P_1 = 22 \text{ kPa } (259 \text{ K} / 220 \text{ K})^{(1.4/(1.4-1))}$$

$$P_1 = 38.94 \text{ kPa}$$

63. **(C)**

The compressor must be analyzed since there is insufficient information to analyze the turbine. The key here is to remember that the magnitude of the turbine work is equal to the magnitude of the compressor work. In a turbojet engine there is no net shaft work; *all* of the work produced by the turbine is utilized by the compressor. Thus the magnitudes of the work are equal. It is essential to realize that the compressor work is in and thus is negative. Since the compressor is ideal, the heat transfer is zero, and assuming that the changes in potential and kinetic energy are zero, the first law reduces to,

$$\dot{m}_1(h_1) = \dot{m}_2(h_{2s}) + \dot{W}_{cv}$$

rearranging and solving for the mass flow rate which is constant from 1 to 2

$$\dot{m} = \dot{W}_{cv} / (h_1 - h_{2s})$$

$$h_1 - h_{2s} = C_p(T_1 - T_{2s})$$

$$\dot{m} = \dot{W}_{cv} / (C_p(T_1 - T_{2s}))$$

Next solve for T_{2s}

$$T_{2s} = 259 \text{ K } (2.0809) = 538.95 \text{ K}$$

$$\dot{m} = -755.15 \text{ kW } ((\text{kJ/s}) / \text{kW}) / (1.005 \text{ kJ/kg K } (259 - 538.95)\text{K})$$

$$\dot{m} = 2.684 \text{ kg/s}$$

64. **(A)**

For constant specific heats, the change in entropy is

$$\Delta s = C_p \ln (T_3 / T_{2s}) - R \ln (P_3 / P_2)$$

For the ideal turbojet it is assumed that the combustor operates at constant pressure ($R \ln (P_3 / P_2) = 0$), thus the change in entropy is

$$\Delta s = C_p \ln (T_3 / T_{2s})$$

$$\Delta s = 1.005 \text{ kJ/kg-K} \ln (1240 \text{ K} / 538.95 \text{ K})$$

$$\Delta s = 0.8374 \text{ kJ/kg-K}$$

65. **(D)**

The turbine must be analyzed to find T_4 which is then used to find T_6. Since the turbine is ideal, the heat transfer is zero. The changes in potential and kinetic energy are assumed to be zero. The resulting first law is

$$\dot{m}_3 h_3 = \dot{m}_4 h_4 + \dot{W}_{cv}$$

with $\dot{m}$ constant and using constant specific heats

$$\dot{m}_3 = \dot{m}_4 = \dot{m}$$

$$h_3 - h_4 = C_p (T_3 - T_4)$$

$$\dot{W}_{cv} / \dot{m} = C_p (T_3 - T_4)$$

and solving for T_4

$$T_4 = T_3 - \dot{W}_{cv} / (\dot{m} C_p)$$

$$T_4 = 1240 \text{ K} - 755.15 \text{ kW} / (2.684 \text{ kg/s } 1.005 \text{ kJ/kg-K})$$

$$T_4 = 960.05 \text{ K}$$

Next, the nozzle must be analyzed using T_4 to find T_6. Since the nozzle is ideal there is no heat transfer and with no shaft work, the work is also zero. The change in potential energy is zero, however the exit velocity necessitates inclusion of kinetic energy terms. The velocity at state 4 (V_4) is assumed small and is neglected. The resulting equation, after dividing out the mass flow rate, is

$$h_4 = h_6 + V_6^2 / 2$$

$$h_4 - h_6 = C_p (T_4 - T_6) = V_6^2 / 2$$

solving for T_6

$$T_6 = T_4 - V_6^2 / 2C_p$$

$$= 960. \text{ K} - (954.8)^2 \text{m}^2/\text{s}^2 / (2 \cdot 1.005 \text{ kJ/kg K})$$

$$(\text{kJ/kg}) / (1000 \text{ m}^2/\text{s}^2)$$

$$T_6 = 506.4 \text{ K}$$

66. **(B)**

A reversible compressor requires the least amount of work. If the compressor is irreversible it will require more work to achieve the same pressure rise. Note that since there is no shaft work delivered from the cycle, then there is no net work to change.

ANSWERS 67–70 refer to the following diagram:

$C_p = 1.005$ kJ/kg-K; $k = 1.4$;

$C_v = 0.718$ kJ/kg-K; $R = 0.287$ kJ/kg-K

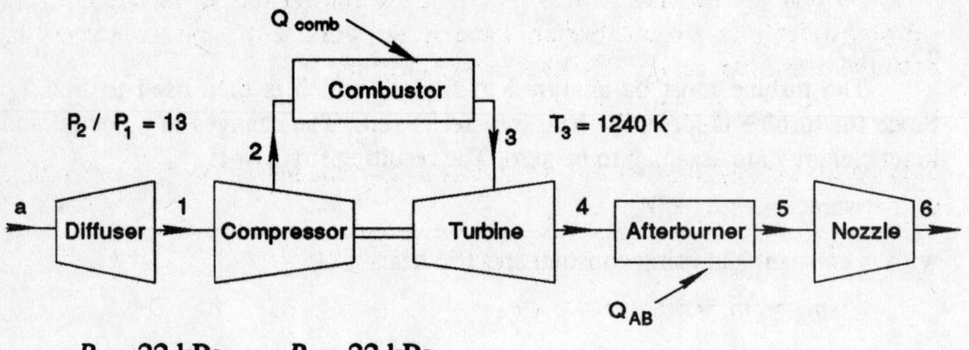

$P_a = 22$ kPa $P_6 = 22$ kPa

$T_a = 220$ K $V_6 = 954.8$ m/s

$T_1 = 259$ K

67. **(A)**

The compressor efficiency compares the isentropic work to the adiabatic actual work. This can be expressed as a comparison of the temperatures,

$$\dot{Q}_{cv})_{act} = \dot{m}(h_1 - h_2)$$

$$\dot{Q}_{cv})_s = \dot{m}(h_1 - h_{2s})$$

$$n_c = \frac{\dot{Q})_s}{\dot{Q})_{act}}$$

$$n_c = \frac{h_1 - h_{2s}}{h_1 - h_2}$$

$$n_c = (T_1 - T_{2s}) / (T_1 - T_2)$$

solving for T_{2s} first using the isentropic relationships

$$T_{2s} = T_1 (P_2 / P_1)^{((k-1)/k)}$$

$$T_{2s} = 259 \text{ K } (13)^{((1.4-1)/1.4)}$$

$$T_{2s} = 538.95 \text{ K}$$

Thus T_2 equals

$$T_2 = T_1 + (T_{2s} - T_1) / n_c$$

$$T_2 = 259 \text{ K} + (538.95 \text{ K} - 259 \text{ K}) / 0.8$$

$$T_2 = 608.94 \text{ K}$$

68. **(B)**

To find the fuel mass flow rate, the heat transfer rate to the air must be calculated first. In the combustor, there is no work term, and no change in potential or kinetic energy. The first law reduces to,

$$\dot{Q}_{cv} + \dot{m}_2 h_2 = \dot{m}_3 h_3$$

Solving for the heat transfer rate and realizing that the mass flow rate is constant ($m_2 = m_3 = m$)

$$\dot{Q}_{cv} = \dot{m}C_p(T_3 - T_2) = (2.8 \text{ kg/s})\ (1.005 \text{ kJ/kg-K})\ (1240 - 608.94) \text{ K}$$

$$\dot{Q}_{cv} = 1775.8 \text{ kW (kJ/s)}$$

Dividing by the lower heating value produces the fuel mass flow rate

$$\dot{m}_{\text{fuel}} = Qcv / LHV$$

$$\dot{m}_{\text{fuel}} = \frac{1775.8 \text{ kJ / s}}{44000 \text{ kJ / kg}} = 0.04036 \text{ kg/s}$$

The lower heating value is the magnitude of the enthalpy of combustion when all the water formed by the combustion is vapor. This value changes as fuels are changed.

69. **(D)**

The key here is to first find the work of the compressor and then realize that the magnitudes of the compressor work and the turbine work are equal. In the compressor, the heat transfer rate is zero and the change in potential and kinetic energy are assumed to be zero. The first law is solved for the work with constant mass flow rate producing,

$$\dot{m}_1 h_1 = \dot{m}_2 h_2 + \dot{W}_{cv}$$

$$\dot{W}_{cv} = \dot{m}Cp(T_1 - T_2) = 2.8 \text{ kg/s } (1.005 \text{ kJ/kg-K } (259 - 608.9)\text{K})$$

$$\dot{W}_{cv})_{comp} = -984.6 \text{ kW}$$

A similar first law around the turbine produces T_4,

$$\dot{W}_{cv})_{tur} = 984.6 \text{ kW (opposite sign of compressor work)}$$

$$T_4 = T_3 - W_{cv} / (mC_p) \text{ (careful with the sign)}$$

$$T_4 = 890.1 \text{ K}$$

70. **(A)**

The drawing of the control volume is critical here. The best way is to include the afterburner and nozzle in the same control volume. The heat transfer occurs only in the afterburner. There is no change in potential energy and no work. The velocity at state 4 (V_4) is assumed to be negligible. The first law reduces to,

$$\dot{Q}_{ab} + \dot{m}_4 h_4 = \dot{m}_6 h_6 + \dot{m}_6 V_6^2 / 2$$

Rearrange to find T_6

$$T_6 = \dot{Q}_{ab} / (\dot{m} C_p) + T_4 - V_6^2 / (2C_p)$$

$$T_6 = 1127 \text{ kJ/s} / (2.8\text{kg/s}) (1.005 \text{ kJ/kg-K}) + 890.1 \text{ K}$$

$$- ((1050)^2 \text{ m}^2/\text{s}^2) / ((1000 \text{ m}^2/\text{s}^2) (2) (1.005 \text{ kJ/kg-K}))$$

$$T_6 = 742.09 \text{ K}$$

Fundamentals of Engineering

A.M. SECTION

Test 2

Fundamentals of Engineering

A.M. SECTION

TEST 2 – ANSWER SHEET

1. A B C D E	26. A B C D E	51. A B C D E
2. A B C D E	27. A B C D E	52. A B C D E
3. A B C D E	28. A B C D E	53. A B C D E
4. A B C D E	29. A B C D E	54. A B C D E
5. A B C D E	30. A B C D E	55. A B C D E
6. A B C D E	31. A B C D E	56. A B C D E
7. A B C D E	32. A B C D E	57. A B C D E
8. A B C D E	33. A B C D E	58. A B C D E
9. A B C D E	34. A B C D E	59. A B C D E
10. A B C D E	35. A B C D E	60. A B C D E
11. A B C D E	36. A B C D E	61. A B C D E
12. A B C D E	37. A B C D E	62. A B C D E
13. A B C D E	38. A B C D E	63. A B C D E
14. A B C D E	39. A B C D E	64. A B C D E
15. A B C D E	40. A B C D E	65. A B C D E
16. A B C D E	41. A B C D E	66. A B C D E
17. A B C D E	42. A B C D E	67. A B C D E
18. A B C D E	43. A B C D E	68. A B C D E
19. A B C D E	44. A B C D E	69. A B C D E
20. A B C D E	45. A B C D E	70. A B C D E
21. A B C D E	46. A B C D E	71. A B C D E
22. A B C D E	47. A B C D E	72. A B C D E
23. A B C D E	48. A B C D E	73. A B C D E
24. A B C D E	49. A B C D E	74. A B C D E
25. A B C D E	50. A B C D E	75. A B C D E

76. Ⓐ Ⓑ Ⓒ Ⓓ Ⓔ	98. Ⓐ Ⓑ Ⓒ Ⓓ Ⓔ	119. Ⓐ Ⓑ Ⓒ Ⓓ Ⓔ
77. Ⓐ Ⓑ Ⓒ Ⓓ Ⓔ	99. Ⓐ Ⓑ Ⓒ Ⓓ Ⓔ	120. Ⓐ Ⓑ Ⓒ Ⓓ Ⓔ
78. Ⓐ Ⓑ Ⓒ Ⓓ Ⓔ	100. Ⓐ Ⓑ Ⓒ Ⓓ Ⓔ	121. Ⓐ Ⓑ Ⓒ Ⓓ Ⓔ
79. Ⓐ Ⓑ Ⓒ Ⓓ Ⓔ	101. Ⓐ Ⓑ Ⓒ Ⓓ Ⓔ	122. Ⓐ Ⓑ Ⓒ Ⓓ Ⓔ
80. Ⓐ Ⓑ Ⓒ Ⓓ Ⓔ	102. Ⓐ Ⓑ Ⓒ Ⓓ Ⓔ	123. Ⓐ Ⓑ Ⓒ Ⓓ Ⓔ
81. Ⓐ Ⓑ Ⓒ Ⓓ Ⓔ	103. Ⓐ Ⓑ Ⓒ Ⓓ Ⓔ	124. Ⓐ Ⓑ Ⓒ Ⓓ Ⓔ
82. Ⓐ Ⓑ Ⓒ Ⓓ Ⓔ	104. Ⓐ Ⓑ Ⓒ Ⓓ Ⓔ	125. Ⓐ Ⓑ Ⓒ Ⓓ Ⓔ
83. Ⓐ Ⓑ Ⓒ Ⓓ Ⓔ	105. Ⓐ Ⓑ Ⓒ Ⓓ Ⓔ	126. Ⓐ Ⓑ Ⓒ Ⓓ Ⓔ
84. Ⓐ Ⓑ Ⓒ Ⓓ Ⓔ	106. Ⓐ Ⓑ Ⓒ Ⓓ Ⓔ	127. Ⓐ Ⓑ Ⓒ Ⓓ Ⓔ
85. Ⓐ Ⓑ Ⓒ Ⓓ Ⓔ	107. Ⓐ Ⓑ Ⓒ Ⓓ Ⓔ	128. Ⓐ Ⓑ Ⓒ Ⓓ Ⓔ
86. Ⓐ Ⓑ Ⓒ Ⓓ Ⓔ	108. Ⓐ Ⓑ Ⓒ Ⓓ Ⓔ	129. Ⓐ Ⓑ Ⓒ Ⓓ Ⓔ
87. Ⓐ Ⓑ Ⓒ Ⓓ Ⓔ	109. Ⓐ Ⓑ Ⓒ Ⓓ Ⓔ	130. Ⓐ Ⓑ Ⓒ Ⓓ Ⓔ
88. Ⓐ Ⓑ Ⓒ Ⓓ Ⓔ	110. Ⓐ Ⓑ Ⓒ Ⓓ Ⓔ	131. Ⓐ Ⓑ Ⓒ Ⓓ Ⓔ
89. Ⓐ Ⓑ Ⓒ Ⓓ Ⓔ	111. Ⓐ Ⓑ Ⓒ Ⓓ Ⓔ	132. Ⓐ Ⓑ Ⓒ Ⓓ Ⓔ
90. Ⓐ Ⓑ Ⓒ Ⓓ Ⓔ	112. Ⓐ Ⓑ Ⓒ Ⓓ Ⓔ	133. Ⓐ Ⓑ Ⓒ Ⓓ Ⓔ
91. Ⓐ Ⓑ Ⓒ Ⓓ Ⓔ	113. Ⓐ Ⓑ Ⓒ Ⓓ Ⓔ	134. Ⓐ Ⓑ Ⓒ Ⓓ Ⓔ
92. Ⓐ Ⓑ Ⓒ Ⓓ Ⓔ	114. Ⓐ Ⓑ Ⓒ Ⓓ Ⓔ	135. Ⓐ Ⓑ Ⓒ Ⓓ Ⓔ
93. Ⓐ Ⓑ Ⓒ Ⓓ Ⓔ	115. Ⓐ Ⓑ Ⓒ Ⓓ Ⓔ	136. Ⓐ Ⓑ Ⓒ Ⓓ Ⓔ
94. Ⓐ Ⓑ Ⓒ Ⓓ Ⓔ	116. Ⓐ Ⓑ Ⓒ Ⓓ Ⓔ	137. Ⓐ Ⓑ Ⓒ Ⓓ Ⓔ
95. Ⓐ Ⓑ Ⓒ Ⓓ Ⓔ	117. Ⓐ Ⓑ Ⓒ Ⓓ Ⓔ	138. Ⓐ Ⓑ Ⓒ Ⓓ Ⓔ
96. Ⓐ Ⓑ Ⓒ Ⓓ Ⓔ	118. Ⓐ Ⓑ Ⓒ Ⓓ Ⓔ	139. Ⓐ Ⓑ Ⓒ Ⓓ Ⓔ
97. Ⓐ Ⓑ Ⓒ Ⓓ Ⓔ		140. Ⓐ Ⓑ Ⓒ Ⓓ Ⓔ

FUNDAMENTALS OF ENGINEERING EXAMINATION

TEST 2

MORNING (AM) SECTION

> **TIME:** 4 Hours
> 140 Questions
>
> **DIRECTIONS:** For each of the following questions and incomplete statements, choose the best answer from the five answer choices.

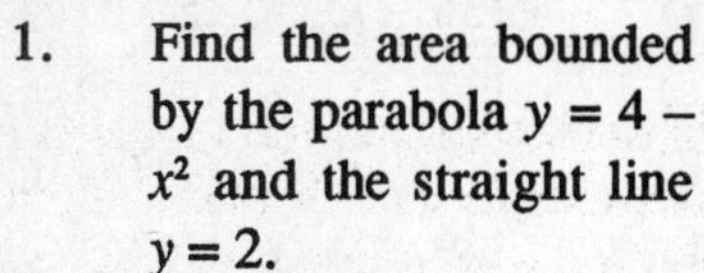

1. Find the area bounded by the parabola $y = 4 - x^2$ and the straight line $y = 2$.

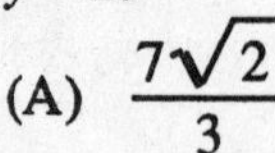

(A) $\frac{7\sqrt{2}}{3}$

(B) $\frac{8\sqrt{2}}{3}$

(C) 0

(D) $3\sqrt{2}$

(E) 4

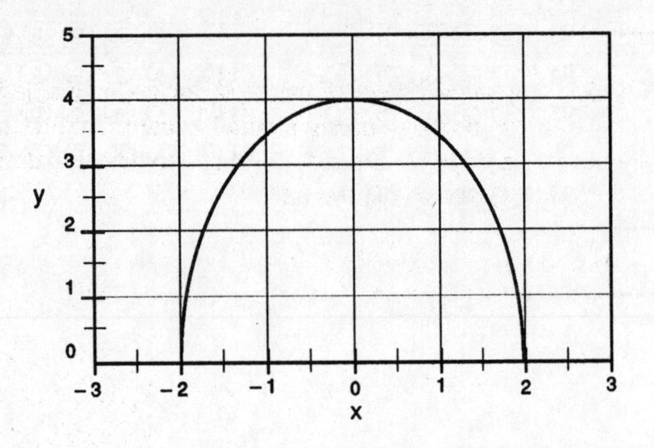

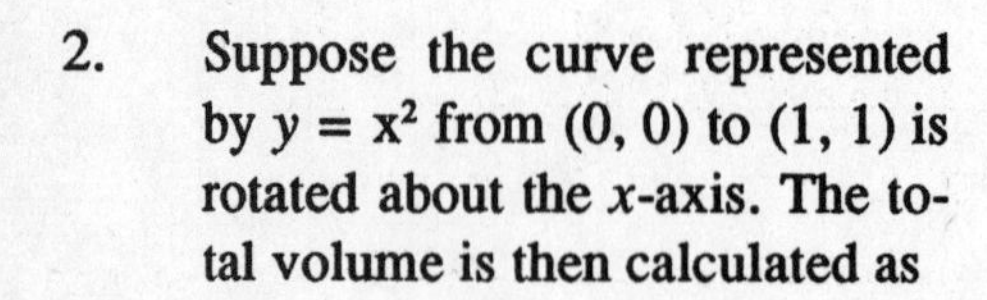

2. Suppose the curve represented by $y = x^2$ from (0, 0) to (1, 1) is rotated about the x-axis. The total volume is then calculated as

(A) 1 / 3

(B) $\pi / 3$

(C) $\pi / 4$

(D) 2 / 3

(E) $\pi / 5$

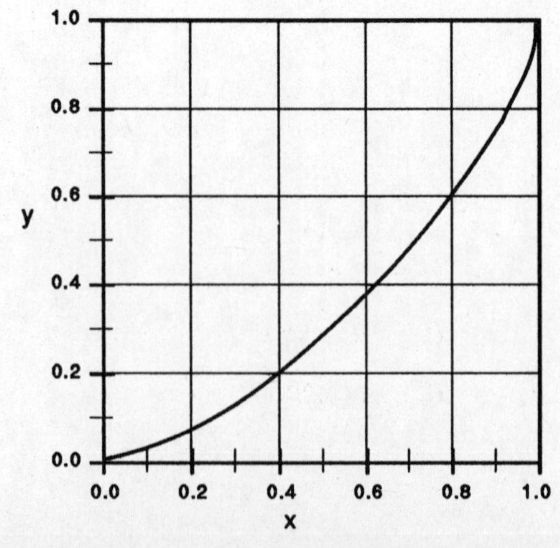

3. Find $\int_0^x 3x\, e^{2x^2}\, dx$.

(A) $\frac{3}{4}\left(e^{2x^2} - 1\right)$

(B) $\frac{3}{4}\, e^{2x^2}$

(C) $3\, e^{2x^2} - 1$

(D) $3e^{2x}$

(E) $3\, e^{x^2}$

4. Consider the following limit

$$\lim_{x \to 0}\left[\frac{\pi \sin x}{\log(1+x)}\right]$$

Its value is equal to

(A) 1

(B) 0

(C) $\pi / \log 2$

(D) π / e

(E) π

5. A square sheet of metal 18 inches on a side is to be used to make an open-top box by cutting a small square from each corner as shown below, then bending up the sides. What should the value of y be so that the volume of the box is maximized?

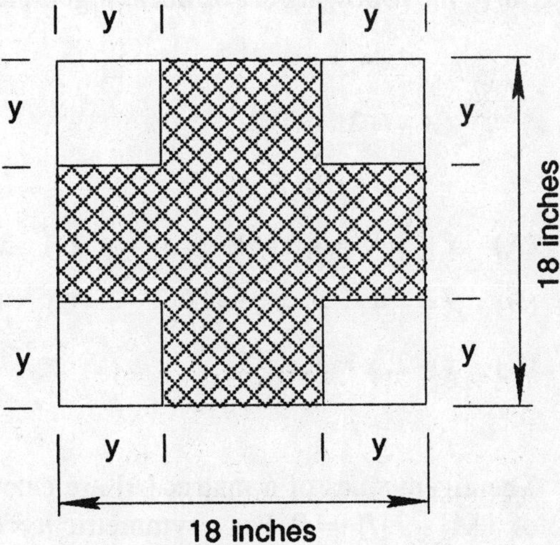

(A) 3 inches

(B) 6 inches

(C) π inches

(D) 2 inches

(E) 4 inches

6. Find the slope of $y = x^{1.5} + \cos \pi x$ at $x = -1$

(A) $1.5j - \pi$
(B) $-1.5 + \pi$
(C) $1.5j + \pi$
(D) $1.5j$
(E) 1.5

7. Find d^2y / dx^2 if $x = 3 - 2z$ and $y = z^2 - 3z^3$

(A) $0.5 + 9z$
(B) $1 - 9z$
(C) $0.5 - 4.5z$
(D) $-1 + 3z^2 / 2$
(E) $3z^2 / 2$

8. The equation of a straight line passing through point (2, 3) and (3, 2) is given by

(A) $x - y = -1$
(B) $x + 2y = 7$
(C) $x - y = 3$
(D) $x + y = 5$
(E) $2x + y = 7$

9. Solve the following set of linear algebraic equations for x, y, and z.

$$x - y = -1$$

$$x + y - 2z = -3$$

$$y + z = 5$$

(A) $x = 3, y = 4, z = 1$
(B) $x = -4, y = -3, z = 8$
(C) $x = 2, y = 3, z = 2$
(D) $x = 1, y = 2, z = 3$
(E) The set has no solution.

10. The eigenvalues of a matrix $[A]$ are calculated by forcing the determinant of $| [A] - \lambda[I] | = 0$. For a symmetric $n \times n$ matrix

(A) There are $(2 \times n)$ eigenvalues
(B) The eigenvalues are unique
(C) The eigenvalues are positive
(D) The eigenvalues are negative
(E) There are (n) eigenvalues which are not necessarily unique.

11. The following matrix $[A]$ has a very special property in that it is equal to its own inverse. Find the determinant of the matrix $2[A]^{10}$.

$$[A]=\begin{bmatrix} 17 & -20 & 8 \\ 40 & -49 & 20 \\ 64 & -80 & 33 \end{bmatrix}$$

(A) 2
(B) −8
(C) 8
(D) −2
(E) Not possible

12. Find the standard deviation of 9, 3, 6, 2, and 10.

(A) $5\sqrt{2}$
(B) $2.5\sqrt{2}$
(C) −0.5
(D) 5.0
(E) $\frac{\sqrt{50}}{2}$

13. How many different groups of six passengers can fit into a four-passenger vehicle?

(A) 30
(B) 60
(C) 120
(D) 15
(E) 360

14. Consider 10 throws of an ordinary coin. The probability for heads or tails is equal to 1/2. What is the probability that exactly 5 heads will turn up?

(A) 0.25
(B) 0.50
(C) 1.000
(D) 0.35
(E) 0.45

15. Solve the differential equation

$$3\frac{dy}{dt}+27y=0 \quad \text{if } y(0)=1.$$

(A) e^{3t}
(B) $2e^{-3t}-1$
(C) e^{-9t}
(D) e^{-27t}
(E) $3e^{-3t}-2$

16. Solve the differential equation

$$\frac{d^2y}{dt^2} - 7\frac{dy}{dt} = 0,\ y(0) = 1 \text{ and } \frac{dy(0)}{dt} = 1$$

(A) $\frac{e^{3.5t}}{7} + \frac{6}{7}$

(B) $\frac{e^{7t}}{7} + \frac{6}{7}$

(C) $\frac{e^{-3.5t}}{7} + \frac{6}{7}$

(D) $\frac{e^{-7t}}{7} + \frac{6}{7}$

(E) $3e^{3t} - 2$

17. Given the differential equation

$$\frac{dy}{dx} - 5\frac{x}{e^y} = 0$$

with $y(1) = 0$, find $y(\sqrt{2})$.

(A) e^5

(B) $e^{2.5}(e^{2.5} - \sqrt{2})$

(C) $e^3 + 5$

(D) $e^5 + 2$

(E) $e^5 - e^{2.5}$

18. Find the center of the circle given by the equation $x^2 + y^2 - 8x + 2y = 9$.

(A) (4, – 0.5)

(B) (– 4, 2)

(C) (– 4, – 1)

(D) (4, – 1)

(E) (8, – 2)

19. The equation $3x^2 + 6xy + 2y^2 - 4y = 10$ represents which conic section?

(A) ellipse

(B) circle

(C) hyperbola

(D) parabola

(E) plane

20. Find the equation of the line normal to the curve $y^2 - 4x + 2y - 3 = 0$ at the point (3, 3). That is, find the tangent at (3, 3), then find the equation of the line that will make a 90 degree angle with that tangent. The normal and the curve are shown below.

(A) $y = 7 - 2x$

(B) $y = 10 - 3x$

(C) $y = 8 + 3x$

(D) $y = 9 - 2x$

(E) $y = 11 - 3x$

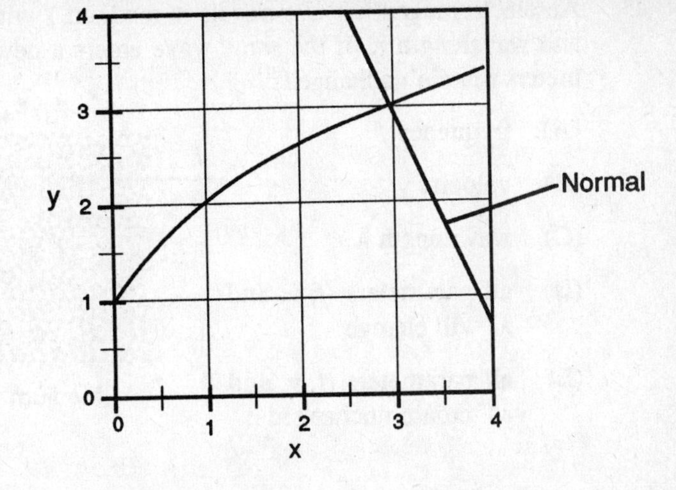

21. Given the DC circuit shown below, which of the following is true?

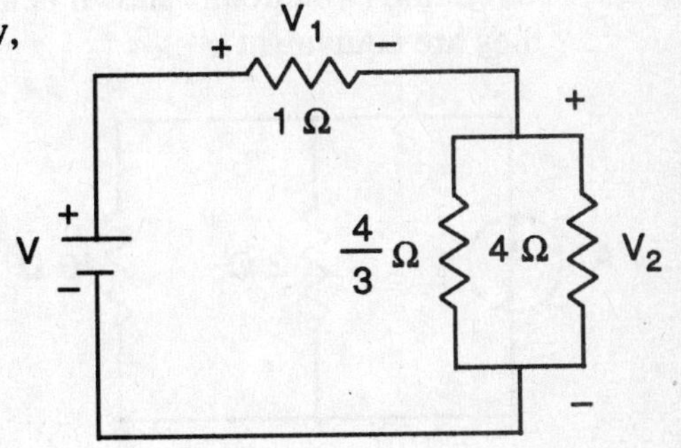

(A) $V_1 = -V_2$

(B) $V_1 = V_2$

(C) $V_1 = 2V_2$

(D) $V_1 = V_2/2$

(E) $V_1 = V^*_2$

22. What should be the values of the capacitor C and the resistor R in order for the circuits (a) and (b) to be equivalent at the frequcney of 15.9 MHz?

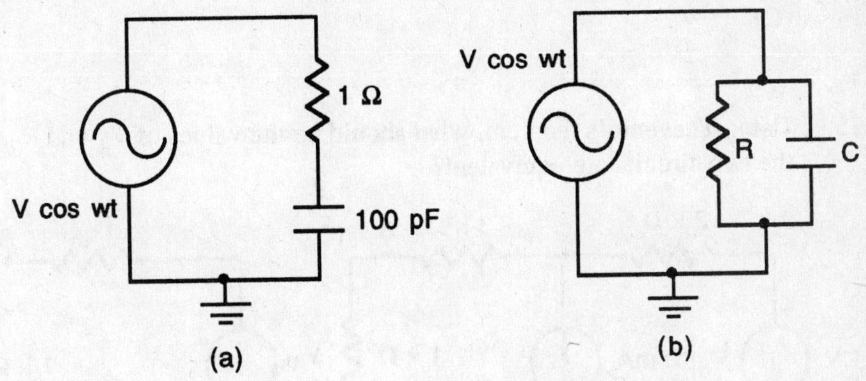

(A) $R = 1\ \Omega, C = 100\ pF$

(B) $R = 10\ k\ \Omega, C = 100\ pF$

(C) $R = 10\ \Omega, C = 1\ pF$

(D) $R = 10\ k\Omega, C = 1\ pF$

(E) impossible

23. An electromagnetic wave travels in medium 1 with frequency f, velocity v, and wavelength λ. If the same wave enters a new medium 2, which parameters remain unchanged?

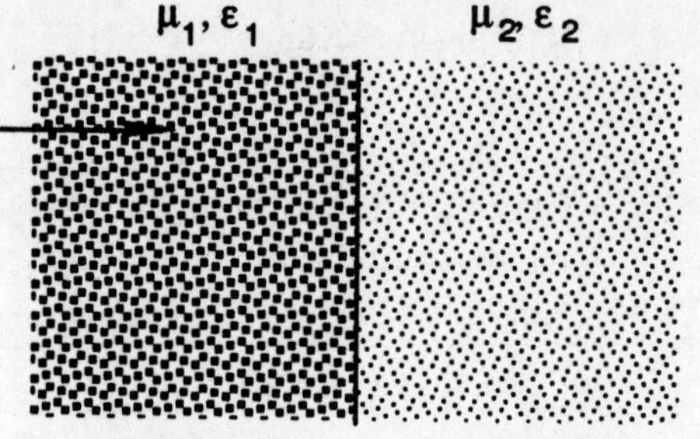

(A) frequency f

(B) velocity v

(C) wavelength λ

(D) all parameters (f, v, and λ) will change

(E) all parameters (f, v, and λ) remain unchanged

24. Given the two circuits shown below, what should be the value of V_s so that they are equivalent?

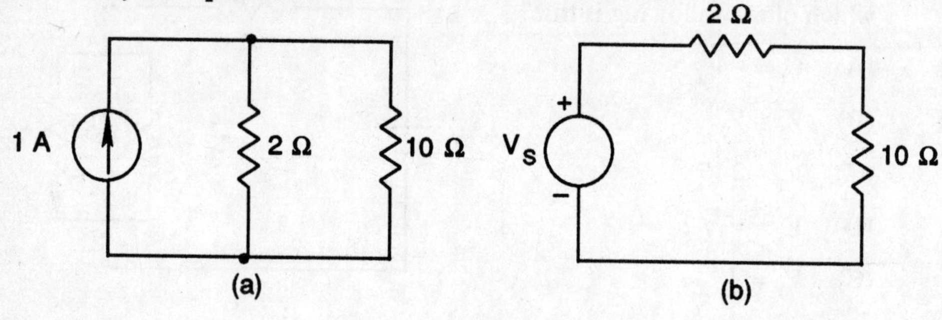

(A) 2 V

(B) 0.5 V

(C) 1 V

(D) 10 V

(E) 0.1 V

25. Using Thevenin's theorem, what should be the values of V_{th} and R_{th} so that the two circuits are equivalent?

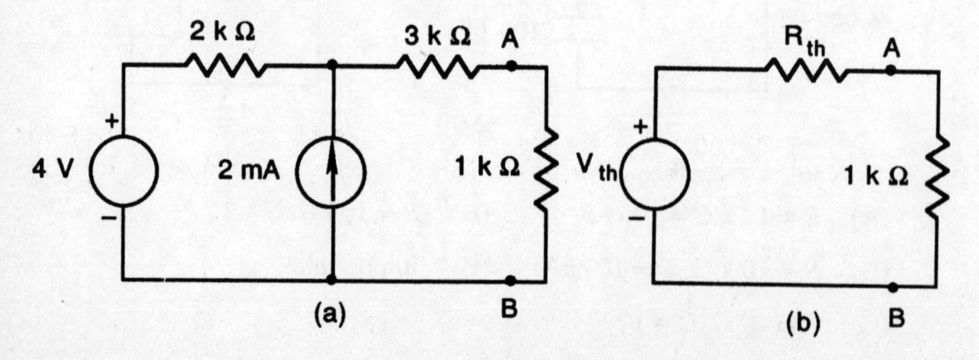

(A) 5 V, 2 K Ω (D) 5 V, 1.2 KΩ

(B) 8 V, 3 KΩ (E) impossible

(C) 8 V, 5 KΩ

26. Using Norton's theorem and the same original circuit (*a*) in Problem 22, what should be the values of I_{nr} and R_{nr} so that the two circuits are equivalent?

(A) 4 mA, 5 KΩ

(B) 1.6 mA, 1.2 KΩ

(C) 4 mA, 1.2 K Ω

(D) 1.6 mA, 5 KΩ

(E) none of the above

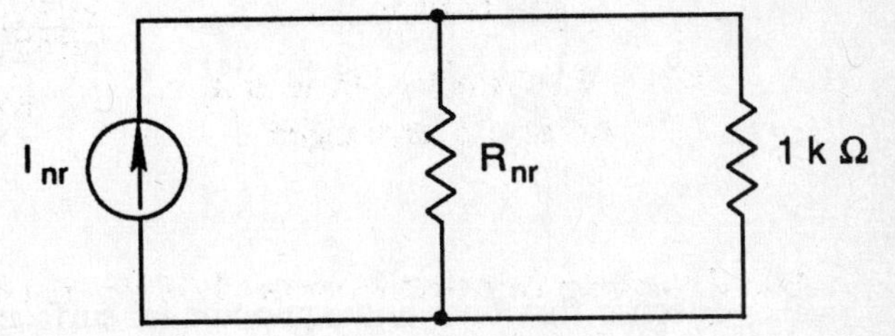

27. Given the passive circuit shown, which of the following relations is true?

(A) $V_1 = 2V_2$

(B) $V_1 = V_2 / 2$

(C) $V_1 = V_2$

(D) $V_1 = -V_2$

(E) none of the above

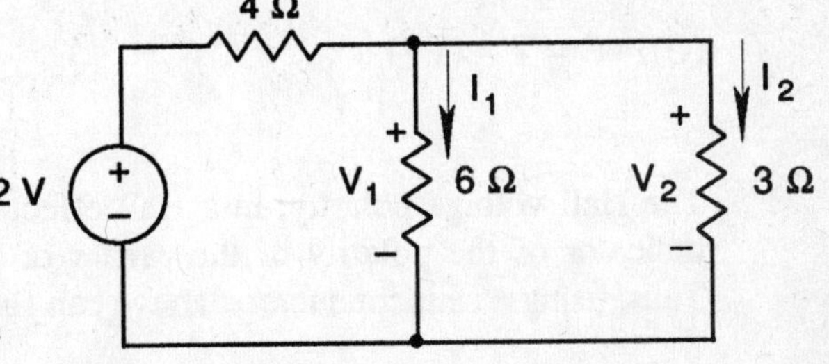

28. For the same circuit in Problem 27, which of the following relations is true?

(A) $I_1 = 2I_2$ (D) $I_1 = -I_2$

(B) $I_1 = I_2 / 2$ (E) none of the above

(C) $I_1 = I_2$

29. The current $i(t)$ is flowing through an inductor L. The voltage across this inductor is given by $v(t)$ as shown below. What is the value of L, knowing that

$$v(t) = L\,\frac{di}{dt}$$

(A) 1/3 H (B) 2 H

(C) 3 H

(D) 0.667 H

(E) impossible

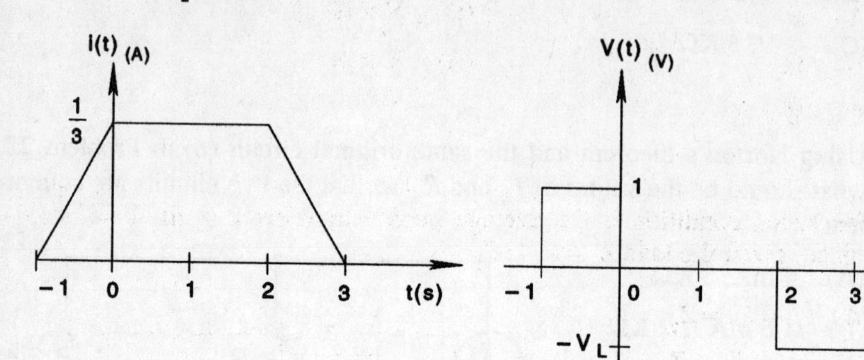

30. Electric flux lines and equipotential surfaces intersect at right angles. This follows from

(A) $V = \nabla \cdot \mathbf{E}$

(B) $V = \nabla(\nabla \cdot \mathbf{E})$

(C) $\mathbf{E} = \nabla \times (\nabla V)$

(D) $\mathbf{E} = -\nabla V$

(E) none of the above

31. The Hall voltage polarity, in a Hall effect experiment, can be used as an indicator of the polarity of the carrier of I in a semiconductor material. Thus, using a semiconductor slab we can find out if it is

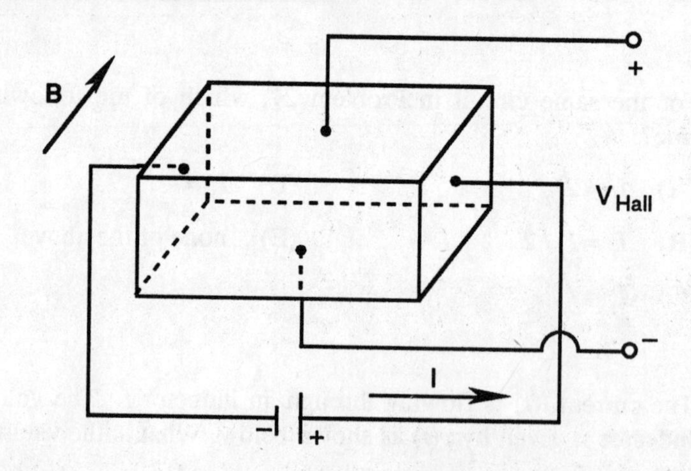

(A) P-type

(B) N-type

(C) metal

(D) P-type or N-type

(E) insulator

32. The power density of an electromagnetic wave is given by $|\mathbf{E}| \cdot |\mathbf{H}|$, then the direction of the power flow is given by

(A) $\mathbf{E}$

(B) $\mathbf{H}$

(C) $\mathbf{E} \times \mathbf{H}$

(D) $\nabla \times \mathbf{E}$

(E) $\nabla \times \mathbf{H}$

33. For which condition is the average power, transferred to the load, maximized, given the load $Z_L = Z_1 \,||\, Z_2$:

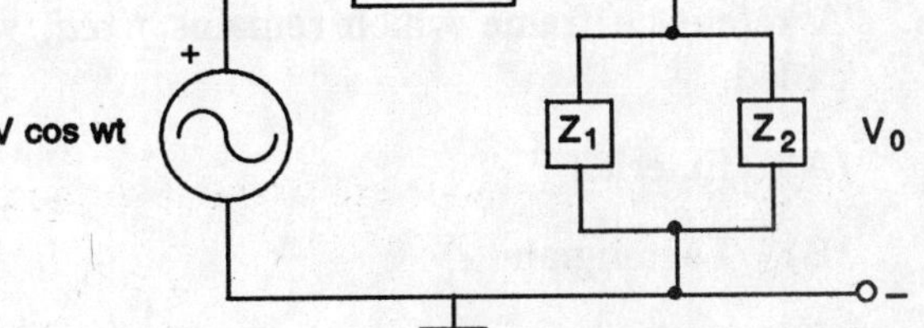

(A) $Z = Z_1 = Z_2$

(B) $Z = \dfrac{Z_1 + Z_2}{2}$

(C) $Z^* = Z_1 + Z_2$

(D) $Z^* = Z_1 \,||\, Z_2$

(E) $Z \cdot Z^* = Z_1 \cdot Z_2$

34. The two circuits given below are equivalent if:

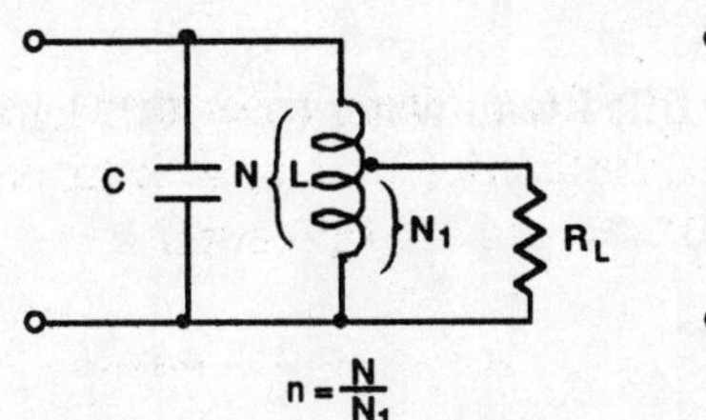

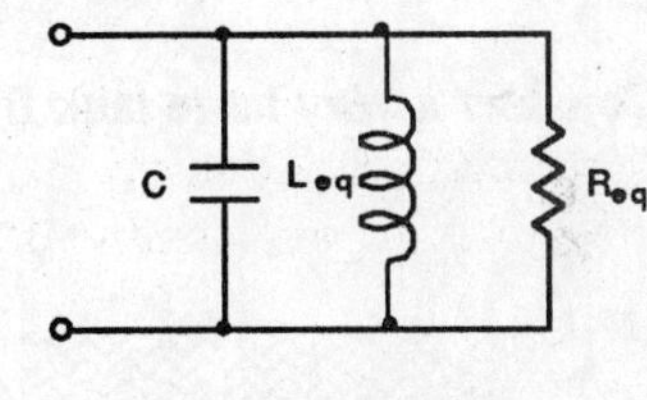

(A) $L_{eq} = n^2 \cdot L,\quad R_{eq} = n^2 \cdot R_L$

(B) $L_{eq} = n^2 \cdot L,\quad R_{eq} = R_L$

(C) $L_{eq} = L / n^2,\quad R_{eq} = n^2 \cdot R_L$

(D) $L_{eq} = L,\quad R_{eq} = n^2 \cdot R_L$

(E) $L_{eq} = L / n^2,\quad R_{eq} = R_L / n^2$

35. A round, flat disk is sliding on a thin film of oil at a velocity V = 10 m/s. The disk is 15 cm in diameter, and the viscosity of the oil is 0.1 N-s/m^2. (See figure on following page.) The drag force of the oil on the plate is

(A) 17.7 N

(B) 70.8 N

(C) 70,000 N (D) 118 N

(E) 0.118 N

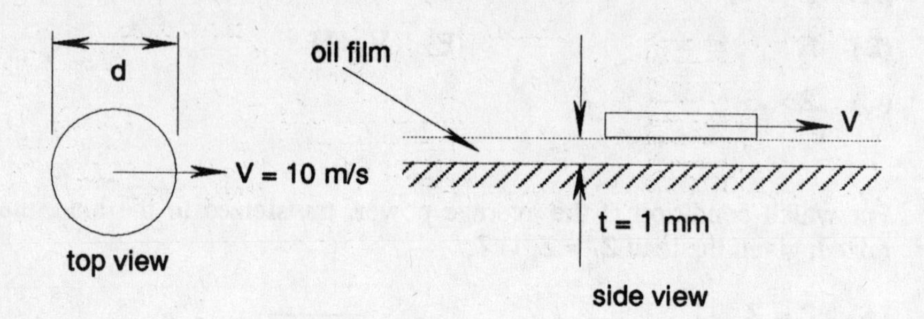

36. A reference frame which remains *fixed*, while fluid flows through it is called:

(A) Eulerian

(B) Lagrangian

(C) Stokesian

(D) A Navier-Stokes Reference Frame

(E) Bernoullian

37. Consider a very large tank filled with water ($\rho = 1000$ kg/m³) as shown.

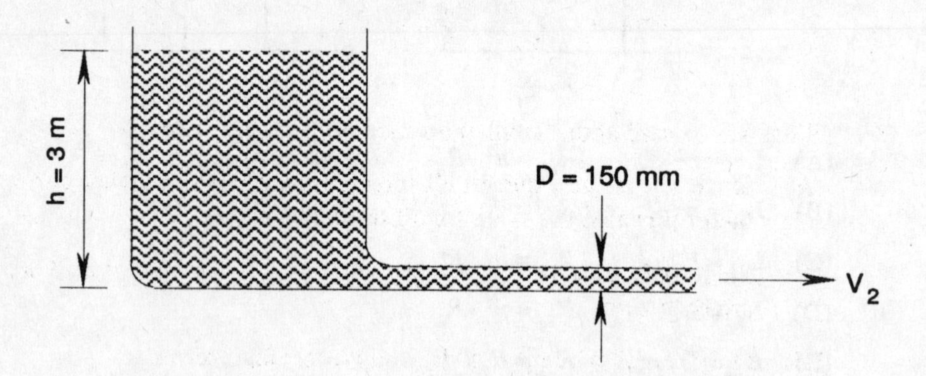

Neglecting friction, what is the steady-state velocity V_2 at the pipe exit, if we neglect the change of water height with time (the tank is extremely large)?

(A) 13.9 m/s (D) 7.67 m/s

(B) infinite, since no friction (E) none of the above

(C) 2.31 m/s

QUESTIONS 38–39 refer to the following:

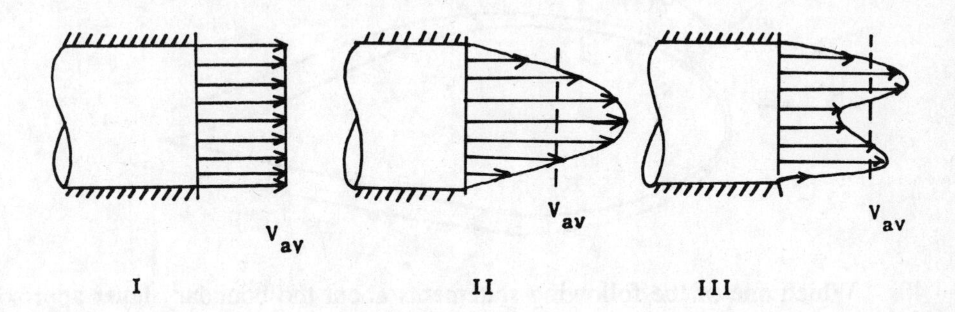

Consider the steady, incompressible flow of water exiting a round pipe. Shown below are three outlet velocity profiles. For all three cases, average velocity, V_{av}, is the same.

38. The volume flow rates, Q_I, Q_{II} and Q_{III} are related as
    (A) $Q_I = Q_{II} = Q_{III}$
    (B) $Q_I > Q_{II} > Q_{III}$
    (C) $Q_I > Q_{II}$, $Q_{II} = Q_{III}$
    (D) $Q_I < Q_{II} < Q_{III}$
    (E) No way to tell from given information

39. What can be said about momentum flux correction factor, β?
    (A) Since V_{av} is the same in all three cases, momentum flux correction factor β is also the same for all three cases.
    (B) Since only case III is non-symmetric, β_{III} is non-zero, but β_I and β_{II} are both zero.
    (C) Only $\beta_I = 0$, while β_{II} and β_{III} are greater than zero.
    (D) $\beta_I = 1.0$, but β cannot be defined for cases II and III.
    (E) $\beta_I = 1.0$, but β_{II} and β_{III} are greater than 1.0.

QUESTIONS 40–41 refer to the diagram below, which represents steady, incompressible flow over a two-dimensional body; a boundary layer coordinate system is sketched on the upper surface:

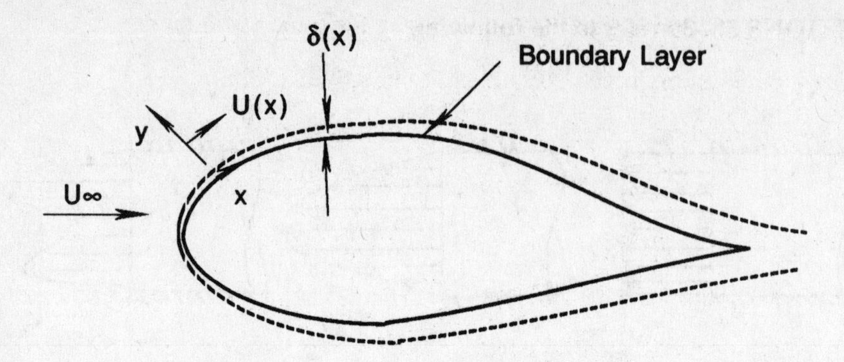

40. Which one of the following statements about the boundary layer approximation is correct?

    (A) Pressure, p, within the boundary layer is constant in the x-direction.

    (B) The rate of change of velocity in the x-direction is much larger than the rate of change of velocity in the y-direction.

    (C) The boundary layer approximation is only valid for very low Reynolds numbers.

    (C) Normal velocity, v, is much smaller than tangential velocity, u, in the boundary layer.

    (E) Boundary layer thickness, δ, does not change in the x-direction unless the boundary layer goes turbulent.

41. If, over a section of the body, pressure, p, increases in the x-direction (i.e., $dp/dx > 0$) in the inviscid flow region outside of the boundary layer, which one of the following statements is correct?

    (A) This condition is called a favorable pressure gradient.

    (B) This condition can be either a favorable or an adverse pressure gradient, depending on the geometry of the body.

    (C) The velocity $U(x)$ just outside of the boundary layer would be decreasing under these conditions.

    (D) In order for dp/dx to be positive, the boundary layer must have separated off the body somewhere upstream.

    (E) Such a condition (i.e., $dp/dx > 0$) is impossible.

42. For steady, incompressible fully-developed flow in a constant-area pipe, friction along the walls of the pipe causes:

(A) a reduction in mass flow rate along the length of the pipe

(B) a reduction in velocity along the length of the pipe

(C) a reduction in static pressure along the length of the pipe

(D) an *increase* in Darcy friction factor along the length of the pipe

(E) all of the above

43. If the atmospheric pressure is measured to be 732 mm of mercury, and the temperature is 20°C, how much gage pressure (in mm of Mercury) would be required to choke the flow of air escaping from a small hole in an automobile tire?

(A) 386 mm Hg

(B) 1386 mm Hg

(C) 2118 mm Hg

(D) 654 mm Hg

(E) Depends on the size of the hole, which is not given. Therefore the question cannot be answered with the information given.

QUESTIONS 44–46 refer to the two pipes shown below. The first one is a round pipe of radius R = 15 mm, and the second one is an annulus with inner radius b and outer radius a.

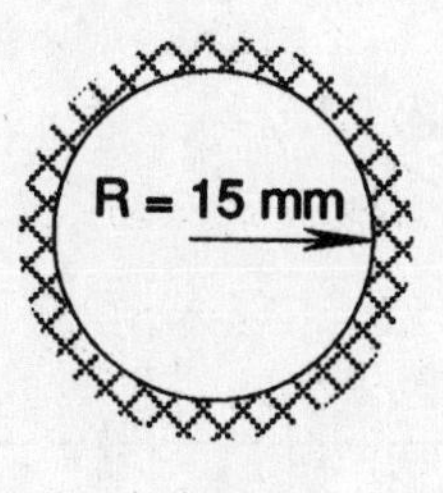

Pipe A

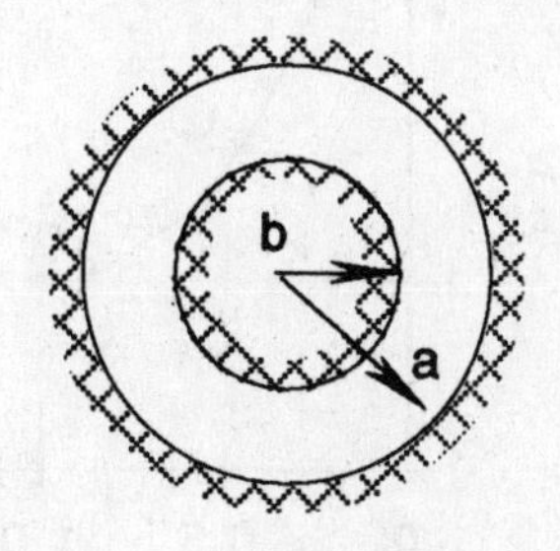

Pipe B

44. If the cross-sectional area of the two pipes is to be identical, and a = 25 mm, b should be

(A) 10 mm

(B) 20 mm

(C) 15 mm

(D) 7.32 mm

(E) 4.47 mm

45. If $a = 25$ mm and $b = 15$ mm, the hydraulic diameter of pipe B is most nearly

(A) 7.5 mm
(B) 10 mm
(C) 15 mm
(D) 20 mm
(E) 30 mm

46. Pipe A is 100 m long, and is to transport water at a volume flow rate $Q = 9.4 \times 10^{-4}$ m³/s. If the friction factor $f = 0.025$, the head loss, h_f, expressed in meters of water is most nearly

(A) 7.5 m
(B) 9.8 m
(C) 12.2 m
(D) 15.0 m
(E) 25.9 m

QUESTIONS 47–48 refer to the following:

The relationship between area ratio A/A^* and Mach number M_a for isentropic compressible flow of a perfect gas in a duct, is sketched below, with labels 1 through 5 representing points along this curve. A^* represents the sonic throat area.

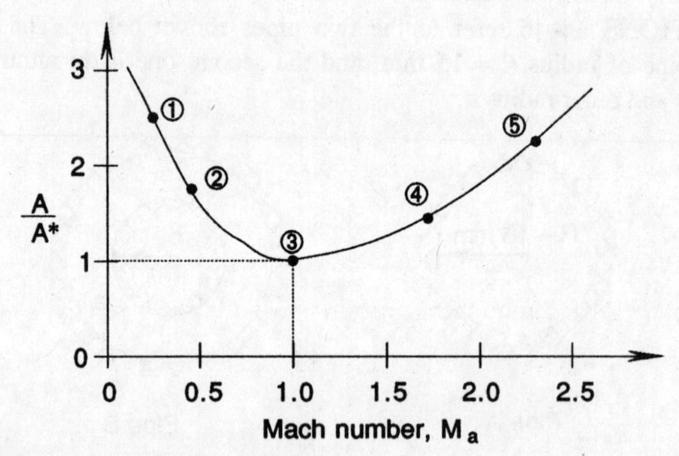

47. In a *diverging* (expanding) section of the duct, if the flow is *supersonic*, which one of the following would be the most accurate qualitative description of an observer travelling with the flow?

(A) The flow moves along the area vs. Mach number curve from point 1 to point 2.

(B) The flow moves along the area vs. Mach number curve from point 2 to point 3.

(C) The flow moves along the area vs. Mach number curve from point 3 to point 4.

(D) The flow moves along the area vs. Mach number curve from point 4 to point 5.

(E) The flow moves along the area vs. Mach number curve from point 5 to point 4.

48. In a *diverging* (expanding) section of the duct, if the flow is *subsonic*, which one of the following would be the most accurate qualitative description of an observer travelling with the flow?

(A) The flow moves along the area vs. Mach number curve from point 1 to point 2.

(B) The flow moves along the area vs. Mach number curve from point 2 to point 3.

(C) The flow moves along the area vs. Mach number curve from point 2 to point 1.

(D) The flow moves along the area vs. Mach number curve from point 3 to point 2.

(E) The flow moves along the area vs. Mach number curve from point 4 to point 5.

49. The vapor dome for water is shown below. Water initially at state point 1 undergoes a process to reach state point 2. All of the following are true EXCEPT:

(A) State 1 is a saturated liquid

(B) State 2 is a saturated vapor

(C) The quality of the mixture decreases from state 1 to state 2

(D) The temperature is constant

(E) The pressure is constant from state 1 to state 2

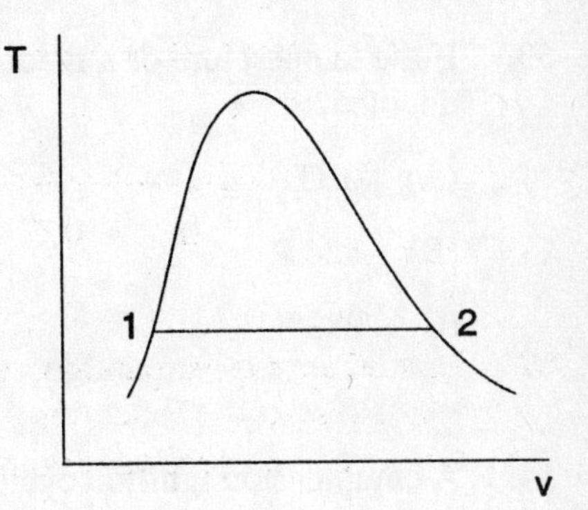

50. A closed system experiences a reversible process where heat rejection is the only energy transfer. The entropy change

(A) must be zero

(B) must be positive

(C) must be negative

(D) cannot be negative due to Second Law requirements.

(E) is equal to the heat transfer

51. Consider an air-water vapor mixture similar to the atmosphere. If the dry-bulb temperature equals the dew point temperature, the relative humidity will be:

(A) 0%

(B) 25%

(C) 50%

(D) 75%

(E) 100%

52. For the reaction

$$CH_4 + (1.5)(2)(O_2 + 3.76N_2) \rightarrow CO_2 + 2H_2O + O_2 + 11.28N_2$$

more air has been supplied than is necessary for complete combustion. The percentage of theoretical air is most nearly:

(A) 300%

(B) 100%

(C) 50%

(D) 150%

(E) 1128%

53. If the temperature of a medium is 0°C, what will the temperature be if it is doubled?

(A) 0°C

(B) 524 R

(C) 273°C

(D) 460 R

(E) 64°F

54. A compression ignition cycle is modeled by which ideal cycle?

(A) Otto cycle

(B) Diesel cycle

(C) Rankine cycle

(D) Brayton cycle

(E) Ericsson cycle

55. An ideal gas is contained in a rigid container. There is no work of a rotating shaft associated with the container. Any heat transfer is a function of

(A) pressure only

(B) volume only

(C) temperature only

(D) There cannot be any heat transfer.

(E) The heat transfer will be equal to the work.

56. In a combustion chamber, fuel is burned to raise the temperature of the medium prior to production of work through an expansion process. The heat generated during the combustion reaction is the

(A) specific heat

(B) heat of vaporization

(C) heat of formation

(D) heat of reaction

(E) heat of fusion

57. There are many types of work associated with energy transfer across a boundary. One form of work is flow work which is described by all of the following EXCEPT:

(A) it is the work which pushes mass into or out of a device

(B) it is necessary to maintain a continuous flow

(C) it is added to the internal energy to obtain enthalpy in the first law for a control volume

(D) it is associated with a closed system

(E) once considered in the enthalpy term, it is no longer accounted for as a work term

58. An open system First Law should be utilized for all the following EXCEPT:

(A) a nozzle

(B) a turbine

(C) a piston-cylinder device with no inlet/exhaust values

(D) a compressor

(E) a pump

59. A piston-cylinder device provides 8 kJ of work to an external device. 2 kg of air are contained inside the cylinder. If the internal energy of the air increases by 2 kJ/kg during the process, the heat transfer is

(A) 12 kJ added
(B) 4 kJ added
(C) zero
(D) 4 kJ rejected
(E) 12 kJ rejected

60. A pump is used to increase the pressure of the water entering the boiler of a steam power cycle. Which statement is true concerning the pump?

(A) The pump produces work.
(B) The pump has no effect on the cycle.
(C) The enthalpy of the water leaving the pump is lower than the enthalpy of the water entering.
(D) The enthalpy of the water leaving the pump is higher than the enthalpy of the water entering.
(E) The enthalpy is constant during the process.

61. An insulated rigid container contains an ideal gas. The container is initially divided in half by a membrane such that one side is a vacuum and the other side contains the ideal gas. The membrane is broken and the gas fills the entire volume. All are true EXCEPT:

(A) the pressure is halved
(B) the volume is doubled
(C) the temperature is quadrupled
(D) the process is irreversible
(E) the internal energy is constant

62. If the gage pressure of a medium is 30 kPa (vacuum) and the atmospheric pressure is 101.3 kPa, then the absolute pressure is:

(A) –131.3 kPa
(B) –71.3 kPa
(C) not valid because gage pressure cannot be negative
(D) +131.3 kPa
(E) +71.3 kPa

63. A stone is thrown from the top of a 200 m building with an initial velocity of 150 m/s at an angle of 30° with the horizontal line. Neglecting the air resistance, determine the maximum height above the ground reached by the stone.

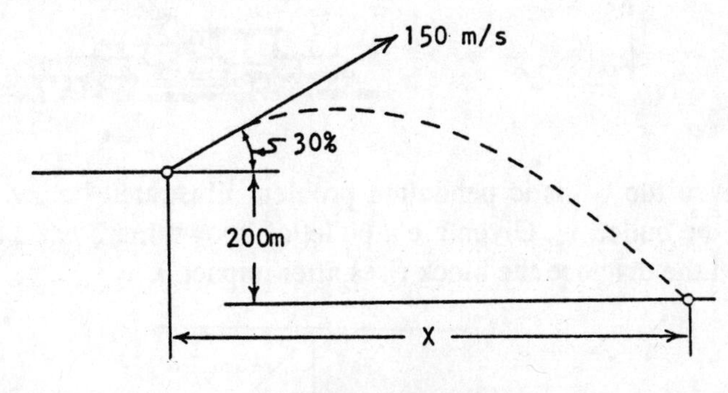

(A) 487 m
(B) 287 m
(C) 87 m
(D) 2289 m
(E) 861 m

64. The outside curve on a highway forms an arc whose radius is 150 ft. If the roadbed is 30 ft. wide and its outer edge is 4 ft. higher than the inner edge, for what speed is it ideally banked?

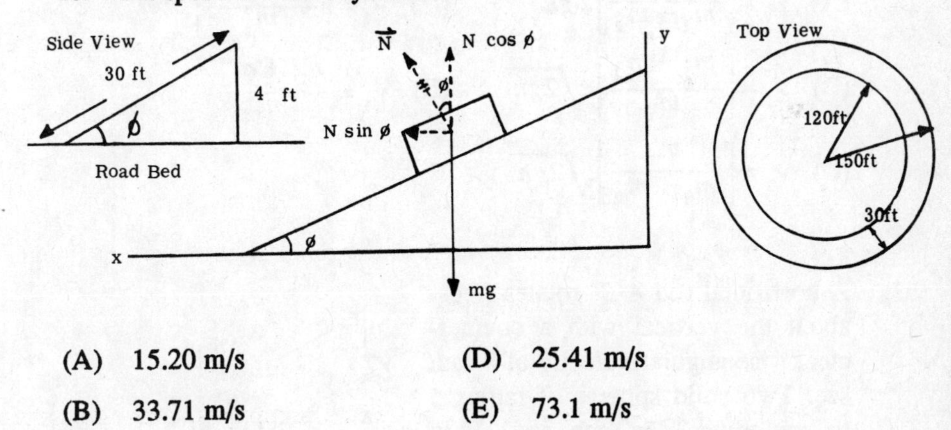

(A) 15.20 m/s
(B) 33.71 m/s
(C) 29.20 m /s
(D) 25.41 m/s
(E) 73.1 m/s

65. A railway gun, initially at rest, whose mass is 70,000 kg fires a 500-kg artillery shell at an angle of 45° and with a muzzle velocity of 200 m/sec. Calculate the recoil velocity of the gun. (See following figure.)

(A) 1.43 m/s
(B) 1.00 m/s
(C) 19,796 m/s
(D) 3.41 m/s
(E) 5,320 m/s

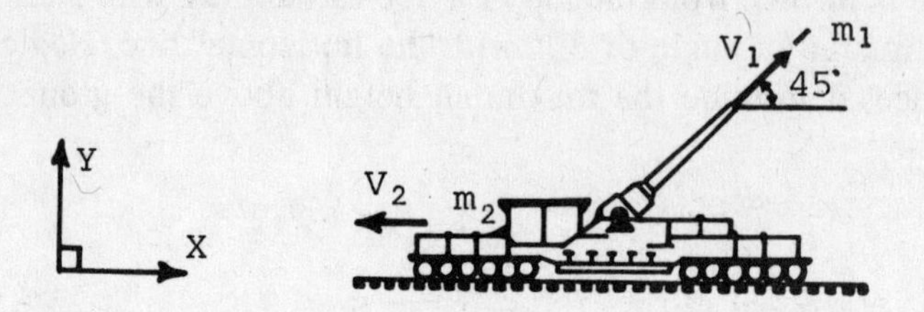

66. Given the ballistic pendulum problem illustrated below, find the velocity of the bullet, v_1. Given are a bullet of known mass m_1, a block of mass m_2, and the distance the block rises after impact h.

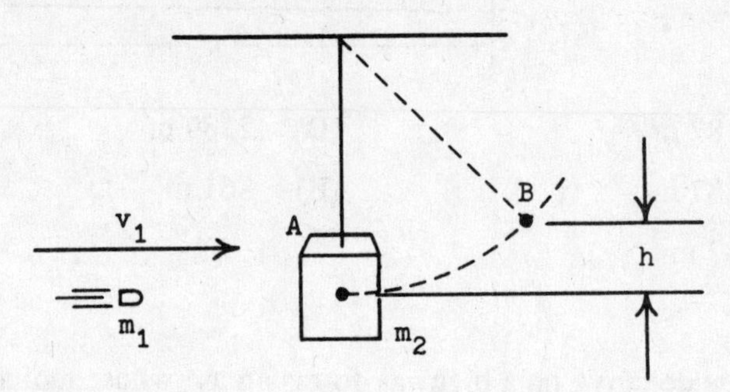

(A) $v_1 = \left[\dfrac{m_1}{m_1 + m_2}\right] 2\,gh$

(B) $v_1 = \left[\dfrac{m_1 + m_2}{m_1}\right] \sqrt{2gh}$

(C) $v_1 = \left[\dfrac{m_1}{m_1 + m_2}\right] \sqrt{2gh}$

(D) $v_1 = \left[\dfrac{m_1 + m_2}{m_1}\right] 2\,gh$

(E) $v_1 = \dfrac{m_1 + m_2}{m_1}$

67. A horizontal rod $A'B'$ rotates freely about the vertical with a counterclockwise angular velocity of 8 rad/sec. Two solid spheres of radius 5 in., weighing 3 lbs each, are held in place at A and B by a cord which is suddenly cut. Knowing that the centroidal moment of inertia of the rod and pivot is $I_r = 0.25$ lb-ft-s², determine the angular velocity of the rod after the spheres have moved to positions A' and B'.

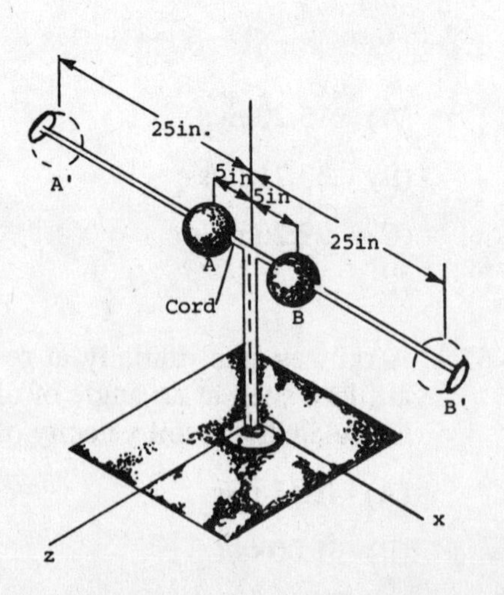

(A) 1.6 rad/s

(B) .441 rad/s

(C) 8.00 rad/s (D) 2.47 rad/s

(E) 2.21 rad/s

68. The system of *A* and *B* and two pulleys *C* and *D* is assembled as shown in the figure. Neglecting friction and the mass of the pulleys, and assuming that the whole system is initially at rest, determine the acceleration of block *A*.

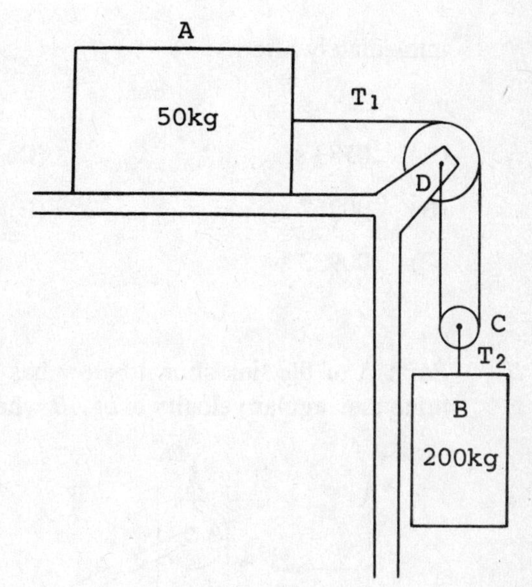

(A) 7.84 m/s²

(B) 3.92 m/s²

(C) 0

(D) 392 m/s²

(E) 784 m/s²

69. Two springs, S_1 and S_2, of negligible mass, with spring constants K_1 and K_2, respectively, are arranged to support a body *A*. In the diagram below, the springs are coupled in "parallel." Determine the equivalent spring constant, k_e, for parallel coupling of springs.

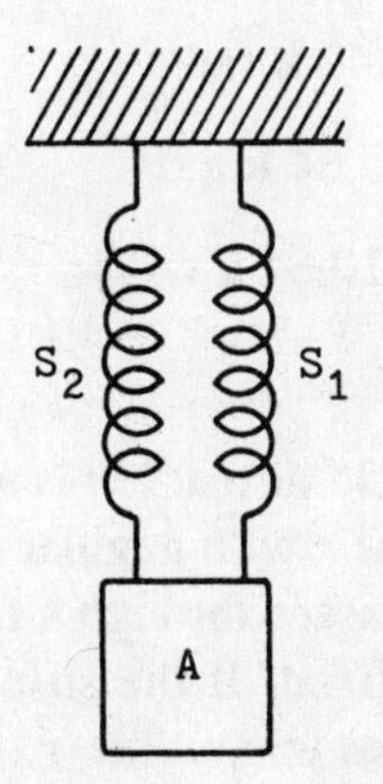

(A) $K_e = K_1K_2$

(B) $K_e = \dfrac{K_1 + K_2}{K_1K_2}$

(C) $K_e = K_1 + K_2$

(D) $K_e = K_1 - K_2$

(E) $K_e = \dfrac{K_1K_2}{K_1 + K_2}$

70. A baseball (mass = .16 kg) is moving 25 m/s when it is hit directly back to the pitcher at a speed of 45 m/s. If the average force exerted by the bat on the ball is 1200 N, how long did the collision last?

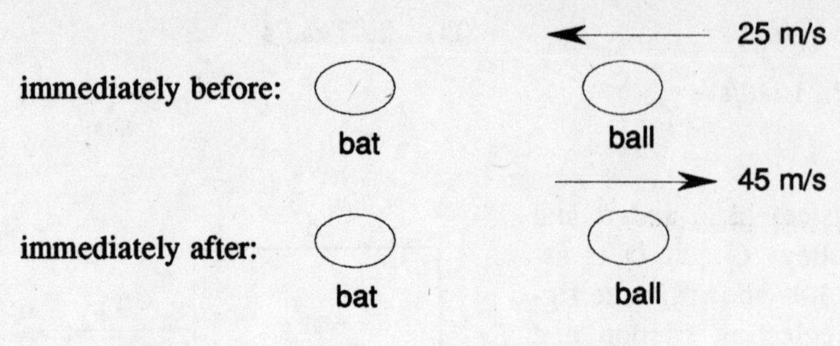

(A) .0583 s

(B) .00267 s

(C) .00933 s

(D) .0167 s

(E) .00321 s

71. Point A of the link shown below has an upward velocity of 3 m/s. Determine the angular velocity ω of *AB* when θ = 20°.

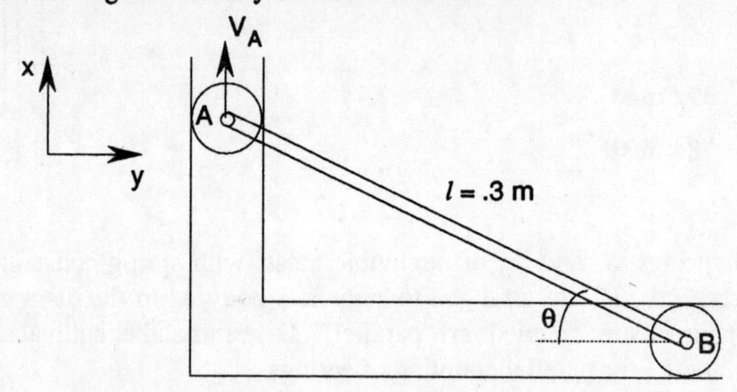

(A) 9.40 **k** rad/s

(B) 10.64 **k** rad/s

(C) –10.64 **k** rad/s

(D) –5.22 **i** rad/s

(E) –9.40 **k** rad/s

72. A particle of mass *m* is attached to the end of a string and moves in a circle of radius *r* with angular velocity ω_0, on a frictionless hortizontal table. The string passes through a frictionless hole in the table and, initially, the other end is fixed. If the string is pulled so that the radius of the circular orbit decreases to a radius, *r*, what is the final angular velocity, ω_f?

(A) $\omega_f = \frac{r_0}{r}\omega_0$

(B) $\omega_f = \omega_0$

(C) $\omega_f = \frac{r}{r_0}\omega_0$

(D) $\omega_f = \left(\frac{r_0}{r}\right)^2 \omega_0$

(E) $\omega_f = \left(\frac{r}{r_0}\right)^2 \omega_0$

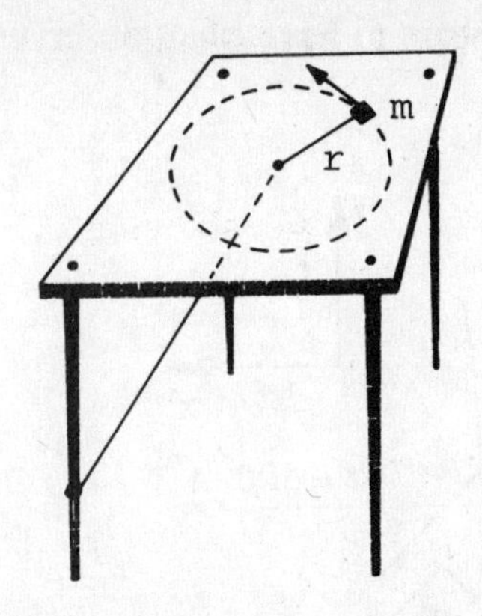

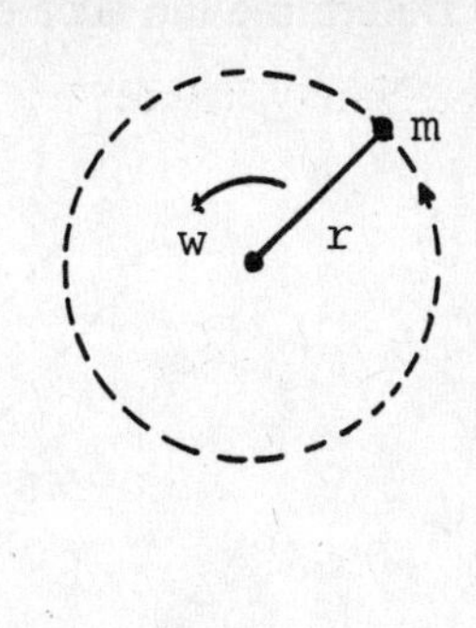

73. Two similar cars, A and B, are connected rigidly together and have a combined mass of 4 kg. Car C has a mass of 1 kg. Initially, A and B have a speed of 5 m/sec and C is at rest as shown in the figure.

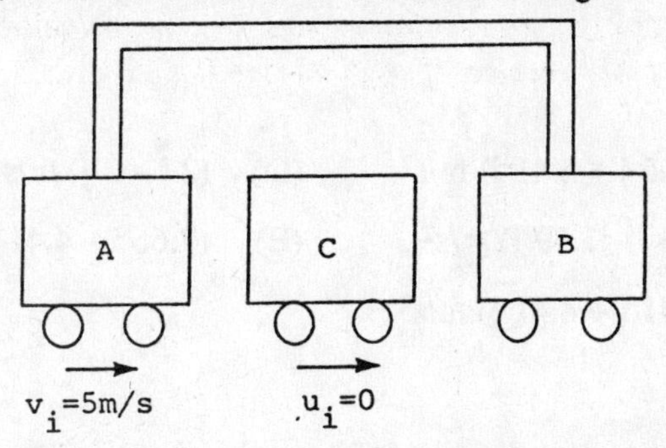

Assuming a perfectly inelastic collision between A and C, the final speed of the system is

(A) 1 m/sec
(B) 2 m/sec
(C) 2.5 m/sec
(D) 3 m/sec
(E) 4 m/sec

74. If the acceleration due to gravity at the surface of the Earth is g, what is the acceleration due to gravity at the surface of Neptune? Neptune's mass is 17 times that of Earth, and its radius is 3.5 times that of Earth.

(A) .206 g
(B) .721 g
(C) 4.86 g
(D) g
(E) 1.39 g

75. At this instant, car A is turning the circular curve at a speed of 25 m/s and is slowing down at a rate of 3 m/s². Car B is speeding up at a rate of 2 m/

s^2. Determine the acceleration car B appears to have to an observer in car A.

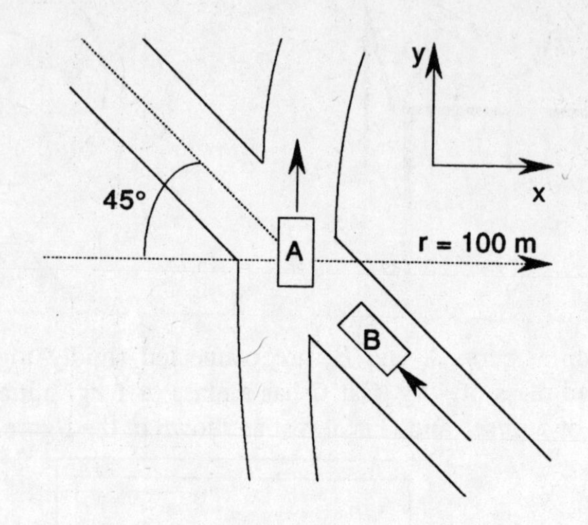

(A) $(-7.66\,\mathbf{i} + 4.41\,\mathbf{j})$ m/s²
(B) $(4.84\,\mathbf{i} - 1.59\,\mathbf{j})$ m/s²
(C) $(-1.41\,\mathbf{i} + 4.41\,\mathbf{j})$ m/s²
(D) $(2\,\mathbf{i} - 3\,\mathbf{j})$ m/s²
(E) $(7.66\,\mathbf{i} - 4.41\,\mathbf{j})$ m/s²

76. The motion of a particle is described by the equation $x(t) = t^3 - 12t^2 - 40t + 60$, where x is measured in meters and t is measured in seconds. Find the velocity when the acceleration of the particle is equal to zero.

(A) – 241 m/s
(B) 32.50 m/s
(C) 0
(D) – 32.50 m/s
(E) – 85.00 m/s

77. The magnitude of the resultant of the force system is:

(A) 140 N
(B) 100 N
(C) 200 N
(D) 280 N
(E) 70 N

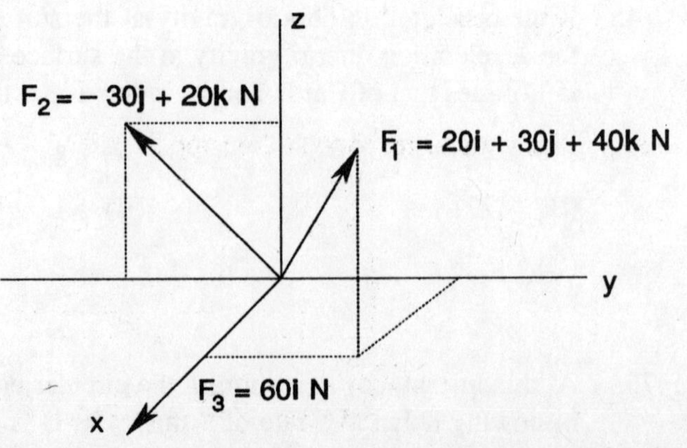

78. The tension T in the pulley system is: (Neglect the friction of the pulley)

(A) W

(B) $W/2$

(C) $W/6$

(D) $W/3$

(E) $W/8$

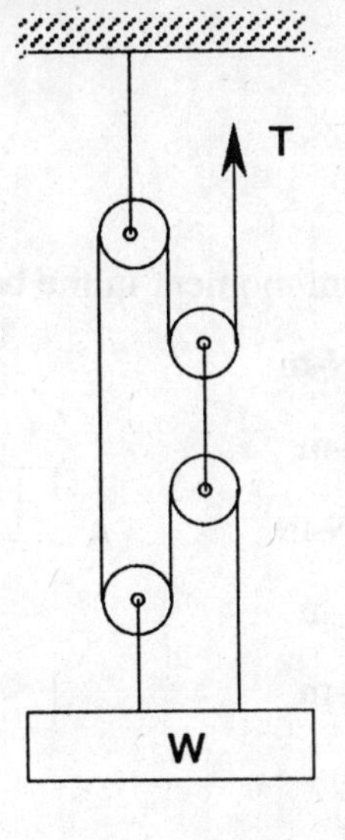

79. The tension in the cable BC is:

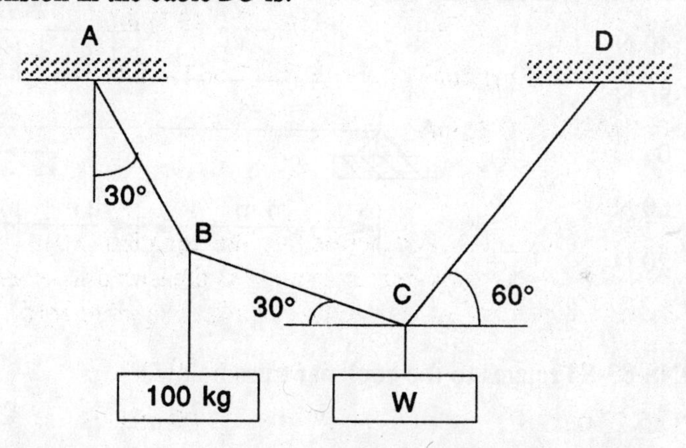

(A) 50 N

(B) 100 N

(C) 981 N

(D) 490 N

(E) 87 N

80. The vertical reaction at the support B is:

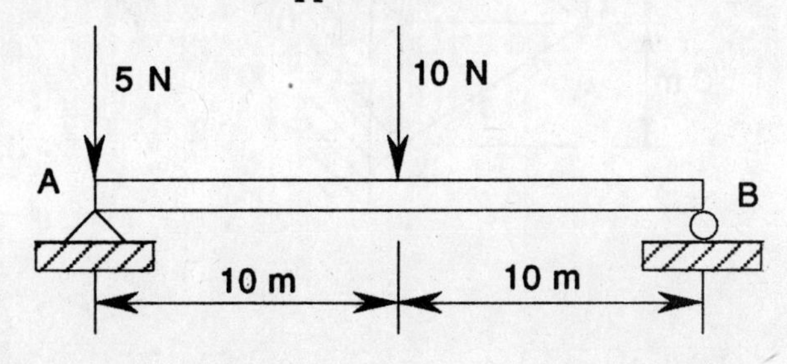

(A) 15 N (D) 0

(B) 10 N (E) 5 N

(C) 7.5 N

81. The maximum moment in the beam is:

(A) 1000 N-m

(B) 500 N-m

(C) 1500 N-m

(D) 100 N-m

(E) 200 N-m

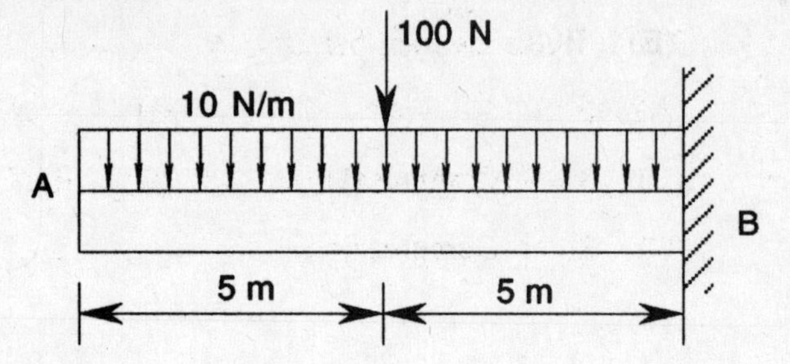

82. The vertical reaction at the support *A* is:

(A) 40 N

(B) 80 N

(C) 0

(D) 60 N

(E) 20 N

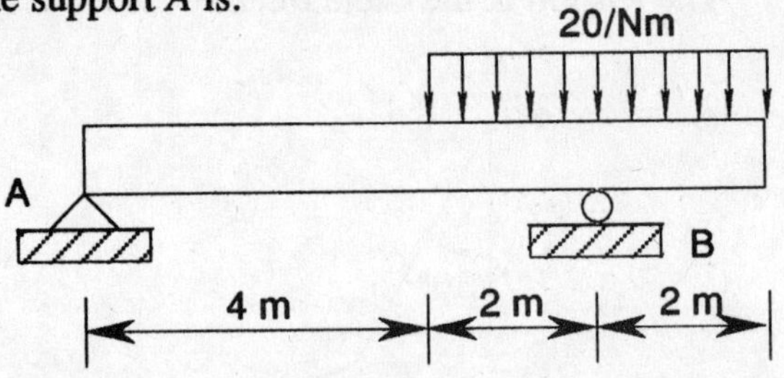

QUESTIONS 83-84 relates to the coplanar truss below.

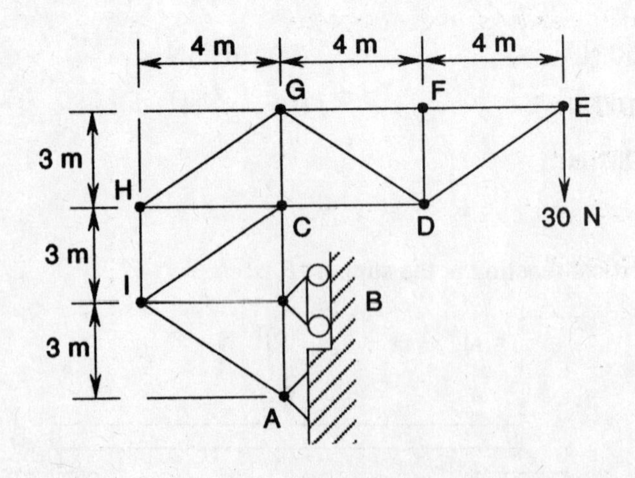

83. The force in the member *GD* is:

(A) 30 N (Compression)
(B) 40 N (Tension)
(C) 50 N (Compression)
(D) 50 N (Tension)
(E) 30 N (Tension)

84. The force in the member DF is:

(A) 0
(B) 30 N (Compression)
(C) 80 N (Compression)
(D) 30 N (Tension)
(E) 80 N (Tension)

85. The coordinates of the centroid of the shaded area is:

(A) (4.0,3.5)
(B) (3.5,4.0)
(C) (3.0,4.0)
(D) (4.0,3.0)
(E) (4.0,4.0)

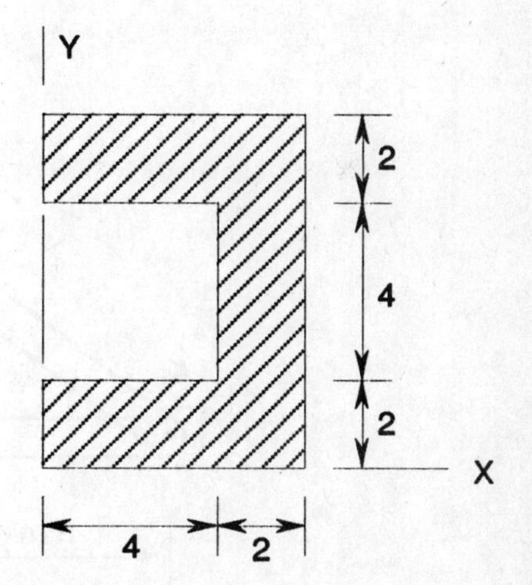

86. The moment of inertia about the centroidal *x*-axis of the composite area is:

(A) 4275 m^4
(B) 8875 m^4
(C) 1440 m^4
(D) 7100 m^4
(E) 6675 m^4

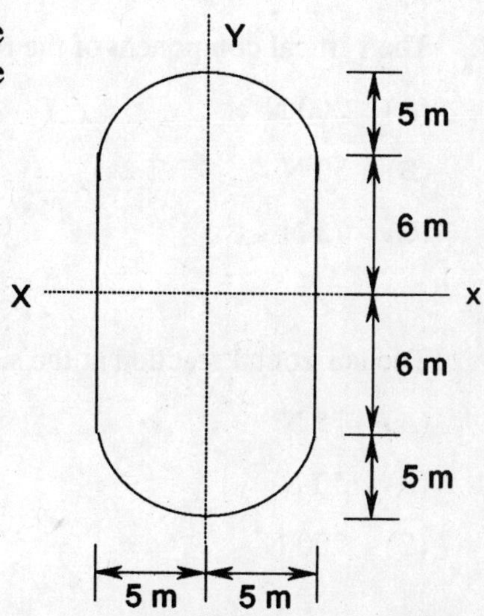

87. The force P which will result in impending motion down the slope is:

(A) 736 N

(B) 981 N

(C) 1128 N

(D) 638 N

(E) 392 N

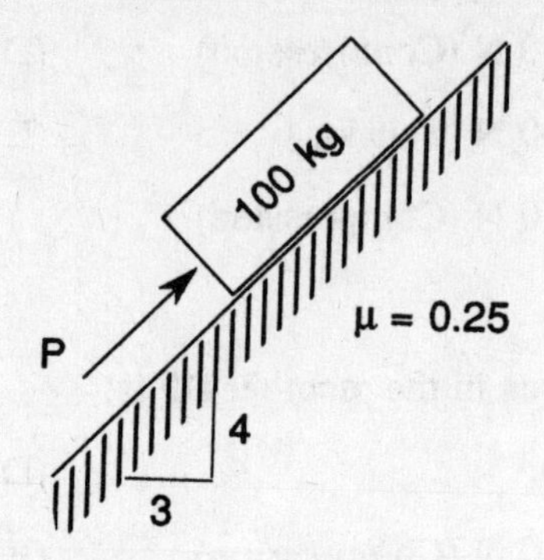

QUESTIONS 88-89 relate to the coplanar frame below:

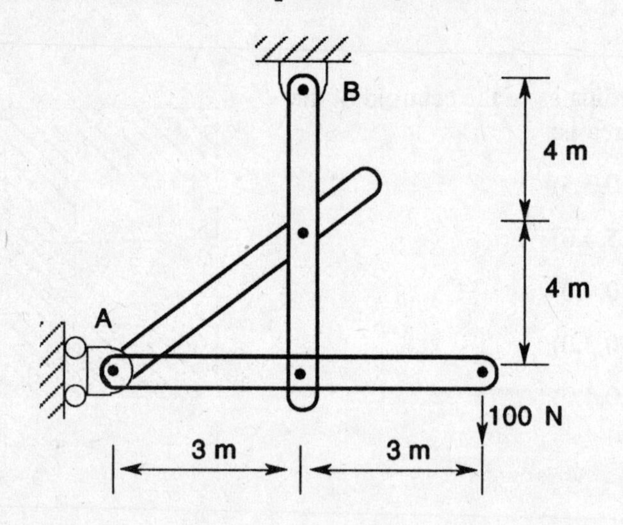

88. The vertical component of the reaction at the support *B* is:

(A) 200 N

(B) 50 N

(C) 75 N

(D) 150 N

(E) 100 N

89. The horizontal reaction at the support *A* is:

(A) 75 N

(B) 50 N

(C) 200 N

(D) 37.5 N

(E) 100 N

90. The horizontal component of the reaction at the support *B* is:

(A) 32 N (rightward)

(B) 32 N (leftward)

(C) 20 N (rightward)

(D) 16 N (leftward)

(E) 20 N (leftward)

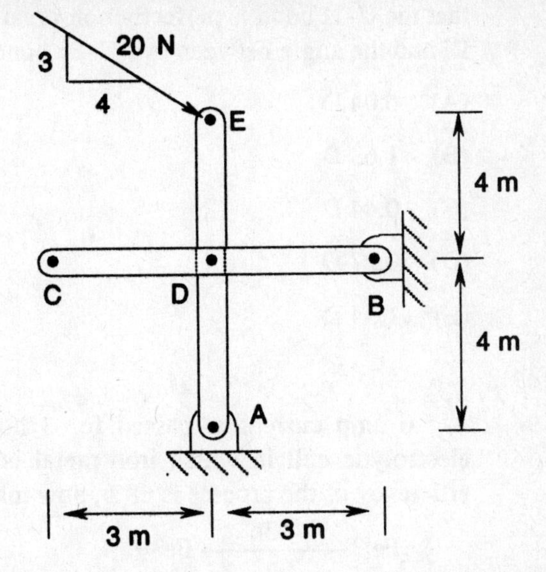

Atomic Weights: N = 14.0 Fe = 56.0

O = 16.0 Cu = 63.5

R, the Gas Constant = 0.082 lit. atm./deg. K mole

91. 0.40 g of a volatile liquid occupies 107.0 ml at 1.00 atm. pressure and 27°C. What is the molecular weight of the liquid?

(A) 8.28

(B) 137

(C) 575

(D) 92

(E) 2232

92. 2.86 g sample of an alloy of gold (Au) and copper (Cu) upon reaction with excess nitric acid (HNO_3) formed 3.75 g of cupric nitrate, $Cu(NO_3)_2$. What is the percent of Au by mass in the alloy?

$$Cu + 4HNO_3 \rightarrow Cu(NO_3)_2 + 2H_2O + 2NO_2$$

(A) 4.44

(B) 1.27

(C) 55.6

(D) 44.4

(E) 95.56

93. Calculate the dipole moment of methylene dibromide (CH_2Br_2) assuming that the C–H bond is perfectly non-polar, the C–Br bond movement is 1.38 D, and the angle between two C–Br bonds is 112°.

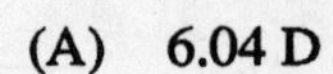

(A) 6.04 D

(B) 1.63 D

(C) 0.44 D

(D) 0.27 D

(E) 1.54 D

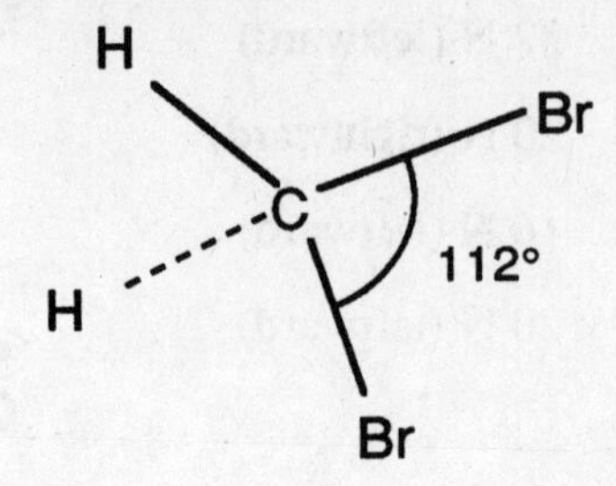

94. A 5.0 amp current is passed for 3 hours and 30 minutes through this electrolytic cell in which iron metal is deposited on the cathode. If the efficiency of the process is 68%, how many g of iron are deposited?

$$Fe^{+3} \xrightarrow{+3e} Fe^{0}$$

(A) 8.26 g

(B) 12.19 g

(C) 1.66 g

(D) 2.44 g

(E) 24.86 g

95. Trinitrotoluene (TNT) is a high explosive. The explosion of TNT can be represented by the equation:

$$2\,C_7H_5(NO_2)_{3(s)} \rightarrow 7C_{(s)} + 7CO_{(g)} + 3N_{2(g)} + 5H_2O_{(g)}$$

How much heat in Kcal will be generated by detonating 3.00 lbs of TNT?

Given: Heats of formation (H_f) of:

$C_7H_5\,(NO_2)_{3(g)}$ = – 87.1 Kcal/mole

$CO_{(g)}$ = – 26.4 Kcal/mole

$H_2O_{(g)}$ = – 57.8 Kcal/mole

Note: Heat of formation of any element by itself is zero.

(A) – 184.8

(B) – 299.6

(C) – 898.8

(D) – 473.8

(E) + 174.2

96. The solubility product (K_{sp}) of silver chromate (Ag_2CrO_4) at 25°C is 8.5 ×

10^{-8}. What is the solubility of silver chromate at 25°C in moles/lit?

$$Ag_2CrO_{4(s)} \rightarrow 2Ag^+_{(aq)} + CrO_4^{-2}{}_{(aq)}$$

(A) 2.83×10^{-7}

(B) 2.55×10^{-7}

(C) 3.49×10^{-3}

(D) 2.77×10^{-3}

(E) 4.4×10^{-3}

97. Many transition metals exhibit more than one valence state in simple reactions because

(A) s and p electrons in the same orbit have very different reactivities.

(B) d electrons from one orbital down are not as reactive as the s electrons in the outer orbit.

(C) the valence states of metals change if the other elements in the molecule change.

(D) it is not always possible to predict how even simple reactions will occur.

(E) valence is determined by probability

98. From the following data regarding decomposition of *A* to form *B* and *C*, which metal(s) act(s) as a catalyst(s)?

	Conc. of *A*	Conc. Metal *X*	Conc. Metal *Y*	Conc. Metal *Z*	Rate of Decomposition of *A* (moles/lit.sec)
Expt. #1	0.020	0.020	0.020	0.020	1.6×10^{-2}
Expt. #2	0.020	0.020	0.040	0.020	1.6×10^{-2}
Expt. #3	0.030	0.020	0.020	0.020	2.4×10^{-2}
Expt. #4	0.030	0.020	0.020	0.030	3.6×10^{-2}
Expt. #5	0.020	0.040	0.040	0.020	1.6×10^{-2}

(A) Metal *X*

(B) Metal *Y*

(C) Metal *Z*

(D) All three metals

(E) None of the metals

99. 0.5 mole of $CO_{(g)}$ and 1.0 mole of $H_2O_{(g)}$ were placed in a 5.0 lit. steel

container at 127°C. When equilibrium was reached, 0.3 mole of CO_2 was formed.

$$CO_{(g)} + H_2O_{(g)} \rightarrow CO_{2(g)} + H_{2(g)}$$

What is the K_c for the reaction?

(A) 1.6

(B) 0.45

(C) 2.22

(D) 0.64

(E) 0.18

100. For this reaction, the equilibrium conditions are:

$$H_{2(g)} + I_{2(g)} \rightarrow 2HI_{(g)} \qquad \Delta H = -7.9 \text{ kJ}$$

(A) Pressure only

(B) Pressure, temperature and concentration

(C) Concentration only

(D) Temperature only

(E) Temperature and concentration

101. Ammonia is prepared by Haber's process:

$$N_{2(g)} + 3H_{2(g)} \rightarrow 2NH_{3(g)} \qquad \Delta H = -5.3 \text{ kJ}$$

In order to maximize the yield of ammonia, one can:

(A) Increase concentrations of hydrogen and ammonia, and decrease the temperature.

(B) Increase concentrations of hydrogen and nitrogen, and decrease the temperature.

(C) Increase concentrations of hydrogen and nitrogen, and increase the temperature.

(D) Increase concentrations of all three gases, and increase the temperature.

(E) Decrease concentrations of all three gases, and decrease the temperature.

102. A 20.0 g sample of fish fillet was found to contain 0.5 mg of mercury. What is the concentration of mercury in the fish fillet in ppm?

(A) 15.0 (D) 75.0

(B) 25.0 (E) 7.5

(C) 350.0

103. What is the actual e.m.f (E_{actual}) of the cell

$Cu^{o} / Cu^{+2} // Ag^{+} / Ag^{o}$

at 25°C if $[Cu^{+2}]$ = 0.1 M and $[Ag^{+}]$ = 0.1 M? The standard reduction potentials of Cu and Ag are as follows:

$Cu^{+2} + 2e \rightarrow Cu^{o}$ + 0.34 volts

$Ag^{+} + 1e \rightarrow Ag^{+}$ + 0.80 volts

(A) 0.43 volts (D) 0.40 volts

(B) 0.46 volts (E) 0.52 volts

(C) 1.14 volts

104. Methanol can be manufactured by the following reaction:

$$CO_{(g)} + 2H_{2(g)} \rightleftharpoons CH_3OH_{(g)}$$

2.5 moles of $CO_{(g)}$ and 3.6 moles of $H_{2(g)}$ were placed in a 10.0 lit. container at 100°C. At equilibrium 1.6 moles of $CH_3OH_{(g)}$ was formed. Calculate the K_p for the reaction.

(A) 4.89 atm (D) 1.46/atm^2

(B) 2.75 atm (E) 1.19/atm^2

(C) 1.22 atm^2

105. A 50,000 lb load is supported by two 1" diameter aluminum rods and a 2" diameter steel rod as shown in the figure below. The length of all three

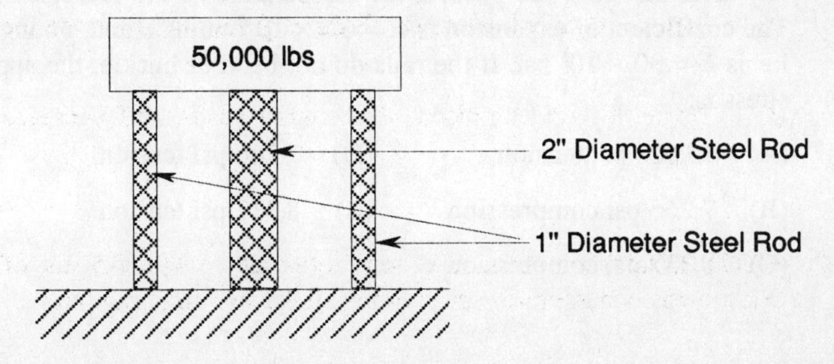

rods is 10" before the load is placed. The load is placed in such a fashion that the deformation of all three rods is the same. Young's Modulus for steel and aluminum is $E_S = 30 \times 10^6$ psi and $E_A = 10 \times 10^6$ psi, respectively. The approximate load carried by the steel rod is:

(A) 16,667 lbs
(B) 25,000 lbs
(C) 42,857 lbs
(D) 33,334 lbs
(E) 12,500 lbs

106. The beams shown in Figures (a) and (b) below have identical cross section and length. The beam in Figure (a) has hinge joint at *A* and roller support at *B*. The beam in Figure (b) has both ends *A* and *B* fixed. For both beams, the concentrated load *P* is applied at the center.

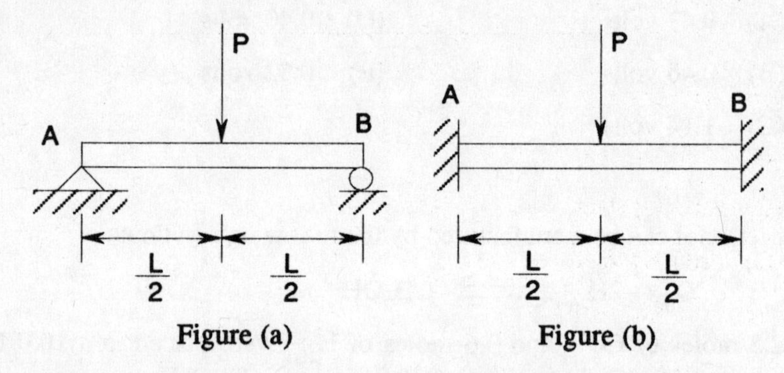

Figure (a) Figure (b)

The load carrying capacity of the beam in Figure (b) compared to the beam in Figure (a) is:

(A) 50% more
(B) 100% more
(C) 200% more
(D) 100% less
(E) Same

107. The gap between two 30 ft. railroad tracks laid at 60°F is 0.1 in. On a hot day after the train has passed, the temperature of the rail rises to 140°F. The coefficient of expansion is $\alpha = 6.5 \times 10^{-6}$ in/in/°F and Young's Modulus is $E = 30 \times 10^6$ psi. If the rails do not bend or buckle, the approximate stress is:

(A) 15,600 psi tension
(B) 7,266 psi compression
(C) 8,337 psi compression
(D) 7,266 psi tension
(E) 8,337 psi tension

108. For the same problem as stated in Problem 107, the actual strain in the 30 ft. rail is approximately

(A) 0.000277 in/in
(B) 0.000242 in/in
(C) 0.000520 in/in
(D) 0.003324 in/in
(E) 0.002900 in/in

109. For a 10 foot long, simply supported beam, the shear diagram is shown. The maximum bending moment is approximately:

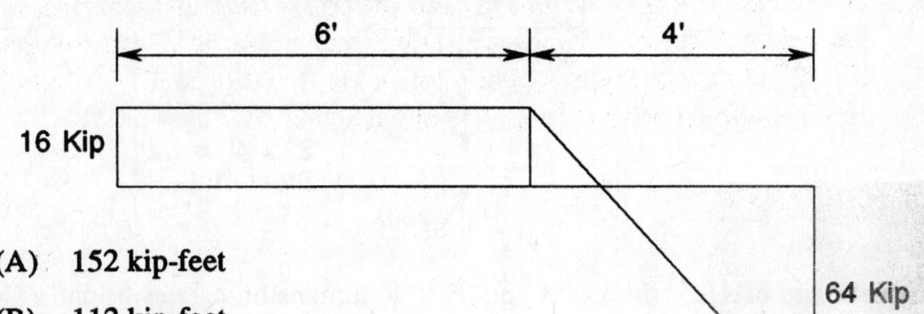

(A) 152 kip-feet
(B) 112 kip-feet
(C) 102 kip-feet
(D) 205 kip-feet
(E) 216 kip-feet

110. A 6" diameter 3" long steel rod is subjected to an axial tensile force of 50,000 lbs. The modulus of elasticity of the material is $E = 30 \times 10^6$ psi and Poisson ratio is $\nu = 0.3$. The actual change in volume of the steel rod is

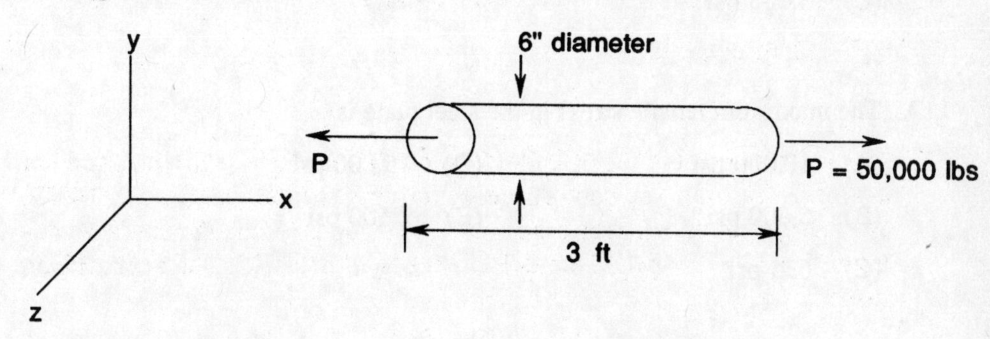

(A) no change
(B) – .0959 cu in
(C) + .0959 cu in
(D) + .024 cu in
(E) – 0.24 cu in

QUESTIONS 111–112 refer to the following diagram.

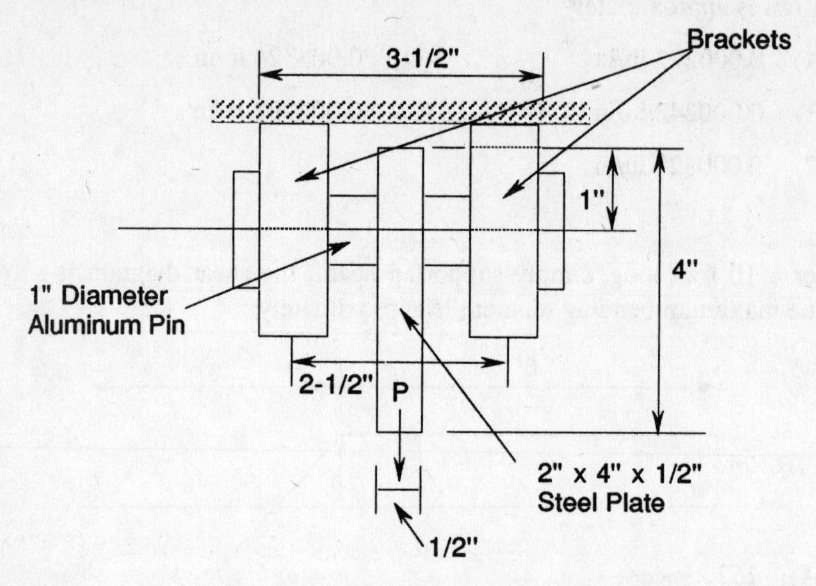

A steel plate of 1/2" thickness and 2" × 4" dimension carries a load P = 5,000 lbs. The plate is attached to a pair of brackets by a round aluminum pin of 1" diameter. The yield stress for tension and shear for steel are 80,000 psi and 50,000 psi respectively and for aluminum they are 40,000 psi and 20,000 psi respectively.

111. The average shear stress in the aluminum pin is approximately

(A) 2,500 psi
(B) 6,370 psi
(C) 3,185 psi
(D) 5,000 psi
(E) 10,000 psi

112. The maximum tensile stress in the steel plate is

(A) 10,000 psi
(B) 5,000 psi
(C) 625 psi
(D) 15,000 psi
(E) 2,500 psi

113. The beam shown below is made up of two different materials with different cross sections at AB and BC. The Young's Modulus for the two sections AB and BC are $E_{AB} = 20 \times 10^6$ psi and $E_{BC} = 30 \times 10^6$ psi, respectively. The cross-sectional moment of inertia for these sections are, $I_{AB} = 0.67$ in^4 and

$I_{BC} = 0.083\ in^4$. Beam deflection at C is approximately given by:

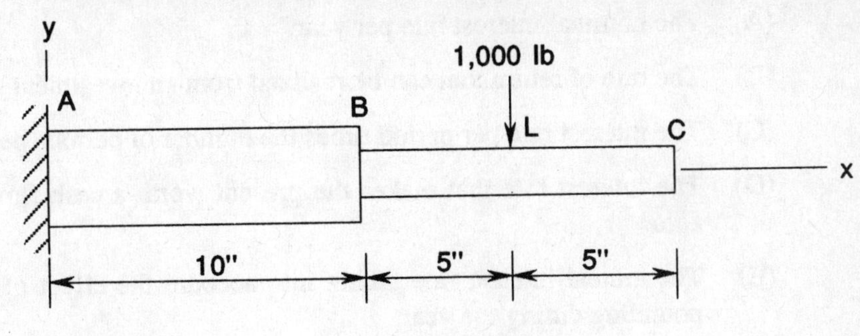

(A) 0.1416 in

(B) 0.1599 in

(C) 0.7129 in

(D) 0.25 in

(E) 0.001 in

114. The cross section of a beam subjected to a moment of $M = 10{,}000$ lb-in is shown. The stress at point A is given by:

(A) 196 psi

(B) 306 psi

(C) 208 psi

(D) 281 psi

(E) 266 psi

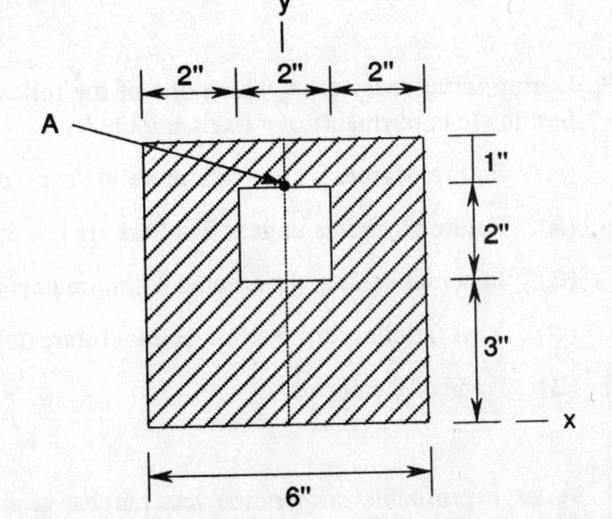

115. A cylindrical steel (yield stress = 30,000 psi), thin walled pressure vessel of 3 ft. internal diameter holds 1000 psi pressure. If the longitudinal stress must be less than 20% of the yield stress, the necessary minimum wall thickness t is given by:

(A) 1.50 in

(B) 3 in

(C) 0.75 in

(D) 3.75 in

(E) 7.0 in

116. Effective interest rate per year is defined as:

(A) The nominal interest rate per year.

(B) The rate of return that can be realized from an investment.

(C) The interest rate per period times the number of periods per year.

(D) The interest rate that makes the present worth a cash flow equal to zero.

(E) The annual interest rate taking into account the effect of any compounding during the year.

117. Given a nominal interest rate per year of 10% and two compounding subperiods per year, what is the effective rate per compounding subperiod?

(A) 0.1%

(B) 20%

(C) 2.5%

(D) 5%

(E) None of the above

118. During inflationary time, which one of the following statements describes best the loan payments of a fixed amount?

(A) Future payments are worth more in year 0 dollars.

(B) Future payments in year 0 dollars are not constant in amount.

(C) In actual dollars, the amount of future payments decreases.

(D) Year 0 dollars are worth less than future dollars.

(E) None of the above.

119. A large profitable corporation has purchased a jet plane for use by its executives. The cost of the plane is $76 million. It has a useful life of 5 years. The estimated resale value at the end of five years is six million dollars. Using the sum-of-years-digit method of depreciation, what is the book value of the jet plane at the end of 3 years?

(A) $14 million

(B) $15.2 million

(C) $20 million

(D) $34 million

(E) $48 million

120. You purchased a lot for building your house four years ago for $20,000. Each year you paid $220 in property taxes. Each year you spent $80 in maintaining the lot, because the town requires that lot be mowed periodically to keep grass and weeds less than 10" in height. Now you are selling the lot and will get $25,000 after deducting selling expense. What is the rate of return you receive on your investment? Assume that the interest is compounded annually.

(A) 10%
(B) 8%
(C) 6%
(D) 4%
(E) 2%

121. You have decided to take a car loan from a bank that charges a nominal rate of 9% compounded monthly. The car costs $16,000 after down payment, including all options, taxes and other expenses. You agree to pay the loan of $16,000 in a series of 60 equal monthly payments. What is the monthly payment?

(A) $212.80
(B) $266.67
(C) $332.80
(D) $1440.00
(E) $1777.78

122. A company is investigating options to increase the warehousing space. The additional space will be needed for the next 10 years. One of the options is described below. "Build a small warehouse now and expand it in three years. It would cost $110,000 to build the small warehouse now. The addition would cost $50,000 at the end of three years. Annual expenses for a warehouse would be $1,000 for the first three years and $2,000 for each year thereafter. The warehouse can be sold for $50,000 at the end of the tenth year." What is the Equivalent Uniform Annual Cost for this option? Use a 12% MARR and annual compounding. Round off to the nearest hundred dollars.

(A) $18,000
(B) $20,500
(C) $24,500
(D) $26,200
(E) $28,000

123. An art museum requires $125,000 every 15 years for major renovation. One patron of the museum has volunteered to set up a trust fund for these expenses. The first renovation has to be 5 years hence. How much money

should the person set aside in the trust fund now? Assume that the funds can be invested to yield a nominal annual interest rate of 8%.

(A) $57,506.50
(B) $124,209.50
(C) $125,608.75
(D) $182,090.60
(E) $256,808.50

124. A small town has been without a physician for a few years. The town has found it difficult to recruit a physician for the local hospital. Therefore the town has decided to set up a fund that will provide a scholarship of $200,000 every ten years to a local young person to become a physician. Each recipient of the scholarship will serve the town until the next recipient is ready. The town can invest the perpetual fund and obtain a constant annual interest of 8%. How much principal would be needed to pay forever the above scholarship every 10 years?

(A) $13,800
(B) $17,480
(C) $172,500
(D) $200,000
(E) $218,500

125. Two different fork lift trucks are being considered for a warehouse. One of the two should be bought to realize savings in material handling operations. The cost and benefit estimates are as follows:

	Truck A	Truck B
Initial Cost	$15,900	$32,100
Uniform Annual Savings	$5,400	$6,300
Useful Lives in Years	5	6

Use a nominal annual interest rate of 10%. Neglecting taxes, what is the annual net benefit advantage of truck A over truck B?

(A) $135.00
(B) $900.00
(C) $1,070.00
(D) $1,206.00
(E) $2,276.00

126. I am building a new house. I have to choose one of the two following alternatives for the roof material. I want a life of 40 years for the roof. *Alternative A:* Use standard asphalt shingles costing $20 per sq. ft. with a

life of 20 years. It is expected to cost $100 per sq. ft. 20 years hence to replace the old one and have additional life of 20 years for the roof. *Alternative B:* Use built-up asphalt shingles costing $40 per sq. ft. with a life of 40 years. I expect to own the house for 40 years and would consider 12% to be a suitable nominal annual interest rate. What is the cost advantage per sq. ft. of alternative A over alternative B? Round off to the nearest dollar. Use present worth analysis.

(A) $5

(B) $10

(C) $15

(D) $20

(E) $25

127. The Miller indices of the plane shown in the accompanying figure are

(A) 112

(B) 100

(C) 221

(D) 111

(E) 010

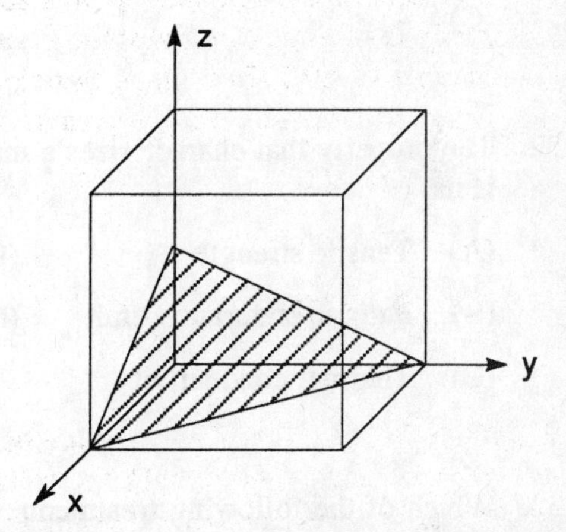

128. A polyester with an elastic modulus of 400,000 psi is reinforced by the addition of 20 percent (by volume) of glass fibers with a modulus of 10 million psi. The elastic modulus of the composite when it is loaded parallel to the glass fibers is

(A) 10 million psi

(B) 400,000 psi

(C) 5.26 million psi

(D) 2.32 million psi

(E) 8.08 million psi

129. Which of the following material properties ia adversely affected by grain refinement?

(A) Tensile strength

(B) Creep resistance

(C) Ductility

(D) Elastic modulus

(E) Yield strength

130. Which of the following classes of materials exhibits a decreasing electrical conductivity with increasing temperature?

(A) Pure ionic materials
(B) Intrinsic semiconductors
(C) Metals
(D) p-type semiconductors
(E) n-type semiconductors

131. How many atoms are in a body centered cubic unit cell?

(A) 2
(B) 1
(C) 4
(D) 8
(E) 9

132. The property that characterizes a material's ability to be drawn into a wire is its

(A) Tensile strength
(B) Fatigue endurance limit
(C) Thermal conductivity
(D) Impact strength
(E) Ductility

133. Which of the following treatments will result in an increase in the fatigue strength of steel?

(A) Annealing
(B) Cold Working
(C) Shot peening
(D) Surface roughening
(E) Surface decarburization

134. Isotopes of an element are atomic species with

(A) the same number of protons and electrons, but differing numbers of neutrons.

(B) the same number of protons, but differing numbers of neutrons and electrons.

(C) the same number of neutrons, but differing numbers of protons and electrons.

(D) the same number valence electrons, but differing numbers of protons, neutrons, and non-valence electrons.

(E) the same numbers of protons, neutrons, and electrons.

135. The separation of isotopes of an element always requires a physical, rather than chemical, separation because

(A) the electron configuration and, hence, chemical reactivity are the same.

(B) the atomic mass, a physical characteristic, constitutes the difference between the isotopes.

(C) differences in isotopes show up as differences in physical properties.

(D) all of the above

(E) none of the above

136. In p-type semiconductors, conductivity is provided by

(A) the motion of electrons.

(B) the thermal excitation of electrons from the valance to conduction band.

(C) the motion of ions.

(D) the motion of holes.

(E) all of the above.

137. The electronic configuration, $1s^2 2s^2 2p^6 3s^2 3p^5$, is that of

(A) an alkali metal

(B) an inert gas

(C) a halogen

(D) a transition metal

(E) an alkaline earth metal

138. A photon is

(A) a neutral proton

(B) a neutral electron

(C) a particle of electro-magnetic radiation

(D) a charged neutron

(E) electromagnetic radiation

139. The measured β radiation coming from a sample of a radioisotope whose decay product is stable decreases by a factor of 10 in 7.5 hours (450 minutes). What is the half life of the isotope to the closest minute?

(A) 225 minutes
(B) 45 minutes
(C) 60 minutes
(D) 337 minutes
(E) 135 minutes

140. An Angstrom unit, commonly used for atomic and sub-atomic distances, is:

(A) 10^{-9} m
(B) 10^{10} m
(C) the distance light travels in 10^{-10} sec.
(D) 10^{-6} m
(E) 10^{-10} m

TEST 2 (AM)

ANSWER KEY

1.	(B)	26.	(D)	51.	(E)	76.	(E)
2.	(E)	27.	(C)	52.	(D)	77.	(B)
3.	(A)	28.	(A)	53.	(C)	78.	(D)
4.	(E)	29.	(C)	54.	(B)	79.	(C)
5.	(A)	30.	(D)	55.	(C)	80.	(E)
6.	(C)	31.	(D)	56.	(D)	81.	(A)
7.	(C)	32.	(C)	57.	(D)	82.	(C)
8.	(D)	33.	(D)	58.	(C)	83.	(D)
9.	(D)	34.	(D)	59.	(A)	84.	(C)
10.	(E)	35.	(A)	60.	(D)	85.	(B)
11.	(C)	36.	(A)	61.	(C)	86.	(D)
12.	(E)	37.	(D)	62.	(E)	87.	(D)
13.	(D)	38.	(A)	63.	(A)	88.	(E)
14.	(A)	39.	(E)	64.	(D)	89.	(D)
15.	(C)	40.	(D)	65.	(B)	90.	(B)
16.	(B)	41.	(C)	66.	(B)	91.	(D)
17.	(E)	42.	(C)	67.	(E)	92.	(C)
18.	(D)	43.	(D)	68.	(A)	93.	(E)
19.	(C)	44.	(B)	69.	(C)	94.	(A)
20.	(D)	45.	(D)	70.	(C)	95.	(C)
21.	(B)	46.	(A)	71.	(C)	96.	(D)
22.	(B)	47.	(D)	72.	(D)	97.	(B)
23.	(A)	48.	(C)	73.	(E)	98.	(C)
24.	(A)	49.	(C)	74.	(E)	99.	(D)
25.	(C)	50.	(C)	75.	(A)	100.	(E)

101. (B)	111. (C)	121. (C)	131. (A)
102. (B)	112. (A)	122. (C)	132. (E)
103. (A)	113. (B)	123. (B)	133. (C)
104. (E)	114. (C)	124. (C)	134. (A)
105. (C)	115. (A)	125. (E)	135. (D)
106. (B)	116. (E)	126. (B)	136. (D)
107. (B)	117. (D)	127. (A)	137. (C)
108. (A)	118. (B)	128. (D)	138. (C)
109. (C)	119. (C)	129. (B)	139. (E)
110. (D)	120. (D)	130. (C)	140. (E)

DETAILED EXPLANATIONS OF ANSWERS
TEST 2

MORNING (AM) SECTION

1. **(B)**

Find the points of intersection between the parabola and the horizontal line. This is accomplished by equating the two equations

$$4 - x^2 = 2$$

Solving for the unknown x-values gives the upper and lower limits of integration. Thus

$$x = \sqrt{2} \quad \text{and} \quad x = +\sqrt{2}$$

The shaded area is calculated by integrating the vertical strip from the lower to the upper limit. The width of the strip is dx and its height is $h = 4 - x^2 - 2 = 2 - x^2$

$$\text{Area} = \int_{-\sqrt{2}}^{\sqrt{2}} h\,dx$$

$$= 2\int_{0}^{\sqrt{2}} h\,dx$$

$$= 2\int_{0}^{\sqrt{2}} (2 - x^2)\,dx$$

$$= 2\left[2x - \frac{x^3}{3}\Big|_{0}^{\sqrt{2}}\right]$$

$$= \frac{8\sqrt{2}}{3}$$

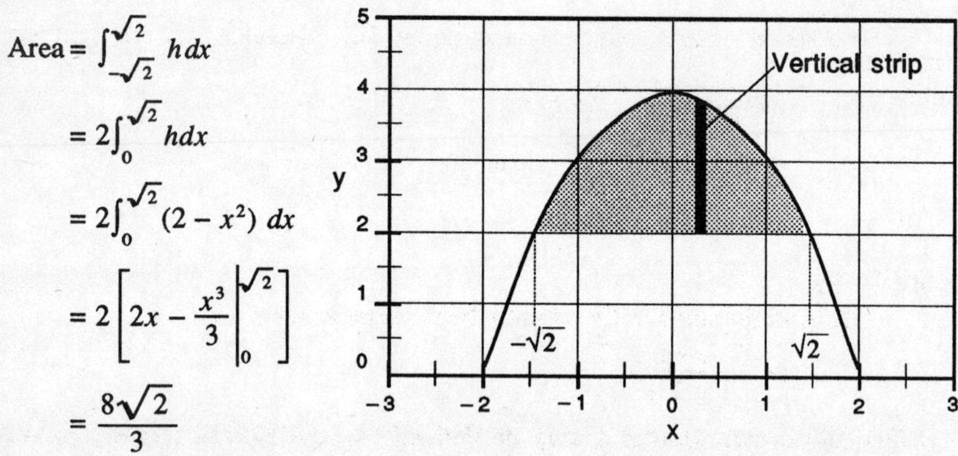

Note that because of symmetry of the shaded area, it is possible to evaluate 1/2 the area, then multiply the results by 2.

2. **(E)**

The volume is calculated by rotating the vertical strip with a radius equal to r as shown in the figure below about the x-axis. Note that the disc has a width of dx. The volume of a disk is given by

$$dv = \pi r^2\, dx = \pi\, y^2 dx = \pi[x^2]^2\, dx = \pi\, x^4\, dx$$

The total volume is then given by integrating from 0 to 1.0 as follows:

$$\text{Volume} = v = \int_0^1 \pi\, x^4\, dx = \frac{\pi x^5}{5}\Big|_0^1 = \frac{\pi}{5}$$

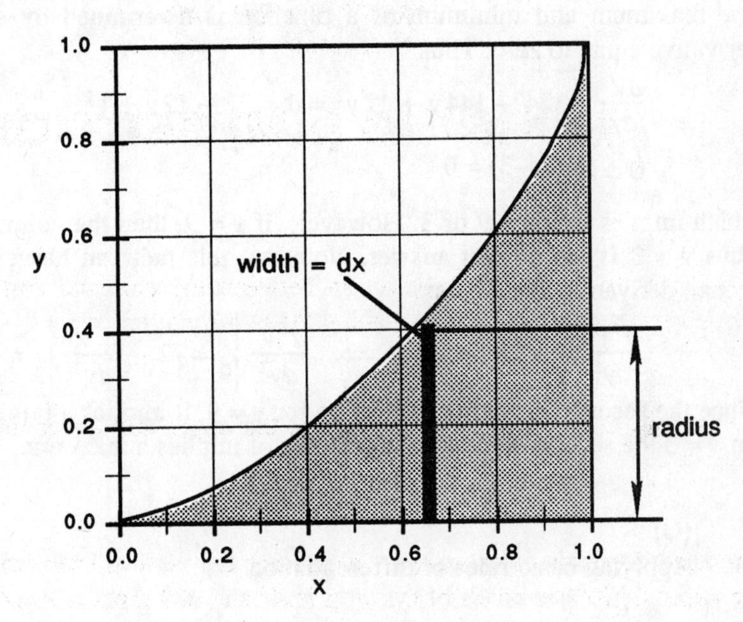

3. **(A)**
This integral is evaluated by making the following algebraic substitutions:

$$\text{if } u = e^{2x^2} \Rightarrow du = 4xe^{2x^2}\, dx$$

$$\int_0^x 3xe^{2x^2}\, dx = \int_0^x \frac{3}{4}\, du = \frac{3}{4}\, u\Big|_0^x = \left(\frac{3}{4}\right)e^{2x^2}\Big|_0^x = \frac{3}{4}\left[e^{2x^2} - 1\right]$$

4. **(E)**
Substituting the limit $x = 0$ into the expression gives

$$\lim_{x\to 0}\left[\frac{\pi \sin x}{\log(1+x)}\right] = \frac{0}{0}$$

When 0/0 occurs, the limit may be defined using L'Hopital's rule. This rule stipulates that the limit may be found by differentiating the numerator and the denominator which gives:

$$\frac{d}{dx}(\pi \sin x) = \pi \cos x \qquad \frac{d}{dx}(\log(1+x)) = \frac{1}{1+x}$$

$$\lim_{x\to 0}\left[\frac{\pi \sin x}{\log(1+x)}\right] = \lim_{x\to 0}\left[\frac{\pi \cos x}{\frac{1}{(1+x)}}\right] = \left[\frac{\pi \cos 0}{\frac{1}{(1+0)}}\right] = \frac{\pi}{1} = \pi$$

Note that ($\pi \sin x$) and $\log(1 + x)$ are differentiated separately.

5. **(A)**

Clearly, the volume of the box is computed as the product of the area (18 – 2y) (18 – 2y) and the height y. That is

$$V = (18 - 2y)^2\, y = 324y - 72y^2 + 4y^3$$

The maximum and minimum of a function is determined by setting the first derivative equal to zero. Thus

$$\frac{dV}{dy} = 324 - 144y + 12y^2 = 0 = 27 - 12y + y^2$$

$$(y - 9)\,(y - 3) = 0$$

which implies that $y = 9$ or 3. However, if $y = 9$, then the volume will be zero. Thus $y = 3$ is the correct answer. Note that this fact can be verified using the second derivative as follows:

$$\frac{d^2V}{dy^2} = -144 + 24y \;\Rightarrow\; \left.\frac{d^2V}{dy^2}\right|_9 = 72 \quad \left.\frac{d^2V}{dy^2}\right|_3 = -72$$

Since the second derivative is positive for $y = 9$, it implies a minimum and since for $y = 3$ the second derivative is negative, it implies a maximum.

6. **(C)**

Applying basic rules of differentiation

$$\frac{dy}{dx} = 1.5x^{0.5} - \pi \sin \pi x$$

$$\frac{dy}{dx} = 1.5(-1)^{0.5} - \pi \cos(-\pi)$$

$$\frac{dy}{dx} = 1.5j + \pi$$

Note that the square root of (– 1) is imaginary number called *j* or *i*.

7. **(C)**

The required derivatives are evaluated using the following notation:

$$\frac{dy}{dx} = y' = \frac{dy/dz}{dx/dz}$$

$$\frac{d^2y}{dx^2} = y'' = \frac{dy'}{dx} = \frac{dy'/dz}{dx/dz}$$

$$y' = \frac{dy/dz}{dx/dz} = \frac{2z - 9z^2}{-2} = -z + \frac{9}{2}z^2$$

$$y'' = \frac{dy'/dz}{dx/dz} = \frac{\frac{d}{dz}\left(-z + \frac{9}{2}z^2\right)}{-2} = \frac{-1 + 9z}{-2} = 0.5 - 4.5z$$

8. **(D)**

The equation of a straight line is given as $y = a + mx$, where a is the intercept and m is the slope. Therefore, substituting the two points through which the straight line is supposed to pass into the equation gives two equations in two unknowns.

$$3 = a + 2m$$
$$2 = a + 3m$$

Solving for the unknowns using Cramer's rule gives

$$a = \frac{\begin{vmatrix} 3 & 2 \\ 2 & 3 \end{vmatrix}}{\begin{vmatrix} 1 & 2 \\ 1 & 3 \end{vmatrix}} = \frac{5}{1} = 5 \quad \text{and} \quad m = \frac{\begin{vmatrix} 1 & 3 \\ 1 & 2 \end{vmatrix}}{1} = \frac{-1}{1} = -1$$

The equation of the line is then given by $y = 5 - x$ or $y + x = 5$.

9. **(D)**

The set of linear algebric equations can be solved easily using the Gauss elimination method. Using R to refer to a row, we have

$$\begin{bmatrix} 1 & -1 & 0 & : & -1 \\ 1 & 1 & -2 & : & -3 \\ 0 & 1 & 1 & : & 5 \end{bmatrix} \xrightarrow{-R_1+R_2} \begin{bmatrix} 1 & -1 & 0 & : & -1 \\ 0 & 2 & -2 & : & -2 \\ 0 & 1 & 1 & : & 5 \end{bmatrix} \xrightarrow{R_2/2}$$

$$\begin{bmatrix} 1 & -1 & 0 & : & -1 \\ 0 & 1 & -1 & : & -1 \\ 0 & 1 & 1 & : & 5 \end{bmatrix} \xrightarrow{-R_2+R_3} \begin{bmatrix} 1 & -1 & 0 & : & -1 \\ 0 & 1 & -1 & : & -1 \\ 0 & 0 & 2 & : & 6 \end{bmatrix}$$

Using back substitutions, gives

$$\left.\begin{aligned} x - y &= -1 \\ y - z &= -1 \\ 2z &= 6 \end{aligned}\right\} \Rightarrow \quad \begin{aligned} x &= -1 + 2 = 1 \\ y &= -1 + 3 = 2 \\ z &= 6/2 = 3 \end{aligned}$$

10. **(E)**

The eigenvalues are a property of any square matrix. For an $n \times n$ matrix, there are n eigenvalues. It makes no difference whether the matrix is symmetrical or asymmetrical, the corresponding eigenvalues are generally unique. However, in some cases, the eigenvalues may not be unique. For the special case described by the determinant

$$|[A] - 1[I]| = 0,$$

the eigenvalues are not unique. Consequently, the correct answer is (E).

11. **(C)**

Since the matrix is equal to its own inverse, then [*A*] [*A*] = [*I*], where [*I*] is the identity matrix. Therefore, raising the matrix to the power of 10 gives the identity matrix as well. The determinant is given by

$$\text{determinant} = 2\,[A]^{10} = 2\begin{bmatrix} 1 & 0 & 0 \\ 0 & 1 & 0 \\ 0 & 0 & 1 \end{bmatrix} = \begin{vmatrix} 2 & 0 & 0 \\ 0 & 2 & 0 \\ 0 & 0 & 2 \end{vmatrix} = 2 \times 2 \times 2 = 8$$

12. **(E)**

The standard deviation is calculated in terms of the average value of the given data as follows:

$$\text{Mean} = \overline{x} = \frac{\sum_{i=1}^{n} x_i}{n} = \frac{\sum_{i=1}^{5} x_i}{5} = \frac{9+3+6+2+10}{5} = 6$$

$$\text{Standard Deviation} = \sigma = \sqrt{\frac{\sum_{i=1}^{n} (x_i - \overline{x})^2}{n-1}} = \sqrt{\frac{\sum_{i=1}^{5} (x_i - \overline{x})^2}{4}}$$

$$\sigma = \left[\frac{(9-6)^2 + (3-6)^2 + (6-6)^2 + (2-6)^2 + (10-6)^2}{4}\right]^{1/2} = \frac{\sqrt{50}}{2}$$

13. **(D)**

This is a combination problem in which $n = 6$ and $r = 4$ where

$$C(n,r) = \frac{n!}{(n-r)!\ r!} = \frac{6!}{(6-4)!\ 4!} = \frac{6 \times 5 \times 4!}{2!\ 4!} = \frac{30}{2 \times 1} = 15$$

Therefore, there are 15 different groups that can be fitted into the four-passenger car.

14. **(A)**

Although one may assume that the answer is 0.50, it is not. In an independent-trials process with two outcomes, the probability of exactly "*r*" successes in "*n*" experiments is given by the following binomial probability expression:

$$b(r;n,p) = \frac{n!}{r!\ (n-r)!} P^r (1-P)^{n-r}$$

For $P = 0.5$, $n = 10$, and $r = 5$, then

$$b(5; 10, 1/2) = \frac{10!}{5!\ (10-5)!}(1/2)^5\ (1-1/2)^{10-5}$$

$$b(5; 10, 1/2) = \frac{10 \times 9 \times 8 \times 7 \times 6 \times 5!}{(5 \times 4 \times 3 \times 2 \times 1)\ 5!}(1/2)^{10} = \frac{30240}{120}\ (0.00097656\ )$$

$$b(5; 10, 1/2) = 0.246 \approx 0.25$$

15. **(C)**

This differential equation is integrated by separating the variables to give the following expression:

$$\frac{dy}{y} = -9\,dt$$

Integrating

$$\ln y = -9t + c_0 \Rightarrow y = e^{-9t} + c$$

The constant of integration "c" is evaluated by substituting the intial condition $y(0) = 1$ into the solution, which gives

$$y(0) = e^0 + c \;\Rightarrow\; 1 = 1 + c \;\Rightarrow\; c = 0$$

Therefore, the solution is $y = e^{-9t}$.

16. **(B)**

This differential equation is integrated by assuming a solution of the form:

$$y = Ge^{st} \quad\Rightarrow\quad \dot{y} = s\,Ge^{st} \quad\Rightarrow\quad \ddot{y} = s^2\,Ge^{st}$$

Substituting into the differential equation and simplifying gives

$$s^2 - 7s = 0 \quad\Rightarrow\quad s = 0 \text{ and } s = 7$$

The solution is of the form

$$y = C_1 + C_2\,e^{7t}$$

The constant C_1 and C_2 are evaluated using the given initial conditions. That is

$$\left.\begin{aligned} y(0) &= 1 = C_1 + C_2 \\ \dot{y}(0) &= 1 = 7C_2 \end{aligned}\right\} \Rightarrow C_1 = 6/7 \text{ and } C_2 = 1/7$$

The solution to the differential equation is then given by

$$y = \frac{6}{7} + \frac{e^{7t}}{7}$$

17. **(E)**

This differential equation is solved by separating the variables then integrating as follows:

$$e^y\,dy = 5x\,dx \quad\Rightarrow\quad e^y = 2.5x^2 + c'$$

Or more simply

$$y = e^{2.5x^2} + c$$

Noting that $e^{c'}$ is a constant equal to c, then introducing the initial condition at x

= 1, we have $y(1) = 0$:

$$y(1) = 0 = e^{2.5(1)} + c \quad \Rightarrow \quad c = -e^{2.5}$$

The solution is then given as

$$y = e^{2.5x} - e^{2.5} \quad \Rightarrow \quad y(\sqrt{2}) = e^{2.5(2)} - e^{2.5} = (e^5 - e^{2.5})$$

18. **(D)**

The equation of a circle with a center at $(-k, -c)$ and a radius r has the following form:

$$(x + k)^2 + (y + c)^2 = r^2$$

"Complete the squares" to convert the given equation

$$x^2 + y^2 - 8x + 2y = 9$$

into the above form.

After grouping the variables:

$$(x^2 - 8x) + (y^2 + 2y) = 9$$

By squaring the half of each coefficient in front of x and y and adding to each side of the equation

$$(x^2 - 8x + 16) + (y^2 + 2y + 1) = 9 + 16 + 1$$

Rewriting the above equation

$$(x - 4)^2 + (y + 1)^2 = 26$$

we see that

$$-k = -4 \quad -c = 1$$

Therefore, the center of the circle is at $(4, -1)$.

19. **(C)**

Algebraic equations of the form $Ax^2 + Bxy + Cy^2 + Dx + Ey + F = 0$ are classified as follows:

Ellipse: $B^2 - 4AC < 0$

Parabola: $B^2 - 4AC = 0$

Hyperbola: $B^2 - 4AC > 0$

In this problem, the term $A = 3$; $B = 6$; and $C = 2$

$$36 - 4(3)(2) = 12 > 0$$

Therefore, the equation represents a hyperbola.

20. **(D)**

Differentiate both sides of the equation with respect to *x*, then solve for *dy / dx*, which gives

$$2y\frac{dy}{dx} - 4 + 2\frac{dy}{dx} = 0 \quad \Rightarrow \quad \frac{dy}{dx} = \frac{2}{1+y}$$

The slope at (3, 3) is computed as

$$\left.\frac{dy}{dx}\right|_{(3,3)} = \frac{2}{1+3} = \frac{1}{2} = m_1$$

The normal will have a slope equal to $m_2 = -1 / m_1 = -2$. The equation of the normal is a straight line of the form $y = a + m_2 x$. The value of a is determined from the point (3, 3) as follows:

$$y = a - 2x \quad \Rightarrow \quad 3 = a - 2(3) \quad \Rightarrow \quad a = 9$$

The equation of the normal is then given as $y = 9 - 2x$.

21. **(B)**

The parallel combination of 4 Ω and $^4/_3$ Ω is 1 Ω. Therefore, the voltage source V will split equally, and $V_1 = V_2 = V / 2$.

22. **(B)**

The two circuits will be equivalent if the impedances, as seen by the AC voltage source in both circuits and at the specified frequency, are the same.

Thus, the series combination of the $R_s = 1\ \Omega$ resistor and $C_s = 100$ pF capacitor, should be the same as the parellel combination of the resistor $R = R_p$ and the capacitor $C = C_p$ at $f = 15.9$ MHz.

After setting up the equations, we solve for R_p and C_p, and find

$$R_p = \frac{R_s^2 + (\omega \cdot C_s)^2}{R_s} = 10\,\text{k}\Omega; \quad C_p = \frac{\dfrac{1}{\omega^2 \cdot C_s}}{R_s^2 + \dfrac{1}{(\omega \cdot C_s)^2}} = 100 \text{ pF}$$

with $\omega = 2 \cdot \pi \cdot f$.

23. **(A)**

In any medium of permittivity constant ε and permeability constant μ, the velocity of any electromagnetic wave in the same medium is given by

$$v = \frac{1}{\sqrt{\mu \cdot \varepsilon}} \text{ (m/sec)}$$

So, if the medium is changed, so does the velocity.

On the other hand, we have a relation that links the three specified parameters, f, v, λ, which is

$$\lambda = \frac{v}{f}$$

The wave cannot be packed on one side of medium 1, but it has to flow evenly. The only way this can happen is if the frequency stays constant and the wavelength changes in such a way that the previous equation is verified.

24. **(A)**

Using the transformation of a current source in parallel with a resistance into a voltage source in series with a resistance, as shown in the figure, we find that V_s should equal $1\ \text{A} \times 2\ \Omega = 2\ \text{V}$.

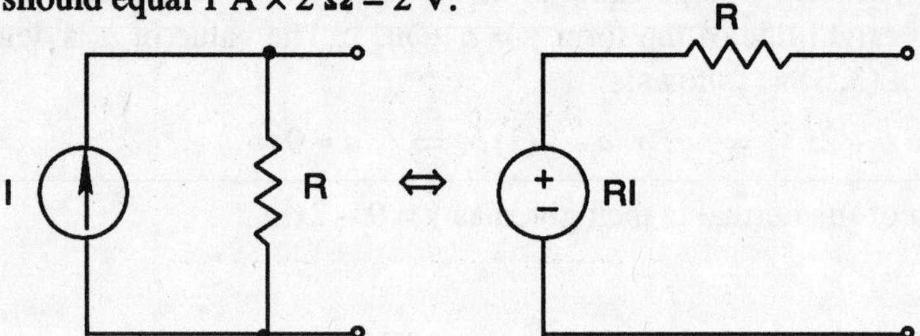

25. **(C)**

Applying Thevenin's theorem, which says that *any linear circuit seen from two points can be reduced into a voltage source in series with a resistance. The voltage source is obtained by open-circuiting the two considered points and computing the voltage between them; the resistance is found by short-circuiting voltage sources and open-circuiting current sources and then grouping all resistances between the two points into a single resistance.*

Thus, open-circuiting *A* and *B* we see a voltage of

$$V_{th} = 4\ \text{V} + 2\ \text{mA} \cdot 2\ \text{K}\Omega = 8\ \text{V}$$

and, short-circuiting the 4 V voltage source and open-circuiting the 2 mA current source, we see in series a resistance of

$$R_{th} = 2\ \text{K}\Omega + 3\ \text{K}\Omega = 5\ \text{K}\Omega$$

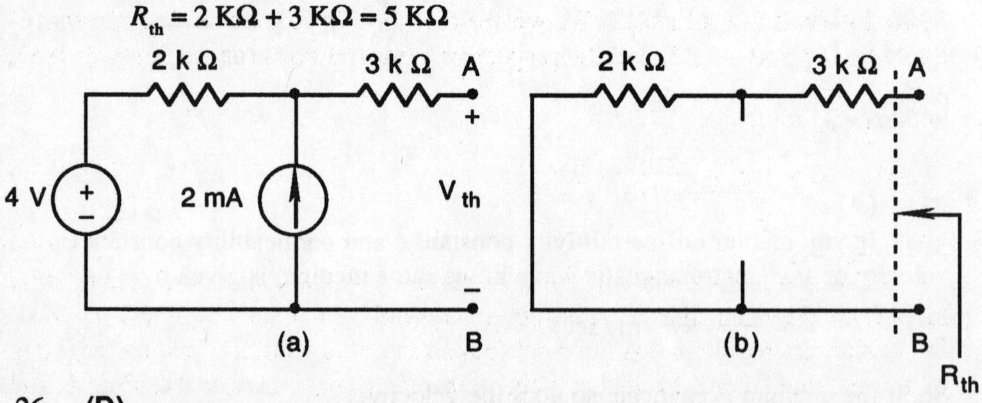

26. **(D)**

Applying Norton's theorem, which says that *any linear circuit seen from two points can be reduced into a current source in parallel with a resistance.*

The current source is obtained by short-circuiting the two considered points and computing the current that flows through them. The resistance is found in exactly the same way as in Thevenin's theorem.

Thus, short-circuiting A and B, we see a current

$$I_{nr} = \left(2\text{ mA} + \frac{4\text{ V}}{2\text{ K}\Omega}\right) \cdot \left(\frac{3\text{ K}\Omega}{3\text{K}\Omega + 2\text{ K}\Omega}\right) = 1.6\text{ mA}$$

and

$$R_{nr} = 5\text{ K}\Omega$$

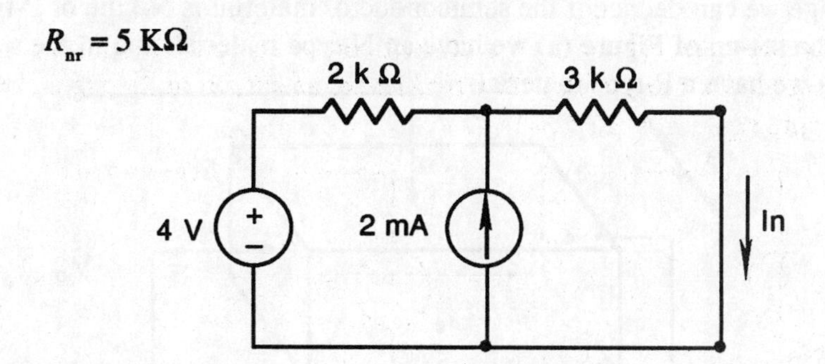

27. **(C)**

Since the two resistances of 6 Ω and 3 Ω are in parallel, the voltages V_1 and V_2 across them are equal.

28. **(A)**

From Problem 27 and the fact that $I = V / R$, we see that

$$I_1 = \frac{V_1}{6\Omega} = \frac{V_2}{2.3\Omega} = I_2/2$$

29. **(C)**

Using the equation given, we deduce that the inductor L will be given by:

$$L = \frac{v(t)}{di/dt}$$

In the intervals (–1, 0) and (2, 3), we see that di/dt is constant and equal respectively to 1/ 3 and – 1 / 3, and the respective values of $v(t)$ in those intervals are 1 and – 1.

Thus,

$$L = \frac{1}{1/3} = \frac{-1}{-1/3} = 3\text{H}.$$

30. **(D)**

The gradient of any surface F is given by $-\nabla F$. Since, the electric flux lines **E** are perpendicular to the equipotential surface V, then we have $\mathbf{E} = -\nabla V$.

Another reason is that the electric intensity at any point is just the negative of the potential gradient at that point; the direction of the electric field is the direction in which the gradient is greatest, or in which the potential changes most rapidly.

The electric field **E** can also be written in terms of the potential V as

$$\mathbf{E} = -\frac{\partial V}{\partial x}\mathbf{i} - \frac{\partial V}{\partial y}\mathbf{j} - \frac{\partial V}{\partial z}\mathbf{k}$$

31. **(D)**

Depending on the set-up of the experiment and the sign of the measured Hall voltage we can deduce if the semiconductor material is N-type or P-type.

In the set-up of Figure (a) we have an N-type material, and in the set-up of Figure (b) we have a P-type material.

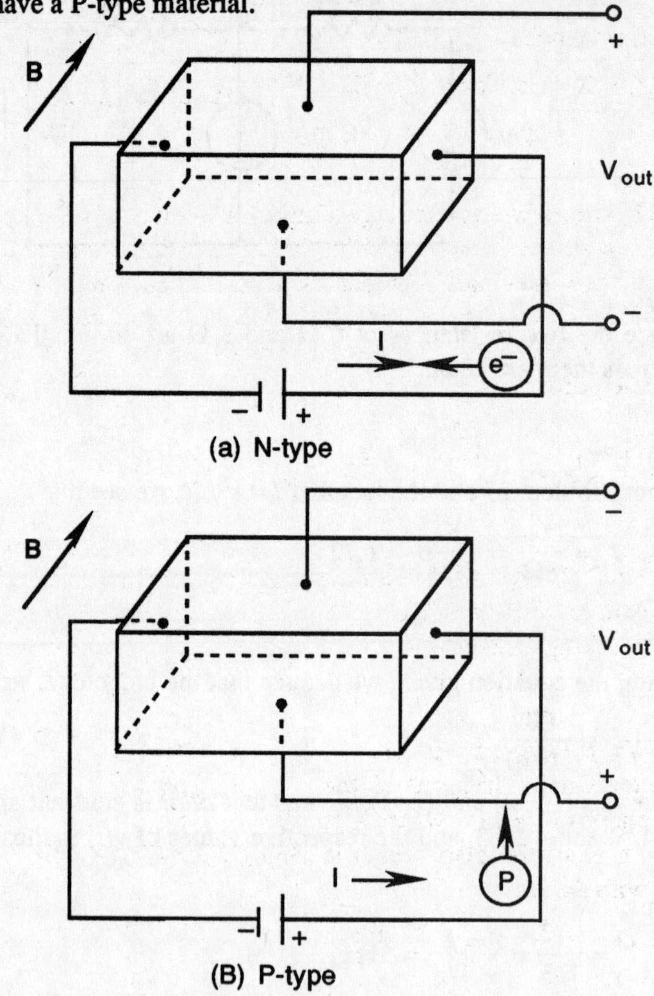

(a) N-type

(B) P-type

32. **(C)**

From electromagnetism courses, we know that the integral of $\mathbf{E} \times \mathbf{H}$ over any closed surface gives the rate of energy flow through the surface. It is seen that the vector

$$\mathbf{P} = \mathbf{E} \times \mathbf{H}$$

has the dimensions of watts per square meter. It is *Poynting's theorem* that the vector product $\mathbf{P} = \mathbf{E} \times \mathbf{H}$ at any point is a measure of the rate of energy flow per unit area at that point. The direction of flow is perpenduclar to **E** and **H** in the direction of the vector **E** × **H**.

33. **(D)**

The condition for maximum power transfer in a series circuit as shown here, is:

$$Z_L = Z_s{}^*$$

In our case, we have:

$$Z_L = Z_1 \,||\, Z_2 \text{ and } Z_s = Z.$$

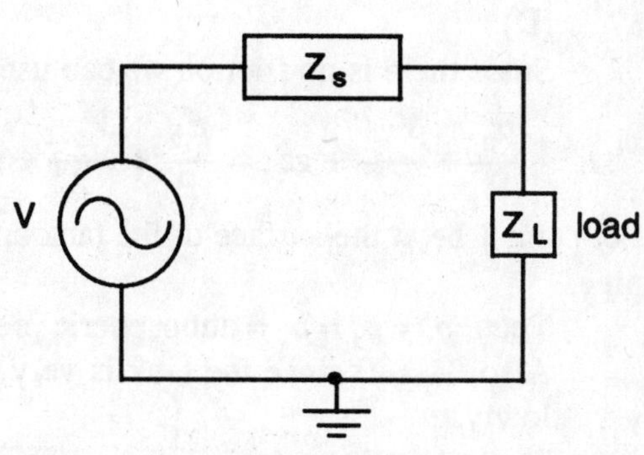

34. **(D)**

In this circuit we have a tapped inductor, which acts like a high frequency transformer, with the primary winding being the whole inductor, and the secondary winding being the part on which the load is connected.

If we had a conventional transformer with N_1 turns in the primary, and N_2 turns in the secondary, then any load Z_L connected across the secondary will be seen from the primary as:

$$(Z_L)_{\text{primary}} = \left(\frac{N_1}{N_2}\right)^2 \cdot Z_L$$

35. **(A)**

Since the oil layer is *thin*, a linear velocity distribution between the ground and the disk may be assumed:

For this simple shear flow,

$$\tau = \mu \frac{du}{dy}$$

where $\mu = Vy/t$. Hence, $\tau = \mu V/t$.

The total drag (friction) of the oil on the plate is $D = \tau \cdot A$ where A is the surface area of the plate which is in contact with the oil.

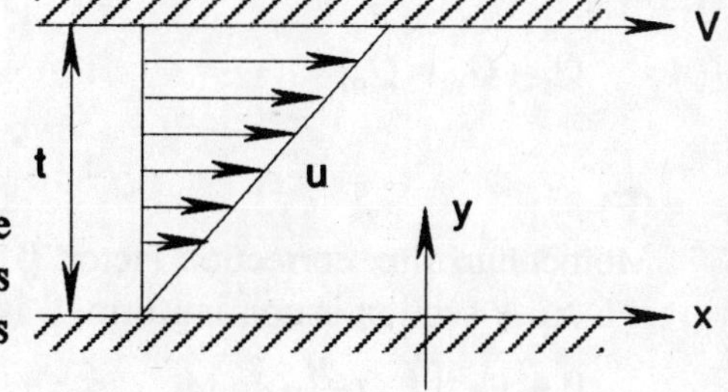

$$\therefore \quad D = \mu \frac{V}{t} A = \mu \frac{V}{t} \frac{\pi d^2}{4}$$

$$= \left(0.1 \frac{N \cdot s}{m^2}\right)\left(10 \frac{m}{s}\right)\left(\frac{1}{1\,mm}\right)\left(\frac{1000\,mm}{m}\right)\left(\frac{\pi}{4}(0.15\,m)^2\right)$$

$$D = 17.7\,N$$

36. **(A)**

The Lagrangian reference frame is one where individual fluid particles are "tagged" and followed. In the Eulerian reference frame, individual particles are not followed; rather, fluid particles flow through the control surface. The three other choices (C), (D), (E) are undefined.

37. **(D)**

Since there is no friction we can use Bernoulli's steady-state equation

$$\frac{p_1}{\rho} + \frac{V_1^2}{2} + gz_1 = \frac{p_2}{\rho} + \frac{V_2^2}{2} + gz_2 \qquad (1)$$

Let point 1 be at the surface of the tank and let point 2 be at the exit plane of the pipe.

Then, $p_1 = p_2 = p_a$ = atmospheric pressure.

Also, $V_1 \approx 0$ since the tank is very large and the surface will move down very slowly.

Thus (1) reduces to

$$V_2 = \sqrt{2g(z_1 - z_2)}$$

$$V_2 = \sqrt{(2)(9.8\ \text{m/s}^2)(3\ \text{m})}$$

$$V_2 = 7.67\ \text{m/s}$$

38. **(A)**

By definition, the average velocity, $V_{av} = Q / A$, where Q = volume flow rate and A = cross-sectional area. Here, it is known that

$$V_{av_{I}} = V_{av_{II}} = V_{av_{III}}$$

Also, the cross-sectional areas are identical for the three cases. Hence it follows that

$$Q_I = Q_{II} = Q_{III}$$

39. **(E)**

Momentum flux correction factor, β, is defined for cases where the velocity over a cross-section is non-uniform. β is defined as

$$\beta = \frac{1}{A} \iint_A \left(\frac{u}{V_{av}}\right)^2 dA$$

From this definition, $\beta = 1.0$ for uniform flow, as in case I. For cases II and III, $\beta > 1.0$, since the velocity profiles are non-uniform. Note that β is always ≥ 1.0.

40. **(D)**

Choices (A), (B), (C), and (E) are invalid for the following reasons:

(A) p is a function of x, but does not change appreciably in the y-direction.

(B) The opposite is true; for example

$$\frac{\partial u}{\partial x} << \frac{\partial u}{dy}.$$

In other words, changes of velocity in the x-direction can be completely neglected with respect to changes in the y-direction.

(C) The opposite is true; the boundary layer approximation is only valid for the very *high* Reynolds numbers.

(E) Boundary layer thickness, δ, is a function of x, (typically δ increases with x).

41. **(C)**

In the inviscid flow outside of the boundary layer, Bernoulli's equation is valid.

$$\frac{p}{\rho} + \frac{1}{2}U^2 + gz = \text{constant}$$

Neglecting gravitational effects, as p increases, U must decrease. Choices (A), (B), (D), and (E) are incorrect for the following reasons:

(A) This condition is not favorable because increasing pressure usually leads to boundary layer separation.

(B) When $dp/dx > 0$, the pressure gradient is unfavorable (i.e. adverse) regardless of the body geometry.

(D) If $dp/dx > 0$, separation is more likely to occur; however, it is possible for the boundary layer to remain attached for small values of dp/dx.

(E) Such a condition is not only possible, it is quite typical for the real portions of bodies in a flow.

42. **(C)**

By definition, "fully developed" pipe flow implies that the velocity profile shape or magnitude does not change with downstream distance along the pipe. Mass flow rate cannot change either since mass must be conserved. Darcy friction factor f, is constant along the entire length of a fully developed pipe flow. Of all the parameters listed in the choices, static pressure is the only one that changes. In other words, a pressure drop is required to "push" the fluid through the pipe as it has to overcome friction.

43. **(D)**

This problem can be represented by a large tank discharging air into the atmosphere:

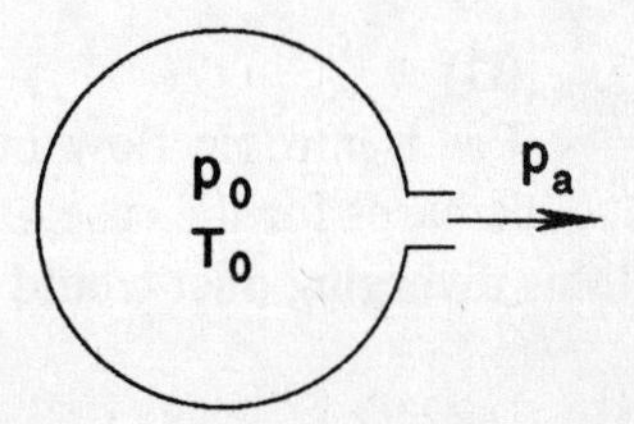

The mass flow of air becomes "choked" when $p_a / p_0 \approx 0.5283$, regardless of the hole size, for standard temperatures.

Here, then

$$p_0 = \frac{p_a}{0.5283} = \frac{(732 \text{ mm Hg})}{0.5283} = 1386 \text{ mm Hg}$$

However, the *gage* pressure is equal to $p_0 - p_a$. i.e.,

$$p_0 - p_a = 1386 - 732 = 654 \text{ mm Hg.}$$

44. **(B)**

Pipe *A* has cross-sectional area $A_A = \pi R^2$

Pipe *B* has cross-sectional area $A_B = \pi a^2 - \pi b^2 = \pi(a^2 - b^2)$

Equating, $a^2 - b^2 = R^2$

or $\quad b = (a^2 - R^2)^{1/2} = (25^2 - 15^2)^{1/2} = 20$ mm.

45. **(D)**

Hydraulic diameter is defined as

$$D_h = \frac{4A}{P}$$

where *A* is the cross sectional area, and *P* is the wetted perimeter. Here

$$A = \pi(a^2 - b^2) \text{ and } P = 2\pi a + 2\pi b = 2\pi(a + b)$$

Thus

$$D_h = \frac{4\pi(a^2 - b^2)}{2\pi(a + b)} = \frac{2(a - b)(a + b)}{(a + b)} = 2(a - b)$$

Substituting $a = 25$ mm and $b = 15$ mm gives $D_h = 20$ mm.

46. **(A)**

Head loss h_f, by definition of friction factor, is

$$h_f = F\left(\frac{L}{D}\right)\left(\frac{V_{av}^2}{2g}\right)$$

Substituting from above, and with $g = 9.8 \text{ m/s}^2$,

$$h_f = (0.025)\left(\frac{100 \text{ m}}{0.03 \text{ m}}\right)\left(\frac{(1.33 \text{ m/s})^2}{2(9.8 \text{ m/s}^2)}\right) = 7.52 \text{ m}$$

47. **(D)**

For isentropic flow in a diverging duct, as area increases, Mach number also increases for the case of supersonic flow. An observer moving with the flow in this diverging duct would see a shift from point 4 to point 5 on the curve.

48. **(C)**

For the case of *subsonic* flow in a diverging duct, Mach number decreases as area increases. In this case the observer would see the flow shift from point 2 to point 1.

49. **(C)**

The quality of a saturated liquid (state 1) is 0.0. The quality of a saturated vapor (state 2) is 1.0. As the process proceeds from state 1 to state 2 the quality will increase. Under the vapor dome, temperature and pressure are not independent and both remain constant unless acted upon by some outside force.

50. **(C)**

In the reversible process there is no entropy generation. Since entropy is a function of absolute temperature (always positive) and the heat transfer, then, if the heat transfer is negative (rejected), the entropy change will be negative. This does not violate the Second Law since the entropy of the universe must be considered to complete the Second Law analysis.

51. **(E)**

On the saturated vapor curve of the Psychrometric chart, the dew point temperature and the dry-bulb temperature are identical. It is at this point that the relative humidity is 100%. The pressure of the vapor mixture, p_v, and the saturation pressure, p_{sat} or p_g, are equal. The ratio of these pressures produces the relative humidity.

52. **(D)**

A stoichiometric relationship would require 100% theoretical air, or exactly the amount needed for complete combustion with no excess. 50% theoretical air would cause incomplete combustion resulting in unburned hydrogen and/or carbon monoxide. 150% theoretical air indicates that there is 50% more oxygen (along with the appropriate amounts of nitrogen) present. It is readily apparent that $2O_2$ are required for complete combustion and that $3O_2$ are present. Thus, there is 50% excess or 150% theoretical air.

53. **(C)**

When temperatures are multiplied, an absolute scale must be utilized. Two times 273 K equals 546 K or 273°C. The other answers were obtained by multiplying temperatures in either °F or °C without converting first to an absolute scale.

54. **(B)**

The Diesel cycle is used to model the ideal compression ignition cycle.

The others are: Otto-spark ignition cycle, Rankine-steam power cycle, Brayton-gas turbine cycle, and Ericsson (like the Carnot cycle with the two isentropic processes replaced by two constant presure regeneration processes).

55. **(C)**

According to the First Law for a closed system, in the absence of work, the heat transfer will be equal to the change in internal energy. Since internal energy is a function of temperature only for an ideal gas, the heat transfer will be a function of temperature only.

56. **(D)**

In exothermic reactions some amount of energy is liberated in the form of heat. This is known as the heat of reaction and is normally associated with combustion processes. The heat of formation is related to the energy absorbed or released as the compound is formed.

57. **(D)**

Flow work is associated with open systems, flow processes, or control volume problems. Flow work is included via pressure and volume terms where their product is added to the internal energy to obtain a new property, enthalpy. In equation form:

$$h = u + pv.$$

Flow work cannot occur in closed systems since no mass enters or leaves the boundaries of the system.

58. **(C)**

The piston-cylinder device is a closed system since no mass flow crosses the boundary. All the other devices are flow devices and are analyzed using a control volume across which mass can flow.

59. **(A)**

To increase the internal energy of 2 kg of air by 2 kJ/kg, 4 kJ of energy are required. Since the work produced is 8 kJ, the total amount of heat required is the sum, or 12 kJ. In equation form from the first law for a closed system

$$Q_{12} = m(u_2 - u_1) + W_{12}, \text{ where } (u_2 - u_1) = 2 \text{ kJ/kg}$$

$$Q_{12} = 2 \text{ kg}(2 \text{ kJ/kg}) + 8 \text{ kJ} = 12 \text{ kJ}$$

60. **(D)**

Water enters the pump at a state point with an associated enthalpy. Work, an energy transfer, is performed on the water to increase the pressure and the temperature. The result is an increase in the enthalpy of the water leaving the pump. A simple open system First Law analysis will verify this conclusion. It must be remembered that the work is added, and that by convention this work has a negative value.

61. **(C)**

This is simply Joule's experiment with the ice bath replaced by the insulation. Since the medium is an ideal gas, if the pressure is halved and the volume doubled, the temperature will remain constant. In equation form, the ideal gas law is:

$$pv = RT.$$

It can be readily seen that the left side remains constant, and since "R" is the gas constant, that the temperature must remain constant. The process is irreversible since work from an external source would be required to obtain the original state. From his experiment Joule concluded that internal energy was a function of temperature only.

62. **(E)**

The pressure equation states that the absolute pressure is the sum of the atmospheric or reference pressure and the gage pressure. In this case, the gage pressure is – 30 kPa, since it is a vacuum pressure. Thus, the absolute pressure is less than the reference pressure.

63. **(A)**

The best way to solve projectile problems is to consider the horizontal and vertical motions separately. The initial velocity given may be decomposed as such:

$$v_{0x} = v_0 \cos \theta$$

$$v_{0x} = 150 \text{ m/s} (\cos 30°) = +129.9 \text{ m/s}$$

$$v_{0y} = v_0 \sin \theta$$

$$v_{0y} = 150 \text{ m/sec} (\sin 30°) = +75 \text{ m/s}.$$

The vertical motion is uniformly accelerated motion. The horizontal motion is uniform motion.

The projectile reaches the maximum height at the momemt when $v_y = 0$. Applying the equation

$$v_y = v_{0y} + at$$

yields $0 = +75 \text{ m/s} - 9.8 \text{ m/s}^2\, t$

or $$t = \frac{-7.5}{-9.8}\, s = 7.7s.$$

Now, substituting this value into the vertical motion equation to solve for y_{max} yields

$$y = v_{0y} + {}^1/_2\, a_y t^2$$

$$y_{max} = +75 \text{ m/s } (7.7 \text{ s}) - 4.9 \text{ m/s}^2 (7.7 \text{ s})^2 = 287 \text{ m}.$$

However, this describes the distance above the roof which the stone reaches rather than its elevation from the ground. The elevation above the ground is found by adding to y_{max} the height of the building. Hence, the greatest elevation

$$= 200 \text{ m} + 287 \text{ m} = 487 \text{ m}.$$

64. **(D)**

We wish to relate the velocity of the car to ϕ, the banking angle. Note that the car is undergoing circular motion, hence its acceleration in the x-direction is

$$a = \frac{v^2}{R},$$

where R is its distance from the center of the circle (see figure). Applying Netwon's Second Law, $F = ma$, to the x component of motion,

$$ma = N \sin \phi$$

But $a = v^2 / R$ and

$$\frac{mv^2}{R} = N \sin \phi \qquad (1)$$

The acceleration of the car in the y-direction is zero, since it remains on the road. Applying the Second Law to this component of motion,

$$N \cos \phi = mg \qquad (2)$$

Dividing (1) by (2),

$$\frac{\frac{mv^2}{R}}{mg} = \frac{N \sin\phi}{N \cos\phi} = \tan\phi$$

Hence $\quad v = \sqrt{Rg \tan\phi}$

Now, note that the width of the road bed is much smaller than the inner radius of the road. Hence, we may approximate R as the inner radius.

$$R \approx 150 \text{ ft}$$

$$v = \sqrt{(150 \text{ ft})(32 \text{ ft/s}^2) \tan\phi}$$

From the figure,

$$\sin\phi = 4 \text{ ft}/30 \text{ ft}$$

$$\cos^2\phi = 1 - \sin^2\phi = \frac{900 \text{ ft}^2}{900 \text{ ft}^2} - \frac{16 \text{ ft}^2}{900 \text{ ft}^2} = \frac{884 \text{ ft}^2}{900 \text{ ft}^2}$$

Hence

$$\cos\phi = \frac{\sqrt{884\ \text{ft}^2}}{30\ \text{ft}}$$

and

$$\tan\phi = \frac{\dfrac{4\ \text{ft}}{30\ \text{ft}}}{\dfrac{\sqrt{884\ \text{ft}^2}}{30\ \text{ft}}} = \frac{4\ \text{ft}}{\sqrt{884\ \text{ft}^2}} = .1345$$

Therefore

$$v = \sqrt{(150\ \text{ft})\ (32\ \text{ft}/\text{s}^2)\ (.1345)}$$

$$v = \sqrt{645.6\ \text{ft}^2/\text{s}^2}$$

$$v = 25.41\ \text{ft}/\text{s}$$

65. **(B)**

Designate the shell with subscript 1 and the gun with subscript 2.

Since there are no external forces acting on the system (gun + shell) in the x-direction, p_x (before) = p_x (after) = p_{1x} (after) + p_{2x} (after). Since p_x (before) = 0, p_{1x} (after) – p_2x (after) = 0, because the gun will move to the left.

$$0 = m_1 v'_{1x} - m_2 (v'_{2x}).$$

So, $\quad m_2 v'_{2x} = m_1 v'_{1x}.$

But, $\quad v'_{1x} = v'_1 \cos 45°.$

Therefore, express p_{1x} (after) as

$$p_{1x}\ (\text{after}) = m_1 v'_1 \cos 45°$$

$$p_{1x}\ (\text{after}) = (500\ \text{kg})\ (200\ \text{m/s})\ (0.707)$$

$$= 7.07 \times 10^4\ \text{kg-m/sec}.$$

This must be numerically equal, but oppositely directed to p_{2x} (after) from equation (1). So, p_{2x} (after) $= m_2 v'_{2x} = -7.07 \times 10^4$ kg-m/sec.

And

$$v'_{2x} = \frac{-7.07 \times 10^4\ \text{kg-m}/\text{sec}}{7 \times 10^4\ \text{kg}}$$

$$v'_{2x} \approx 1\ \text{m}/\text{s}$$

Ignore the vertical component of the recoil momentum due to the extremely large value of the earth's mass compared to the railway gun.

66. **(B)**

The ballistic pendulum problem naturally divides into two parts of analysis: The totally inelastic collision when the bullet imbeds itself into the block, and the rise of the bullet and the block together due to the velocity imparted by the collision.

The collision is inelastic, so we are restricted to the always-applicable conservation of momentum equation.

$$m_1 v_1 + m_2 v_2 = m_1 V_1 + m_2 V_2$$

(before collision) (after collision)

For the ballistic pendulum, $v_2 = 0$ and $v_1 = v_2$ since the bullet imbeds itself in the block. So

$$m_1 v_1 = (m_1 + m_2) V. \qquad (1)$$

Now we must determine V by consideration of the rise of the pendulum. The equation used now is conservation of energy

$$KE_i + PE_i = KE_f + PE_f$$

At the top of the rise, the system is not moving, so $KE_f = 0$. We use the available arbitrariness to set $PE_i = 0$. Therefore, we have

$$\tfrac{1}{2} (m_1 + m_2) V^2 = (m_1 + m_2)\, gh. \qquad (2)$$

Equations (1) and (2) contain all the analyses needed for this problem. Solving for v_1 yields

$$v_1 = \left(\frac{m_1 + m_2}{m_1}\right) \sqrt{2gh}. \qquad (3)$$

Just to put some perspective on this highly practical equation, we will provide some pertinent data: m_1 weighs 0.10 lb, m_2 weighs 25 lb and $h = 4$ in.

$$v_1 = \left(\frac{0.10/32.2 + 25/32.2}{0.10/32.2}\right) \sqrt{2 \cdot 32.2 \cdot 4/12}$$

$$= \left(\frac{25.1}{0.10}\right) \sqrt{21.4} = 1163 \text{ ft/sec.}$$

67. **(E)**

It is clear that the final angular velocity may be found by using conservation of total angular momentum since there are no external forces involved. The total angular momentum is

$$L = I_{\text{rod}}\, \omega + 2I_{\text{sphere}}\, \omega + 2\, (m_{\text{sphere}}\, r^2)\, \omega.$$

The last two terms follow from the parallel axis theorem. Setting $L_i = L_f$,

$$I_r \omega_i + 2I_s \omega_i + 2m_s r_i^2\, \omega_i = I_r \omega_f + 2I_s \omega_f + 2m_s r_f^2 \omega_f$$

$$(I_r + 2I_s + 2m_s r_i^2)\, \omega_i = (I_r + 2I_s + 2m_s r_f^2) \omega_f \qquad (1)$$

We are given $I_r = 0.25$ lb·ft·sec², $m_s = 3$ lb, $r_i = {}^5/_{12}$ ft and $r_f = {}^{25}/_{12}$ ft. We compute

$$I_s = \frac{2}{5} m_s a^2 = \frac{2}{5}\left(\frac{3 \text{ lb}}{(32.2 \text{ ft/sec}^2)}\right)\left(\frac{5}{12} \text{ ft}\right)^2$$

$$= 0.00645 \text{ lb} \cdot \text{ft} \cdot \text{sec}^2,$$

and

$$m_s r_i^{\,2} = \left(\frac{3}{32.2}\right)\left(\frac{5}{12}\right)^2 = 0.01617 \text{ lb} \cdot \text{ft} \cdot \text{sec}^2$$

$$m_s r_f^{\,2} = \left(\frac{3}{32.2}\right)\left(\frac{25}{12}\right)^2 = 0.4043 \text{ lb} \cdot \text{ft} \cdot \text{sec}^2.$$

Substituting these values into equation (1) yields:

$$[0.25 + 2\,(.00645) + 2\,(0.1617)]\omega_i$$

$$= [(0.25 + 2\,(.00645) + 2\,(.4043)]\omega_f$$

$$\omega_f = \frac{0.2963}{1.0715}\,\omega_i = .2765\,(8 \text{ rad/ s}) = 2.21 \text{ rad/ sec.}$$

68. **(A)**

Since the pulleys are massless, the cord ACD has the same tension throughout, say T_1. The tension in cord BC can be called T_2. The effect of having the cord double back around pully C produces

$$d_B = {}^1/_2\, d_A, \tag{1}$$

where d_B and d_A are the distances moved, respectively, by blocks B and A. Then differentiating twice

$$a_B = {}^1/_2\, a_A \tag{2}$$

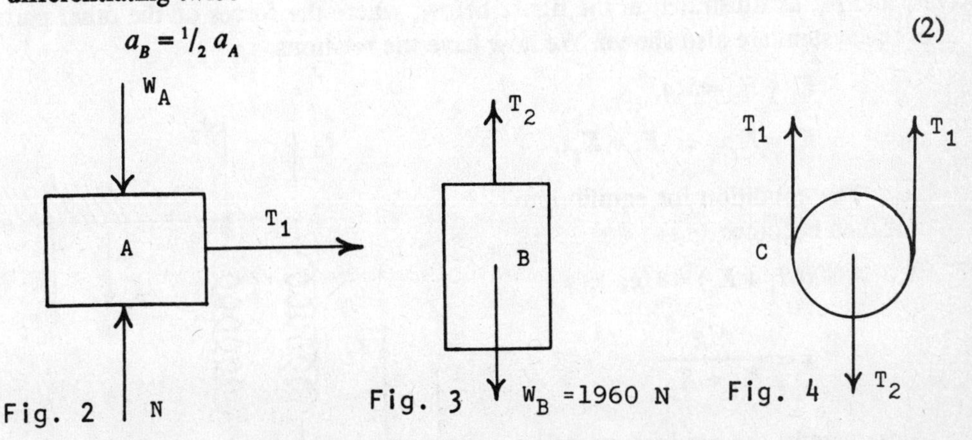

Now we shall analyze blocks A, B, and pulley C as free bodies. The free body diagram of block A is given in Figure 2. W_A and N, the weight and normal force respectively, cancel, so from Newton' s Second Law we have

$$T_1 = 50\, a_A. \tag{3}$$

The free body diagram of block B is in Figure 2.

$$W_B = mg = (200 \text{ kg})\,(9.8 \text{ m/s}^2) = 1960 \text{ N}.$$

Again using Newton's Law

$$W_B - T_2 = 1960\,N - T_2 = (200\text{ kg})\,a_B \tag{4}$$

Pulley C (Figure 3) is massless, therefore the sum of the forces on it must be zero.

$$2T_1 - T_2 = 0 \tag{5}$$

Equations (2), (3), (4) and (5) now form a system of 4 equations.

Substituting for a_A from (2) into (3) yields T_1 = (100 kg) a_B. Using this and equation (4) to substitute into (5) yields

$$(200\text{ kg})\,a_B - (1960\text{ N} - (200\text{ kg})\,a_B) = 0$$

$$(200\text{ kg})\,a_B + (200\text{ kg})\,a_B = 1960\text{ N}$$

$$a_B = \frac{1960\text{ N}}{500\text{ kg}} = 3.92\text{ m/s}^2. \tag{6}$$

Putting (6) back into (2) gives us

$$a_A = 7.84\text{ m/s}^2 \tag{7}$$

69. **(C)**

In a parallel coupling, the extensions of the two springs are the same:

$$x_1 = x_2 = x.$$

Body A is now acted upon by three forces, the weight and the two spring forces F_1 and F_2, as illustrated in the figure below, where the forces on the other parts of the system are also shown. We now have the relations

$$F_1 + F_2 = Mg,$$

and $\quad F_1 = K_1 x \;,\; F_2 = K_2 x.$

The condition for equilibrium of A then becomes

$$x(K_1 + K_2) = Mg,$$

$$x = \frac{Mg}{K_1 + K_2}.$$

Consequently the equivalent spring constant in this case becomes

$$K_e = K_1 + K_2.$$

In the special case when $K_1 = K_2 = K$, $K_e = 2K$.

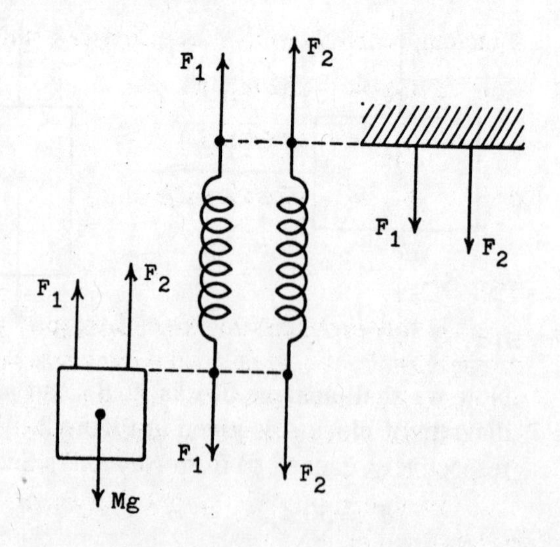

70. **(C)**

The true length of the collision can be found using the impulse-momentum theorem. Where

$$\Delta \mathbf{p} = \mathbf{F}\Delta t$$

or

$$\Delta t = \frac{\Delta \mathbf{p}}{\mathbf{F}}$$

Since we are working in one direction, the vector notation may be dropped:

$$\Delta t = \frac{\Delta p}{F} = \frac{.16 \text{ kg}[45 \text{ m/s} - (-25 \text{ m/s})]}{1200 \text{ N}}$$

$$\Delta t = .00933 \ s.$$

71. **(C)**

To find ω_{AB}, use the equation of relative velocity between ends A and B:

$$\mathbf{V}_A = \mathbf{V}_B + \mathbf{V}_{A/B}.$$

We know that

$$\mathbf{V}_{A/B} = \omega_{AB} \times \mathbf{r}_{A/B} = \omega_{AB}\,\mathbf{k} \times (-.3 \cos 20° \,\mathbf{i} + .3 \sin 20° \,\mathbf{j})\text{M}$$

$$\mathbf{V}_B = -\mathbf{V}_B\,\mathbf{i}$$

$$\mathbf{V}_A = 3\,\mathbf{j} \text{ m/s}$$

Therefore

$$3\,\mathbf{j} \text{ m/s} = -V_B\,\mathbf{i} + \omega_{AB}\,\mathbf{k}x\,(-.282\,\mathbf{i} + .103\,\mathbf{j})\text{M}$$

$$3\,\mathbf{j} \text{ m/s} = -V_B\,\mathbf{i} - .282\,\omega_{AB}\,\mathbf{j}\,\text{M} - .103\,\omega_{AB}\,\mathbf{i}\,\text{M}$$

Matching coefficients of the respective **i** and **j** terms, we get

$$3 \text{ m/s} = -.282\,\omega_{AB}\,\text{M}$$

$$\omega_{AB} = -10.64 \text{ rad/s}$$

or $$\omega_{AB} = -10.64\,\mathbf{k} \text{ rad/s}.$$

72. **(D)**

In this problem, no external torques act on the particle, therefore, angular momentum is conserved. Angular momentum, L, equals $\mathbf{r} \times m\mathbf{v}$ and, for this problem, $L = mr^2\omega$ since the motion is circular and $r\omega = \mathbf{v}$.

If the strings length is altered from r_0 to r, the angular rotation changes from ω_0 to ω.

However, $L = L_0$ (angular momentum is conserved) and it follows that $mr^2\omega = mr_0^{\,2}\,\omega_0$.

Rearranging the above expression and dividing by m yields

$$\omega_f = \frac{r_0^2\omega_0}{r^2} = \left(\frac{r_0}{r}\right)^2 \omega_0.$$

73. **(E)**

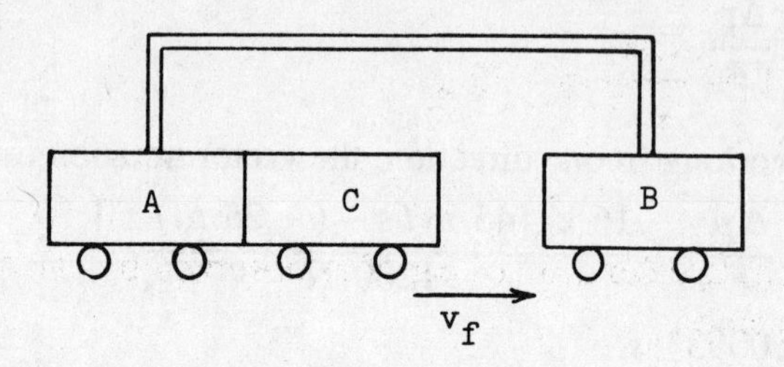

A perfectly inelastic collision means that the two colliding bodies stick together and move with the same velocity after the collision, as shown in the figure. From the Principle of Conservation of Linear Momentum, we may write

Total Momentum Before Collision = Total Momentum After Collision.

Thus,

$$4 \text{ kg } (v_i) + 0 = (4 \text{ kg} + 1 \text{ kg})(v_f). \quad (1)$$

Solving for the final velocity, v_f:

$$v_f = \frac{4 \text{ kg}}{5 \text{ kg}} v_i = \frac{4 \text{ kg}}{5 \text{ kg}} (5 \text{ m/sec}) = 4 \text{ m/sec}.$$

74. **(E)**

Using the universal law of gravitation,

$$F = \frac{Gm_1m_2}{r^2}$$

For the earth's surface:

$$F = m_1 g = \frac{Gm_e m_1}{r_e^2}$$

$$g = \frac{Gm_e}{r_e^2}$$

For Neptune's surface:

$$F = m_1 g_n = \frac{Gm_n m_1}{r_n^2}$$

With

$$m_n = 17m_e, \quad r_n = 3.5\, r_e$$

$$g_n = \frac{Gm_n}{r_n^2}$$

$$g_n = \frac{G(17\, m_e)}{(3.5 r_e)^2}$$

$$g_n = \frac{17 G m_e}{12.25 r_e^2}$$

$$g_n = 1.39 \frac{G m_e}{r_e^2} = 1.39\ g.$$

75. **(A)**

To find the acceleration of car B as observed by car A we use the relative acceleration equation

$$\mathbf{a}_B = \mathbf{a}_A + \mathbf{a}_{B/A}$$

From the problem statement, the acceleration of car B is

$$\mathbf{a}_B = 2\,(-\cos 45°\ \mathbf{i}\ \ \sin 45°\ \mathbf{j})\ \text{m/s}^2$$

$$\mathbf{a}_B = (-1.41\ \mathbf{i} + 1.41\ \mathbf{j})\ \text{m/s}^2$$

The acceleration of car A is:

$$\mathbf{a}_A = \frac{(25\ \text{m/s})^2}{100\ \text{m}}\ \mathbf{i} - 3\ \text{m/s}\ \mathbf{j}$$

$$\mathbf{a}_A = (6.25\ \mathbf{i} - 3\mathbf{j})\ \text{m/s}^2$$

Solving for $\mathbf{a}_{B/A}$,

$$\mathbf{a}_{B/A} = \mathbf{a}_B - \mathbf{a}_A$$

$$\mathbf{a}_{B/A} = (-1.41\ \mathbf{i} + 1.41\ \mathbf{j})\ \text{m/s}^2 - (6.25\ \mathbf{i} - 3\ \mathbf{j})\ \text{m/s}^2$$

$$\mathbf{a}_{B/A} = (-7.66\ \mathbf{i} + 4.41\ \mathbf{j})\ \text{m/s}^2.$$

76. **(E)**

With $x(t) = t^3 - 12t^2 - 40t + 60$, we realize that

$$v(t) - \dot{x}(t) = \frac{dx}{dt} = 3t^2 - 24t - 40$$

$$a(t) = \frac{dv}{dt} = \ddot{x}(t) = \frac{d^2x}{dt^2} = 6t - 24$$

Setting $a(t) = 0$.

$$6t - 24 = 0$$

$$t = 3.$$

Substituting $t = 3$ into the velocity function,

$$v(3) = 3(3)^2 - 24(3) - 40 = -85\ \text{m/s}.$$

77. **(B)**

The resultant vector, $\mathbf{R} = \Sigma F_i$

$$\mathbf{R} = (20\,\mathbf{i} + 30\,\mathbf{j} + 40\,\mathbf{k})\text{N} + (-30\,\mathbf{j} + 20\,\mathbf{k})\text{N} + (60\,\mathbf{i})\,\text{N}$$

$$\mathbf{R} = (80\,\mathbf{i} + 60\,\mathbf{k})\,\text{N}$$

Magnitude of $R = \sqrt{(80)^2 + (60)^2} = 100$ N.

78. **(D)**

The tension T in the main cable is same at any section. If we take a section below the top pulley, then

$$\Sigma F_y = 0, \quad T + T + T - W = 0$$

$$T = \frac{W}{3}.$$

79. **(C)**

Weight of 100 kg mass

$= 100 \text{ kg} \times 9.81 \text{ m/s}^2 = 981$ N

By analyzing joint B:

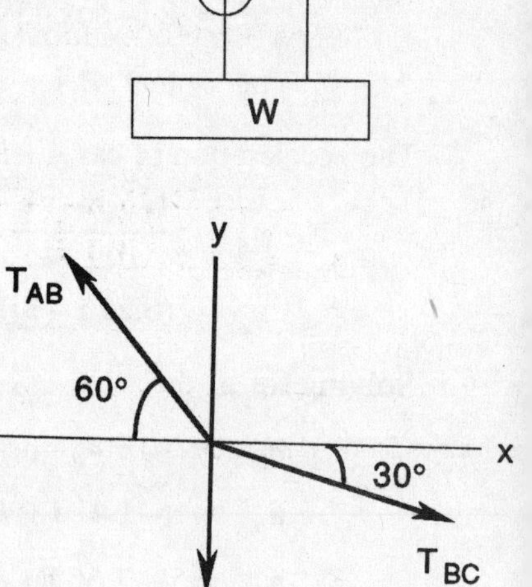

$$\Sigma F_x = 0, \quad -T_{AB} \cos 60° + T_{BC} \cos 30° = 0$$

$$T_{AB} = 1.732\, T_{BC}$$

$$\Sigma F_y = 0, \quad +T_{AB} \sin 60° - T_{BC} \sin 30° - 981 \text{ N} = 0$$

$$(1.732\, T_{BC})(0.866) - 0.5\, T_{BC} - 981\text{N} = 0$$

$$T_{BC} = 981 \text{ N}$$

80. **(E)**

Use the fundamental equations of equilibrium

$$\Sigma F_x = 0 \quad \Sigma F_y = 0 \quad \Sigma M = 0$$

to find reaction force at B.

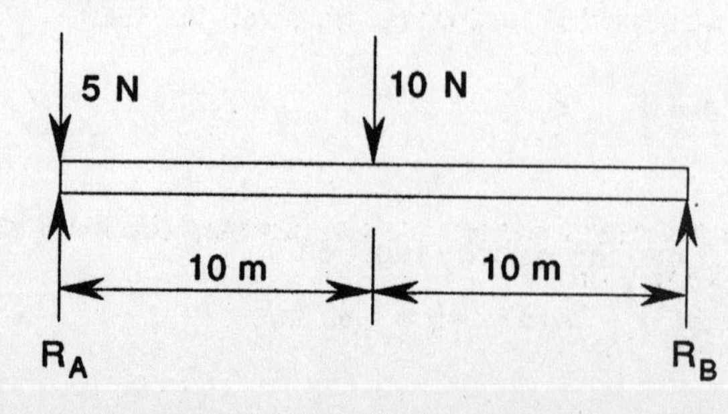

Note: Since there are no horizontal external forces acting on the beam, horizontal reaction forces at A and B equal zero. Otherwise, we would have to use $\Sigma F_x = 0$.

Summing the vertical forces

$$R_A - 5\text{ N} - 10\text{ N} + R_B = 0$$

Summing the moments about A (counterclockwise direction is positive)

$$-(10\text{ N} \times 10\text{ M}) + (R_B \times 20\text{ M}) = 0$$

$$R_B = 5\text{ N}$$

Note: Since there are only two unknown forces and one may be eliminated by taking the sum of the moments about A, the last equation alone was sufficient to solve the problem. However, this may not always be the case.

81. **(A)**

The bending moment diagram for the loaded beam is:

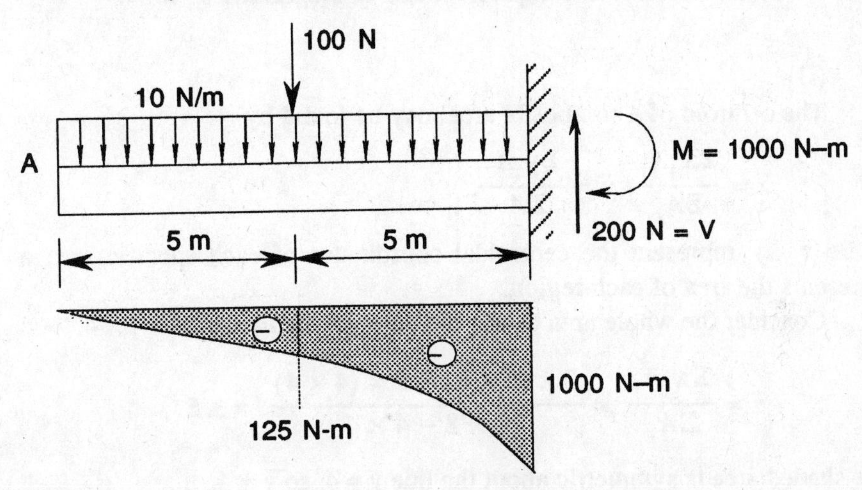

From the bending moment diagram, the maximum bending moment is 1000 N-M.

82. **(C)**

The load 20 N/M is equally distributed to the right and to the left of the support at B. So, support B will take care of all the load. The reaction at A will be zero.

83. **(C)**

If we take a vertical section passing through the member GD and consider the left side of the section,

$$\Sigma F_y = 0, \quad {}^3/_5\, F_{GD} - 30\text{ N} = 0$$

$F_{GD} = 50$ N (Tension)

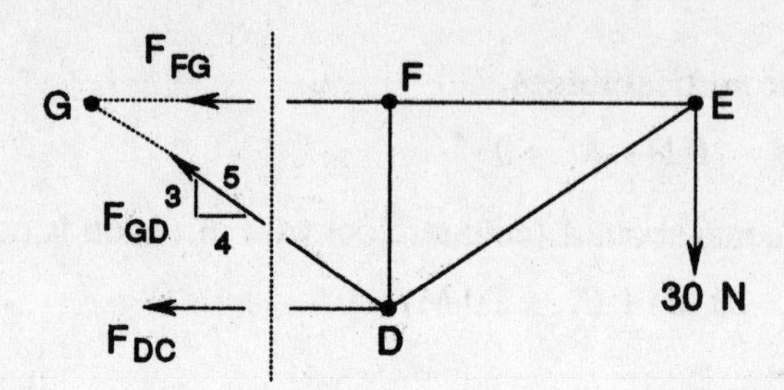

84. **(C)**

By taking moment about the point G, considering clockwise positive,

$$F_{DC} * 3 + 30 * 8 = 0$$

$$F_{DC} = -80 \text{ N}$$

Force in the member DC is compressive and of magnitude of 80 N.

85. **(B)**

The centroid of a composite area may be found by

$$\bar{x} = \frac{\Sigma x_i A_i}{\Sigma A_i} \quad \bar{y} = \frac{\Sigma y_i A_i}{\Sigma A_i}$$

where $\bar{x}_i$, $\bar{y}_i$ represent the centroidal coordinates of each specific area and A_i represents the area of each region.

Consider the whole area 6×8, and subtract the inside area 4×4,

$$\bar{x} = \frac{\Sigma x_i A_i}{\Sigma A_i} = \frac{3 \times (6 \times 8) - 2 \times (4 \times 4)}{6 \times 8 - 4 \times 4} = 3.5.$$

The shaded area is symmetric about the line $y = 4$, so $\bar{y} = 4$.

The centroid is at (3.5, 4).

86. **(D)**

I_x for a composite area about its centroidal axis may be found from

$$I_x = \Sigma I_{x_i} + \Sigma A_i d_i^2$$

where I_{x_i} = centroidal moment of inertia for given area

A_i = area of given area

d_i = distance between centroid of given area and centroidal axis of composite area

Redraw the composite area and divide into separate areas:

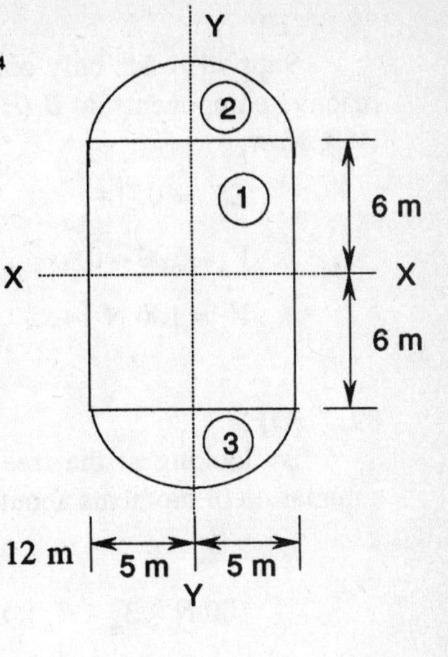

For area ①

$$I_{x_1} = \frac{bh^3}{12} = \frac{(10\text{ m})(12\text{ m})^3}{12} = 1440\text{ m}^4$$

$A_1 = (12\text{ m})(10\text{ m}) = 120\text{ m}$

$d_1 = 0$ M (centroid of area 1 coincides with centroidal axis)

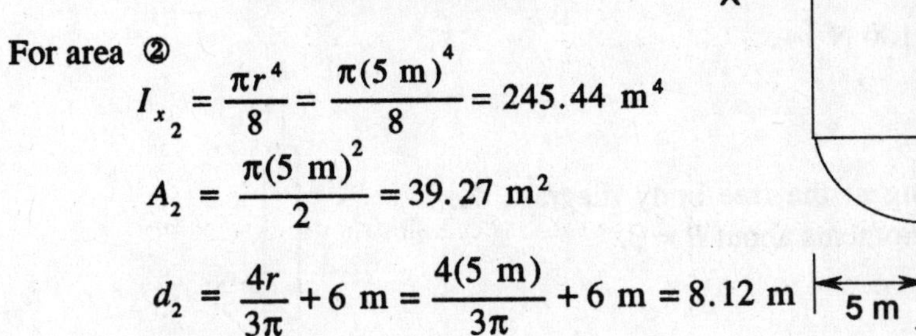

For area ②

$$I_{x_2} = \frac{\pi r^4}{8} = \frac{\pi (5\text{ m})^4}{8} = 245.44\text{ m}^4$$

$$A_2 = \frac{\pi (5\text{ m})^2}{2} = 39.27\text{ m}^2$$

$$d_2 = \frac{4r}{3\pi} + 6\text{ m} = \frac{4(5\text{ m})}{3\pi} + 6\text{ m} = 8.12\text{ m}$$

For area ③

Since Area ① and ③ are identical

$I_{x_3} = 245.44\text{ m}^4$

$A_3 = 39.27\text{ m}^2$

$d_3 = 8.12\text{ m}$

Therefore $I_x = [1440\text{ m}^4 + 0] + 2[245.44\text{ m}^4 + 39.27\text{ m}^2\,(8.12\text{m})^2] = 7109\text{ m}^4$.

87. **(D)**

The weight of the block = 100 kg × 9.81 m/s² = 981 N acting in the downward direction. By taking sum of all forces along the *x*-diorection equal to zero,

$\Sigma F_x = 0.$

$P + F - {}^4/_5 * 981\text{ N} = 0$ (1)

and $\Sigma F_y = 0, \quad N - {}^3/_5 \times 981\text{ N} = 0$

$N = 588.6\text{ N}$

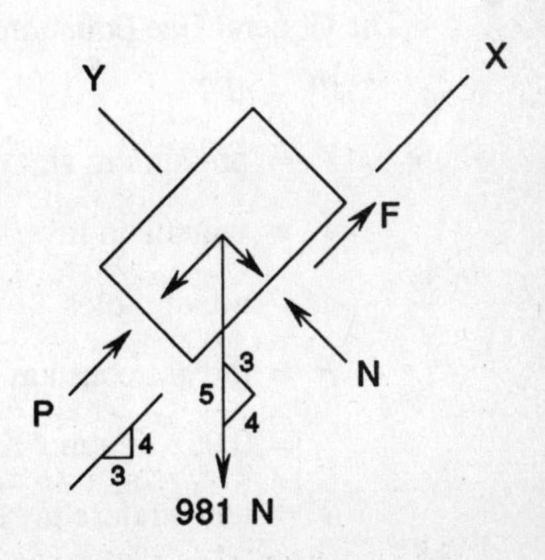

The frictional force, $F = \mu N = 0.25 * 588.6$ N, $F = 147.15$ N. Using equation (1),

$P = {}^4/_5 * 981\text{ N} - F = {}^4/_5 \times 981\text{ N} - 147.15\text{ N}$

$P = 637.65\text{ N}$

88. (E)

Support A has only one horizontal reaction, and the vertical at B has two reactive components: at B (H_B) a horizontal reaction, and a vertical reaction at B (V_B). Now,

$$\Sigma F_V = 0, \uparrow +,$$

$$V_A - 100 = 0$$

$$V_A = 100 \text{ N} \uparrow +.$$

89. (D)

By looking at the free body diagram the summation of moments about $B = 0$,

$$\Sigma M_B = 0$$

$$100 \text{ N} \times 3_m - R_A \times 8 = 0$$

$$R_A = 37.5 \rightarrow +$$

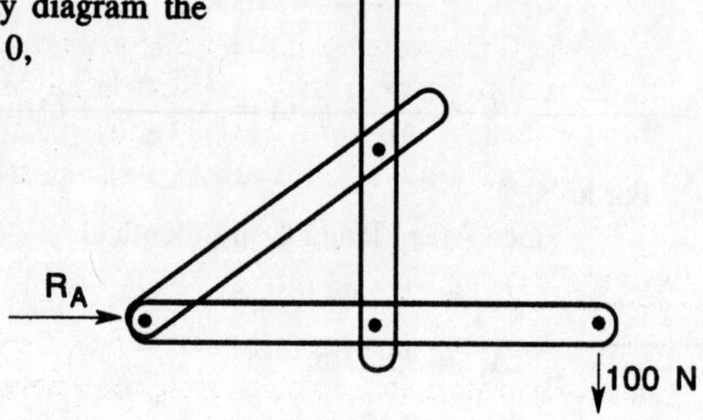

90. (B)

The member CB is a two force member, so the reaction at B is only horizontal reaction.

$$\Sigma M_A = 0,$$

$$4/_5 \times 20 \times 8 - H_B \times 4 = 0$$

$$H_B = 32 \, N \leftarrow +.$$

91. (D)

The General Gas Equation can be used to solve this problem

$$PV = nRT$$

where P = pressure in atm

V = volume in lit

n = no. of moles

R = the gas constant

= 0.082 lit atm./°K· mole

T = temperature in °K

Since n, the no. of moles

$$= \frac{\text{mass in g}}{\text{molecular wt.}},$$

the above equation can be written as,

$$PV = \frac{gRT}{\text{M.W.}}$$

Substituting

$$(1.0 \text{ atm})\ (0.107 \text{ lit}) = \frac{(0.4 \text{ g})\ (0.082 \text{ lit atm}/°\text{K Mole})\ (300°\text{K})}{\text{M.W.}}$$

M.W. = 92.0 g/mole.

92. **(C)**

Gold is a noble metal. Even concentrated acids, such as nitric, hydrochloric or sulfuric, do not attack it. Only a mixture of conc. hydrochloric and conc. nitric acids in a 3 : 1 ratio (called "aqua-regia") can dissolve gold.

In solving this problem, therefore, the assumption is that only copper metal from the alloy is reacted upon by nitric acid.

$$Cu + 4\,HNO_3 \rightarrow Cu(NO_3)_2 + 2H_2O + 2NO_2$$

Using atomic wts., the molecular wt of $Cu(NO_3)_2$ is calculated to be 187.5 g/mole. By setting the stoichiometric ratio between Cu and $Cu(NO_3)_2$, the mass of copper in the sample of alloy can be calculated as follows:

$$3.75 \text{ g } Cu(NO_3)_2 \times \frac{63.5 \text{ g Cu}}{187.5 \text{ g } Cu(NO_3)_2} = 1.27 \text{ g Cu}$$

Therefore, the mass of gold in the sample is

2.86 g alloy – 1.27 g. Cu = 1.59 g Au

Percent of Au in alloy

$$= \frac{1.59 \text{ g Au}}{2.86 \text{ g alloy}} \times 100$$

$$= 55.6\% \text{ Au.}$$

93. **(E)**

By convention, the direction of dipole is from δ + to δ–. The resultant of two C–Br moments is exactly in the middle of two C–Br bonds, at an angle of 56° from each C–Br bond. The contribution of each C–Br bond to the resultant can be calculated by using vector analysis. Contribution of each C–Br bond to the resultant

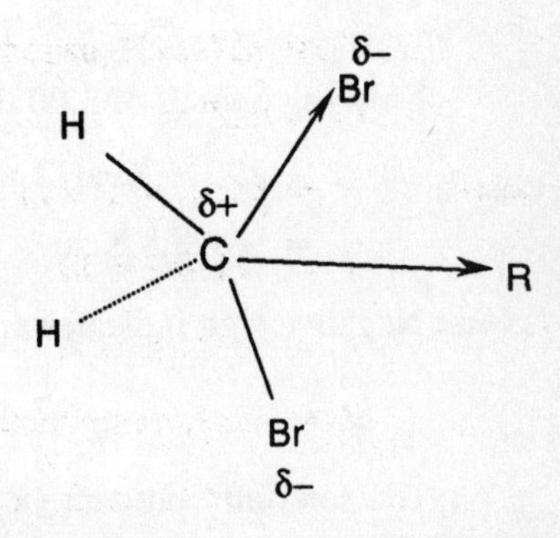

= bond moment of C–Br bond × cos 56°

= 1.38 D × 0.56 = 0.77 D

Contribution of the other C–Br bond to the resultant is also 0.77 D in the same direction.

Therefore, the total resultant, which is the (total) dipole moment of CH_2Br_2 is 0.77 D + 0.77 D = 1.54 D.

94. **(A)**

Let us first express the amount of electricity passed in Faradays.

Coulombs passed = amp × sec = 5.0 amp × 12,600 secs = 63,000

$$63,000 \text{ coulombs} \times \frac{1 \text{ faraday}}{96,500 \text{ coulombs}} = 0.65 \text{ Faraday}$$

$Fe^{+3} + 3e \rightarrow Fe^{\circ}$

One mole of Fe^{+3} with 3 moles of electrons (or 3 moles of Faraday) forms one mole of Fe metal. Therefore, the equivalent wt of Fe° in this reaction

$$= \frac{\text{Atomic wt of Fe}}{3 \text{ Faraday}} = \frac{56}{3} = 18.67 \text{ g / Faraday}$$

That is one Faraday of electricity will theoretically deposit 18.67 g of Fe metal.

0.65 Faraday will deposit 12.14 g of Fe metal at 100% efficiency.

At 66% efficiency, 68% of 12.14 g = 8.26 g of Fe metal will be deposited.

95. **(C)**

The Heat of Reaction (ΔH), also called the Change in Enthalpy, is equal to the total enthalpy of all the products minus the total enthalpy of all the reactants.

$$\Delta H_{\text{reaction}} = H_{\text{products}} - H_{\text{reactants}}$$

$$= [7\, H_f C_{(s)} + 7\, H_f CO_{(g)} + 3\, H_f N_{2(g)} + 5\, H_f H_2O_{(g)}]$$

$$- [2\, H_f TNT_{9(s)}]$$

$$= [7 \text{ (zero)} + 7\,(-26.4) + 3 \text{ (zero)} + 5(-57.8)]$$

$$- [2(-87.1)]$$

$$= -299.6 \text{ Kcal per two moles of TNT}$$

Now solve for H of 3.00 lbs of TNT.

$$-299.6 \text{ Kcal/2} \times \frac{1}{227} \times \frac{1362}{3 \text{ lb}}$$

$$= -898.8 \text{ Kcal.}$$

Note: Negative sign indicates an exothermic reaction.

96 **(D)**

The solubility product (K_{sp}) is used to express solubilities of very slightly soluble substances. The K_{sp} of a substance is equal to the multiplication of molar concentrations of each of the ions the substance produces in solution raised to the power which is equal to the coefficient of that ion.

$$Ag_2CrO_{4(s)} \rightarrow 2\,Ag^+_{(aq)} + CrO_{4(aq)}^{-2}$$

This equation shows that when one mole of $AgCrO_4$ (solid) goes into aqueous solution, it forms 2 moles of Ag^+ and one mole of CrO_4^{-2}. The expression for the K_{sp} of Ag_2CrO_4 is

$$K_{sp} = [Ag^+]^2\,[CrO_4^{-2}] = 8.5 \times 10^{-8}$$

Therefore, the solubility of Ag_2CrO_4 in mole/lit is equal to the conc. of CrO_4^{-2} produced in the solution.

If x moles of Ag_2CrO_4 go in the solution, $2x$ moles of Ag^+ and x moles of CrO_4^{-2} are formed.

Substituting in the above expression for K_{sp},

$$8.5 \times 10^{-8} = (2x)^2\,(x) = 4x^3$$

$$x = 2.77 \times 10^{-3} \text{ moles/lit}$$

The solubility of Ag_2CrO_4 is 2.77×10^{-3} moles/lit.

97. **(B)**

The metals that exhibit multiple valence states lie principally in GroIb through VIIIb of the Periodic Table. These mtals all have two (or sometines one) "s" electrons in the outer orbit, and "d" electrons are added one orbit down with each successive element is formed. These "d" electrons have significantly different reactivity than the "s" electrons in the outer orbit and lead to multiple valence states depending on whether or not they react.

Answer (A) is incorrect. None of the transition metals have both "s" and "p" electrons in the outer orbit.

Answer (C) is incorrect. Often a transition metal will exhibit more than one valence state with a same element — e.g., copper un curpous chloride (CuCl) and cupric chloride ($CuCL_2$).

Answers (D) and (E) are incorrect. Reactivity of metals in simple reactions is quite predictable and are not governed by probability.

98. **(C)**

When the conc. of a metal is changed, there should be a change in the rate of reaction if the metal is acting as a catalyst. If changing the conc. of a metal does not change the rate of reaction, it is not acting as a catalyst.

In order to determine if a metal is acting as a catalyst, for comparison select two experiments in which the only difference is different concentrations of a given metal. If the rate is higher, with higher conc. of the metal, it is acting as a catalyst.

To evaluate the effect of metal Y, compare expt. Nos. 1 and 2. There is a higher conc. of Metal Y in expt. 2 than in expt. 1, but the reaction rates for both experiments are the same. Therefore, metal Y does not act as a catalyst. In the same manner, comparing expt. Nos. 2 and 5, we find that metal *X* does not act as a catalyst. Comparing expt. Nos. 3 and 4, we find that metal *Z* does act as a catalyst.

99. **(D)**

The K_c for a reaction is its equilibrium constant involving concentration in moles/lit or Molarity (M). In order to calculate K_c, we must know the M of all reactants and products at equilibrium.

$$CO_{(g)} + H_2O_{(g)} \rightleftharpoons CO_{2(g)} + H_{2(g)}$$

From the equation, we see that one mole of CO reacts with one mole of H_2O to form one mole of CO_2 and one mole of H_2. Therefore, if 0.3 mole of CO_2 is present at equilibrium, the moles of CO at equil. are

(0.5 – 0.3) = 0.2 mole, moles of H_2O at equil. are (1.0 – 0.3) = 0.7

and moles of H_2 at equil. are 0.3.

Now let's calculate equilibrium concentrations of each.

$$[CO] = \frac{0.2 \text{ mole}}{5.0 \text{ lit}} = 0.04 \text{ M}$$

$$[H_2O] = \frac{0.7 \text{ mole}}{5.0 \text{ lit}} = 0.14 \text{ M}$$

$$[CO_2] = \frac{0.3 \text{ mole}}{5.0 \text{ lit}} = 0.06 \text{ M}$$

$$[H_2] = \frac{0.3 \text{ mole}}{5.0 \text{ lit}} = 0.06 \text{ M}$$

The expression for K_c for this reaction is:

$$K_c = \frac{[CO_2]\,[H_2]}{[CO]\,[H_2O]} = \frac{(0.06)\,(0.06)}{(0.04)\,(0.14)} = 0.64$$

100. **(E)**

Le Chatelier's Principle states that if any change is made in the conditions of equilibrium, the equilibrium will shift in such a way as to nullify or cancel the effect of change.

There are three possible conditions of equilibrium.. i) Concentration, ii) pressure and iii) temperature.

i) Concentration is always a condition of equil. If conc. of any or all reactants is increased, or the conc. of any or all products is decreased (e.g. by removing the product from equil.). The equilibrium will shift to the right, i.e. more of the products will be formed.

ii) Pressure is a condition of equilibrium if there are a different number of moles of gases in a balanced equation for the reaction.

iii) Temp. is a condition of equilibrium if a reaction is either exothermic or endothermic. If a reaction, as written from left to right, is exothermic, decreasing the temp. of reaction (by absorbing heat), will shift the equil. more to the right.

Looking at the reaction in this problem, the conditions of equil are conc., and temp.

101. **(B)**

Conditions for this equilibrium are concentration, pressure (because we have 4 moles of gases on the left and 2 moles of gases on the right), and temp. (because the reaction is exothermic).

The correct choice, therefore, is (B). viz. in order to maximize the yield of NH_3, increase conc. of H_2 and N_2 and decrease the temp.

102. **(B)**

Firstly, we must express the total amount of sample (fish fillet) and the contaminant (mercury) using the same units. 20.0 g = 20,000 mg = 2.0×10^4 mg. Now set up the simple proportion

$$\frac{0.5 \text{ mg Hg}}{2.0 \times 10^4 \text{ mg total}} = \frac{x \text{ mg Hg}}{1.0 \times 10^6 \text{ mg total}}$$

$$x = 25.0 \text{ ppm}.$$

103. **(A)**

The standard reduction potential for

$$Ag^+ + 1\,e \rightarrow Ag^\circ$$

is more positve (+0.80 volts) than for

$$Cu^{+2} + 2\,e \rightarrow Cu^\circ$$

(+0.34 volts). This means that silver has greater tendency for reduction than copper. Therefore, in this galvanic cell, copper will be oxidized and silver will be reduced.

$Cu^\circ + 2\,e \rightarrow Cu^{+2}$ Half equation for oxidation.

$Ag^+ + 1\,e \rightarrow Ag^\circ$ Half equation for reduction.

Using the electron balance method, we get the balanced equation:

$$Cu^\circ + 2\,e \rightarrow Cu^{+2}$$

$$2\,Ag^+ - \;\; 2\,e \rightarrow 2\,Ag^\circ$$

$$Cu_{(s)}{}^\circ + 2\,Ag^+_{(aq)} \rightarrow Cu^{+2}_{(aq)} + Ag_{(s)}{}^\circ \qquad (1)$$

The standard electrode potential,

$$E° = E_{reduction} - E_{oxidation}$$
$$= (+0.80 \text{ volts}) - (+0.34 \text{ volts}) = 0.46 \text{ volts}$$

The actual e.m.f. (E_{actual}) can be calculated using the Nearnst Equation.

$E_{actual} = E° - RT \ln K$, where R = Gas constant

T = temp. in °K

K = equilibrium expression for the reaction

Looking at balanced equation (1),

$$K = \frac{[Cu^{+2}]}{[Ag^{+}]^2}$$

At 25° C. and using log to the base of 10, the Nearnst Equation can be written as:

$$E_{actual} = E° - \frac{0.0592}{n} \log K,$$

where n = No. of moles of electrons transferred in a balanced equation.

In this reaction $n = 2$. Substituting in the equation,

$$E_{actual} = E° - \frac{0.0592}{n} \log \frac{[Cu^{+2}]}{[Ag^{+}]^2}$$

$$= 0.56 - \frac{0.0592}{2} \log \frac{[0.1]}{[0.1]^2}$$

$$= 0.43 \text{ v.}$$

104. **(E)**

A balanced equation shows ratio by moles between reactants and products. For each mole of $CH_3OH_{(g)}$ formed, one mole of $CO_{(g)}$ and two moles of $H_{2(g)}$ are consumed. We can therefore calculate the moles of each substance at equilibrium.

	Initial Moles	Moles at Equilibrium
$CH_3OH_{(g)}$	0.0	1.6
$CO_{(g)}$	2.5	0.9
$H_{2(g)}$	3.6	0.4

The partial pressure of each gas at equilibrium can now be calculated using the General Gas Equation.

$$PV = nRT \;,\; \text{or } P = \frac{nRT}{V}$$

Where, P = pressure in atm

V = volume in lit

n = number of moles

R = General Gas Constant

= 0.082 lit atm/deg mole

T = temp. in °K

Substituting

$$p\text{CH}_3\text{OH} = \frac{(1.6)\ (0.082)\ (373)}{10.0}$$

$$= 4.89 \text{ atm}$$

Similarly, pCO = 2.75 atm and pH_2 = 1.22 atm. The expression for K_p for this reaction is:

$$K_P = \frac{(p\text{CH}_3\text{OH})}{(p\text{CO})\ (pH_2)^2} = \frac{4.89 \text{ atm}}{(2.75 \text{ atm})\ (1.22 \text{ atm})^2}$$

$$= 1.19/\text{atm}^2$$

105. **(C)**

P_1 and P_3 are forces generated by two aluminum rods with corresponding deformation, Δ_1 and Δ_3.

P_2 is the force generated by the steel rod in the middle with corresponding deformation, Δ_2.

From equilibrium equation:

$$P_1 + P_2 + P_3 = 50{,}000 \qquad (1)$$

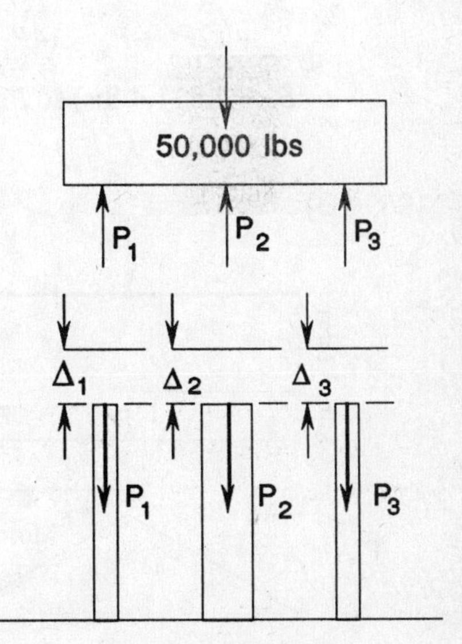

Deformations

This is a statically indeterminate problem since the equilibrium equation has too many unknowns. Therefore, deformation must be considered.

$$\Delta_1 = \Delta_2 = \Delta_3;$$

since $\sigma = P/A$; $\varepsilon = \Delta/L$ and $\sigma = E\varepsilon$ gives

$$\frac{P}{A} = E\frac{\Delta}{L} \text{ or; } \Delta = \frac{PL}{AE}$$

We have:

$$L_1 = L_2 = L_3 = 10 \text{ in}$$

$$A_1 = A_3 = \pi(^1/_2 \text{ in})^2 \text{ and } A_2 = \pi(1 \text{ in})^2$$

$$E_1 = E_3 = E_{AL} = 10 \times 10^6 \text{ psi and } E_2 = E_s = 30 \times 10^6 \text{ psi.}$$

Considering $\Delta_1 = \Delta_3$ and substituting above values,

$$\frac{P_1(10 \text{ in})}{\pi\left(\frac{1}{2} \text{ in}\right)^2 \times (10 \times 10^6 \text{ psi})} = \frac{P_3(10 \text{ in})}{\pi\left(\frac{1}{2} \text{ in}\right)^2 \times (10 \times 10^6 \text{ psi})}$$

gives $P_1 = P_3$. (2)

This could also have been obtained by observing symmetry. Consider $\Delta_1 = \Delta_2$ and substituting values of A, E and L,

$$\frac{P_1(10 \text{ in})}{\pi\left(\frac{1}{2} \text{ in}\right)^2 \times (10 \times 10^6 \text{ psi})} = \frac{P_2(10 \text{ in})}{\pi(1 \text{ in})^2 \times (30 \times 10^6 \text{ psi})}$$

gives $P_2 = 12P_1$. (3)

Substituting (2) and (3) in (1), we get

$$P_1 + 12P_1 + P_1 = 14P_1 = 50{,}000 \text{ lb.}$$

$$P_1 = 3{,}571.4 \text{ lbs; so } P_2 = 12P_1 = 12(3{,}571.4) = 42{,}857 \text{ lbs.}$$

106. **(B)**

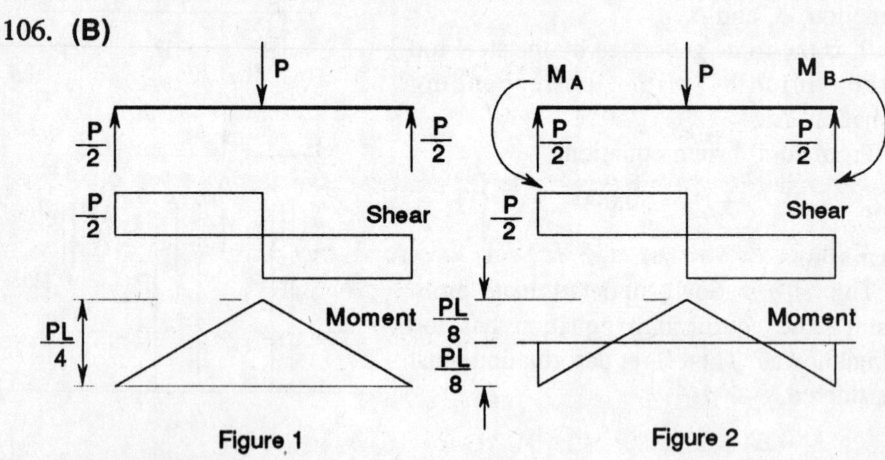

Figure 1 Figure 2

For Beam in Figure 1:

Between $x = 0$ to $L/2$ (left side)

$$V_x = \frac{P}{2}; \text{since } \frac{dM}{dx} = V$$

$$M = \int V dx = \frac{P}{2}x + c$$

at $x = 0$, $M = 0$ giving $c = 0$ so

$$M = \frac{P}{2}\, x,$$

giving maximum moment at $x = {}^{L}/_{2}$

$$M_{max} = \frac{P}{2}\left(\frac{L}{2}\right) = \frac{PL}{4}$$

For Beam in Figure 2:

Between $x = 0$ to $L/2$ (left side)

$$V_x = \frac{P}{2} \text{ giving } M = \frac{P}{2}\, x + c$$

at $x = 0$, $M = M_A$ giving $c = -M_A$ so

$$M = \frac{P}{2}\, x - M_A.$$

In this moment equation the slope of moment line is $P/2$ and intercept is $-M_A$. Maximum moment is at $x = 0$ and $x = L/2$

$$M_{max}\Big|_{x=0} = -\frac{PL}{8}$$

$$M_{max}\Big|_{x=\frac{L}{2}} = \frac{PL}{8}$$

Since maximum moment of Case 2 is half that of Case 1, therefore, the beam in Case 2 will support twice the load, hence, 100% more.

107. **(B)**

If the 30 ft long rail was allowed to expand freely, then total elongation would be

$$\Delta = \alpha\,(\Delta T)\,L = 6.5 \times 10^{-6}\ \text{in/in°F}\ (140°\text{F} - 60°\text{F})\ (30\ \text{ft} \times 12\ \text{in/ft})$$

$$= 0.1872 \text{ inches.}$$

Since the rail can grow freely up to 0.1 in, no stress will develop up to that growth.

Compressive stress will develop due to restriction of $(0.1872 - 0.1) = 0.0872$ in growth. The equivalent stress for this growth is:

corresponding strain is $\varepsilon = \Delta/L = 0.0872\ \text{in} / (30\ \text{ft} \times 12\ \text{in/ft})$

$= 0.0002422$

$\sigma = E\varepsilon = 30 \times 10^{6}\ \text{psi} \times 0.0002422 / (30\ \text{ft} \times 12\ \text{in/ft})$

$= 7267$ psi (compression)

108. **(A)**

The actual elongation is 0.1 in. So the actual strain is

$$\varepsilon = \Delta/L = 0.1 \text{ in} / (30 \text{ ft} \times 12 \text{ in/ft}) = 0.000277.$$

Note: In thermal stress problems, the formula $\sigma = E\varepsilon$ does not use actual strain. As, for example, in the previous problem, it used an equivalent strain of 0.0002422, which resulted from restricting the growth of 0.0872 in. A rod that grows freely has zero stress but finite strain. Similarly, a rod that is totally restricted to grow has zero strain but finite stress.

109. **(C)**

The beam is simply supported, so the bending moment at the ends are zero. Since

$$\frac{dM}{dx} = V, \ M \text{ is maximum where } \frac{dM}{dx} = 0 \text{ or } V = 0.$$

So we locate the zero shear point as follows:

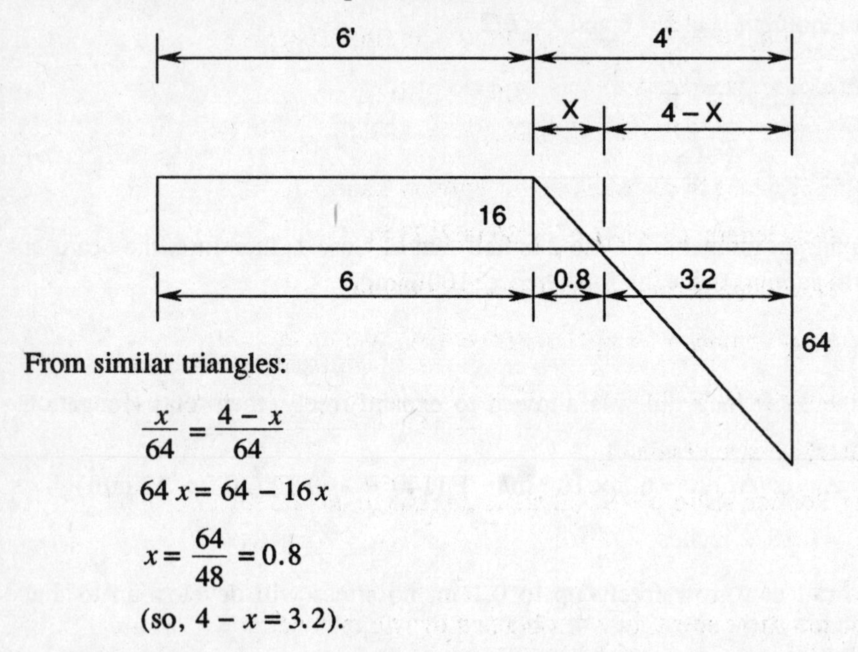

From similar triangles:

$$\frac{x}{64} = \frac{4-x}{64}$$

$$64\,x = 64 - 16x$$

$$x = \frac{64}{48} = 0.8$$

$$(\text{so, } 4 - x = 3.2).$$

Since moment is zero at the ends, the moment at zero shear point is given by the area of the shear diagram to the left (or to the right) of the zero shear point. Note:

$$dM = Vdx; \int_0^M dM = \int_0^x Vdx;$$

hence, increase in moment = area of shear diagram.

Area of shear diagram to the left

$$= (16 \text{ k})(6 \text{ ft}) + {}^1/_2 (16 \text{ k})(0.8 \text{ ft}) = 102.4 \text{ k-ft or}$$

Area of shear diagram to the right

$$= {}^1/_2 (64 \text{ k})(3.2 \text{ ft}) = 102.4 \text{ k-ft}.$$

110. **(D)**

Original volume of Rod

$$V_0 = \pi R^2 L = \pi(3 \text{ in})^2 \ (3 \text{ ft} \times 12 \text{ in/ft}) = 1017.36 \text{ cu in}$$

Axial Strain

$$\varepsilon_{xx} = \frac{\sigma_{xx}}{E} = \frac{P}{AE} = \frac{50,000 \text{ lb}}{\pi(3 \text{ in})^2 (30 \times 10^6 \text{ psi})} = 58.98 \times 10^{-6} \text{ in/in}$$

There is no stress in Y and Z direction. Hence,

$$\varepsilon_{yy} = \varepsilon_{zz} = -\upsilon\varepsilon_{xx} = -0.3\ (58.98 \times 10^{-6} \text{ in/in}) = -17.69 \times 10^{-6} \text{ in/in.}$$

Changes in dimensions are obtained from the definition of strain as follows:

$$\varepsilon_{xx} = \frac{\Delta L}{L} = \frac{\Delta L}{36 \text{ in}} = 58.98 \times 10^{-6} \text{ in/in; } \Delta L = 2123 \times 10^{-6} \text{ in}$$

(+ indicates increase in length due to tension.)

$$\varepsilon_{yy} = \varepsilon_{zz} = \frac{\Delta D}{D} = -17.69 \times 10^{-6} \text{ in/in; } D = 6 \text{ in}$$

Therefore change in diameter

$$\Delta D = (-17.69 \times 10^{-6} \text{ in/in})\ (6 \text{ in}) = -106.14 \times 10^{-6} \text{ in.}$$

(– indicates decrease in diameter due to Poisson Ratio effect.)

New length $L' = L + \Delta L = 36.002123$ in;

New diameter = 6 – .000106 = 5.99989 in

New volume $V' = \pi(D'/2)^2(L') = 1017.384$ cu in

Change in volume = $V' - V_0$ = 1017.384 – 1017.360 = +0.024 cu in

(+ indicates increase in volume)

Note: For Poisson Ratio $\upsilon = 0.5$, there is no change in volume.

111. **(C)**

The pin shear stress may be obtained by two methods:

1. The total load of 5,000 lb is supported by two cross sectional areas a and b as shown in Figure (a).

   Net shear area $= 2\left[\frac{\pi}{4} d^2\right] = 2\left[\frac{\pi}{4} (1 \text{ in})^2\right] = 1.57 \text{ in}^2$

   Pin shear stress $\tau = \frac{5000 \text{ lb}}{1.57 \text{ in}^2} = 3185 \text{ psi}$

2. The approximate free body diagram of the pin is shown in Figure (b). From symmetry, R on both the left and right end of the pin is 2,500 lbs. Considering the pin as a beam, maximum shear force is 2,500 lbs. Hence, max shear stress is

$$\frac{2500 \text{ lb}}{(\pi/4)\,(1 \text{ in})^2} = 3185 \text{ psi}$$

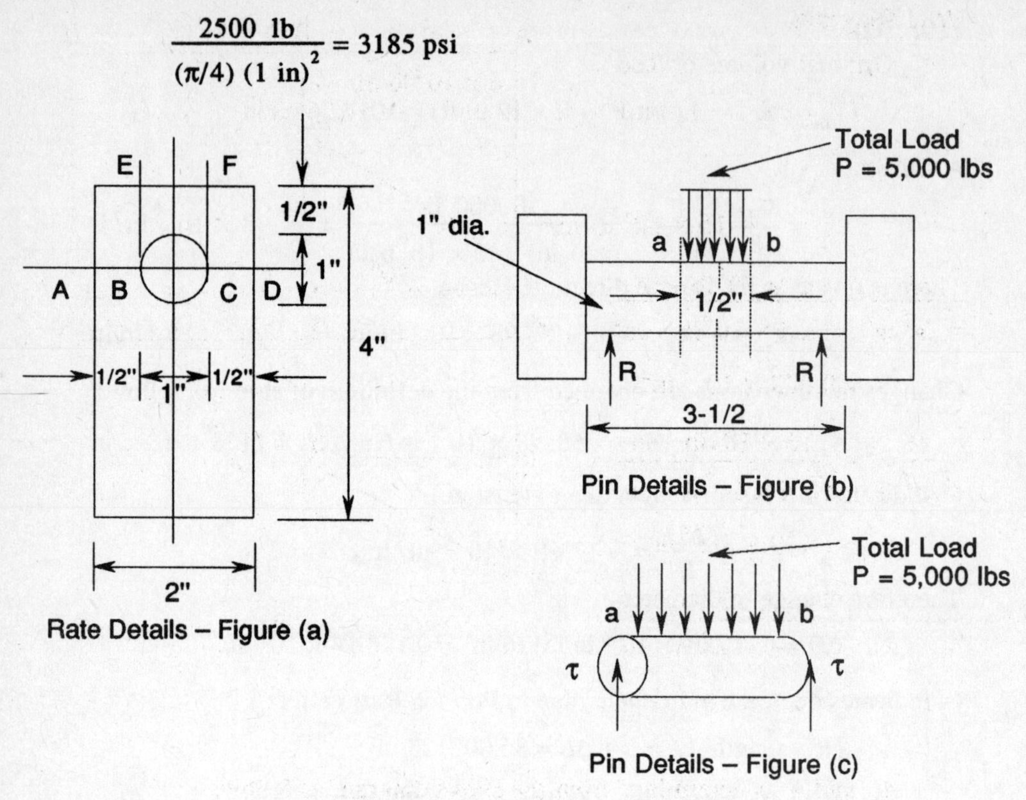

112. **(A)**

Maximum tensile stress will be acting at minimum plate area (section *AB* and *CD* at Figure (A) in Answer 111).

Max Plate Tensile Stress

$$= \frac{5000 \text{ lb}}{2(1/2 \text{ in} \times 1/2 \text{ in})} = 10,000 \text{ psi}$$

113. **(B)**

Since *EI* is variable along the length the "Moment-Area" method is employed.

y_c = Deflection at $C = A_{EI}\overline{X}_c$,

where A_{EI} = Area of M/EI vs x diagram left of point C

$\overline{X}_c$ = Distance of centroid of A_{EI} measured from C.

Moment Diagram (See following Page)

$$E_{AB}I_{AB} = 20 \times 10^6 \text{ psi} \times 0.67 \text{ in}^4$$
$$= 13.4 \times 10^6 \text{ lb-in}^2$$

$$E_{BC}I_{BC} = 30 \times 10^6 \text{ psi} \times 0.083 \text{ in}^4$$
$$= 2.49 \times 10^6 \text{ lb-in}^2$$

***M/EI* Diagram**

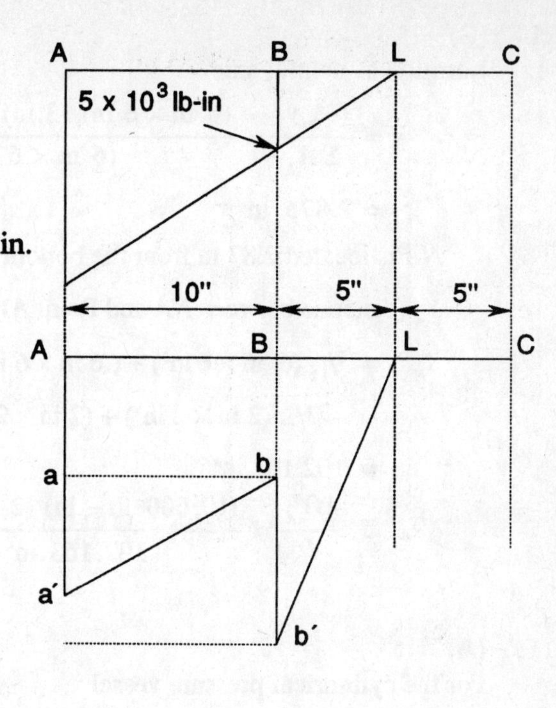

15 × 10³ lb-in.

$$\frac{M}{E_{AB}I_{AB}} = \frac{5 \times 10^{3}\,\text{lb} - \text{in}}{13.4 \times 10^{6}\,\text{lb} - \text{in}^{2}}$$

$$= 0.373 \times 10^{-3}\ \text{in}^{-1}$$

$$\frac{M}{E_{AB}I_{AB}} = \frac{15 \times 10^{3}\,\text{lb} - \text{in}}{13.4 \times 10^{6}\,\text{lb} - \text{in}^{2}}$$

$$= 1.119 \times 10^{-3}\,\text{in}^{-1}$$

$$\frac{M}{E_{BC}I_{BC}} = \frac{5 \times 10^{3}\,\text{lb} - \text{in}}{2.49 \times 10^{6}\,\text{lb} - \text{in}^{2}}$$

$$= 2.008 \times 10^{-3}\ \text{in}^{-1}$$

A_{EI} and $\overline{X}_c$ is determined from the above diagram as follows:

A_1 = $ABba = 0.373 \times 10^{-3}\ \text{in}^{-1} \times 10\ \text{in} = 3.73 \times 10^{-3};$

x_1 = $10\ \text{in} + 5\ \text{in} = 15\ \text{in}$

A_2 = $aba' = {}^{1}/_{2}\,(1.119 - .373) \times 10^{-3}\,(10\ \text{in}) = 3.73 \times 10^{-3};$

x_2 = $10\ \text{in} + {}^{2}/_{3}\,(10\ \text{in}) = 16.67\ \text{in}$

A_3 = $BLb' = {}^{1}/_{2}\,(2.008 \times 10^{-3}\ \text{in}^{-1})\,(5\ \text{in}) = 5.02 \times 10^{-3}.$

x_3 = $5\ \text{in} + {}^{2}/_{3}\,(5\ \text{in}) = 8.33\ \text{in}$

A_{EI} = $A_1 + A_2 + A_3 = 12.48 \times 10^{-3}$

$$\overline{X}_c = \frac{\Sigma A_i X_i}{\Sigma A_i}$$

$$= \frac{3.73 \times 10^{-3}(15) + 3.73 \times 10^{-3}(16.67) + 5.02 \times 10^{-3}(8.33)}{12.48 \times 10^{-3}}$$

$$= 12.816\ \text{in}$$

$\overline{Y}_c$ = $12.48\,(10^{-3})\,(12.816\ \text{in}) = 159.9 \times 10^{-3}\ \text{in}$

$\overline{Y}_c$ = $0.1599\ \text{in}.$

114. **(C)**

Location of neutral axis

$$\bar{y} = \frac{\Sigma A_i y_i}{\Sigma A_i} = \frac{(6\text{ in} \times 6\text{ in})(3\text{ in}) - (2\text{ in} \times 2\text{ in})(3\text{ in} + 1\text{ in})}{(6\text{ in} \times 6\text{ in}) - (2\text{ in} \times 2\text{ in})}$$

$$= 2.875\text{ in}$$

NA is located 2.87 in from the bottom surface.

Y_A (distance between *NA* and Point *A*) = 5 in – 2.875 in = 2.125 in.

$$I_{NA} = {}^1/_{12}(6\text{ in} \times 6\text{ in}^3) + (6\text{ in} \times 6\text{ in})(3\text{ in} - 2.875\text{ in})^2$$

$$- [{}^1/_{12}(2\text{ in} \times 2\text{ in}^3) + (2\text{ in} \times 2\text{ in})(4\text{ in} - 2.875\text{ in})^2]$$

$$= 102.163\text{ in}^4$$

$$\sigma_A = \frac{MY_A}{I} = \frac{(10{,}000\text{ lb} - \text{in})(2.125\text{ in})}{102.163\text{ in}^4} = 208\text{ psi}$$

115. **(A)**

For the cylindrical pressure vessel

$$\sigma_{long} = \frac{P_r}{Zt} = \frac{(1000\text{ psi})(36\text{ in}/2)}{Zt}$$

Since $\sigma_{long} = 0.2\,\sigma_{yield}$

$$\frac{(1000\text{ psi})(18\text{ in})}{2t} = 0.2\,(30{,}000\text{ psi})$$

$$t = \frac{(18\text{ in})(1000\text{ psi})}{0.4\,(30{,}000\text{ psi})} = 1.5\text{ in.}$$

116. **(E)**

Effective interest rate per year is the annual interest rate taking into account the effect of any compounding during the year.

117. **(D)**

Effective interest rate per compounding subperiod is

$$i = \frac{r}{m}$$

Here $r = 10\%$ and $m = 2$. Therefore $i = {}^{10\%}/_2 = 5\%$.

118. **(B)**

The inflation may vary. Therefore, the future payments in year 0 dollars will not be constant and will change with changes in inflation.

119. **(C)**

Year	Book Value Before Depreciation Change	SOYD Depreciation For The Year	Book Value After Depreciation Change
1	76.00	5/15 (76–6) = 23.33	52.67
2	52.67	4/15 (76 – 6) = 18.67	34.00
3	34.00	3/15 (76 – 6) = 14.00	20.00
4	20.00	2/15 (76 – 6) = 9.33	10.67
5	10.67	1/15 (76 – 6) = 4.67	6.00

Note: The book values and depreciation in the above table are in millions of dollars.

120. **(D)**

Draw the cash-flow diagram.

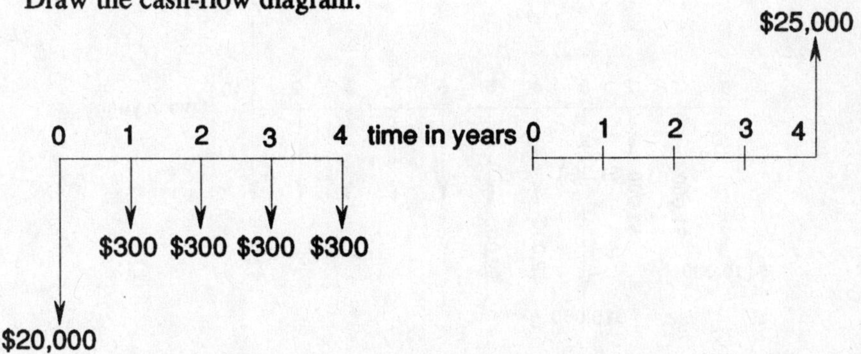

i = interest rate per year compounded annually

The rate of return must be found by a trial-and-error process as shown below. If the rate of return is i%, then the present worth of revenue should equal the present worth of expenses.

$$\underbrace{\$20,000 + \$300\ (P/A,\ i\%,4)}_{\text{Present worth of expenses}} = \underbrace{\$25,000(P/F,\ i\%,4)}_{\text{Present worth of revenue}}$$

Compute the left-hand-side (LHS) and right-hand-side (RHS) of the above equation for different values of i using interest tables for annual compounding. The value of i that makes LHS = RHS is the rate of return on the investment.

Interest Rate (%)	LHS ($)	RHS ($)	RHS– LHS ($)
3	21,115.00	22,212.50	1,097.50
4	21,088.00	21,370.00	282.00
5	21,063,00	20,567.00	– 496.00

The rate of return is about 4%.

121. **(C)**

Note that the payments are paid monthly and the interest is compounded monthly. Interest rate per month is 9/12% or 3/4%.

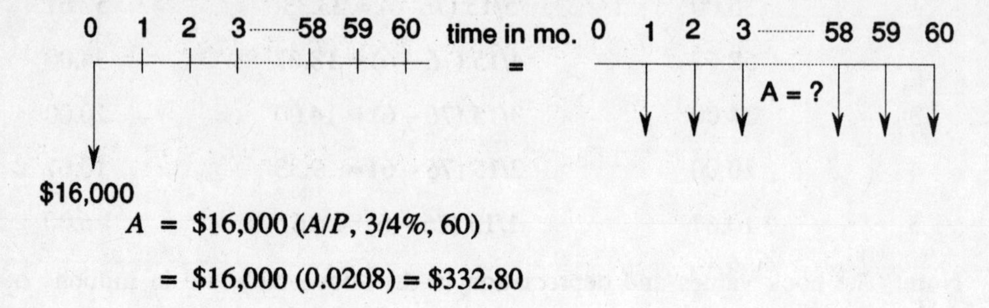

$$A = \$16{,}000\ (A/P,\ 3/4\%,\ 60)$$
$$= \$16{,}000\ (0.0208) = \$332.80$$

122. **(C)**

Draw the cash-flow diagram

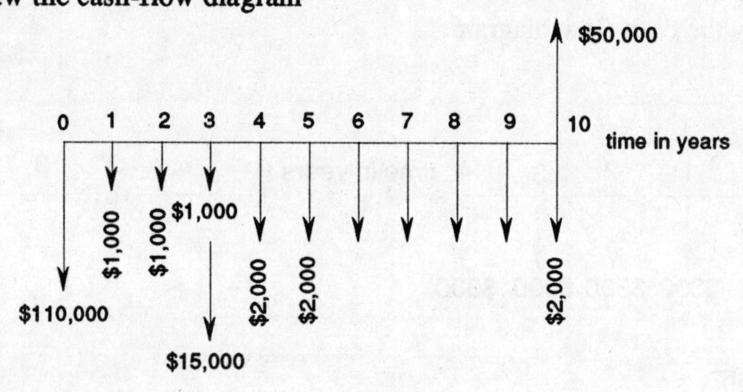

First find the Future Worth of all cash-flow.

$$FW = \$110{,}000\ (F/P,\ 12\%,\ 10) + \$1{,}000\ (F/A,\ 12\%,\ 10)$$
$$+ \$50{,}000\ (F/P,\ 12\%,\ 7) + \$1{,}000\ (F/A,\ 12\%,\ 7) - \$50{,}000$$
$$= \$110{,}000\ (3.106) + \$1{,}000\ (17.549) + \ (2.211)$$
$$+ \$1{,}000\ (10.089) - \$50{,}000$$
$$= \$341{,}660 + \$17{,}549 + \$110{,}550 + \$10{,}089 - \$50{,}000$$
$$= \$429{,}848$$

Convert Future Worth of $429,848 at the end of ten years to Equal Uniform Annual Cost (EUAC)

$$EUAC = FW\ (A/F,\ 12\%,\ 10)$$
$$= \$429{,}848\ (0.0057)$$
$$= \$24{,}501 \text{ or } \$24{,}500$$

Note: You may also find the Present Worth (PW) of all cash-flow and then convert *PW* to *EUAC*.

$$
\begin{aligned}
PW &= \$110{,}000 + \$1{,}000\,(P/A, 12\%, 10) + [\$50{,}000 \\
&\quad + \$1{,}000\,(P/A, 12\%, 7)](P/F, 12\%, 3) - \$50{,}000\,(P/F, 12\%, 10) \\
&= \$110{,}000 + \$1{,}000\,(5.6502) + [\$50{,}000 + \$1{,}000\,(4.5638)] \\
&\quad (0.7118) - \$50{,}000\,(0.322) \\
&= \$110{,}000 + \$5{,}650.20 + [\$50{,}000 + 4{,}563.80]\,(0.7118) - \$16{,}100 \\
&= \$138{,}388.71 \\
EUAC &= PW(A/P, 12\%, 10) \\
&= \$138{,}388.71\,(0.177) \\
&= \$24{,}494.80 \text{ or } \$24{,}500
\end{aligned}
$$

123. (B)

First consider $t = 5$ years. Find the equivalent uniform annual cash flow for the following diagram:

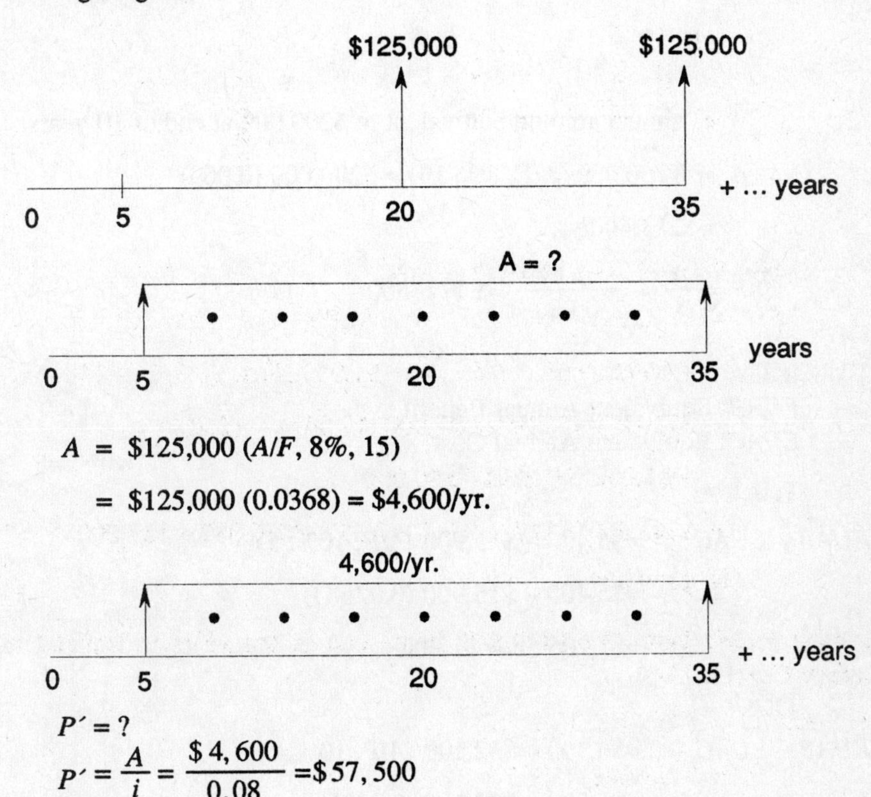

$$
\begin{aligned}
A &= \$125{,}000\,(A/F, 8\%, 15) \\
&= \$125{,}000\,(0.0368) = \$4{,}600/\text{yr.}
\end{aligned}
$$

$P' = ?$

$$P' = \frac{A}{i} = \frac{\$4{,}600}{0.08} = \$57{,}500$$

Now consider the following:

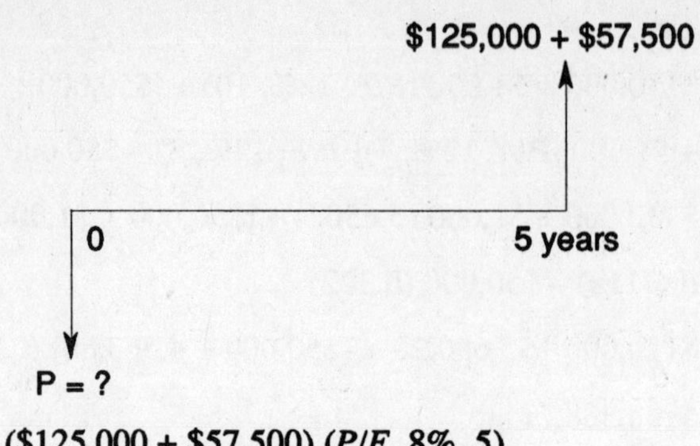

$$P = (\$125{,}000 + \$57{,}500)\ (P/F, 8\%, 5)$$
$$= 182{,}500\ (0.6806) = \$124{,}209.50$$

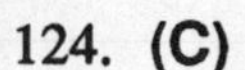
124. **(C)**

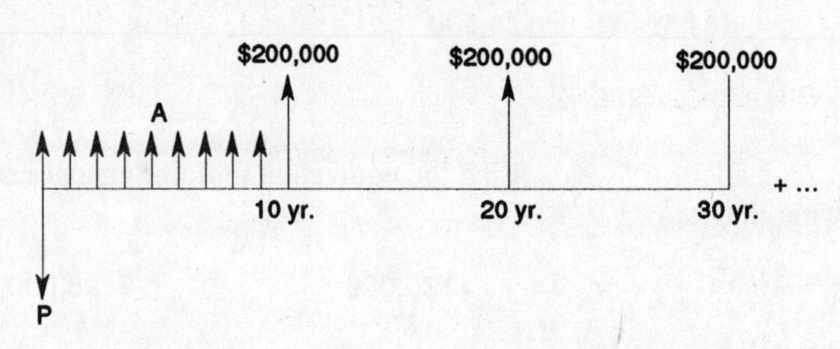

A : annual amount equivalent to $200,000 at end of 10 years.

$$A = \$200{,}000\ (A/F, 8\%, 10) = \$200{,}000\ (0.069)$$
$$= \$13{,}800\text{/yr.}$$
$$P = \frac{A}{i} = \frac{\$\,13{,}800}{0.08} = \$\,172{,}500$$

125. **(E)**
EUAB: Equivalent Annual Benefit
EUAC: Equivalent Annual Cost

Truck *A*:

$$EUAB - EUAC = \$5{,}400 - \$15{,}900\ (A/P, 10\%, 5)$$
$$= \$5{,}400 - \$15{,}900\ (0.02683)$$
$$= \$1{,}205.58 \approx \$1{,}206$$

Truck *B*:

$$EUAB - EUAC = \$6{,}300 - \$32{,}100\ (A/P, 10\%, 6)$$
$$= \$6{,}300 - \$32{,}100\ (0.2296)$$
$$= -\$1{,}070.16 \approx -\$1{,}070$$

Advantage of Truck *A* over Truck *B* = $1,205.58 – (–$1,070.16) = $2,276.

126. **(B)**

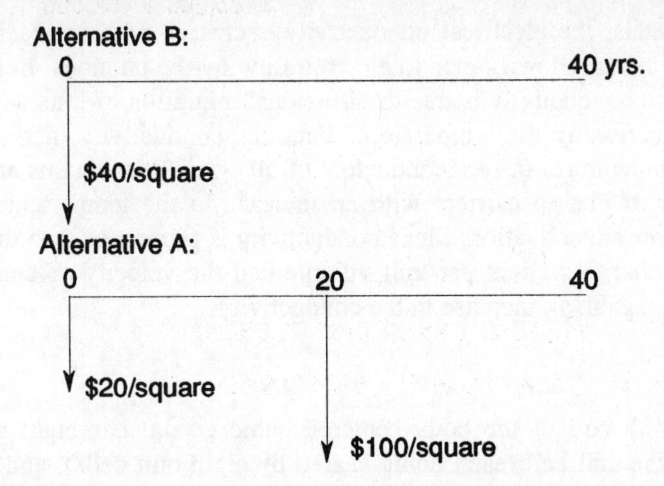

$$PW \text{ (Cost / Square)} = \$20 + \$100\,(P/F, 12\%, 20)$$

$$= \$30.37 \text{ or } \$30$$

Cost advantage per square of alternative *A* over alternative *B*

= $40 – $30

= $10.

127. **(A)**

The intercepts made by the plane with the *x, y, z* axes are 1, 1, $^1/_2$ respectively. Hence the indices of the plane are 112.

128. **(D)**

In a fiber reinforced material which is loaded in a direction parallel to that of the fibers, the elastic modulus of the composite is given by the rule of mixtures as $E_c = (1 - f)E_m + fE_f$ where f is the volume fraction of the fibers, E_m is the elastic modulus of the matrix, and E_f is the elastic modulus of the fibers. Application of this formula yields for the elastic modulus of the composite a value of 2.32 million psi.

$$E_c = (1 - .2)\ 400{,}000 \text{ psi} + .2\ (10{,}000{,}000 \text{ psi}) = 2.32 \times 10^6 \text{ psi}$$

129. **(B)**

Tensile and yield strengths increase with decreasing grain size according to the Hall-Petch relationship. Impact strength and ductility also increase with grain refinement. Creeep resistance decreases with decreasing grain size due to increased diffusion flow contributing to several mechanisms of creep.

130. **(C)**

In metals, the electrical conductivity decreases with increasing temperature due to increased resistance to electron flow by the phonons. In ionic materials, electrical conductivity is due to diffusional migration of ions which is accelerated by increasing the temperature. Thus the conductivity increases with increasing temperature. In semiconductors of all types, one obtains an increase in the number of charge carriers with an increase in the temperature due to the increased thermal activation. Since conductivity is proportional to the product of number of charge carriers per unit volume and the velocity of charge carriers, one obtains, again, an increase in the conductivity.

131. **(A)**

The unit cell of the body centered cubic crystal has eight atoms in the corners of the unit cell (each being shared by eight unit cells); and one atom at the center of the cube which belongs entirely to the unit cell. Thus the total number of atoms belonging to one unit cell is $1 + {}^{8}/_{8} = 2$.

132. **(E)**

In wire drawing, the material is plastically deformed to obtain a reduction in the cross sectional area. The material property that characterizes this is the ductility. None of the other properties in the list relate to this.

133. **(C)**

The fatigue strength relates to the ease with which fatigue cracks can initiate and propagate in a given alloy. In most cases, fatigue cracks originate at the surface of the part. Factors which improve the fatigue strength include surface strengthening, removing surface roughness, putting compressive residual stresses on the surface and reducing the crack propagation rate in the bulk material. All treatments except shot peening will have the effect of reducing the fatigue strength from these considerations.

134. **(A)**

This question tests knowledge of the structure of atoms and the difference between isotopes of the elements. The correct answer is (A). The element is determined by the number of protons in the nucleus (which, by the requirement of electrical neutrality, equals the number of electrons). The atomic weight is changed by differing numbers of neutrons. Answer (B) is incorrect. Electrical neutrality requirements dictate that the number of protons and electrons must be equal. Answer (C) is incorrect. If the number of protons and electrons change, the element also changes. Answer (D) is incorrect. The valence electrons are a function of the identity of the element and do not change with different isotopes of the same element. Answer (E) is incorrect. If the number of protons, electrons, and neutrons are the same, the species is the same isotope of an element.

135. **(D)**

This question tests knowledge of isotopes and how isotopes of an element differ. Isotopes of an element have an identical electronic configuration and, hence, excluding extremely light isotopes such as hydrogen and deuterium, have identical chemical properties. Separation, therefore, requires procedures that take advantage of differences in atomic weight since there are no chemical differences. The correct answer is (D). Answers (A), (B), and (C) are all correct statements about isotopes.

136. **(D)**

This question tests knowledge of the structure of semiconductors and the mechanism of conduction in semiconducting materials. It also tests knowledge of the difference between p-type and n-type semiconductors. P-type semiconductors are those in which conductivity is provided by positive (p-type) carriers. Holes (or the absence of electrons in the valence band) are the positive carriers. Hence, the correct answer is (D). Answer (A) is incorrect. The motion of electrons provides conductivity in n-type semiconductors. Electrons are negative or n-type carriers. Answer (B) is incorrect. The thermal excitation of electrons from the valence to the conduction band provides conductivity in intrinsic semiconductors. Impurity semiconductors (those doped with either n-type or p-type impurities) do not rely on thermal excitation from the valence to the conduction band as a mechanism for conductivity. Answer (C) is incorrect. Ions do not provide for the conductivity in semiconducting materials. Inasmuch as answers (A), (B), and (C) are incorrect, answer (E) is also incorrect.

137. **(C)**

This question tests knowledge of the structure of the electronic configuration of matter and the relationship of electronic configuration to chemical reactivity. In the configuration above, the first and second electronic orbits are filled. The third electronic orbit is one electron short of being filled — i.e. if the third orbit contained 6 p electrons, instead of only 5, it would be filled. Elements needing only one p electron to complete the outer orbit are the halogens. The correct answer, therefore, is (C). Answer (A) is incorrect. The alkali metals are characterized by an ns^1 electron in the nth or outer orbit. Answer (B) is incorrect. The inert gases are characterized by ns^2np^6 electrons in the nth or outer orbit. Answer (D) is incorrect. The transition metals are characterized by ns^2 electrons in the nth or outer orbit and d electrons in the (n-1) orbit. Answer (E) is incorrect. The alkaline earth metals are characterized by ns^2 electrons in the nth or outer orbit.

138. **(C)**

This question tests knowledge of the nature of sub-atomic particles and their relationship to light. The correct answer is (C). A quantum amount of light or light particle is called a photon. It has both momentum and energy. Answer

(A) is incorrect. There is no such thing as a neutral proton. A proton has a charge of +1. Answer (B) is also incorrect since there is no such thing as a neutral electron. An electron has a charge of –1. Answer (D) is incorrect. Neutrons are not charged, but are neutral. Answer (E) is incorrect. Although light is sometimes described as electromagnetic waves, when it is described in terms of photons, it is considered to be particulate in nature. Light can be described as either particulate or wave-like.

139. **(E)**

This question tests knowledge of the rate of decay of radioisotopes and what is meant by the half life of an isotope. The half life is the time required for one-half of the nuclides in a sample of a radioisotope to decompose. Since the decay products are stable, all of the b radiation comes from the parent isotope and is proportional to the number of nuclides present. The number of nuclides is given by the equation:

$$N = N_o e^{-kt}$$

where N_o is the number of nuclides at zero time and k is a constant. To determine k, observe that $N = 0.5N_o$ when $t = t_{1/2}$.

$$0.5 = \exp(-kt_{1/2})$$

$$k = \ln(2)/t_{1/2}$$

Therefore:

$$0.1\ N_o = N_o \exp(-(\ln(2)\ t/t_{1/2}) \text{ where } t = 450 \text{ min.}$$

$$t_{1/2} = -\ln(2)\ (450 \text{ min.})/\ln(0.1) = 135.46 \text{ minutes}$$

Therefore, the correct answer is (E). Answer (A) is incorrect; it is one-half the total time for a ten-fold decay. Answer (B) is incorrect; it is one-tenth the total time for a ten-fold decay. Answer (C) is incorrect; it is obtained by dividing the time in minutes by the time in hours. Answer (D) is incorrect; it is obtained by taking 75% of the total observed time for a ten-fold decay.

140. **(E)**

This question tests knowledge of an Angstrom unit — a very basic unit in specifying molecular and atomic distances. The correct answer is (E), 10^{-10} m. Answer (A) is incorrect, since 10^{-9} m is a nanometer. Answer (B) is incorrect. The sign in the exponent is incorrect. Answer (C) is incorrect. The distance traveled by light in 10^{-10} sec. can be calculated as follows:

$$d = (3 \times 19^9 \text{ m/sec})(10^{-10} \text{ sec}) = 0.3 \text{ m} = 30 \text{ cm}$$

Answer (D) is incorrect, since 10^{-6} m is a micrometer or micron.

Fundamentals of Engineering

P.M. SECTION

Test 2

Fundamentals of Engineering

P.M. SECTION

TEST 2 – ANSWER SHEET

1. A B C D E
2. A B C D E
3. A B C D E
4. A B C D E
5. A B C D E
6. A B C D E
7. A B C D E
8. A B C D E
9. A B C D E
10. A B C D E
11. A B C D E
12. A B C D E
13. A B C D E
14. A B C D E
15. A B C D E
16. A B C D E
17. A B C D E
18. A B C D E
19. A B C D E
20. A B C D E
21. A B C D E
22. A B C D E
23. A B C D E
24. A B C D E
25. A B C D E
26. A B C D E
27. A B C D E
28. A B C D E
29. A B C D E
30. A B C D E
31. A B C D E
32. A B C D E
33. A B C D E
34. A B C D E
35. A B C D E
36. A B C D E
37. A B C D E
38. A B C D E
39. A B C D E
40. A B C D E
41. A B C D E
42. A B C D E
43. A B C D E
44. A B C D E
45. A B C D E
46. A B C D E
47. A B C D E
48. A B C D E
49. A B C D E
50. A B C D E
51. A B C D E
52. A B C D E
53. A B C D E
54. A B C D E
55. A B C D E
56. A B C D E
57. A B C D E
58. A B C D E
59. A B C D E
60. A B C D E
61. A B C D E
62. A B C D E
63. A B C D E
64. A B C D E
65. A B C D E
66. A B C D E
67. A B C D E
68. A B C D E
69. A B C D E
70. A B C D E

FUNDAMENTALS OF ENGINEERING EXAMINATION

TEST 2

AFTERNOON (PM) SECTION

TIME: 4 Hours
70 Questions

DIRECTIONS: For each of the following questions and incomplete statements, choose the best answer from the five answer choices. You must answer all questions.

QUESTIONS 1–10 refer to the following problem.

Questions 1–10 relate to the phase diagram shown below. Assume equilibrium conditions unless specified otherwise. α, β and δ are solid solutions.

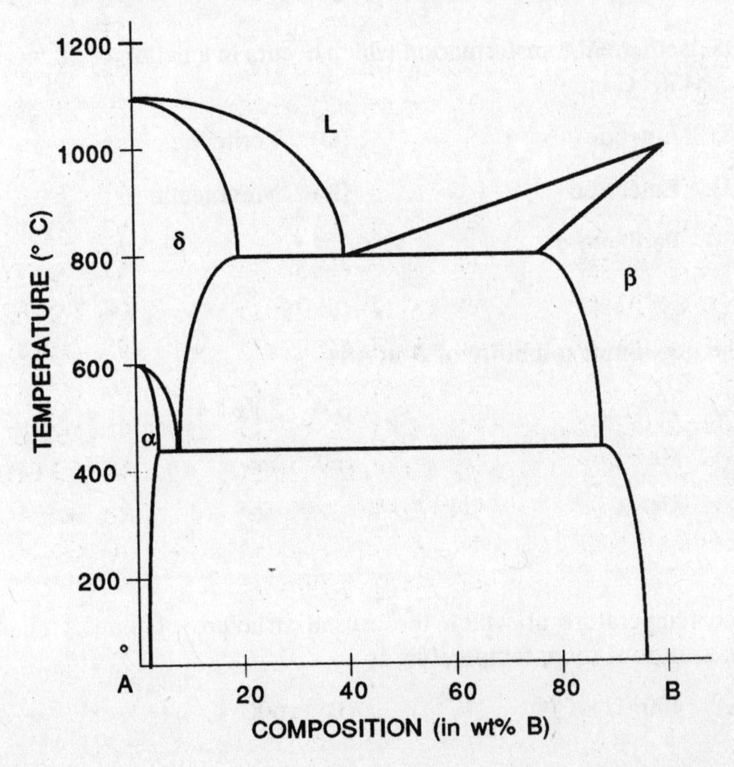

1. The lowest temperature at which liquid will be present in an alloy of 60 weight percent *B is*

   (A) 800° C
   (B) 450° C
   (C) 950° C
   (D) 1100° C
   (E) 870° C

2. The quantity of α at room temperature in 80 kilograms of an alloy of 48 weight percent *B* is most nearly

   (A) 48 kilograms
   (B) 52 kilograms
   (C) 39.6 kilograms
   (D) 36.7 kilograms
   (E) 41.6 kilograms

3. The composition of β in an alloy of 48 weight percent *B* at 700° C is

   (A) 80% *B*
   (B) 90% *B*
   (C) 14% *B*
   (D) 48% *B*
   (E) 84% *B*

4. The isothermal transformation which occurs in an alloy of 48 weight percent *B* at 450° *C* is

   (A) Eutectic
   (B) Eutectoid
   (C) Peritectoid
   (D) Peritectic
   (E) Monotectic

5. The maximum solubility of *B* in *A* is

   (A) 40%
   (B) 80%
   (C) 6%
   (D) 20%
   (E) 1%

6. The temperature at which the crystal structure of pure *A* changes upon heating from room temperature is

   (A) 450° C
   (B) 600° C

(C) 800° C

(D) 1200° C

(E) 1000° C

7. The percentages of phases present in the eutectic alloy at 500° C is most nearly

(A) 33.3% δ and 66.7% β

(B) 40% δ and 60% β

(C) 60% δ and 40% β

(D) 66.7% α and 33.3% β

(E) 15% α and 85% *L*

8. The solubility limit of *A* in *B* at 450° C is

(A) 6%

(B) 90%

(C) 2%

(D) 98%

(E) 10%

9. How many degrees of freedom exist for an alloy containing 30 weight percent *B* at temperatures less than 400° C at a constant pressure of 1 atmosphere?

(A) 1

(B) 2

(C) 3

(D) 4

(E) 0

10. The percentage of eutectic liquid in an alloy of 60 weight percent *B* at 800° C is

(A) 50%

(B) 40%

(C) 80%

(D) 60%

(E) 0%

QUESTIONS 11–20 refer to the following problem.

The beam *ABCDEF* below is loaded with 10 kips (k) at *B* and 20 kips at *D*. A uniformly distributed load of 2 kips per linear foot (k/ft) is spread over the entire beam. The beam has the *T* shape with dimensions as shown. The 30 kip axial force is applied at the centroid of the section.

Units:	k	= kips	in	= inches
	ft	= feet	lb	= pounds

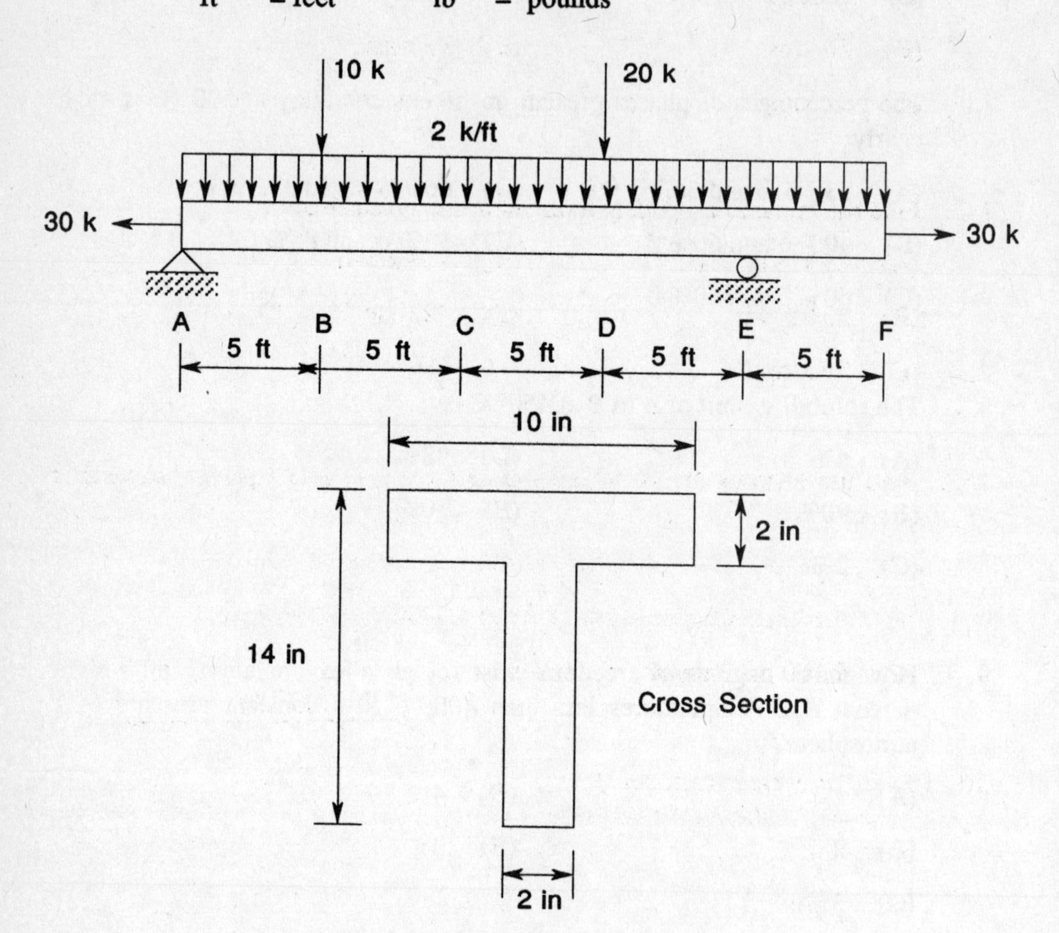

11. Determine the location of the neutral axis of the beam section.

(A) 4.23 inches from the top of the beam

(B) 4.82 inches from the top of the beam

(C) 5.23 inches from the top of the beam

(D) 7.00 inches from the top of the beam

(E) 8.00 inches from the top of the beam

12. Determine the moment of inertia about the neutral axis of the *T*-section.

(A) 44 in^4

(B) 295 in^4

(C) 829 in^4

(D) 1135 in^4

(E) 2287 in^4

13. Find the maximum shear force in the loaded beam.

(A) 10 kips
(B) 20 kips
(C) 31 kips
(D) 39 kips
(E) 49 kips

14. Find the maximum bending moment in the given beam.

(A) 25 kip-ft
(B) 131 kip-ft
(C) 144 kip-ft
(D) 163 kip-ft
(E) 625 kip-ft

15. Find the bending stress on the top of the beam at B. (psi = pounds per square inch)

(A) 0 psi
(B) 9,160 psi Compression
(C) 9,160 psi Tension
(D) 17,440 psi Compression
(D) 17,440 psi Tension

16. Find the horizontal shear stress at $m - m$ at the right hand side of section E.

(A) 0 psi
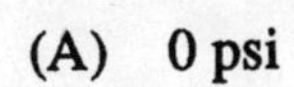
(B) 9 psi
(C) 46 psi
(D) 227 psi
(E) 341 psi

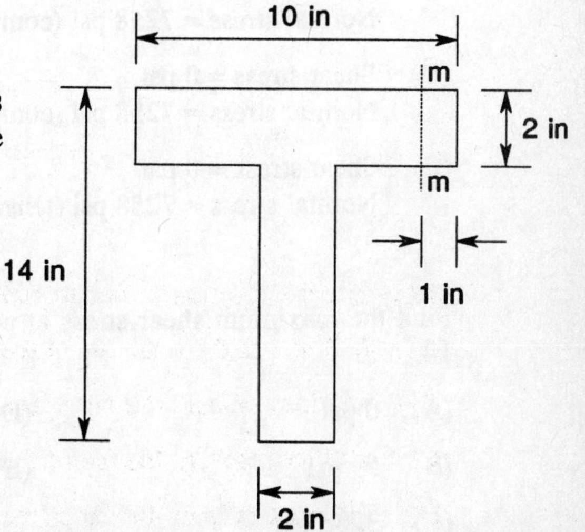

17. Find the maximum horizontal shear stress at the section on the left hand side of section B.

(A) 0 psi
(B) 216 psi
(C) 483 psi
(D) 724 psi
(E) 1081 psi

18. Find stresses at point *n* which is located at the section on the right hand side of *B* and 1 inch from the top of the beam.

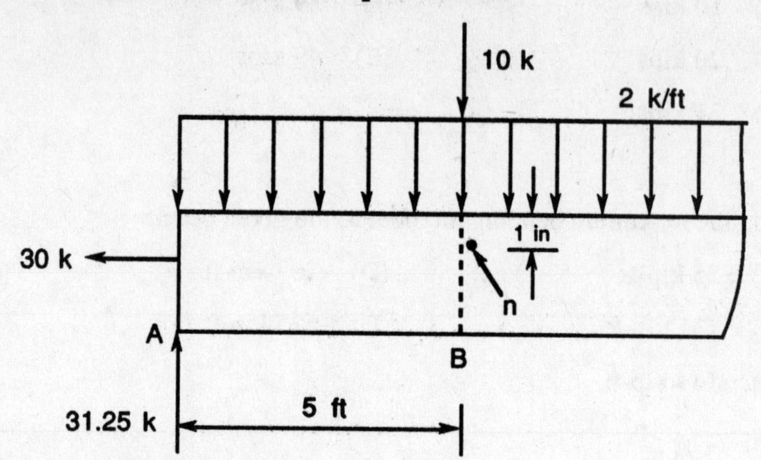

(A) Shear stress = 59 psi
Normal stress = 6576 psi (tension)

(B) Shear stress = 59 psi
Normal stress = 6576 psi (compression)

(C) Shear stress = 59 psi
Normal stress = 7258 psi (compression)

(D) Shear stress = 0 psi
Normal stress = 7258 psi (compression)

(E) Shear stress = 0 psi
Normal stress = 7258 psi (tension)

19. Find the maximum shear stress at point *n* which was defined in problem 18.

(A) 0 psi
(B) 59 psi
(C) 3288 psi
(D) 6576 psi
(E) 7258 psi

20. Find the maximum tensile stress at point *m* which is located at the N.A. of the section on the right hand side of *B*. (See figure following)

(A) 0 psi
(B) 666 psi
(C) 682 psi
(D) 890 psi
(E) 1007 psi

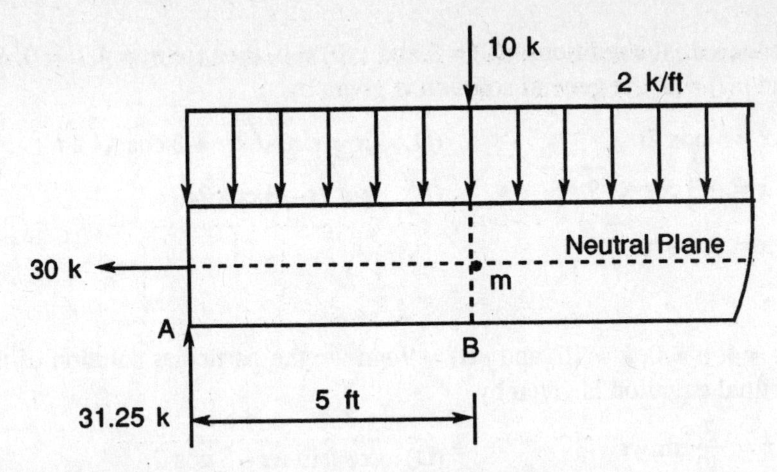

QUESTIONS 21–30 refer to the following problem.

The motion of the mass-dashpot-spring system shown below is given by the following ordinary differential equation:

$$m\ddot{x} + c\dot{x} + kx = p(t)$$

Where x is the displacement, t is time, m is the mass, c is the damping, k is the spring stiffness, and $p(t)$ is the external force.

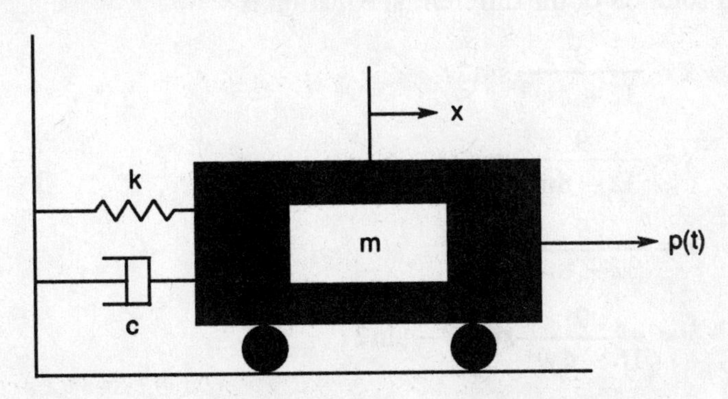

21. For $c = 0$ and $p(t) = 0$, the differential equation is called

    (A) homogeneous
    (B) unstable
    (C) stiff
    (D) nonlinear
    (E) nonhomogeneous

22. If $p(t) = \pi$, then the differential equation is called

    (A) homogeneous
    (B) unstable
    (C) trigonometric
    (D) nonlinear
    (E) nonhomogeneous

23. Given the initial conditions $x(0) = 3$ and $x(0) = 0$, then for $m = 4$, $c = 0$, $k = 16$, and $p(t) = 0$, the general solution is given by

    (A) $x = 3\cos 2t$

    (B) $x = -3\cos\sqrt{2}\,t$

    (C) $x = 2\cos\sqrt{2}\,t$

    (D) $x = \sin\sqrt{2}\,t + 3\cos\sqrt{2}\,t$

    (E) $\sin 2t - 3\cos 2t$

24. For $m = 4$, $c = 0$, $k = 16$, and $p(t) = 9\sin wt$, the particular solution of the differential equation is given by

    (A) $x = \dfrac{3}{49}\sin wt$

    (B) $x = -\dfrac{3}{65}\cos 2t$

    (C) $x = \dfrac{1}{20}\sin wt$

    (D) $x = \sin wt - 3\cos 2t$

    (E) $\dfrac{9}{16 - 4w^2}\sin wt$

25. For $m = 4$, $c = 0$, $k = 16$, and $p(t) = 9\sin wt$, and zero initial conditions, the total solution of the differential equation is

    (A) $x = \dfrac{9}{32 - 8w^2}\sin 2t$

    (B) $x = \dfrac{9}{32 - 8w^2}\cos 2t$

    (C) $x = \dfrac{9}{32 - 8w^2}\sin wt$

    (D) $x = \dfrac{9}{16 - 4w^2}\sin wt - \sin 2t$

    (E) $x = \dfrac{9}{16 - 4w^2}\left[\sin wt - \dfrac{w}{2}\sin 2t\right]$

26. The solution obtained for question 25 suggests that resonance will occur at what frequency of excitation.

    (A) $w = 0$

    (B) $w = 7$

    (C) $w = 4$

    (D) $w = 7$

    (E) $w = 2$

27. Give the initial conditions $x(0) = 1$ and $x(0) = 2$, then for $m = 4$, $c = 0$, $k = 16$, and $p(t) = 2$, the general solution is given by

(A) $x = \frac{1}{8}[\sin 2t + \cos 2t]$

(B) $x = \frac{7}{8}[\cos 2t]$

(C) $x = \frac{1}{8}[\cos 2t]$

(D) $x = \frac{1}{8}[\sin 2t - 7\cos 2t]$

(E) $x = \frac{1}{8}[1 + 8\sin 2t + 7\cos 2t]$

28. For $m = 4$, $c = 0$, $k = 16$, and $p(t) = e^{-\pi t}$, the particular solution is given by

(A) $x = \dfrac{e^{-\pi t}}{16 + 4\pi}$

(B) $x = \dfrac{e^{-\pi t}}{16 + 4\pi^2}$

(C) $x = \dfrac{e^{-\pi t}}{16 - 4\pi}$

(D) $x = \dfrac{e^{-\pi t}}{-16 + 4\pi^2}$

(E) $x = \dfrac{\pi e^{-\pi t}}{16 + 4\pi}$

29. Assuming the $m - c - k$ is initially at rest, the total solution for the differential equation described in problem 28 is given by

(A) $x = \dfrac{1}{16 + 4\pi^2}\left[e^{-\pi t} + \dfrac{\pi}{2}\sin 2t - \cos 2t\right]$

(B) $x = \dfrac{1}{16 + 4\pi}\left[e^{-\pi t} + \dfrac{1}{2}\sin 2t - \cos 2t\right]$

(C) $x = \dfrac{1}{16 + 4\pi^2}[e^{-\pi t} + \sin 2t - \cos 2t]$

(D) $x = \dfrac{1}{16 - 4\pi^2}\left[-\dfrac{\pi}{2}\sin 2t - \cos 2t\right]$

(E) $x = \dfrac{1}{16 + 4\pi^2}\left[\dfrac{\pi}{2}\sin 2t - \cos 2t\right]$

30. For $m = 4$, $c = 1$, $k = 16$, $p(t) = 0$, and initial conditions $x(0) = 1$ and $x(0) = 0$, the total solution is given by

(A) $x = e^{-t/8}\left[\dfrac{1}{\sqrt{255}}\sin\dfrac{\sqrt{255}}{8}t + \cos\dfrac{\sqrt{255}}{8}t\right]$

(B) $x = e^{-t/8}\left[\dfrac{8}{\sqrt{255}}\sin\dfrac{\sqrt{255}}{8}t + \cos\dfrac{\sqrt{255}}{8}t\right]$

(C) $x = e^{-t/8}\left[\sin\frac{1}{4}t + \cos\frac{1}{4}t\right]$

(D) $x = e^{-t/8}\left[\frac{1}{2}\sin 2t + \cos 2t\right]$

(E) $x = e^{-8t}\left[\frac{1}{2}\sin\frac{1}{2}t - \cos\frac{1}{2}t\right]$

QUESTIONS 31–40 refer to the following problem.

Given the fourth order polynomial and the parabola as shown below:

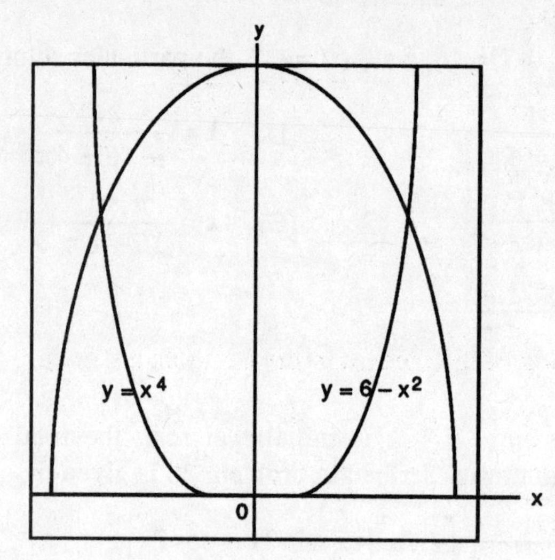

Figure is not to scale

31. Find the points of intersection between the two curves.

(A) $(\pm 1.5, 4)$

(B) $(\pm 1.4, 4)$

(C) $(\pm\sqrt{2}, 4)$

(D) $(\pm\frac{2\sqrt{5}}{3}, 4)$

(E) $(\pm\frac{3\sqrt{3}}{4}, 4)$

32. Determine the area enclosed by the two functions.

(A) $\frac{56}{15}\sqrt{3}$

(B) $\frac{136}{15}\sqrt{2}$

(C) 0

(D) $\frac{112}{15}\sqrt{2}$

(E) $\frac{136}{18}\sqrt{3}$

33. Find the y–coordinate of the center of the area enclosed by the fourth order polynomial $y = x^4$ and a horizontal line drawn at $y = 4$.

(A) $\frac{3}{2}\sqrt{3}$ (D) $\frac{9}{5}\sqrt{2}$

(B) $\frac{4}{3}\sqrt{2}$ (E) $\frac{8}{5}\sqrt{3}$

(C) $\frac{20}{9}$

34. Determine the volume, if the area enclosed by the two curves is rotated about the x–axis.

(A) $\frac{144}{45}\pi\sqrt{2}$ (D) $\frac{2132}{45}\pi$

(B) $\frac{2432}{45}\pi\sqrt{2}$ (E) $\frac{256}{45}\pi\sqrt{2}$

(C) $\frac{1256}{45}\sqrt{2}$

35. Find the equation of the tangent to the parabola at (2, 2).

(A) $y = 10 - 2x$ (D) $y = 10 - 4x$

(B) $y = 8 - 4x$ (E) $y = 10 + 4x$

(C) $y = 8 + 2x$

36. Find the equation of the normal to the fourth order polynomial at point (1, 1).

(A) $y = \frac{5}{4} - \frac{3x}{4}$ (D) $y = 4 - \frac{4x}{5}$

(B) $y = \frac{5}{4} - \frac{x}{4}$ (E) $y = \frac{1}{4} - \frac{5x}{4}$

(C) $y = \frac{4}{5} - \frac{4x}{5}$

37. Use the trapezoidal rule of integration to approximate the arc length of the fourth order polynomial between points (0, 0) and (1, 1). Assume an interval of integration equal to 0.25.

(A) 1.41 (D) 1.96

(B) 1.03 (E) 1.73

(C) 1.66

38. Find the arc length of the parabola between points (0, 6) and (2, 2). Use trigonometric substitution to evaluate the resulting integral. Also, note that the integral of ($\sec^3 \theta \, d\theta$) is equal to [$\sec \theta \tan \theta + \ln (\sec \theta + \tan \theta)$]/2.

(A) 4.95
(B) 4.00
(C) 4.65
(D) 5.15
(E) 4.35

39. Determine the volume swept out when the region bounded by the parabola $y = 6 - x^2$ and the line $x = 0$ is rotated about the line $x = \sqrt{6}$.

(A) 6π
(B) 2π
(C) $\pi\sqrt{3}$
(D) $\pi\sqrt{6}$
(E) π

40. Determine the surface area generated when the parabola above the line where it intersects the function $y = x^4$ is rotated about the y–axis.

(A) $\frac{\pi}{3}$
(B) $\frac{13\pi}{3}$
(C) $\frac{\pi}{2}$
(D) $\pi\sqrt{2}$
(E) $\frac{\pi\sqrt{2}}{2}$

QUESTIONS 41–50 refer to the following problem.

A consumer is using a P_{av} = 11 kW with PF = 1.0 and V_{rms} = 220 V, assume 0.1Ω per-phase transmission line resistance.

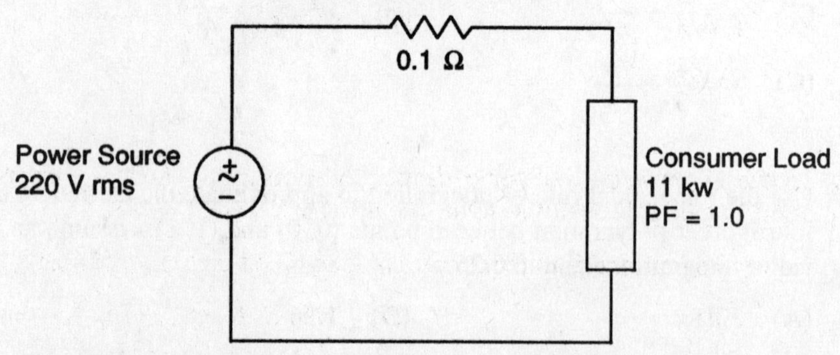

41. What is the *rms* current that flows through the load and the transmission lines?

(A) 100 A
(B) 50 A
(C) 70.71 A
(D) 35.35 A
(E) none of the above

42. How much are the per-phase line losses?

(A) 1 kW
(B) 500 W
(C) 250 W
(D) 750 W
(E) none of the above

43. In order to supply 11 kW to the consumer, how much total power should the power company generate?

(A) 11.5 kW
(B) 12 kW
(C) 11.75 kW
(D) 11.25 kW
(E) none of the above

44. What should be the *rms* voltage at the power company?

(A) 235 V
(B) 215 V
(C) 225 V
(D) 320 V
(E) 200 V

45. What is the percentage of the total generated power that the consumer would be billed?

(A) 91.7 %
(B) 100 %
(C) 93.6 %
(D) 84.6 %
(E) 97.7 %

Now, let the hypothesize another consumer, also requiring 11 kW, but at a PF angle of 60° lagging through the same line resistance.

46. How much current does the power company feed through this load?

(A) 50 A
(B) 57.73 A
(C) 100 A
(D) 115.47 A
(E) none of the above

47. How much are the total line losses (assuming the same transmission line resistance)?

(A) 3 kW
(B) 1.5 kW
(C) 1 kW
(D) 2.67 kW
(E) none of the above

48. What is the percentage that the customer is billed-out-of the actual energy generated?

(A) 95.6%
(B) 80.5%
(C) 94.3%
(D) 78.6%
(E) none of the above

49. How much should the transmission line resistance of 0.1 Ω be reduced or increased by, so that the power paid by the second consumer is increased to 90% of the generated power, assuming 11 kW power consumption and *PF* angle of 60° lag?

(A) – 0.159 Ω
(B) + 1.59 Ω
(C) – 0.059 Ω
(D) + 2.2 Ω
(E) + 0.03 Ω

50. If the resistance of the transmission line is kept constant at 0.1 Ω and the power consumption at 11 kW, what should the *PF* become so that the power efficiency of 90% in problem 49 is obtained?

(A) 57.66°
(B) 38.36°
(C) 84.53°
(D) none of the above
(E) impossible

QUESTIONS 51–60 refer to the following problem.

XYZ is a chemical manufacturing company based in the northeastern US. The company can obtain loans at 10% financing, compounded annually. However, for environmental projects the company can enjoy government secured loans at 6% interest compounded annually. XYZ uses Sum-of-Years-Digits (SOYD) depreciation for tax purposes and operates on a before tax Minimum Attractive

Rate of Return (MARR) of 20%. XYZ pays tax as a major corporation; i.e., 50% tax rate.

51. XYZ engineers propose to purchase machinery for $25,000. This machinery will last for 5 years and will have a salvage value of $5,000. This decision will earn the company a net annual profit of $8,000. The annual worth of this investment using the MARR of 20% is:

(A) $671.90
(B) $8,359.49
(C) $312.41
(D) $8,000
(E) $5,000

52. If this machinery was for an environmental project, compute the capital recovery amount at the given 6% interest compounded annually.

(A) $312
(B) $8,359
(C) $671
(D) $5,048
(E) $7,687

53. XYZ engineers are now given an alternative machine to purchase. Below are the profiles of both alternatives:

Alternative	A	B
First Cost	$25,000	$50,000
Annual Benefit	8,000	8,500
Salvage Value	5,000	5,500
Useful Life	5 years	10 Years

Neglecting taxes, what is the net present worth of the cost advantage of selecting alternative A using the company's MARR of 20%?

(A) $14,785
(B) $13,475.5
(C) $11,475.5
(D) $1,309.5
(E) $5,785

54. Finally the company uses machine A to produce units of specialty polymer which yields a profit contribution of $3.00 per unit. How many units of this polymer does XYZ have to make in order to break even? Use the company's MARR and an analysis period of 10 years.

(A) 2000 units
(B) 2563 units
(C) 2653 units
(D) 2365 units
(E) 3652 units

55. What is the percent rate of return for alternative B, assuming that the salvage value is now zero?

(A) 20%
(B) 6%
(C) 11%
(D) 10%
(E) 12%

56. XYZ has a plant that produces 100 million lbs of cellulose ester per annum and sells it for 50 cents/lb. The direct fixed capital for the plant is $70 million. The plant has a zero salvage value at the end of its useful life of 10 years. For this plant, XYZ is constrained to use the straight line method of depreciation. The cost of manufacturing, including depreciation, is 22 cents/lb, and the general expenses are 8 cents/lb. XYZ's annual income tax payments for this plant in millions of dollars is:

(A) 5
(B) 10
(C) 15
(D) 20
(E) 25

57. In problem 56, the annual profit before taxes in millions of dollars is:

(A) 5
(B) 10
(C) 15
(D) 20
(E) 25

58. In problem 56, the annual profit after taxes in millions of dollars is:

(A) 5
(B) 10
(C) 15
(D) 20
(E) 25

59. XYZ is considering paving the company's entrance road with either asphalt or concrete. Concrete costs $15,000/mile and lasts 20 years. If asphalt lasts 10 years, how much will it cost (per mile) if annual mainte-

nance for both roads are \$500/mile? Minimum financing is available at 12% per annum.

(A) \$11,665
(B) \$11,524
(C) \$11,347
(D) \$11,344
(E) \$11,112

60. If the inflation rate were 5%, what will be XYZ's apparent interest rate, if its real borrowing interest rate is 10%?

(A) 15.00%
(B) 15.50%
(C) 5.00%
(D) 10.00%
(E) 2.00%

QUESTIONS 61–70 refer to the following problem.

A U-tube mercury manometer (ρ_{Hg} = 13,580 kg/m³), a pitot static probe, and several static pressure taps are used to measure the pressure difference between stagnation pressure and static pressure in a water tunnel. (ρ_{water} 1000 kg/m³, and g = 9.8 m/s².

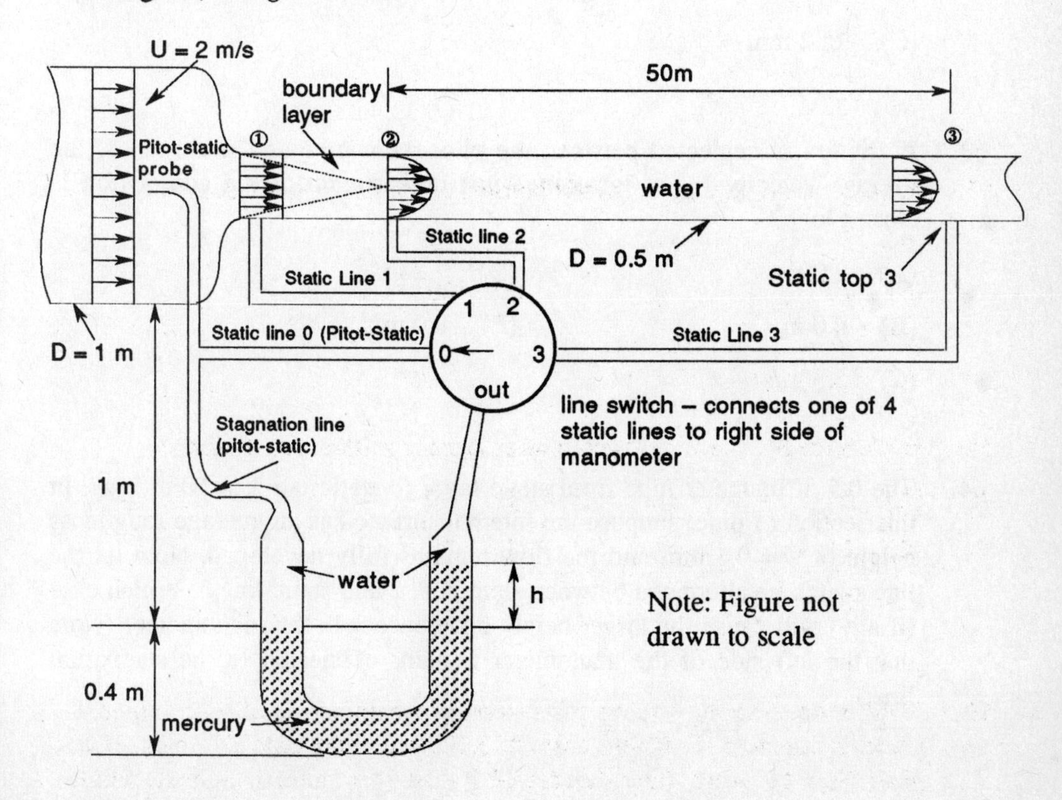

The line switch allows selection of one of the four static lines to be connected to the right side of the manometer. The left side of the manometer is permanently connected to the stagnation line. Assuming a uniform velocity $U = 2$ m/s across the 1 m diameter portion of the test section, and assuming the pitot-static tube is so small it does not significantly affect the flow, answer the following questions:

61. Due to the flow of water around the pitot-static probe, there will be a difference between stagnation pressure p_0 and static pressure p at the probe. This pressure difference $(p_0 - p)$ will be closest to

(A) 136 N/m^2 (D) 1000 N/m^2

(B) 223 N/m^2 (E) 2000 N/m^2

(C) 500 N/m^2

62. With the line switch connecting static line 0 to the manometer, the height difference h between the right side and left side of the manometer will be closest to

(A) 12.8 mm (D) 19.3 mm

(B) 15.0 mm (E) 22.1 mm

(C) 16.2 mm

63. If friction is neglected between the pitot-static tube and static tap 1, the average velocity V_{av} at location 1 just downstream of the contraction is closest to

(A) 2.0 m/s (D) 7.2 m/s

(B) 4.0 m/s (E) 8.0 m/s

(C) 5.3 m/s

64. The 0.5 m diameter tube from static tap 2 to static tap 3 is 50 m long. In this section of pipe, suppose the internal surface has an average roughness height of $\varepsilon = 0.5$ mm and the flow remains fully developed. Now let the line switch be alternated between static line 2 and static line 3. Which case (if any) will cause the larger height difference h in the manometer? (Note that the left side of the manometer remains connected to the stagnation line.)

(A) h will be LARGER when static line 2 is connected

(B) h will be LARGER when static line 3 is connected

(C) h will be identical for both cases

(D) the answer depends on Reynolds number. For low Re, h will be larger when static line 2 is connected. For high Re, h will be larger when static line 3 is connected.

(E) The answer depends on the viscosity of the water. If the water has a high viscosity, h will be larger when static line 2 is connected. On the other hand, if the water viscosity is small, h will be larger when static line 3 is connected.

65. With an average pipe roughness of $\varepsilon = 0.5$ mm and assuming the kinematic viscosity ν of the water is 1.0×10^{-6} m^2/s, the friction factor f from position 2 to position 3 in the pipe will be closest to

(A) 0.0093
(B) 0.0129
(C) 0.0137
(D) 0.0197
(E) 0.0205

66. Again consider the entrance region from static tap 1 to static tap 2. What will happen to static pressure p?

(A) it will INCREASE with downstream distance

(B) it will DECREASE with downstream distance

(C) it will STAY THE SAME with downstream distance

(D) the answer depends on the Reynolds number. For low Reynolds numbers, p will increase; for high Reynolds numbers, p will decrease.

(E) The answer depends on the viscosity of the water. If the water is cold, and very viscous, p will decrease rapidly. If the water is hot, with a low viscosity, p will remain constant through the entire section from 1 to 2.

67. How will the static pressure vary from position 2 to position 3?

(A) it will INCREASE with downstream distance

(B) it will DECREASE with downstream distance

(C) it will STAY THE SAME with the downstream distance

(D) the answer depends on the Reynolds number. For low Reynolds numbers, p will increase; for high Reynolds numbers, p will decrease.

(E) The answer depends on the viscosity of the water. If the water is cold, and very viscous, p will decrease rapidly. If the water is hot, with a low viscosity, p will remain constant through the entire section from 2 to 3.

68. How will stagnation pressure p_0 vary from position 2 to position 3?

(A) it will INCREASE with downstream distance

(B) it will DECREASE with downstream distance

(C) it will STAY THE SAME with the downstream distance

(D) the answer depends on the Reynolds number. For low Reynolds numbers, p will increase; for high Reynolds numbers, p will decrease.

(E) The answer depends on the viscosity of the water. If the water is cold, and very viscous, p will decrease rapidly. If the water is hot, with a low viscosity, p will remain constant through the entire section from 2 to 3.

69. How will centerline velocity U vary from position 2 to position 3? (You may assume the water temperature remains constant throughout the system)

(A) it will INCREASE with downstream distance

(B) it will DECREASE with downstream distance

(C) it will STAY THE SAME with the downstream distance

(D) the answer depends on the Reynolds number. For low Reynolds numbers, p will increase; for high Reynolds numbers, p will decrease.

(E) The answer depends on the viscosity of the water. If the water is cold, and very viscous, p will decrease rapidly. If the water is hot, with a low viscosity, p will remain constant through the entire section from 2 to 3.

70. Again with an average pipe roughness of $\varepsilon = 0.5$ mm and assuming $\nu = 1.0 \times 10^{-6}$ m^2/s, the frictional head loss of h_f from position 2 to position 3 in the pipe will be closest to

(A) 3.03 m of water

(B) 3.03 m of mercury

(C) 4.47 m of water

(D) 6.43 m of water

(E) 9.80 m of water

TEST 2 (PM)

ANSWER KEY

1.	(A)	21.	(A)	41.	(B)	60.	(B)
2.	(C)	22.	(E)	42.	(C)	61.	(E)
3.	(E)	23.	(A)	43.	(C)	62.	(C)
4.	(B)	24.	(E)	44.	(C)	63.	(E)
5.	(D)	25.	(E)	45.	(C)	64.	(B)
6.	(B)	26.	(E)	46.	(C)	65.	(D)
7.	(C)	27.	(E)	47.	(C)	66.	(B)
8.	(E)	28.	(B)	48.	(D)	67.	(B)
9.	(A)	29.	(A)	49.	(C)	68.	(B)
10.	(A)	30.	(A)	50.	(B)	69.	(C)
11.	(B)	31.	(C)	51.	(C)	70.	(D)
12.	(C)	32.	(B)	52.	(D)		
13.	(D)	33.	(C)	53.	(A)		
14.	(D)	34.	(B)	54.	(B)		
15.	(B)	35.	(D)	55.	(C)		
16.	(C)	36.	(B)	56.	(B)		
17.	(E)	37.	(C)	57.	(D)		
18.	(B)	38.	(C)	58.	(B)		
19.	(C)	39.	(A)	59.	(C)		
20.	(E)	40.	(B)				

DETAILED EXPLANATIONS OF ANSWERS

TEST 2

AFTERNOON (PM) SECTION

1. **(A)**

The 60% alloy has its last liquid solidify at the eutectic temperature of 800° C. Below this temperature, the alloy is completely solid.

2. **(C)**

Application of lever rule at room temperature for the 48% alloy yields for the fraction of α, f_α, in the alloy the following:

$$f_\alpha = (48 - 1) / (96 - 1) = 0.495$$

The amount of α in 80 kg of the alloy is 80 × 0.495 = 39.6 kg.

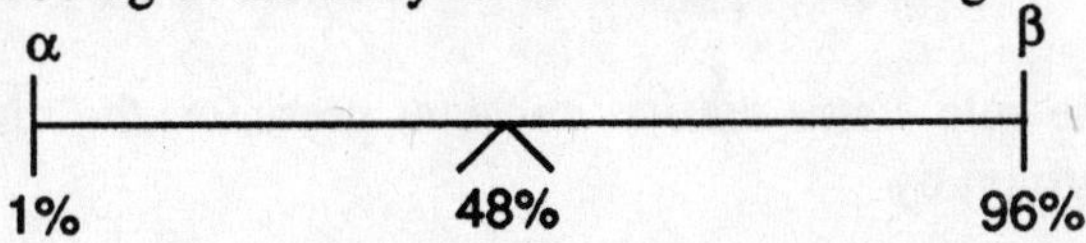

3. **(E)**

The horizontal line drawn at 700° C intersects the solvus line for β at a composition of 84% *B*. Thus this is the composition of β at 700° C.

4. **(B)**

The transformation which occurs at 450° C is $\delta \rightarrow \alpha + \beta$. This is a reaction in which one solid phase transforms into two different solid phases. Thus, it is the eutectoid transformation.

5. **(D)**

The maximum solubility of *B* in *A* is given by the maximum horizontal distance to which the phase field of a terminal *A*–rich solid solution extends. From the given phase diagram, it is seen that the phase field of α extends to 20% *B* on the composition axis at a temperature of 800° C and hence this the maximum solubility of *B* in *A*.

6. **(B)**

Pure *A* exists as the α phase at temperatures below 600° C and it exists as the δ phase in the temperature range of 600° C – 1200° C. Thus the change from α to δ in pure *A* occurs at 600° C.

7. **(C)**

The eutectic alloy has a composition of 40% *B*. Application of lever rule for this alloy at 500° C yields

$$\text{percent } \delta = \frac{89.5 - 40}{89.5 - 7} \times 100 = 60\%$$

$$\text{percent } \beta = \frac{40 - 7}{89.5 - 7} \times 100 = 40\%$$

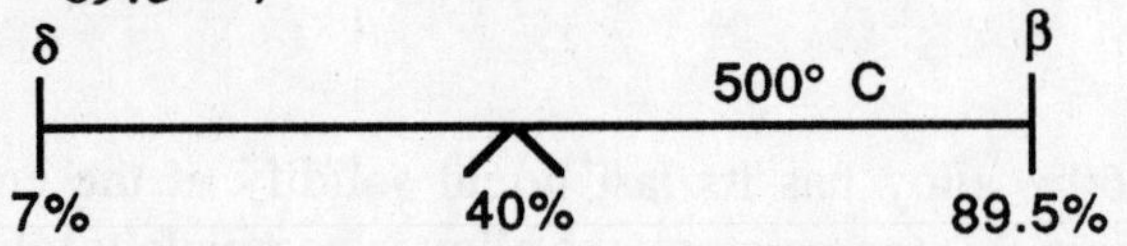

δ β
500° C
7% 40% 89.5%

8. **(E)**

The solubility limit of *A* in *B* is given as the maximum amount of *A* that can be dissolved in the terminal solid solution of *B*, namely β. At 450° C, this is obtained by the intersection point of the horizontal line at 450° C and the solvus line of the β phase. This occurs at a composition of 90% *B*, 10% *A*. Thus the solubility limit of *A* in *B* at 450° C is 10%.

9. **(A)**

The phase rule states that, at constant pressure, the number of degrees of freedom, *F*, is given by

$$F = C - P + 1$$

where *C* is the number of components and *P* is the number of phases. In this case, there are two components (*A* and *B*). For the 30% alloy at temperatures below 400, the number of phases is two (α and β). Thus there is one degree of freedom per the phase rule.

10. **(A)**

The amount of eutectic liquid can be calculated by applying the lever rule to the *L* + β region of the eutectic temperature (800° C). This yields

$$\text{percent } L = \frac{80 - 60}{80 - 40} \times 100 = 50\%$$

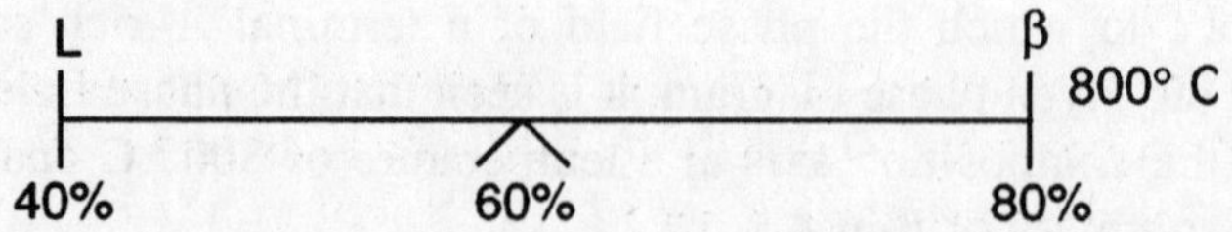

L β
800° C
40% 60% 80%

11. **(B)**

The neutral axis (N.A.) of a section is the centroidal axis of the section. The neutral axis is the plane where the bending stress is zero. In order to determine the location of the neutral axis, we must first locate the centroid of the section ($\bar{y}$).

First, the subsection is divided into two rectangular sections; A_1 and A_2, as shown.

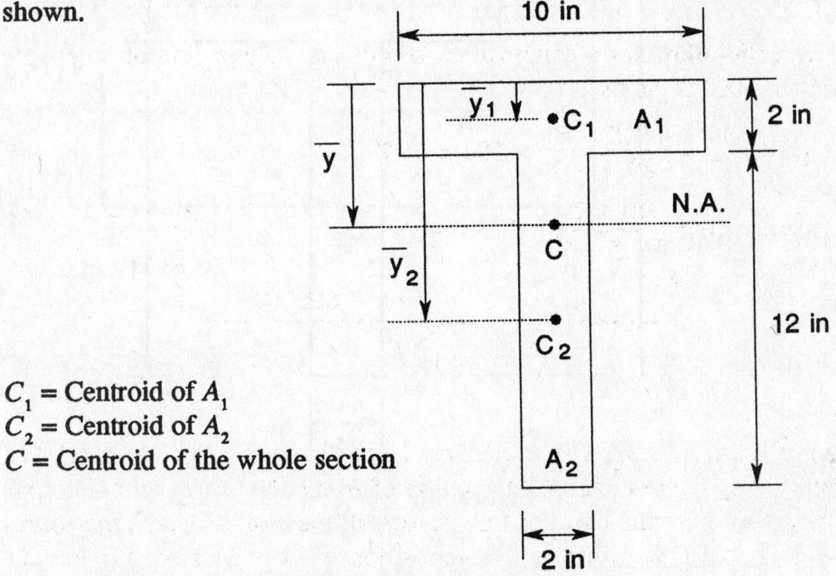

C_1 = Centroid of A_1
C_2 = Centroid of A_2
C = Centroid of the whole section

The calculation is summarized in the table below.

Area (A_i)	Distance to Centroid ($\overline{Y}_i$)	(A_i) ($\overline{Y}_i$)
$A_1 = 10\text{ in} \times 2\text{ in} = 20\text{ in}^2$	$\overline{Y}_1 = 1\text{ in}$	$A_1\overline{Y}_1 = 20\text{ in}^2 \times 1\text{ in} = 20\text{ in}^3$
$A_2 = 12\text{ in} \times 2\text{ in} = 24\text{ in}^2$	$\overline{Y}_2 = 18\text{in}$	$A_2\overline{Y}_2 = 24\text{ in}^2 \times 8\text{ in} = 192\text{ in}^3$
$\Sigma(A_i) = 44\text{ in}^2$		$\Sigma(A_i)(\overline{Y}_i) = 212\text{ in}^3$

Note: Σ = summation.

Therefore,

$$\overline{Y} = \Sigma(A_i)(\overline{Y}_i) / \Sigma(A_i) = \frac{(212\text{ in}^3)}{(44\text{ in}^2)} = 4.82\text{ in}$$

The neutral axis is 4.82 inches from the top of the beam. (B) is the correct answer.

12. **(C)**

From problem 11, we have already known the location of the neutral axis (N.A.). Again, the section is divided into two areas; A_1 and A_2, as shown in Figure (a) on the following page.

For a rectangular section, the moment of inertia about the centroidal axis can be computed from: (Figure (b))

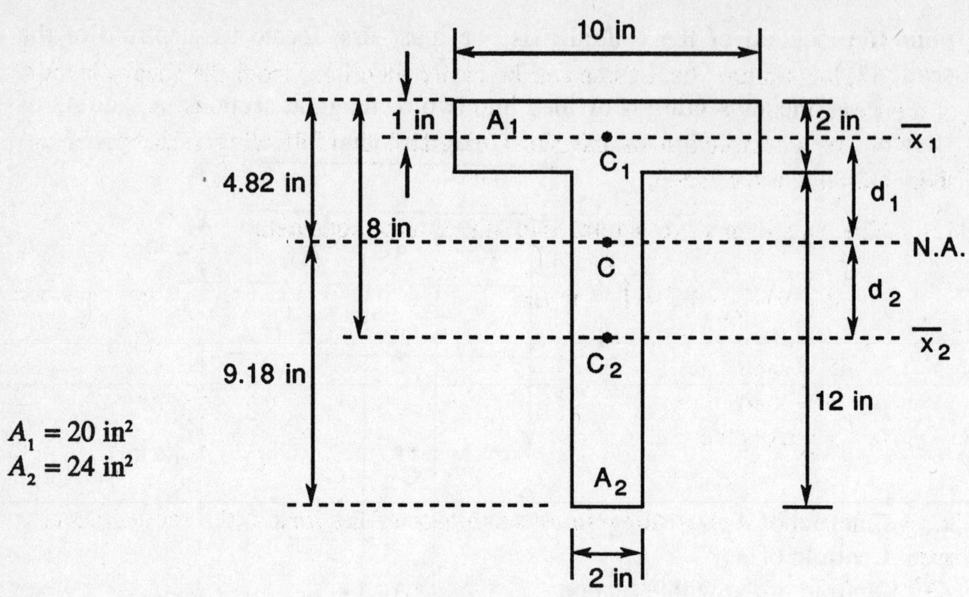

Figure (a)

$$I_{\bar{x}} = \frac{1}{12} bh^3$$

Figure (b)

In this problem, the neutral axis is neither the centroidal axis of A_1, $(\overline{X}_1)$ nor A_2 $(\overline{X}_2)$. The moment of inertia about the neutral axis can be computed by the Parallel Axis Theorem, which is illustrated below:

A = Area
C = Centroid
$\bar{x}$ = Centroidal Axis
x = Axis which is parallel to $\bar{x}$
d = distance between x and $\bar{x}$

$$I_x = I_{\bar{x}} + Ad^2$$

Therefore,

$$I = [I_{\bar{x}_1} + A_1 d_1^2] + [I_{\bar{x}_2} + A_2 d_2^2]$$

$$= [\tfrac{1}{12}(10\text{ in})(2\text{ in})^3 + (20\text{ in}^2)(4.82\text{ in} - 1\text{ in})^2]$$

$$+ [\tfrac{1}{12}(2\text{ in})(12\text{ in})^3 + (24\text{ in}^2)(8\text{ in} - 4.82\text{ in})^2]$$

$$= 829.21\text{ in}^4.$$

13. **(D)**

The maximum shear force can be easily identified from the shear diagram of the given beam.

In the construction of the shear diagram, the following relations of the shear and load are utilized:

1. The change in shear = total load in the range considered

$$\Delta V = V_2 - V_1 = \int_{x_1}^{x_2} w\, dx$$

V = shear
w = load
x = distance

2. The slope of shear diagram = magnitude of the load at the section considered.

$$\frac{dV}{dx} = w$$

Load is considered positive, when it points upward.

Reactions:

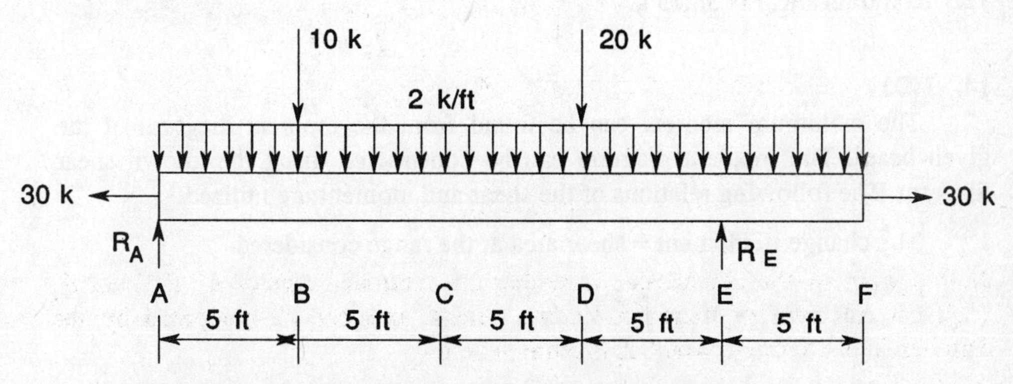

ΣM about $A = 0$, using clockwise direction as positive:

$$(10\text{ k})(5\text{ ft}) + (2 \times 25\text{ k})(12.5\text{ ft}) + (20\text{ k})(15\text{ ft}) - (R_E\text{k})(20\text{ ft}) = 0$$

$$R_E = 48.75\text{ k}$$

ΣM about $E = 0$, using clockwise direction as positive:

$$(R_A\text{ k})(20\text{ ft}) - (10\text{ k})(15\text{ ft}) - (20\text{ k})(5\text{ ft}) - (2 \times 25\text{ k})(7.5\text{ ft}) = 0$$

$$R_A = 31.25\text{ k}$$

Check: ΣF in the vertical direction = 0, using upward direction as positive:

$(31.25\text{ k}) - (10\text{ k}) - (20\text{ k}) - (2 \times 25\text{ k}) + (48.75\text{ k}) = 0.$ O.K.

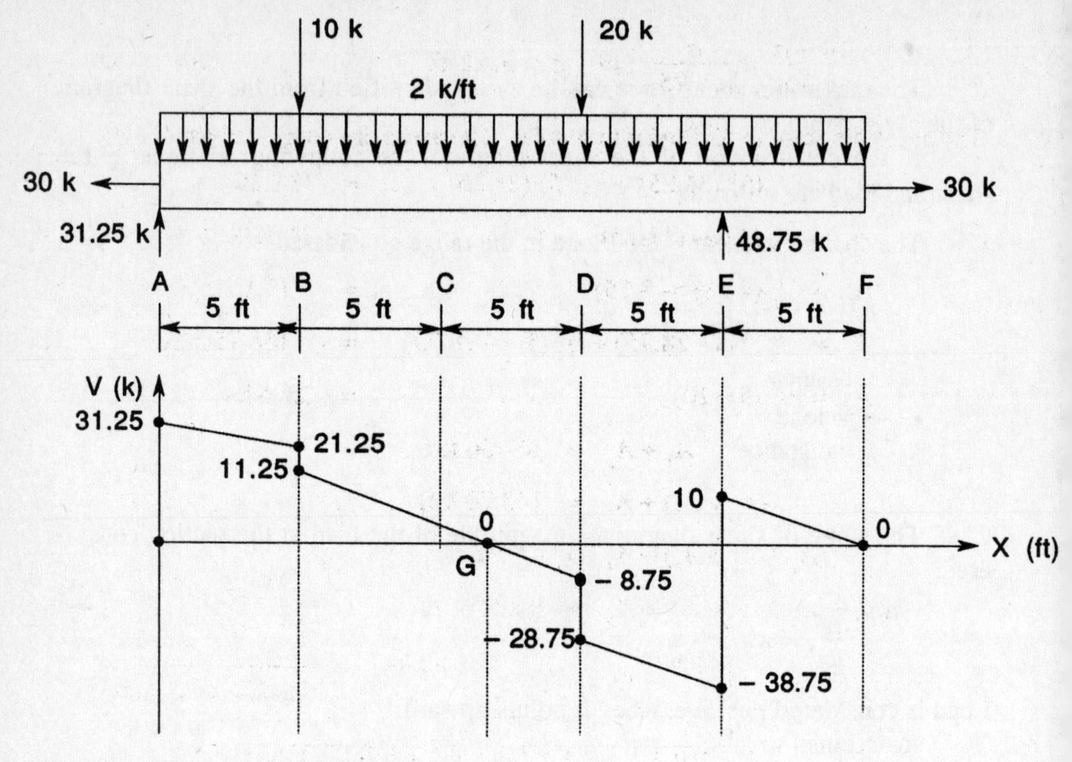

The maximum shear is 38.75 k.

14. **(D)**

The maximum moment can be found from the moment diagram of the given beam. The moment diagram can be constructed using the known shear diagram. The following relations of the shear and moment are utilized:

1. The change in moment = shear area at the range considered.

$$\Delta M = M_2 - M_1 = \int_{x_1}^{x_2} V \, dx$$

V = shear
M = moment
x = distance

2. The slope of moment diagram = magnitude of the shear at the section considered.

$$\frac{dM}{dx} = V$$

From the shear diagram, the shear is zero at point G. When the shear is zero, the slope of the moment diagram is zero. When the slope is zero, the moment is either a maximum or minimum point. It is, therefore, necessary to locate point G. By using similar triangles, distance a can be found:

$$\frac{10-a}{8.75}=\frac{a}{11.25} \quad \Rightarrow \quad a=5.62 \text{ ft}$$

The shear diagram is divided into 5 areas. The areas are computed as:

$$A_1 = {}^1/_2\,(5)\,(31.25) + {}^1/_2\,(5)\,(21.25) = 131.25 \text{ k-ft}$$

$$A_2 = {}^1/_2\,(5.62)\,(11.25) = 31.62 \text{ k-ft}$$

$$A_3 = {}^1/_2\,(4.38)\,(-8.75) = -19.16 \text{ k-ft}$$

$$A_4 = {}^1/_2\,(5)\,(-28.75) + {}^1/_2\,(5)\,(-38.75) = -168.75 \text{ k-ft}$$

$$A_5 = {}^1/_2\,(5)\,(10) = 25 \text{ k-ft}$$

$$A_1 + A_2 = 162.86 \text{ k-ft}$$

$$(A_1 + A_2) + A_3 = 143.70 \text{ k-ft}$$

$$(A_1 + A_2 + A_3) + A_4 = -25 \text{ k-ft}$$

$$(A_1 + A_2 + A_3 + A_4) + A_5 = 0 \text{ k-ft}$$

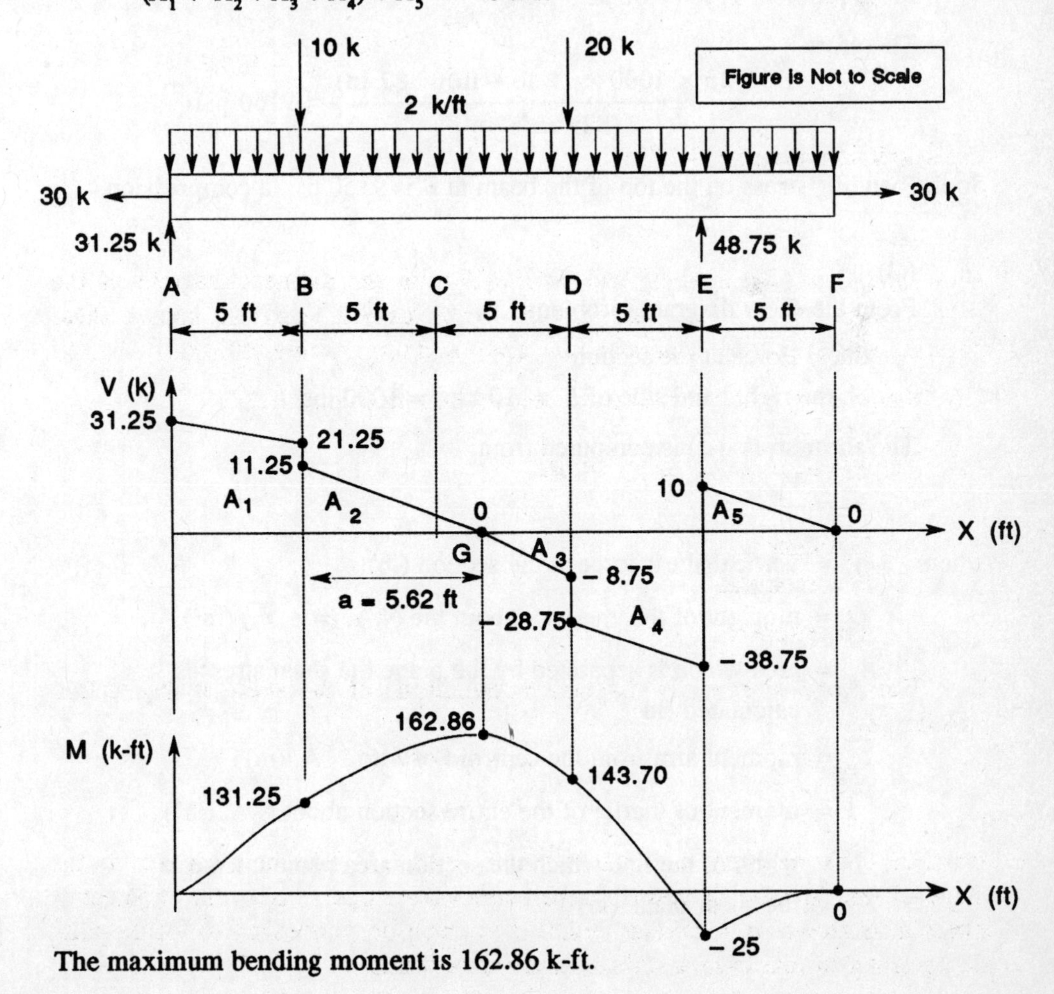

The maximum bending moment is 162.86 k-ft.

15. **(B)**

From the moment diagram in problem 14, the moment at *B* is +131.25 kip-ft. The positive moment causes the beam to bend such that the compression zone is on the top and the tension zone is on the bottom. Since we want to determine the stress at the top, the stress must be "Compression." Therefore, answers (C) and (E) may be eliminated immediately.

The value of the bending stress (σ) can be found from:

$$\sigma = My / I$$

where M = bending moment (lb-in)

y = distance from the N.A. to the desired point (in)

I = moment of inertia about the N.A. (in^4)

From problem 14 , $M = 131.25$ k-ft $= 131.25 \times 1000 \times 12$ lb-in.

From problem 12, $I = 829\ in^4$

From problem 1, $y = 4.82$ in

Therefore,

$$\sigma = \frac{(131.25 \times 1000 \times 12\ \text{lb} - \text{in})(4.82\ \text{in})}{(829\ \text{in}^4)} = 9160\ \text{psi}$$

So the bending stress on the top of the beam at *B* is 9160 psi in compression.

16. **(C)**

From the shear diagram (problem 13),

Shear Force at the section
on the right hand side of E = 10 kips = 1000 lbs

The shear stress ($\overline{\sigma}$) is computed from:

$$\overline{\sigma} = \frac{VQ}{Ib}$$

where V = vertical shear force at the section (lb)

Q = moment of the area, A_F, about the N.A. ($= A_F\overline{Y_F}$) (in^3)

A_F = area which is separated by the plane the shear stress is calculated (in^2)

$\overline{Y_F}$ = moment arm from the centroid of A to N.A. (in)

I = moment of inertia of the entire section about N.A. (in^4)

b = width of the line which the section area plane intersects the shear plane (in)

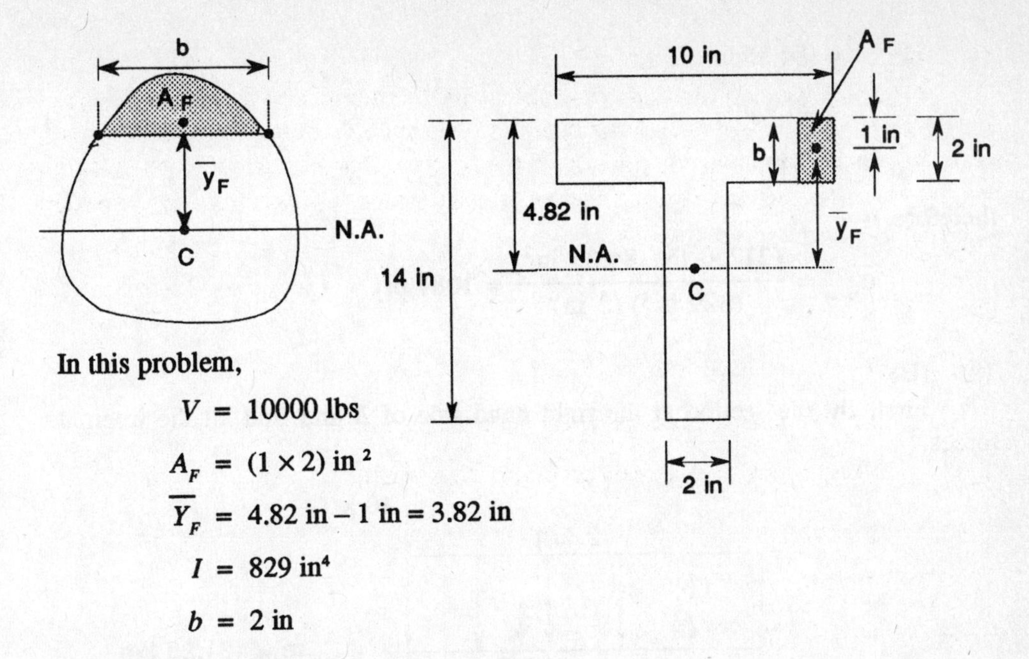

In this problem,

$$V = 10000 \text{ lbs}$$

$$A_F = (1 \times 2) \text{ in}^2$$

$$\overline{Y}_F = 4.82 \text{ in} - 1 \text{ in} = 3.82 \text{ in}$$

$$I = 829 \text{ in}^4$$

$$b = 2 \text{ in}$$

therefore,

$$\overline{\sigma} = \frac{(10000 \text{ lb})(1 \times 2 \times 3.82 \text{ in}^3)}{(829 \text{ in}^4)(2 \text{ in})} = 46 \text{ psi}$$

17. **(E)**

From the shear diagram (problem 13),

Shear Force at the section on the left hand side of B = 21.25 kips = 21,250 lbs

The shear stress (σ) is computed from:

$$\overline{\sigma} = \frac{VQ}{IB}$$

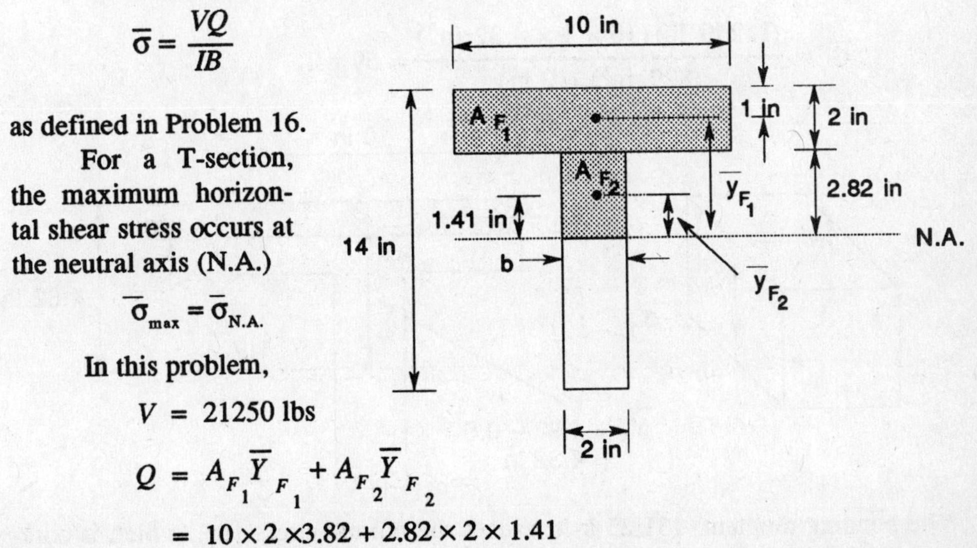

as defined in Problem 16.

For a T-section, the maximum horizontal shear stress occurs at the neutral axis (N.A.)

$$\overline{\sigma}_{max} = \overline{\sigma}_{N.A.}$$

In this problem,

$$V = 21250 \text{ lbs}$$

$$Q = A_{F_1}\overline{Y}_{F_1} + A_{F_2}\overline{Y}_{F_2}$$

$$= 10 \times 2 \times 3.82 + 2.82 \times 2 \times 1.41$$

$$= 84.35 \text{ in}^3$$

$$I = 829 \text{ in}^4$$

$$b = 2 \text{ in}$$

therefore,

$$\bar{\sigma}_{max} = \frac{(21250 \text{ lb})(84.35 \text{ in}^2)}{(829 \text{ in}^4)(2 \text{ in})} = 1081 \text{ psi}$$

18. **(B)**

First, cut the section at the right hand side of *B* and find all the internal forces.

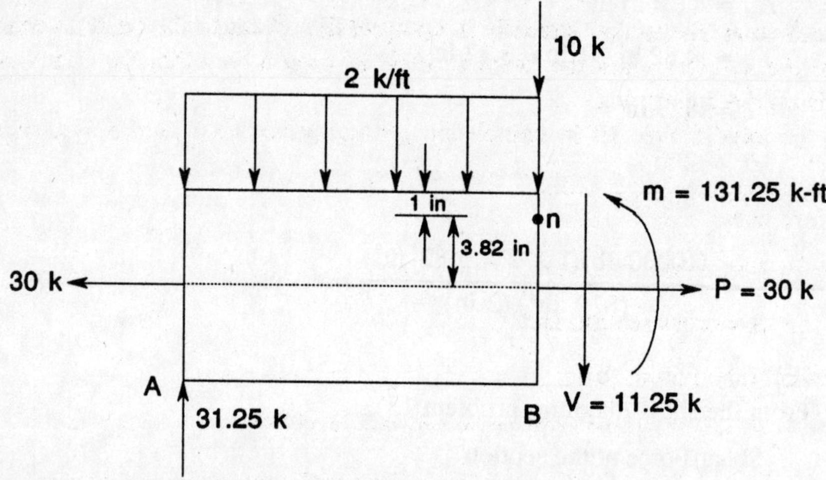

The shear force, 11.25 k, causes the shear stress ($\bar{\sigma}_1$), which is computed from (all terms are defined in problem 16):

$$\bar{\sigma}_1 = \frac{(11250 \text{ lb})(10 \times 1 \times 4.32 \text{ in}^3)}{(829 \text{ in}^4)(10 \text{ in})} = 59 \text{ psi}$$

10 in

A_F

1 in

4.82 in

$\bar{\sigma}_1$

n

N.A.

$\bar{y}_F = 4.82 - 0.5$
$= 4.32$ in

The bending moment, 131.25 k-ft, causes the normal stress (σ_1), which is computed from (all terms are defined in problem 15): (see Figure (a))

$$\sigma_1 = \frac{(131.25 \times 1000 \times 12 \text{ lb-in})(3.82 \text{ in})}{(829 \text{ in}^4)} = 7258 \text{ psi (compression)}$$

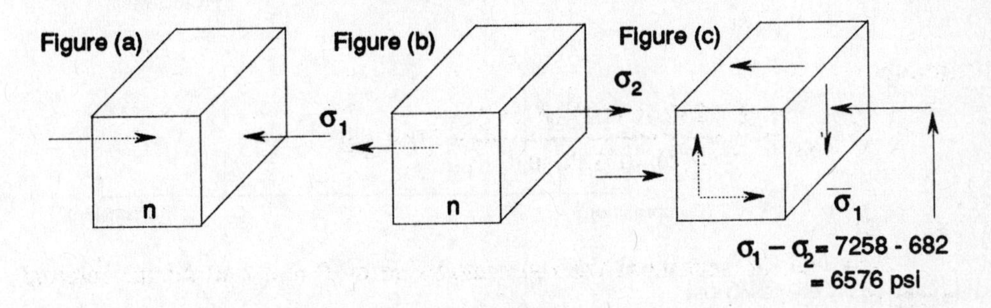

We know that the normal stress is in compression because the bending moment is positive, which causes the beam to bend in such a way that the compression zone is on the upper part.

the axial force, 30 k, causes the normal stress (σ_2), which is computed from:

$$\sigma = P / A$$

where P = axial force,

A = cross section area

Therefore, (see Figure (b))

$$\sigma_2 = (30000 \text{ lb}) / (44 \text{ in}^2) = 682 \text{ psi (tension)}$$

The stress is in tension because the force 30 k pulls the section and causes tensile stress.

The final state of stress at point n is then (see Figure (c))

$$\sigma_1 - \sigma_2 = 7285 - 682 = 6576 \text{ psi}$$

The shear stress = 59 psi
The normal stress = 6576 psi (compression)

19. **(C)**
From problem 18, we know the state of stress at point n:

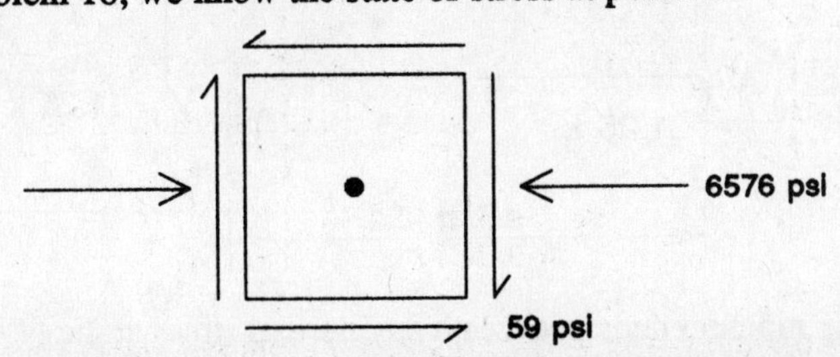

When the point is rotated to a new orientation, the stresses change accordingly.

At one orientation, the shear stress is maximum. This maximum shear stress can be found using the Mohr's Circle Method.

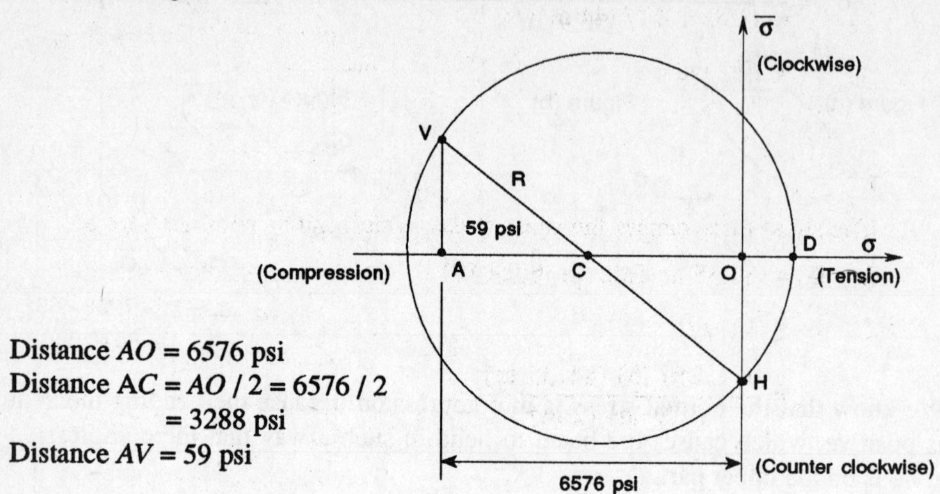

Distance $AO = 6576$ psi
Distance $AC = AO / 2 = 6576 / 2$
$= 3288$ psi
Distance $AV = 59$ psi

The radius of the Mohr's Circle,

$$R = CV = \sqrt{AV^2 + AC^2}$$

$$R = \sqrt{3288^2 + 59^2} = 3288.53 \text{ psi}$$

The maximum shear stress $= R = 3288.53$ psi.

20. **(E)**

The internal forces at the required section can be determined from the following free-body diagram.

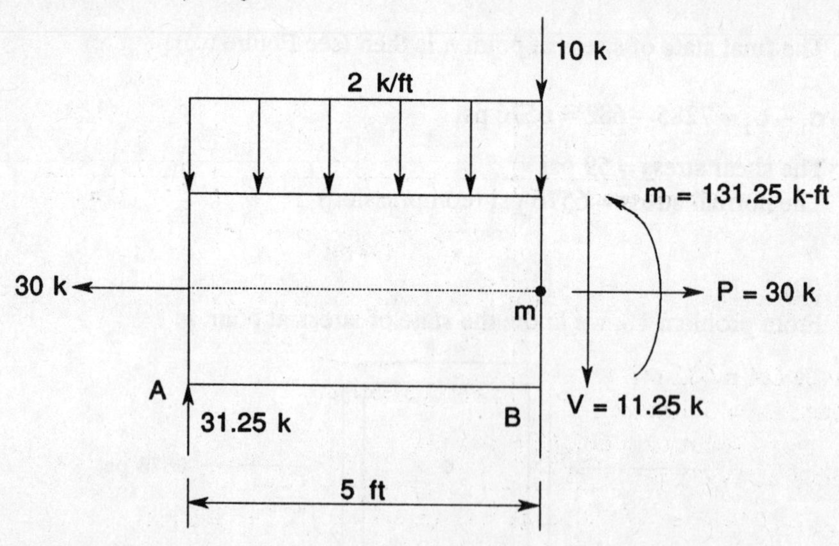

The bending moment does not create any normal stress at the N.A.

The axial force creates the tensile stress, which can be computed from:

$$\sigma = P/A$$
$$= (30000\ \text{lb}) / (44\ \text{in}^2)$$
$$= 682\ \text{psi}$$

σ

The shear force causes the shear stress, which can be computed from:

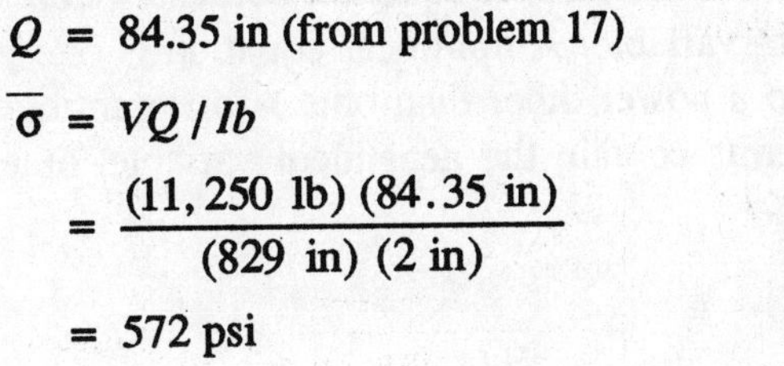

$$Q = 84.35\ \text{in (from problem 17)}$$
$$\overline{\sigma} = VQ/Ib$$
$$= \frac{(11,250\ \text{lb})\ (84.35\ \text{in})}{(829\ \text{in})\ (2\ \text{in})}$$
$$= 572\ \text{psi}$$

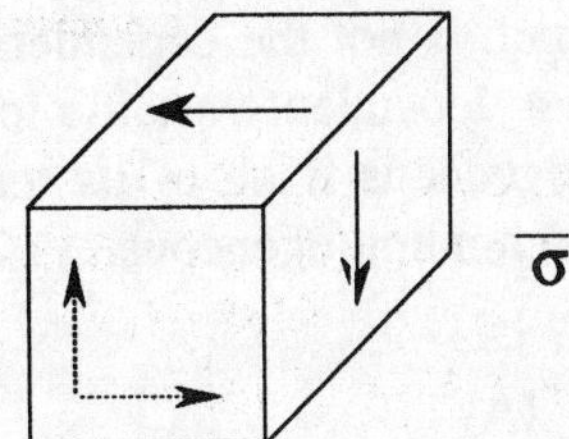

The state of stress at point m is:

The maximum tensile stress can be found using the Mohr's Circle Method.

682 psi

572 psi

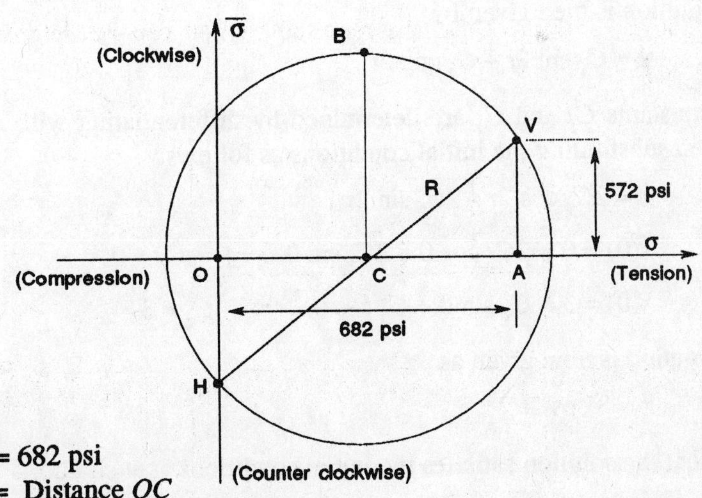

Distance $OA = 682$ psi
Distance CA = Distance OC
$= OA / 2 = 682 / 2 = 341$ psi
Distance $AV = 572$ psi
Radius $= CV = R = \sqrt{CA^2 + AV^2} = \sqrt{341^2 + 572^2} = 666$ psi
Maximum tensile stress $= OB = OC + R = 341 + 666 = 1007$ psi
The maximum tensile stress = 1007 psi

21. **(A)**

This is a homogeneous differential equation because each term contains the dependent variable. A nonlinear equation is one which contains a dependent variable to a power other than one. A differential equation is homogeneous if all of its terms contain the dependent variable, otherwise it is called nonhomogeneous.

22. **(E)**

This is a nonhomogeneous differential equation because it contains a term π which is not the dependent variable. A nonlinear equation is one which contains a dependent variable to a power other than one. A differential equation is homogeneous if all of its terms contain the dependent variable, otherwise it is called nonhomogeneous.

23. **(A)**

Substituting into the differential equation gives

$$4\ddot{x} + 16x = 0$$

Assuming a solution of the form

$$x = Ge^{st} \quad \Rightarrow \quad \dot{x} = sGe^{st} \quad \Rightarrow \quad \ddot{x} = s^2 Ge^{st}$$

Substituting into the differential equation, then solving for s gives

$$4s^2 + 16 = 0 \quad \Rightarrow \quad s = \pm 2j$$

The solution is then given by

$$x = C_1 \sin 2t + C_2 \cos 2t$$

The constants C_1 and C_2 are determined by differentiating with respect to time (t), then substituting the initial conditions as follows:

$$\dot{x} = 2C_1 \cos 2t - 2C_2 \sin 2t$$

$$\dot{x}(0) = 0 = 2C_1 \cos 0 - 2C_2 \sin 0 \quad \Rightarrow \quad C_1 = 0$$

$$x(0) = 3 = C_1 \sin 0 + C_2 \cos 0 \quad \Rightarrow \quad C_2 = 3$$

The solution is now given as

$$x = 3 \cos 2t$$

Note that the solution satisfies the initial conditions as it should.

24. **(E)**

Substituting into the differential equation gives

$$4\ddot{x} + 16x = 9 \sin wt$$

Assuming a solution of the form

$$x = G \sin wt \quad \Rightarrow \quad \dot{x} = wG \cos wt \quad \Rightarrow \quad \ddot{x} = -w^2 G \sin wt$$

Substituting into the differential equation, then solving for G gives

$$G = \frac{9}{16 - 4w^2}$$

The particular solution is then given by

$$x = \frac{9}{16 - 4w^2} \sin wt$$

25. **(E)**

Given the particular solution in problem 24, the complimentary solution is given in the form outlined earlier in problem 23. That is

$$x = C_1 \sin 2t + C_2 \cos 2t$$

The total solution is equal to the sum of the particular and complimentary solutions. This is given as

$$x = x_{\text{part.}} + x_{\text{comp.}} = \frac{9}{16 - 4w^2} \sin wt + C_1 \sin 2t + C_2 \cos 2t$$

and

$$\dot{x} = \frac{9}{16 - 4w^2} \cos wt + 2C_1 \cos 2t - 2C_2 \sin 2t$$

Substituting the initial conditions into the above gives

$$x(0) = 0 = \frac{9}{16 - 4w^2} \sin 0 + C_1 \sin 0 + C_2 \cos 0 \quad \Rightarrow \quad C_2 = 0$$

$$x(0) = 0 = \frac{9}{16 - 4w^2} \cos 0 + 2C_2 \cos 0 \quad \Rightarrow \quad C_1 = \frac{1}{2}\left[\frac{-9w}{16 - 4w^2}\right]$$

The total solution is given by

$$x = \frac{9}{16 - 4w^2} \sin wt + \frac{1}{2}\left[\frac{-9w}{16 - 4w^2}\right] \sin 2t$$

Which simplifies to

$$x = \frac{9}{16 - 4w^2}\left[\sin wt = \frac{w}{2} \sin 2t\right]$$

Note that the solution is not equal to the particular solution even though the initial conditions are equal to zero.

26. **(E)**

The solution obtained in problem 25 suggests that $m - c - k$ system will become unstable when the frequency w reaches a specific value. This is true in that the solution contained the following term:

$$\frac{9}{16-4w^2}$$

The implication is that if $4w^2 = 16$, then the displacement will be infinite! Therefore, for $w = 2$, resonance will occur.

27. **(E)**

Substituting into the differential equation gives

$$4\ddot{x} + 16x = 2$$

Note that in this case, we are asked to find the total solution. Assuming a particular solution of the form

$$x = G \quad \Rightarrow \quad \ddot{x} = 0$$

Substituting into the differential equation, then solving for G gives

$$4(0) + 16G = 2 \quad \Rightarrow \quad G = {}^1/_8$$

The particular solution is then given by

$$x_p = {}^1/_8$$

The complimentary solution is of the form given in problem 23. Hence

$$x_c = C_1 \sin 2t + C_2 \cos 2t$$

The total solution is then given as the sum of $x = x_p + x_c$

$$x = {}^1/_8 + C_1 \sin 2t + C_2 \cos 2t$$

The constants C_1 and C_2 are determined using the given initial conditions. Thus

$$\dot{x} = 2C_1 \cos 2t - 2C_2 \sin 2t$$

$$x(0) = 1 = {}^1/_8 + C_1 \sin 0 + C_2 \cos 0 \quad \Rightarrow \quad C_2 = {}^7/_8$$

$$\dot{x}(0) = 2 = 2C_1 \cos 0 - 2C_2 \sin 0 \quad \Rightarrow \quad C_1 = 1$$

The total solution is then given as

$$X = {}^1/_8 + \sin 2t + {}^7/_8 \cos 2t$$

28. **(B)**

Substituting into the differential equation gives

$$4\ddot{x} + 16x = e^{-\pi t}$$

Assuming a solution of the form

$$x = Ge^{-\pi t} \quad \Rightarrow \quad \dot{x} = -\pi Ge^{-\pi t} \quad \Rightarrow \quad \ddot{x} = \pi^2 Ge^{-\pi t}$$

Substituting into the differential equation, then solving for G gives

$$4\pi^2 G + 16G = 1 \quad \Rightarrow \quad G = \frac{1}{4\pi^2 + 16}$$

The particular solution is then given by

$$x = \frac{e^{-\pi t}}{4\pi^2 + 16}.$$

29. **(A)**

Having determined the particular solution in problem 28, the complimentary solution is given by

$$x_c = C_1 \sin 2t + C_2 \cos 2t$$

Note that this is the case because $m = 4$ and $k = 16$. The total solution is then given as the sum of $x = x_p + x_c$

$$x = \frac{e^{-\pi t}}{4\pi^2 + 16} + C_1 \sin 2t + C_2 \cos 2t$$

The constants C_1 and C_2 are determined using the given initial conditions. Thus, differentiating first, we have

$$\dot{x} = \frac{-\pi e^{-\pi t}}{4\pi^2 + 16} + 2C_1 \cos 2t - 2C_2 \sin 2t$$

$$x(0) = 0 = \frac{e^0}{4\pi^2 + 16} + C_1 \sin 0 + C_2 \cos 0 \quad \Rightarrow \quad C_2 = \frac{-1}{4\pi^2 + 16}$$

$$\dot{x}(0) = 0 = \frac{-\pi e^0}{4\pi^2 + 16} + 2C_1 \cos 0 - 2C_2 \sin 0 \quad \Rightarrow \quad C_1 = \frac{1}{2}\,\frac{\pi}{4\pi^2 + 16}$$

The total solution is then given as

$$x = \frac{e^{-\pi t}}{4\pi^2 + 16} + \frac{1}{2}\,\frac{\pi}{4\pi^2 + 16} \sin 2t - \frac{1}{4\pi^2 + 16} \cos 2t$$

Which simplifies to

$$x = \frac{1}{4\pi^2 + 16}\left[e^{-\pi t} + \frac{\pi}{2} \sin 2t - \cos 2t\right]$$

30. **(A)**

Substituting into the differential equation gives

$$4\ddot{x} + \dot{x} + 16x = 0$$

Assuming a solution of the form

$$x = Ge^{st} \quad \Rightarrow \quad \dot{x} = sGe^{st} \quad \Rightarrow \quad \ddot{x} = s^2 Ge^{st}$$

Substituting into the differential equation gives

$$4s^2 + s + 16 = 0$$

This equation is solved for s using the quadratic formula, which gives

$$s = \frac{-1 \pm \sqrt{1-4(4)(16)}}{2(4)} = -\frac{1}{8} \pm \frac{\sqrt{255}}{8} j$$

The solution is then given by

$$x = e^{-t/8}\left[C_1 \sin \frac{\sqrt{255}}{8} t + C_2 \cos \frac{\sqrt{255}}{8} t\right]$$

The constants C_1 and C_2 are determined by differentiating with respect to time (t), then substituting the initial conditions as follows:

$$\dot{x} = -\frac{1}{8} e^{-t/8}\left[C_1 \sin \frac{\sqrt{255}}{8} t + C_2 \cos \frac{\sqrt{255}}{8} t\right]$$

$$+ e^{-t/8}\left[\frac{\sqrt{255}}{8} C_1 \cos \frac{\sqrt{255}}{8} t - \frac{\sqrt{255}}{8} C_2 \sin \frac{\sqrt{255}}{8} t\right]$$

$$x(0) = 1 = e^0[C_1 \sin 0 + C_2 \cos 0] \quad \Rightarrow \quad C_2 = 1$$

$$\dot{x}(0) = 0 - \frac{1}{8}[C_2] + \left[\frac{\sqrt{255}}{8} C_1\right] \quad \Rightarrow \quad C_1 = \frac{1}{\sqrt{255}}$$

The solution is now given as

$$x = e^{-t/8}\left[\frac{1}{\sqrt{255}} \sin \frac{\sqrt{255}}{8} t + \cos \frac{\sqrt{255}}{8} t\right].$$

31. **(C)**

This is accomplished by equating the two equations, then solving for the unknown x–values.

$$x^4 + x^2 - 6 = 0 = (x^2 - 2)(x^2 + 3)$$

Which gives

$$x^2 = 2 \text{ or } x = \pm\sqrt{2} \text{ and } x^2 = -3 \text{ or } x = \pm\sqrt{3}\, j.$$

The imaginary solution is not needed. The corresponding functional value to the real x–values is calculated as follows:

$$f(x) = x^4 = (\pm\sqrt{2})^4 = 4$$

The points of intersection are then given as

$$(\sqrt{2}, 4) \text{ and } (\sqrt{2}, 4) \text{ or more simply } (\pm\sqrt{2}, 4).$$

32. **(B)**

The shaded area shown below is determined by integrating the strip of width dx and of height $y_1 - y_2$ as follows:

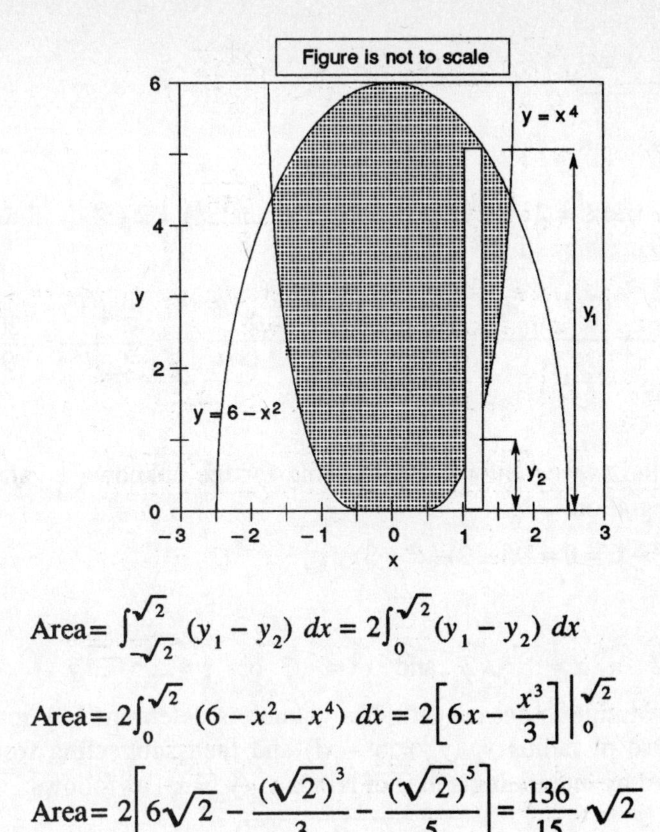

$$\text{Area} = \int_{-\sqrt{2}}^{\sqrt{2}} (y_1 - y_2)\,dx = 2\int_0^{\sqrt{2}} (y_1 - y_2)\,dx$$

$$\text{Area} = 2\int_0^{\sqrt{2}} (6 - x^2 - x^4)\,dx = 2\left[6x - \frac{x^3}{3}\right]\Bigg|_0^{\sqrt{2}}$$

$$\text{Area} = 2\left[6\sqrt{2} - \frac{(\sqrt{2})^3}{3} - \frac{(\sqrt{2})^5}{5}\right] = \frac{136}{15}\sqrt{2}$$

33. **(C)**

The center of the shaded area shown below is determined by using the strip of width *dy* and of length *L* as follows:

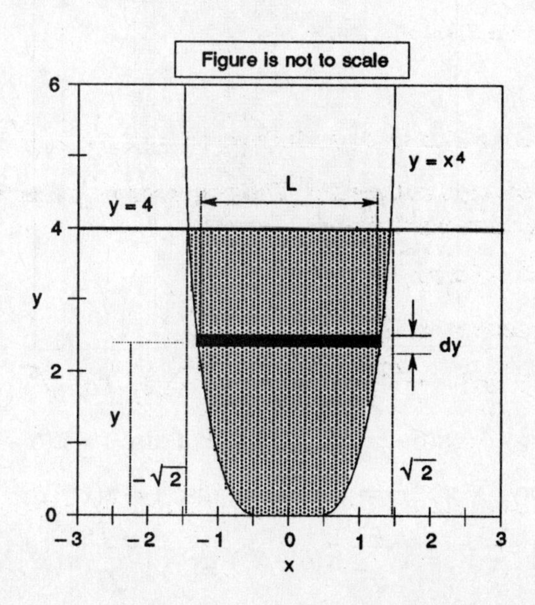

$$\bar{y} = \frac{\int_0^4 y\, dA}{\int_0^4 dA}$$

Where $dA = Ldy$ and $L = 2x = 2y^{1/4}$ which implies that $dA = 2\, y^{1/4}\, dy$. Substituting into the above expression gives

$$\bar{y} = \frac{\int_0^4 y\,(2y^{1/4})\,dy}{\int_0^4 2y^{1/4}dy} = \frac{\int_0^4 y^{5/4}dy}{\int_0^4 y^{1/4}dy} = \left.\frac{\frac{4}{9}y^{9/4}}{\frac{4}{5}y^{5/4}}\right|_0^4 = \frac{\frac{4}{9}4^{9/4}}{\frac{4}{5}4^{5/4}} = \frac{20}{9}$$

34. **(B)**

Equating the two equations, then solving for the unknown x-values yield the limits of integration.

$$x^4 + x^2 - 6 = 0 = (x^2 - 2)\,(x^2 + 3)$$

Which gives

$$x^2 = 2 \text{ or } x = \pm\sqrt{2} \text{ and } x^2 = -3 \text{ or } x = \pm\sqrt{3}\,j.$$

The imaginary solution is not needed. The volume is determined by integrating the area of a disc of radius y_1 ($y_1 = 6 - x^2$) and then subtracting from it the volume generated by integrating a disc of radius y_2 ($y_2 = x^4$) as follows:

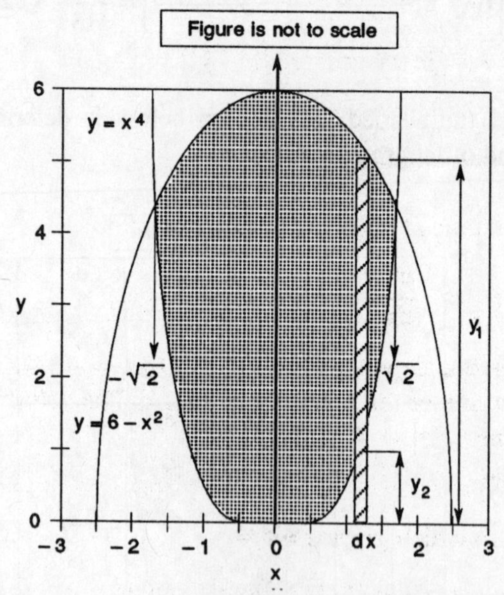

Area of disc 1 = $\pi y_1^2 = \pi(6 - x^2)^2 \Rightarrow$ Volume of disc 1 = $\pi(6 - x^2)^2\, dx$

Area of disc 2 = $\pi y_2^2 = \pi(x^4)^2 \Rightarrow$ Volume of disc 1 = $\pi(x^4)^2\, dx$

$$V = V_1 - V_2 = \int_{-\sqrt{2}}^{\sqrt{2}} \pi(6 - x^2)^2 dx - \int_{-\sqrt{2}}^{\sqrt{2}} \pi(x^4)^2 dx$$

Note that because of symmetry, it is possible to simplify the integral as follows:

$$V = V_1 - V_2 = \int_{-\sqrt{2}}^{\sqrt{2}} \pi(6 - x^2)^2 dx - \int_{-\sqrt{2}}^{\sqrt{2}} \pi(x^4)^2 dx$$

$$V = 2\pi\left[36x - \frac{12}{3}x^3 + \frac{1}{5}x^5\right]\Bigg|_0^{\sqrt{2}} - 2\pi\left[\frac{1}{9}x^9\right]\Bigg|_0^{\sqrt{2}}$$

$$V = \frac{2432}{45}\pi\sqrt{2}$$

35. **(D)**

Find the derivative, then evaluate it at (2, 2). This gives the slope of the line at that point as follows:

$$\frac{dy}{dx} = -2x \Rightarrow \frac{dy}{dx}\bigg|_{2,2} = -2(2) = -4 = m$$

The equation of a straight line is given, in general by the following expression:

$$y = a + mx$$

Substituting the $x = 2$ and $y = 2$ and noting that the slope is $m = -4$ permits the determination of the intercept a as

$$2 = a - 4(2) \quad \Rightarrow \quad a = 10$$

The equation of the tangent at (2, 2) is then given by

$$y = 10 - 4x$$

36. **(B)**

Solve for the slope at point (1, 1) as follows:

$$\frac{dy}{dx} = 4x^3 \Rightarrow \frac{dy}{dx}\bigg|_{(1,1)} = 4(1)^3 = 4 = m_1$$

The normal will have a slope equal to $m_2 = -1/m_1 = -1/4$. The equation of the normal is a straight line of the form $y = a + m_2 x$. The value of the intercept a is determined using the point (1, 1) as follows:

$$y = a - x/4$$

$1 = a - 1/4$ which gives $a = 5/4$.

The equation of the normal at (1, 1) is then given as

$$y = 5/4 - x/4$$

37. **(C)**

The arc length of the fourth order polynomial between points (0, 0) and

(1, 1) is calculated using the following integral:

$$\text{Arc Length} = I = \int_0^1 \sqrt{1 + \left(\frac{dx}{dy}\right)^2}\, dx$$

$$\frac{dy}{dx} = 4x^3$$

$$I = \int_0^1 \sqrt{1 + (4x^3)^2}\, dx = \int_0^1 f(x)\, dx$$

Using the Trapezoidal Rule, the integral $f(x)$ is evaluated as follows:

$$I = \frac{0.25}{2}\,[f(0) + 2f(0.25) + 2f(0.5) + 2f(0.75) + f(1.0)]$$

Where

$$f(0) = \sqrt{1 + 16(0)^6} = 1.0$$

$$f(0) = \sqrt{1 + 16(0.25)^6} = 1.0019512$$

$$f(0) = \sqrt{1 + 16(0.50)^6} = 1.1180340$$

$$f(0) = \sqrt{1 + 16(0.75)^6} = 1.9615423$$

$$f(0) = \sqrt{1 + 16(1.00)^6} = 4.1231056$$

Substituting into the expression, then evaluating gives

$$I = 1.6607701$$

38. **(C)**

The arc length of the parabola between points (0, 6) and (2, 2) is evaluated using the following integral.

$$\text{Arc Length} = I = \int_0^2 \sqrt{1 + \left(\frac{dx}{dy}\right)^2}\, dx$$

$$\frac{dy}{dx} = -2x$$

Substituting into the expression yields the following integral:

$$I = \int_0^2 \sqrt{1 + (-2x)^2}\, dx = \int_0^2 \sqrt{1 + 4x^2}\, dx$$

The integral can not be directly evaluated. Instead, the following trigonometric substitutions are made:

Let $x = {}^1/_2 \tan\theta \quad \Rightarrow \quad dx = {}^1/_2 \sec^2\theta\, d\theta$

Substituting into the integral gives

$$I = \int_0^2 \sqrt{1+4x^2}\,dx = \int_0^2 \sqrt{1+4(\tfrac{1}{2}\tan\theta)^2}\,(\tfrac{1}{2}\sec^2\theta\,d\theta)$$

$$I = \int_0^2 \sqrt{1+\tan^2\theta}\,(\tfrac{1}{2}\sec^2\theta\,d\theta) = \int_0^2 (\tfrac{1}{2}\sec^3\theta\,d\theta)$$

$$I = \tfrac{1}{2}\,[\sec\theta\tan\theta + \ln(\sec\theta + \tan\theta)]/2$$

However, since

$$\sec\theta = \sqrt{1+4x^2} \quad \text{and} \quad \tan\theta = 2x,$$

then

$$I = \tfrac{1}{4}[2x\sqrt{1+4x^2} + \ln(\sqrt{1+4x^2} + 2x)]\Big|_0^2 = 4.65$$

39. **(A)**

The axis about which the parabola is rotated is shown in the following figure at $x = \sqrt{6}$.

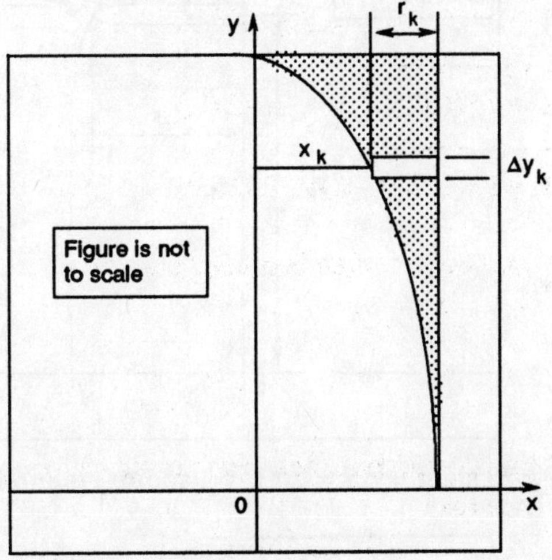

The volume of the *k-th* strip is approximated as

$$\Delta V_k = \pi(\sqrt{6} - x_k)^2 \Delta y_k$$

Noting that the limits of integration is between 0 and 6, the total volume is given by

$$V = \int_0^6 \pi(\sqrt{6} - x)^2\,dy = \int_0^6 \pi(6 - 2\sqrt{6}\,x + x^2)dy$$

However, $x = \sqrt{6-y}$

$$V = \int_0^6 \pi(6 - 2\sqrt{6}\sqrt{6-y} + (6-y)\;dy$$

$$= \pi \int_0^6 [12 - 2\sqrt{6}\sqrt{6-y} - y]\, dy$$

$$V = \pi \left[12y + 2\sqrt{6}\, \frac{(6-y)^{3/2}}{3/2} - \frac{y^2}{2} \right]\Bigg|_0^6 = 6\pi$$

40. **(B)**

The two curves' intersection is determined by equating the two equations, then solving for the unknown x–values.

$$x^4 + x^2 - 6 = 0 = (x^2 - 2)\,(x^2 + 3)$$

Which gives $x^2 = 2$ or $x = \pm\sqrt{2}$

Note that the imaginary roots are not needed. The corresponding functional value to the real x–values is calculated as follows: The surface area is computed for the *k-th* slant using the following figure:

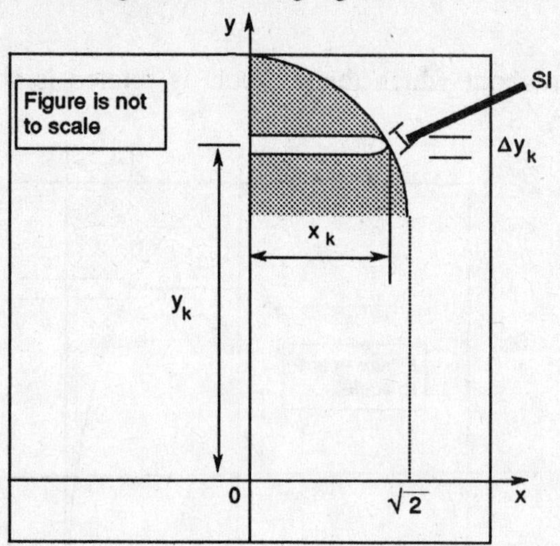

Surface area for the *k-th* slant = $\Delta S = 2\pi$ (avg. radius) (slant height)

$$\Delta S = 2\pi x_k \sqrt{(\Delta x_k)^2 + (\Delta y_k)^2}$$

The total surface area is then given as

$$S = \int_0^{\sqrt{2}} 2\pi x \sqrt{1 + \left(\frac{dy}{dx}\right)^2}\, dx$$

Where $dy / dx = -2x$. Substituting into the integral gives

$$S = \int_0^{\sqrt{2}} 2\pi x \sqrt{1 + (-2x)^2}\, dx = 2\pi \int_0^{\sqrt{2}} x\sqrt{1 + 4x^2}\, dx$$

$$S = 2\pi \left[\frac{1}{8}\, \frac{(1+4x^2)^{3/2}}{3/2} \right]\Bigg|_0^{\sqrt{2}} = \frac{13}{3}\pi$$

41. **(B)**

In any AC circuit, the average power P_{av} through any load is given in terms of the peak voltage V across the load, the peak current I flowing through the load, and the power factor PF by

$$P_{av} = {}^1/_2 \cdot V \cdot I \cdot PF = V_{rms} \cdot I_{rms} \cdot PF$$

Thus,

$$I_{rms} = \frac{P_{av}}{V_{rms} \cdot PF} = \frac{11000\ W}{220\ V \cdot 1.0} = 50\ A\ rms$$

42. **(C)**

The per-phase transmission line loses P_{loss}, or ohmic losses, through any resistance R are given in terms of the current I flowing through R *by*

$$P_{loss} = R \cdot I^2_{rms} = 0.1 \cdot (50)^2 = 250\ \text{W}$$

Note that since the per-phase load and the resistance R are in series, the current flowing through them is the same.

43. **(C)**

The power company should generate the power P_g to feed the load and the line losses. Since the power factor (PF) of the load is 1.0, then we have P_g given by

$$P_g = 3 \cdot P_{loss} + P_{load} = 3.\cdot 250 + 11000 = 11.75\ \text{kW}$$

44. **(C)**

The *rms* voltage drop across phase *a* transmission line resistance is

$$0.1\ \Omega \cdot 50\ \text{A} = 5\ \text{V}\ rms$$

Since the per-phase load and transmission line resistance are in series, and the voltages across each has a 0° phase-angle, the phase *a* source voltage will be given by

$$V_g = 220\ \text{V} + 5\ \text{V} = 225\ \text{V}\ rms$$

45. **(C)**

The power company generates 11.75 W and the consumer takes only 11kW, so the percentage that the consumer will be billed with is

$$\frac{11}{11.75} \cdot 100 = 93.6\%$$

46. **(C)**

Using the same argument as in problem 41, we find that current I is given by

$$I_{rms} = \frac{11000\ W}{220\ V \cdot \cos 60°} = \frac{11000}{220 \cdot 0.5} = 100\ A$$

47. **(C)**

Similarly to problem 2, the total line losses are given by

$$P_{loss} = 3 \cdot (0.1\ \Omega \cdot 100^2) = 3\ \text{kW}$$

48. **(D)**

The actual energy generated at the power company is

$$P_g = 11000 + 3000 = 14\ \text{kW}$$

So, the percentage of the generated power that the second consumer will be billed with is

$$\frac{11}{14} \cdot 100 = 78.6\%$$

49. **(C)**

If the percentage in problem 48 becomes 90, this means that the total generated power P'_g became

$$P'_g = \frac{11000}{0.9} = 12.22\ kW$$

Since the consumption at the load is kept constant, the per-phase transmission line losses are given by

$$P'_{loss} = \frac{(12.22 - 11)}{3} = 406.7\ kW$$

and the current flowing through the load and the transmission line resistance is the same as before. Thus, the per-phase transmission line resistance has become

$$R' = \frac{406.7}{100^2} = 0.041\ \Omega$$

and the original resistance R has to be decreased by

$$R' - R = 0.041 - 0.1 = -0.059\ \Omega$$

50. **(B)**

Similarly, we get P'_g = 12.22 kW, and P'_{loss} = 406.7 W, but since in this case the transmission line resistance is kept constant, the current value becomes I' given by

$$I' = \sqrt{406.7/0.1} = 63.77\ A$$

Hence, the power factor PF' becomes

$$PF' = \frac{11000}{63.77 \cdot 220} = 0.78$$

and corresponding to an angle of 38.36° lag.

51. **(C)**

Given: Annual worth / capital recovery problem

(i)	$P = \$25{,}000$	(iv)	$n = 5$ years
(ii)	$S = \$5{,}000$	(v)	Interest Rate (MARR) = 20%
(iii)	$A = \$8{,}000$		

Required: Calculate Annual Worth (*AW*) of the investment.

Cash Flow Diagram:

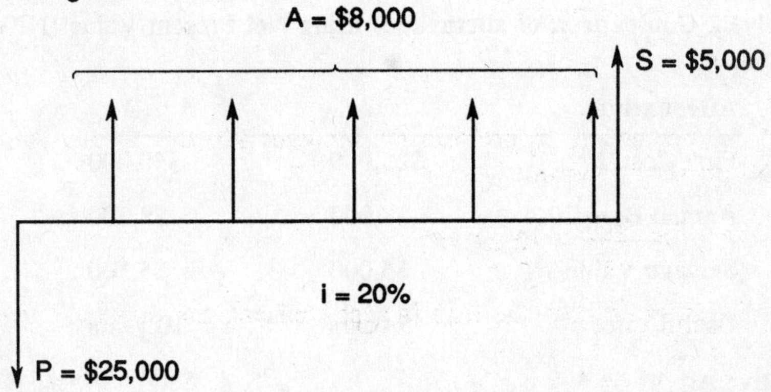

Solution:

$$\text{Annual Worth } (AW) = \left\{\begin{matrix}\text{Annual Revenue} - \text{Expenses} \\ \text{(Annual Net Profit)}\end{matrix}\right\} - \left\{\begin{matrix}\text{Capital Recovery} \\ \text{Amount}\end{matrix}\right\}$$

$AW\,(20\%) = \{A - 0\} - \{P(A/P, 20\%, 5) - S(A/F, 20\%, 5)\}$

Using Tables = \$8,000 – {\$25,000 (0.3344) – \$5,000 (0.1344)}

Solving = \$8,000 – (\$8,359.49 – \$671.90) = \$312.41

52. **(D)**

Given: Capital Recovery Problem

(i) Initial Cost, $P = \$25{,}000$

(ii) Estimated Salvage Value, $F = \$5{,}000$

(iii) Estimated Service Life, $n = 5$ *years*

(iv) Interest Rate (government secured) = 6%

Required: Calculate Capital Recovery, which equals the annual equivalent cost less the annual equivalent salvage value.

Solution:

$$\text{Capital Recovery, } CR(i) = P(A/P, i, n) - S(A/F, i, n)$$

$$CR(6\%) = \$25{,}000\ (A/P, 6, 5) - \$5{,}000\ (A/F, 6, 5)$$

From the tables

$$CR(6\%) = \$25{,}000\ (.2374) - \$5{,}000\ (.1774)$$

$$CR(6\%) = \$5{,}935 - \$887$$

$$CR(6\%) = \$5,048$$

53. **(A)**

Given: Comparison of alternatives using Net Present Value (NPV) analysis technique.

Alternative	A	B
First Cost, *P*	$25,000	$50,000
Annual Benefit, *A*	$8,000	$8,500
Salvage Value, *S*	$5,000	$5,500
Useful Life, *n*	5 years	10 years

Alternative A;

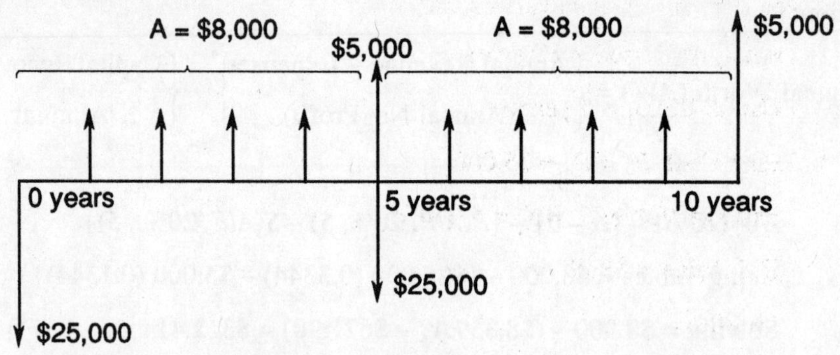

$$\begin{aligned} NPV, A(20\%) &= -\$25{,}000 - (\$25{,}000 - \$5{,}000)\ (P/F, 20\%, 5) \\ &\quad + \$5{,}000\ (P/F, 20\%, 10) + \$8{,}000\ (P/A, 20\%, 10) \\ &= \$\{-25{,}000 - 20{,}000\ (.4019) + 5{,}000\ (.1615) \\ &\quad + 8{,}000(4.1925)\} \\ &= \$(25{,}000 - 8{,}038 + 807.5 + 33{,}540) \\ &= +\$1{,}309.5 \end{aligned}$$

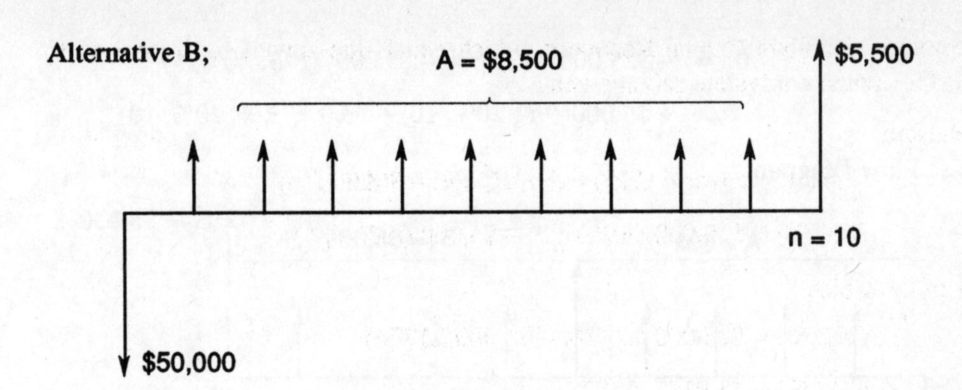

$$NPV, \text{B}(20\%) = -\$50{,}000 + \$5{,}500\,(P/F, 20\%, 10)$$
$$+ \$8{,}500(P/A, 20\%, 10)$$
$$= \$\{-50{,}000 + 5{,}500\,(.1615) + 8{,}500(4.1925)\}$$
$$= \$(-50{,}000 + 888.25 + 35{,}636.25)$$
$$= -\$13{,}475.5$$

So *NPV* of cost advantage of A over B

$$= \$1{,}309.5 - (-13{,}475.5)$$
$$= \$14{,}785$$

54. **(B)**

Given: Breakeven Analysis Problem

(i) Analysis Period = 10 years

(ii) Useful life, $n = 5$ years

(iii) First Cost, $P_1 = P_2 = 25{,}000$

(iv) Salvage Value, $S_1 = S_2 = 5{,}000$

(v) Unit Price = $3.00

(vi) $I = 20\%$

Required: Number of units produced at breakeven point.

Solution: At the breakeven point, Net Present Value, N*PV* = 0. Write an expression for the *NPV* of alternative *A*, assuming *X* = number of units produced, and set equal to zero.

$$NPV, \text{A}\ (20\%) = 0 = -P_1 - P_2(P/F, i, n) + S_1(P/F, i, n)$$
$$+ S_2\ (P/F, i, 10) + \$3 \times (P/A, i, 10)$$

$$0 = -\$25{,}000 - (25{,}000 - \$5{,}000)\,(P/F, 20\%, 5)$$
$$+ \$5{,}000(P/F, 20\%, 10) + \$3.0 \times (P/A, 20\%, 10)$$

Cash Flow Diagram:

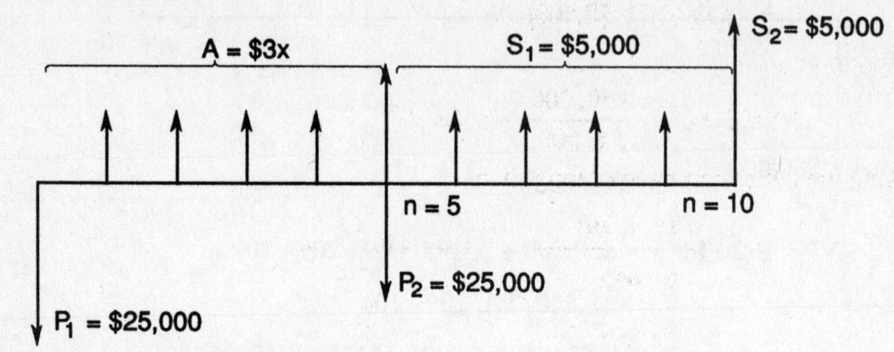

Using the tables to find factors

$$0 = -\$25{,}000 - \$20{,}000(.4019) + \$5{,}000(.1615)$$
$$+ \$3.0 \times (4.1925)$$
$$0 = -\$25{,}000 - \$8.038 - \$807.5 + \$12.58(x)/\text{unit}$$
$$x = \$32{,}230.5/(\$12.58/\text{unit})$$
$$x = 2{,}563 \text{ units}$$

55. **(C)**

Given: Rate of Return / Net Present Value Problem (modification of problem 53).

(i) First Cost, $p = \$50{,}000$

(ii) Annual Benefit, $A = \$8{,}500$

(iii) Useful Life, $n = 10$ years

(iv) Salvage value, $S = 0$

Required: Rate of Return for Alternative B, when salvage value, $S = 0$.

Cash Flow Diagram: A = $8,500

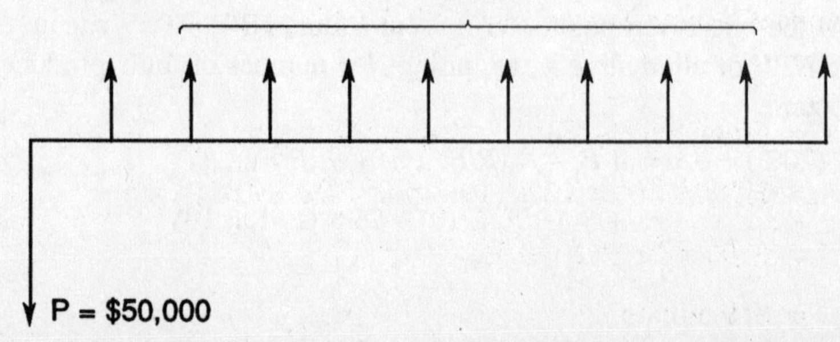

Solution: To find rate of return, derive an expression for the Net Present Value, *NPV*, of alternative B and set it equal to zero. Then solve, through trial and error, for interest rate.

$$NPV\text{, B}(i) = 0 = -P + A(P/A, i, n)$$

$$0 = -\$50{,}000 + \$8{,}500(P/A, i, 10)$$

Solve for factor

$$(P/A, i, 10) = \frac{\$\,50,000}{\$\,8,500} = 5.8824$$

Using the tables and interpolating gives

(P/A, i, 10)	i
6.1446	10%
5.8824	X%
5.6502	12%

$$\frac{x-10}{12-10} = \frac{5.8824-6.1446}{5.6502-6.1446} = \frac{-.2622}{-.4944} = .5303$$

$$x = 10 + 2(.5303) = 11 \quad \dot{z} = 11\%$$

56. **(B)**

Given: Continuous Cash Flow Analysis and Tax Problem.

(i) Direct Fixed Capital (*DFC*) = \$70 m

(ii) Salvage = 0

(iii) Life, $n = 10$ years

(iv) Straight Line, S.L. depreciation

(v) Cost of Manufacture *COM* = \$0.22/lb (Includes Depreciation)

(vi) General Expenses, G = \$0.08/lb

(vii) Sales = \$.50/lb

(viii) Tax Rate = 50%

(ix) Annual Production = 100 m lb

Required: Annual Income Tax paid by XYZ for this plant.

Solution:

$$\begin{aligned} \text{Tax} &= \text{(Total Sales – Total Expenses) (Tax Rate)} \\ &= \$[100 \times 10^6 \times 0.5 - \{100 \times 10^6(0.22 + 0.08)\}]\ 0.5 \\ &= \$100 \times 10^6\ (0.5 - 0.3)\ 0.5 \\ &= \$10 \text{ million} \end{aligned}$$

57. **(D)**

Given: Same information as in problem 56.

Required: Annual Profit Before Taxes

Solution:

Annual Profit Before Taxes = Total Sales – Total Expenses

$$\text{Profit} = \text{Sales} - (COM + G)$$
$$= \$100 \times 10^6 [0.5 - (0.22 + 0.08)]$$
$$= \$20 \text{ million.}$$

58. **(B)**

Given: Extension of problem 56.

Required: Annual Profit After Tax.

Solution:

Profit After Tax (1 – Tax Return) (Profit Before Taxes)

or = Profit Before Taxes – Taxes

Now from problems 56 and 57,

Profit Before Taxes = \$20 million

Tax computed = \$10 million

∴ Profit After Taxes = (\$20 – 10) million

= \$10 million.

or = (1 – Tax Rate) {Total Sales – Total Expenses}

$$= (1 - 0.5)\ \{100 \times 10^6 \times 0.5 - 100 \times 10^6\ (0.22 + 0.08)]$$

= \$10 million.

59. **(C)**

Given: Breakeven or equivalence problem.

Using concrete is equivalent to using asphalt. Since maintenance cost is the same for both materials we may compare only the capital costs distributed over life of each project.

i.e., $15{,}000/\text{mile}\ (A/P, 12\%, 20) = P_{As}(A/P, 12\%, 10)$

where P_{As} = maximum cost of asphalt road

Using Tables,

P_{A_s} = \$15,000/mile (0.1339) / 0.1770

= \$11,347/mile

60. **(B)**
Given: Inflation / Effective Interest Rate / Combined.

Interest Rate Problem:

(i) Real Interest Rate = 10%

(ii) Inflation Rate = 5%

Required: Compute the combined rate.

Solution: Combined rate = $i + j + ij$

where i = interest rate

j = inflation rate

Combined rate

= 0.10 + 0.05 + (0.10) (0.05)

= 0.155

= 15.5%

61. **(E)**
For a small pitot-static tube, Bernoulli's equation is valid between the stagnation point and the static taps in the probe

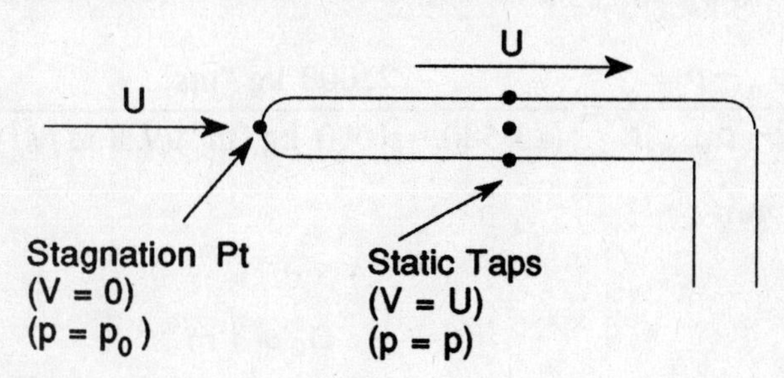

Neglecting gravity over the tiny probe,

$$p_0 + \tfrac{1}{2}\rho V_0^2 = p + \tfrac{1}{2}\rho U^2$$

or $$p_0 - p = \tfrac{1}{2}\rho U^2 = \tfrac{1}{2}(1000\text{ kg/m}^3)(2\text{ m/s})^2 = 2{,}000\text{ kg/ms}^2$$

$$p_0 - p = 2{,}000\text{ N/m}^2.$$

62. **(C)**
With the line switch connecting static line 0 (i.e., the line from the static

pressure of the pitot-static tube), the manometer will measure the pressure difference between p_0 and p.

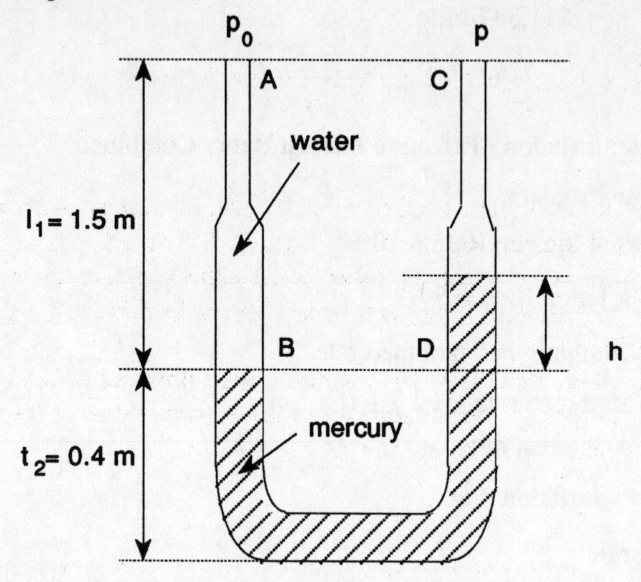

An equivalent system is drawn above. Using the manometer (fluid statics) relationships,

from A to B, $\quad p_B = p_0 + \rho_{\text{water}}\, g l_1.$

from C to D, $\quad p_D = p + \rho_{\text{water}}\, g(l_1 - h) + \rho_{\text{Hg}}\, gh.$

But since a continuous line can be drawn through static mercury from B to D, p_B must equal p_D. Hence

$$p_0 = p + (\rho_{\text{Hg}} - \rho_{\text{water}})\, gh$$

or

$$h = \frac{p_0 - p}{(\rho_{\text{Hg}} - \rho_{\text{water}})g} = \frac{2{,}000 \text{ kg/ms}^2}{(13{,}580 - 1000 \text{ kg/m}^3)(9.8 \text{ m/s}^2)} = 0.0162 \text{ m}$$

$$h = 16.2 \text{ mm}.$$

63. **(E)**

D0 = 1 m
D1 = 0.5 m
U0= 2 m/s
U1= ?

Neglecting friction between the probe and the static tap location 1, conservation of mass can be applied, assuming uniform velocity at location 1.

$$\rho_0 U_0 A_0 = \rho_1 U_1 A_1$$

But $\rho_0 = \rho_1$ for water which is incompressible in this problem. Also,

$$A_0 = {}^{\pi}/_4\, D_0^2 \text{ and } A_1 = {}^{\pi}/_4\, D_1^2$$

Thus,

$$U_1 = U_0 \frac{D_o^2}{D_1^2} = (2\ \text{m/s})\left(\frac{1^2}{0.5^2}\right) = 8.0\ \text{m/s} = U_1$$

64. **(B)**

In the long straight section of pipe, pressure must *decrease* due to frictional losses, even though the flow is fully developed, which means that the velocity profile is constant. In other words, the pressure must be greater upstream than downstream in order to "push" the water through the pipe, having to overcome frictional forces.

Thus, when static line 3 is connected, the pressure difference between the left and right sides of the manometer will be *greater* than when static line 2 is connected. Hence h will be *greater* for case 3.

65. **(D)**

To find the friction factor f from 2 to 3, first calculate the Reynolds number of the pipe flow.

$$R_e = \frac{UD}{\nu} = \frac{(8\ \text{m/s})\,(0.5\ \text{m})}{1.0 \times 10^{-6}\ \text{m}^2/\text{s}} = 4 \times 10^6$$

The roughness factor

$$\varepsilon / D = 0.5\ \text{mm} / 500\ \text{mm} = 0.001$$

From the Moody chart, $f \approx 0.0197$.

66. **(B)**

In the entrance region, as U increases along the centerline, p must decrease by Bernoulli's equation along the inviscid centerline. This is shown below:

$$p + {}^1/_2\, \rho\, U^2 + \rho g z = \text{constant}$$

Along the centerline, z = constant and therefore

$$p + {}^1/_2\, \rho\, U^2 = \text{constant}$$

Thus as U goes up, p must go down.

67. **(B)**

Static pressure must *decrease* between 2 and 3 in order to overcome the frictional losses. This pressure drop is what "pushes" the water through the pipe, overcoming friction.

68. **(B)**

Stagnation pressure only remains constant in isentropic flow. Here, since there is a frictional head loss, stagnation pressure must also *decrease* in the pipe.

69. **(C)**

For fully developed pipe flow, by definition, the velocity profile shape *does not change*. Thus, centerline velocity U must remain constant, even though pressure is dropping along the length of the pipe.

70. **(D)**

To find h_f, the frictional head loss from 1 to 2, use the formula from the Moody chart, i.e.

$$h_f = f\left(\frac{L}{D}\right)\left(\frac{V_{av}^2}{2g}\right)$$

where L = pipe length = 50 m

D = pipe diameter = 0.5 m

V_{av} = average velocity = 8 m/s

g = 9.8 m/s²

Hence,

$$h_f = (0.0197)\left(\frac{50}{0.5}\right)\left(\frac{8^2}{2(9.8)}\right) = 6.43 \text{ m of water.}$$

Fundamentals of Engineering

A.M. SECTION

Test 3

Fundamentals of Engineering

A.M. SECTION

TEST 3 – ANSWER SHEET

1. (A) (B) (C) (D) (E)
2. (A) (B) (C) (D) (E)
3. (A) (B) (C) (D) (E)
4. (A) (B) (C) (D) (E)
5. (A) (B) (C) (D) (E)
6. (A) (B) (C) (D) (E)
7. (A) (B) (C) (D) (E)
8. (A) (B) (C) (D) (E)
9. (A) (B) (C) (D) (E)
10. (A) (B) (C) (D) (E)
11. (A) (B) (C) (D) (E)
12. (A) (B) (C) (D) (E)
13. (A) (B) (C) (D) (E)
14. (A) (B) (C) (D) (E)
15. (A) (B) (C) (D) (E)
16. (A) (B) (C) (D) (E)
17. (A) (B) (C) (D) (E)
18. (A) (B) (C) (D) (E)
19. (A) (B) (C) (D) (E)
20. (A) (B) (C) (D) (E)
21. (A) (B) (C) (D) (E)
22. (A) (B) (C) (D) (E)
23. (A) (B) (C) (D) (E)
24. (A) (B) (C) (D) (E)
25. (A) (B) (C) (D) (E)
26. (A) (B) (C) (D) (E)
27. (A) (B) (C) (D) (E)
28. (A) (B) (C) (D) (E)
29. (A) (B) (C) (D) (E)
30. (A) (B) (C) (D) (E)
31. (A) (B) (C) (D) (E)
32. (A) (B) (C) (D) (E)
33. (A) (B) (C) (D) (E)
34. (A) (B) (C) (D) (E)
35. (A) (B) (C) (D) (E)
36. (A) (B) (C) (D) (E)
37. (A) (B) (C) (D) (E)
38. (A) (B) (C) (D) (E)
39. (A) (B) (C) (D) (E)
40. (A) (B) (C) (D) (E)
41. (A) (B) (C) (D) (E)
42. (A) (B) (C) (D) (E)
43. (A) (B) (C) (D) (E)
44. (A) (B) (C) (D) (E)
45. (A) (B) (C) (D) (E)
46. (A) (B) (C) (D) (E)
47. (A) (B) (C) (D) (E)
48. (A) (B) (C) (D) (E)
49. (A) (B) (C) (D) (E)
50. (A) (B) (C) (D) (E)
51. (A) (B) (C) (D) (E)
52. (A) (B) (C) (D) (E)
53. (A) (B) (C) (D) (E)
54. (A) (B) (C) (D) (E)
55. (A) (B) (C) (D) (E)
56. (A) (B) (C) (D) (E)
57. (A) (B) (C) (D) (E)
58. (A) (B) (C) (D) (E)
59. (A) (B) (C) (D) (E)
60. (A) (B) (C) (D) (E)
61. (A) (B) (C) (D) (E)
62. (A) (B) (C) (D) (E)
63. (A) (B) (C) (D) (E)
64. (A) (B) (C) (D) (E)
65. (A) (B) (C) (D) (E)
66. (A) (B) (C) (D) (E)
67. (A) (B) (C) (D) (E)
68. (A) (B) (C) (D) (E)
69. (A) (B) (C) (D) (E)
70. (A) (B) (C) (D) (E)
71. (A) (B) (C) (D) (E)
72. (A) (B) (C) (D) (E)
73. (A) (B) (C) (D) (E)
74. (A) (B) (C) (D) (E)
75. (A) (B) (C) (D) (E)

76. Ⓐ Ⓑ Ⓒ Ⓓ Ⓔ
77. Ⓐ Ⓑ Ⓒ Ⓓ Ⓔ
78. Ⓐ Ⓑ Ⓒ Ⓓ Ⓔ
79. Ⓐ Ⓑ Ⓒ Ⓓ Ⓔ
80. Ⓐ Ⓑ Ⓒ Ⓓ Ⓔ
81. Ⓐ Ⓑ Ⓒ Ⓓ Ⓔ
82. Ⓐ Ⓑ Ⓒ Ⓓ Ⓔ
83. Ⓐ Ⓑ Ⓒ Ⓓ Ⓔ
84. Ⓐ Ⓑ Ⓒ Ⓓ Ⓔ
85. Ⓐ Ⓑ Ⓒ Ⓓ Ⓔ
86. Ⓐ Ⓑ Ⓒ Ⓓ Ⓔ
87. Ⓐ Ⓑ Ⓒ Ⓓ Ⓔ
88. Ⓐ Ⓑ Ⓒ Ⓓ Ⓔ
89. Ⓐ Ⓑ Ⓒ Ⓓ Ⓔ
90. Ⓐ Ⓑ Ⓒ Ⓓ Ⓔ
91. Ⓐ Ⓑ Ⓒ Ⓓ Ⓔ
92. Ⓐ Ⓑ Ⓒ Ⓓ Ⓔ
93. Ⓐ Ⓑ Ⓒ Ⓓ Ⓔ
94. Ⓐ Ⓑ Ⓒ Ⓓ Ⓔ
95. Ⓐ Ⓑ Ⓒ Ⓓ Ⓔ
96. Ⓐ Ⓑ Ⓒ Ⓓ Ⓔ
97. Ⓐ Ⓑ Ⓒ Ⓓ Ⓔ
98. Ⓐ Ⓑ Ⓒ Ⓓ Ⓔ
99. Ⓐ Ⓑ Ⓒ Ⓓ Ⓔ
100. Ⓐ Ⓑ Ⓒ Ⓓ Ⓔ
101. Ⓐ Ⓑ Ⓒ Ⓓ Ⓔ
102. Ⓐ Ⓑ Ⓒ Ⓓ Ⓔ
103. Ⓐ Ⓑ Ⓒ Ⓓ Ⓔ
104. Ⓐ Ⓑ Ⓒ Ⓓ Ⓔ
105. Ⓐ Ⓑ Ⓒ Ⓓ Ⓔ
106. Ⓐ Ⓑ Ⓒ Ⓓ Ⓔ
107. Ⓐ Ⓑ Ⓒ Ⓓ Ⓔ
108. Ⓐ Ⓑ Ⓒ Ⓓ Ⓔ
109. Ⓐ Ⓑ Ⓒ Ⓓ Ⓔ
110. Ⓐ Ⓑ Ⓒ Ⓓ Ⓔ
111. Ⓐ Ⓑ Ⓒ Ⓓ Ⓔ
112. Ⓐ Ⓑ Ⓒ Ⓓ Ⓔ
113. Ⓐ Ⓑ Ⓒ Ⓓ Ⓔ
114. Ⓐ Ⓑ Ⓒ Ⓓ Ⓔ
115. Ⓐ Ⓑ Ⓒ Ⓓ Ⓔ
116. Ⓐ Ⓑ Ⓒ Ⓓ Ⓔ
117. Ⓐ Ⓑ Ⓒ Ⓓ Ⓔ
118. Ⓐ Ⓑ Ⓒ Ⓓ Ⓔ
119. Ⓐ Ⓑ Ⓒ Ⓓ Ⓔ
120. Ⓐ Ⓑ Ⓒ Ⓓ Ⓔ
121. Ⓐ Ⓑ Ⓒ Ⓓ Ⓔ
122. Ⓐ Ⓑ Ⓒ Ⓓ Ⓔ
123. Ⓐ Ⓑ Ⓒ Ⓓ Ⓔ
124. Ⓐ Ⓑ Ⓒ Ⓓ Ⓔ
125. Ⓐ Ⓑ Ⓒ Ⓓ Ⓔ
126. Ⓐ Ⓑ Ⓒ Ⓓ Ⓔ
127. Ⓐ Ⓑ Ⓒ Ⓓ Ⓔ
128. Ⓐ Ⓑ Ⓒ Ⓓ Ⓔ
129. Ⓐ Ⓑ Ⓒ Ⓓ Ⓔ
130. Ⓐ Ⓑ Ⓒ Ⓓ Ⓔ
131. Ⓐ Ⓑ Ⓒ Ⓓ Ⓔ
132. Ⓐ Ⓑ Ⓒ Ⓓ Ⓔ
133. Ⓐ Ⓑ Ⓒ Ⓓ Ⓔ
134. Ⓐ Ⓑ Ⓒ Ⓓ Ⓔ
135. Ⓐ Ⓑ Ⓒ Ⓓ Ⓔ
136. Ⓐ Ⓑ Ⓒ Ⓓ Ⓔ
137. Ⓐ Ⓑ Ⓒ Ⓓ Ⓔ
138. Ⓐ Ⓑ Ⓒ Ⓓ Ⓔ
139. Ⓐ Ⓑ Ⓒ Ⓓ Ⓔ
140. Ⓐ Ⓑ Ⓒ Ⓓ Ⓔ

FUNDAMENTALS OF ENGINEERING EXAMINATION

TEST 3

MORNING (AM) SECTION

TIME: 4 Hours
140 Questions

DIRECTIONS: For each of the following questions and incomplete statements, choose the best answer from the five answer choices.

1. **A** = **i** + 2**j** is a two-component vector. Another vector **B**, which is perpendicular to **A** and whose component in the **i** direction is – 6.2 is

(A) – 62**i** – 4.7**j**

(B) – 6.2**i** – 1.8**j**

(C) – 6.2**i** + 1.8**j**

(D) – 6.2**i** + 3.1**j**

(E) – 6.2**i** + 4.7**j**

2. The differential equation

$$3\frac{d^2y}{dt^2} + t\frac{dy}{dt} - 4y = \sin(2t) \text{ is}$$

(A) first-order, linear, and homogeneous.

(B) second-order, linear, and homogeneous.

(C) first-order, nonlinear, and homogeneous

(D) second-order, nonlinear, and nonhomogeneous.

(E) second-order, linear, and nonhomogeneous.

3. The value of the limit

$$\lim_{x \to 0} \frac{x + \tan x}{\sin 3x} \text{ is}$$

(A) 1

(B) 2

(C) 1/2 (D) 2/3

(E) 2/9

4. The length of each side of a regular pentagon inscribed in a unit circle is

(A) 0.26 (D) 1.18

(B) 0.59 (E) 1.78

(C) 0.78

5. The product of the complex numbers $(2 + 4i)$ and $(1 - 7i)$ is

(A) $30 - 10i$ (D) $3 + 19i$

(B) $22 + 8i$ (E) $10 - 24i$

(C) $17 + 24i$

6. The values of x and y which satisfy

$4x + 3y = 17$

$x - 5y = -13$ are

(A) $x = 2, y = 3$ (D) $x = -1, y = 2$

(B) $x = 4, y = 7$ (E) $x = 5, y = 11$

(C) $x = -3, y = 3$

7. Let $x = 2$ be an initial approximation of the root to the expression

$x^3 + 4x^2 + 2x - 2 = 0$

The next approximation of the root as calculated by Newton's method is

(A) 2.72 (D) 0.92

(B) 4.77 (E) 1.13

(C) 2.04

8. The concentration of a particular pesticide decays exponentially. One-half of the initial amount of the pesticide decomposes in thirty years. How many years are required for 90% of the pesticide to decompose?

(A) 21

(B) 32

(C) 37

(D) 48

(E) 62

9. The value of

$$\int_1^3 \frac{4x\ dx}{6+x^2} \text{ is}$$

(A) 0.16

(B) 0.58

(C) 1.20

(D) 1.52

(E) 2.99

10. The total differential of

$$F(x, y, z) = (1 + e^{xz}) \sin 4y \text{ is}$$

(A) $z \sin(4y)\, dx + 2 \cos(4y) e^{xy}\, dy + x\, e^{xz} \sin(4y)\, dz$

(B) $z\, e^{xz} \sin(4y)\, dx + 4 \cos(4y)\, (1 + e^{xy})\, dy + x\, e^{xz} \sin(4y)\, dz$

(C) $4 \sin(4y)\, (1 + e^{xy})\, dy + 4x\, e^{xz} \cos(4y)\, dz$

(D) $x\, e^{xz} \cos(4y)\, dx + 16 \cos(4y)\, (1 + e^{xy}) dy + 4x\, e^{xz} \sin(4y)\, dz$

(E) $2\, z\, e^{xz} \cos(4y)\, dx + 2 \cos(4y)\, (1 + 2e^{xy}) dy + y\, e^{xz} \sin(4y)\, dz$

11. The partial fraction expansion of

$$\frac{s-3}{(s+2)\,(s+3)\,(s+4)} \text{ is } \frac{A}{s+2} + \frac{B}{s+3} + \frac{C}{s+4}$$

The values of A, B, and C are

(A) $-3, -5/2, 7$

(B) $-6, 4, 5$

(C) $2, -5, 7$

(D) $-6, -5/2, -7/2$

(E) $-6, 6, -3$

12. Measurements X have a normal distribution, which has a mean of 16 and a standard deviation of 2. The probability of measuring a value of X between 15 and 19 is

(A) $\dfrac{1}{\sqrt{2\pi}} \displaystyle\int_{0.5}^{1.5} e^{-Z^2/2}\, dZ$

(B) $\dfrac{1}{\sqrt{2\pi}} \displaystyle\int_{-0.5}^{1.5} e^{-Z^2/2}\, dZ$

(C) $\frac{1}{\sqrt{2\pi}} \int_{15}^{19} e^{-z^2/2}\, dZ$

(D) $\frac{1}{\sqrt{2\pi}} \int_{-1}^{3} e^{-z^2/2}\, dZ$

(E) $\frac{1}{\sqrt{2\pi}} \int_{0}^{16} e^{-z^2/2}\, dZ$

13. The equation of a straight line which passes through $x = 3$, $y = 1$ and $x = 10$, $y = 15$ is

(A) $y = 2x - 5$

(B) $y = 2x + 5$

(C) $y = 4x - 5$

(D) $y = 2x + 10$

(E) $y = 4x + 10$

14. The value of x which provides the minimum y in the function

$$y = 2x^3 - 24x + 14 \text{ is}$$

(A) -1

(B) -2

(C) 1

(D) 2

(E) 4

15. Forty electrical engineers, twenty chemical engineers, thirty mechanical engineers, and ten civil engineers attend a banquet. A television station randomly selects four engineers to interview. What is the probability that someone from all four disciplines will be interviewed?

(A) 4 / 100

(B) 198,360 / 94,109,400

(C) 240,000 / 100,000,000

(D) 240,000 / 94,109,400

(E) 224,120 / 100,000,000

16. Rotate the curve defined by $x^2 + y^2 = 1$ around the y–axis. The volume is

(A) $\pi/6$

(B) $\pi/3$

(C) $\pi/2$

(D) $4\pi/3$

(E) 2π

17. A set of linear equations is written as

$$\mathbf{Ax} = \mathbf{b}$$

where **A** is the matrix of constant coefficients

x is the vector of unknown values

b is the vector of known constants

Let $\mathbf{A}^T$ represent the transpose of **A**

$\mathbf{A}^{-1}$ represent the inverse of **A**

The vector **x** can be determined by

(A) $\mathbf{x} = \mathbf{A}^{-1}\,\mathbf{b}$

(B) $\mathbf{x} = \mathbf{A}^{T}\,\mathbf{b}$

(C) $\mathbf{x} = \mathbf{b}\,\mathbf{A}^{-1}$

(D) $\mathbf{x} = \mathbf{b}\,/\mathbf{A}$

(E) $\mathbf{x} = \mathbf{b}\,\mathbf{A}^{T}$

18. What is the radius of the circle described by

$$x^2 - 2x + y^2 - 2y - 4 = 0\ ?$$

(A) 1

(B) 2.2

(C) 2.7

(D) 3

(E) 4.2

19. Find the area of the region bounded by

$$x = 0$$

$$y = x + 1$$

$$y = .5\,x^2$$

(A) 0.23

(B) 0.48

(C) 0.99

(D) 1.25

(E) 1.87

20. The tangent to the curve

$y = x\,e^x$ at $x = 0.8$ is

(A) 0.8

(B) 1.8

(C) 3.6

(D) 4.0

(E) 5.2

21. In the circuit shown here, the capacitor had been charged to a voltage of 100 V before it was connected to a 100 K resistor to discharge. The voltage decreased to 36.8 V in 0.2 sec. The value of the capacitance is

(A) 2 μF

(B) 1 μF

(C) 20 μF

(D) 4 μF

(E) 0.5 μF

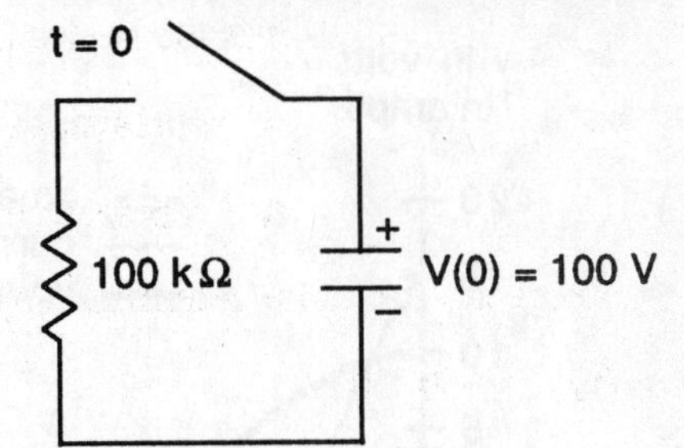

22. In the circuit shown below, before the switch is closed at time $t = 0$, no energy was stored either in the capacitor nor in the inductor. Immediately after closing the switch, the current in the 3 Ω resistor is given by:

(A) 2.4 A

(B) 4.0 A

(C) 0 A

(D) 10.0 A

(E) 3.3 A

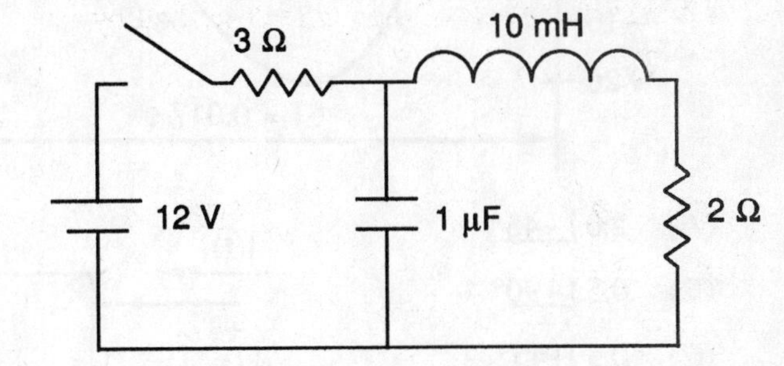

23. A resistor, an inductor and a capacitor are connected in parallel to an AC source, and the values of the inductance L and the capacitance C have been adjusted so that the circuit is in resonance, the current i is:

(A) a maximum

(B) a minimum

(C) lags the voltage by 90°

(D) leads the voltage by 90°

(E) is zero

24. In the circuit shown here the voltage across the 4 Ω resistor is given by:

(A) $80\angle 53^\circ$

(B) $80\angle -37^\circ$

(C) $80\angle 0^\circ$

(D) $18.71\angle 53^\circ$

(E) $18.71\angle -53^\circ$

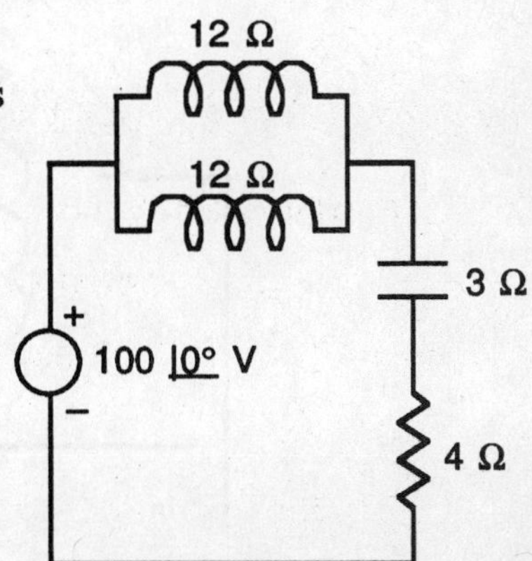

25. The voltage $v(t)$ and current $i(t)$ at the two terminals of a passive element have been recorded and are shown below. The impedance of this element is:

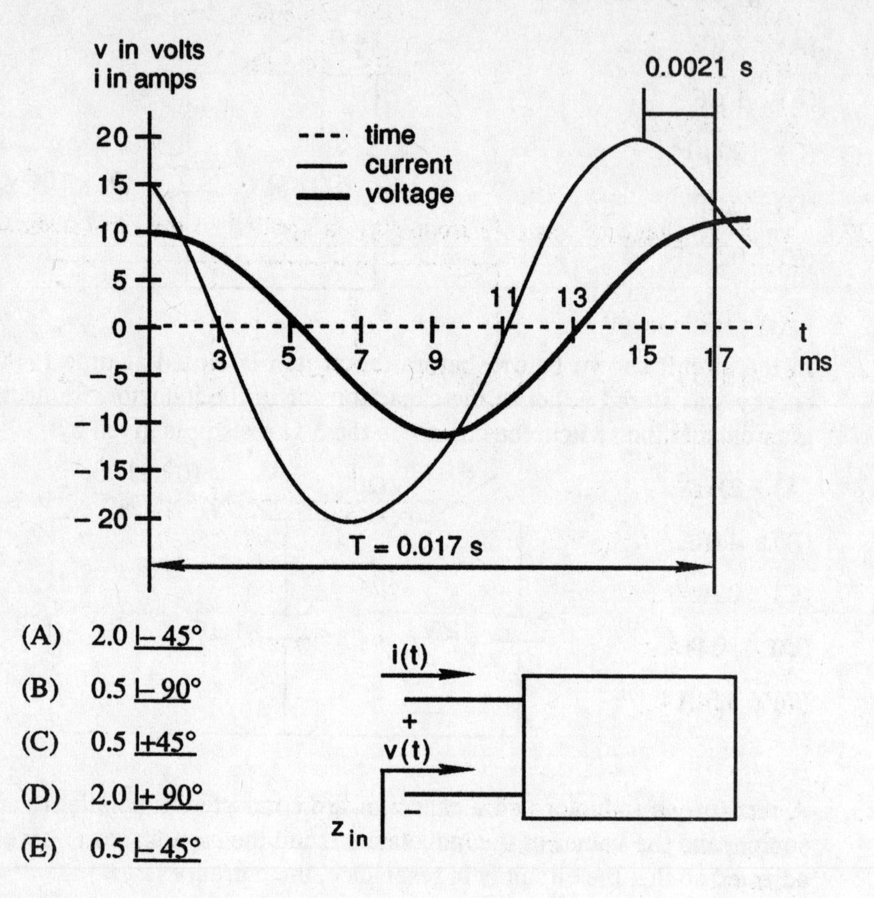

(A) 2.0 $\angle -45°$

(B) 0.5 $\angle -90°$

(C) 0.5 $\angle +45°$

(D) 2.0 $\angle +90°$

(E) 0.5 $\angle -45°$

26. As shown in the circuit below a 5Ω resistor and a 12 Ω inductor are connected in series.

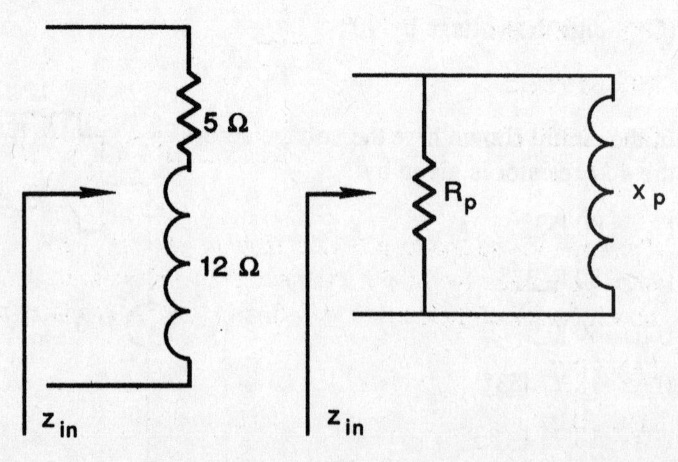

The value of the resistor R_p in a parallel combination shown above with the same input impedance is:

(A) 5.4 Ω (D) 33.8 Ω

(B) 12.0 Ω (E) 14.1 Ω

(C) 13.5 Ω

27. An AC voltage of variable frequency is applied to the circuit shown below:

$v(t) = 120 \cos \omega t$

$R = 10\ k\Omega$

$L = 10$ mH,

$C = 50\ \mu F$.

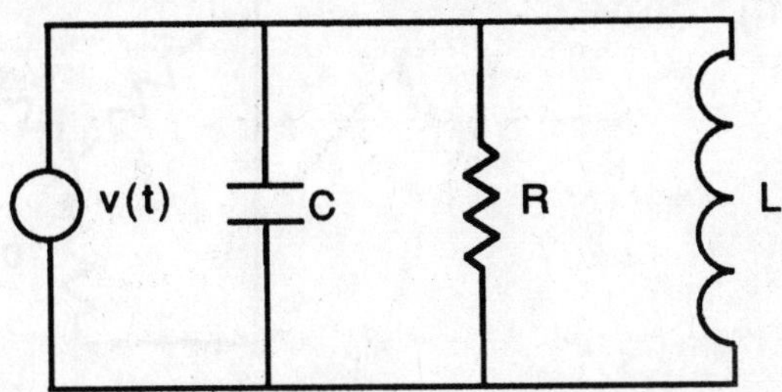

The angular frequency ω at which this circuit is resonant is given by:

(A) $1000\ s^{-1}$ (D) $0.0007 s^{-1}$

(B) $1414\ s^{-1}$ (E) $700 s^{-1}$

(C) $14.14\ s^{-1}$

28. In the circuit shown below the switch is closed at $t = 0$. The transient component of the current after the switch closes has a frequency of oscillation of:

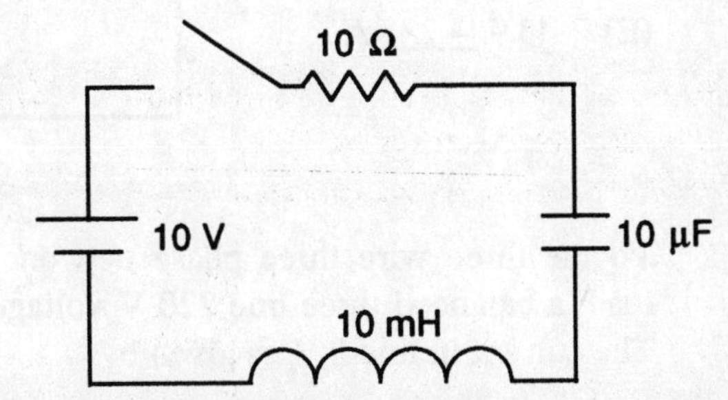

(A) 497 Hz

(B) 6240 Hz

(C) 1000 Hz

(D) 159 Hz

(E) 2350 Hz

29. Two one phase loads, a 5 KW resistor and a 10 KVA load with a 0.707 lagging power factor are connected in parallel to a 120 V source. The power factor of the current supplied to the two loads by the source is

(A) 0.5 (D) 0.46

(B) 0.71 (E) 0.86

(C) 0.67

30. The circuit below shows a 10 Ω per phase resistive Y connected balanced load and a 17.32 Ω per phase capacitive Δ connected balanced load supplied by a 3 phase, 220 V_1 network. The current supplied by the load is

(A) 50.86 A

(B) 25.43 A

(C) 43.99 A

(D) 60.12 A

(E) 30.06 A

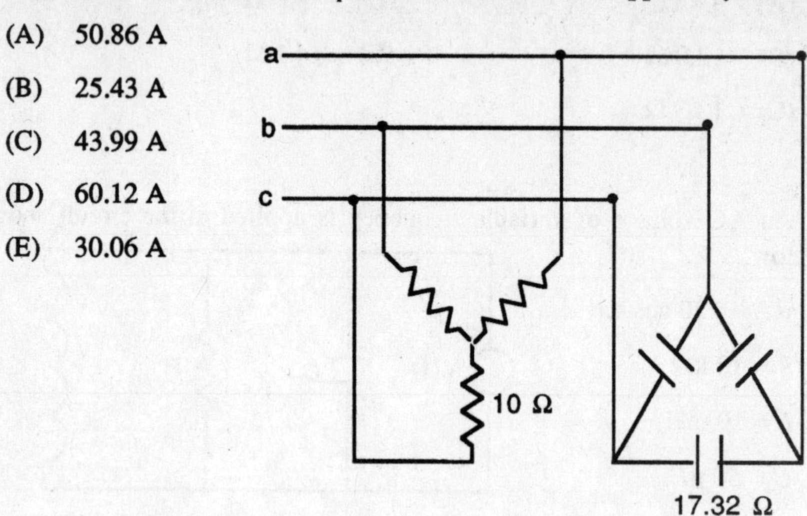

31. In the circuit shown here, a capacitive load of 0.12 Ω is connected to an AC source of 220 $\angle 90^\circ$ V using an ideal transformer with a turn ratio of 1:10. The primary current I is given by the phasor

(A) 16.9 $\angle 157^\circ$ A

(B) 16.9 $\angle 23^\circ$ A

(C) 43.9 $\angle 157^\circ$ A

(D) 43.9 $\angle 23^\circ$ A

(E) 43.9 $\angle -23^\circ$ A

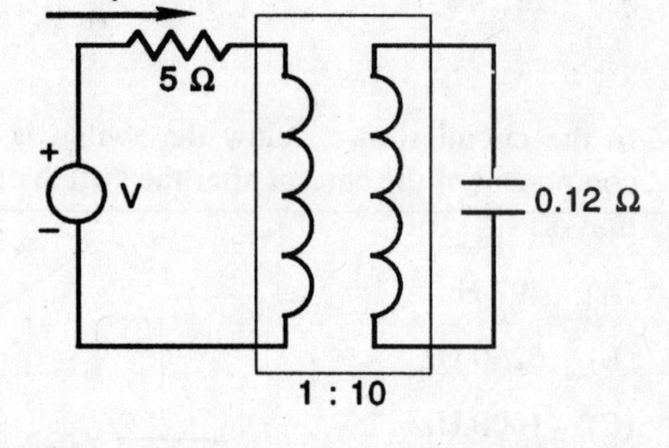

32. To the three wire three phase network shown here with an unbalanced load, a balanced three line 220 V voltage of positive sequence is applied. The current in line b, I_b is given by:

(A) 62.3 A

(B) 75.5 A

(C) 55.5A

(D) 105.7 A

(E) 104.6 A

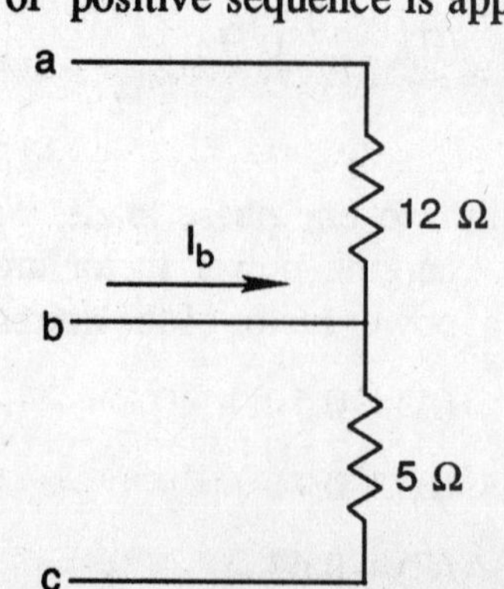

33. In the circuit shown here a balanced three phase voltage of positive sequence and value 220 V is applied to a balanced load and to a second load connected between phases a and b. The current in phase a is given by:

(A) 69.9 A

(B) 88.0 A

(C) 50.8 A

(D) 67.2 A

(E) 55.7 A

34. An unbalanced Y connected load is suplied by a 4-wire network from a three phase balanced voltage source with the phase sequence ... $a\ b\ c\ a\ b$ ... . To find the three phase currents I_a, I_b, I_c and the current in the neutral wire I_n, how many loop equations must be solved?

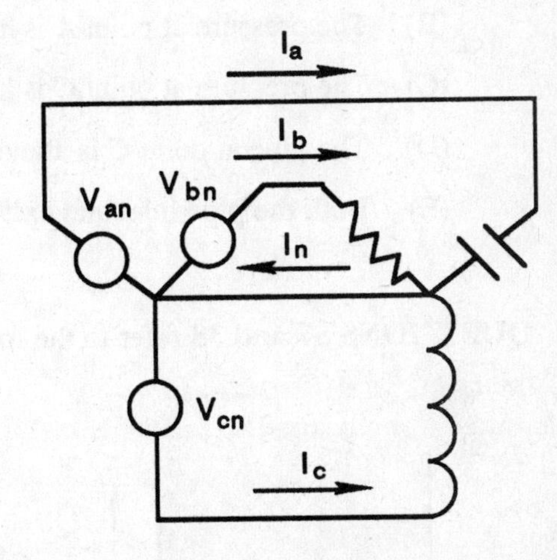

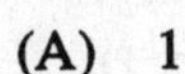

(A) 1

(B) 3

(C) 4

(D) 2

(E) 5

35. The resistance of a *Newtonian* fluid to a constant shearing force shows what general characteristics?

(A) Rate of deformation decreases with time.

(B) Rate of deformation is independent of shear force.

(C) Rate of deformation is proportional to the square of the shear force.

(D) Rate of deformation is directly proportional to shear force.

(E) Rate of deformation decreases with increasing shear force.

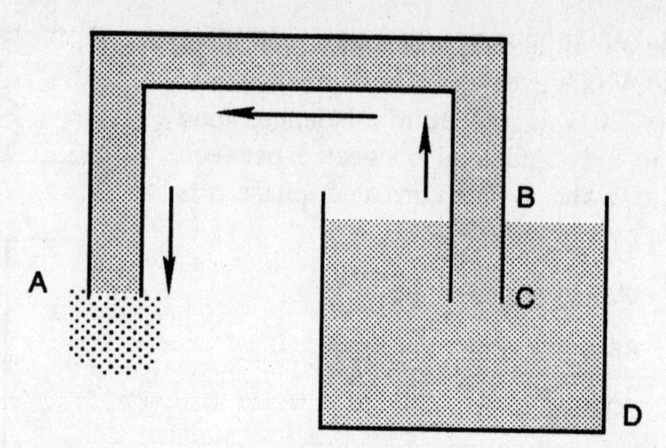

36. In the figure above, water is flowing out of the tank through the tube. Gravity is acting downward. Set up in this fashion, liquid will flow out of the tube because?

    (A) The pressure at point *A* is higher than the pressure at point *C*.

    (B) The pressure at point *A* is higher than the pressure at point *B*.

    (C) The pressure at point *C* is higher than the pressure at point *A*.

    (D) The pipe at point *C* is above point *D*.

    (E) Both the pipe inlet and exit are above point *D*.

QUESTIONS 37 and 38 refer to the following figure:

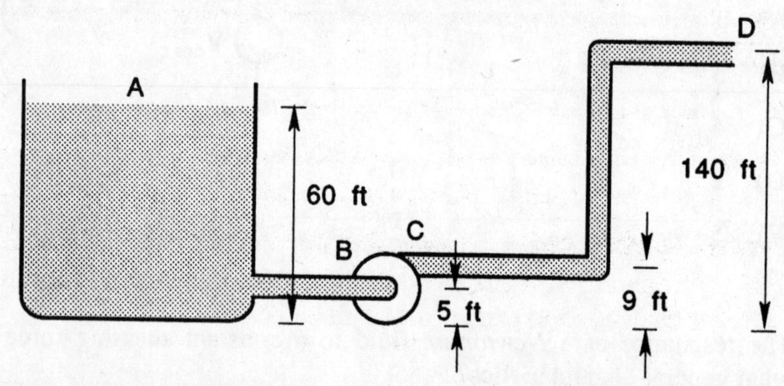

Note: The density of water is 62.27 lbm/ft³. All piping diameters are 1 ft.

37. If the pump in the figure provides 100 ft of head, what is the discharge velocity at point *D* in ft/sec? Neglect friction losses in the piping.

    (A) 1.8 ft/sec
    (B) 35.9 ft/sec
    (C) 54.2 ft/sec
    (D) 74 ft/sec
    (E) 1023 ft/sec

38. What is the pressure at the inlet to the pump if the discharge velocity is measured at 50 ft/sec?

(A) – 14.7 psig
(B) – 1.6 psig
(C) 0.3 psig
(D) 7.0 psig
(E) 102.2 psig

39. A Pitot-static tube is placed into an air duct and the deflection of the gage is 1.6 inches of water. What is the velocity of air in the duct? Assume that the density of air is 0.075 lbm/ft^3.

(A) 84 ft/sec
(B) 309.2 ft/sec
(C) 24.1 ft/sec
(D) 1133 ft/sec
(E) 6.4 ft/sec

40. Consider the hydraulic jump which exists at the outflow of a dam. The outflow is controlled so that the depth of the fluid at the outflow is constant. An increase in the volumetric flow rate through the dam will have what effect on the hydraulic jump?

(A) Increase the height difference across the jump and move its location upstream.

(B) Decrease the height difference across the jump and move its location upstream.

(C) Increase the height difference across the jump and move its location downstream.

(D) Decrease the height difference across the jump and move its location downstream.

(E) Volumetric flow changes will have no effect on the size and location of the hydraulic jump.

41. Which of the following instruments for measuring fluid velocities does not need to be calibrated?

(A) Orifice Meter
(B) Pitot-Static Tube
(C) Hot-Wire Anemometer
(D) Weir
(E) Venturi Tube

42. A submarine whose length is 120 ft is to be towed at 3.5 ft/sec. A model submarine (length 4 ft) is towed in a towing tank and requires a force of 200 lbf to pull it at a dynamically similar velocity. Estimate what force will be required to tow the full size submarine.

(A) 84 lbf
(B) 309.2 lbf
(C) 24.1 lbf
(D) 1133 lbf
(E) 146.9 lbf

43. A stream of water 1 in with a diameter (density = 62.27 lbm/ft^3) impinges on a bathroom scale which measures a force of 250 lbf. What is the velocity of the stream?

(A) 153.7 ft/sec
(B) 24.3 ft/sec
(C) 24.9 ft/sec
(D) 190.6 ft/sec
(E) 6.4 ft/sec

44. In compressible fluid mechanics, a shock wave is considered a "normal" shock wave when:

(A) The flow remains supersonic on both sides of the shock wave.
(B) The angle between the direction of flow and the wave is 90°.
(C) The shock wave occurs in air.
(D) The shock wave occurs at an oblique angle to the flow.
(E) The shock wave does not produce any viscous dissipation.

45. Capillarity, or the rise or fall of a liquid in a thin tube is primarily controlled by which fluid property?

(A) density
(B) viscosity
(C) surface tension
(D) temperature
(E) electrical conductivity

46. A sphere of constant density and 4 in diameter floats in water (density of water is 62.27 lbm/ft^3) such that the upper hemisphere is above the water level. What is the weight of the sphere?

(A) 0.6 lbf
(B) 1.2 lbf
(C) 16 lbf
(D) 32.2 lbf
(E) 106 lbf

47. An engineer needs to deliver 150 ft^3/min of water in a round pipe in such a way that the flow in the pipe remains laminar. If the critical Reynolds number is 2,000, what diameter of pipe is necessary to insure laminar flow? (Note: the kinematic viscosity of water is 4.75×10^{-5} ft^2/sec). Assume that the flow in the pipe will be uniform.

(A) 1.2 ft
(B) 3.8 ft
(C) 21.5 ft
(D) 33.5 ft
(E) 106 ft

48. Cavitation in fluid mechanics refers to

(A) the separation of the air flow behind a wing.
(B) the oscillations of a plate due to turbulent fluctuations.
(C) the liquid to gaseous phase transition of a fluid due to low pressure.
(D) the condensation of vapor into liquid due to high pressure.
(E) the violation of the "no-slip" condition on a wall.

49. A polytropic process is one in which the functional relationship between pressure P and volume V is given by the equation

$$PV^n = \text{Const.}$$

The exponent n may possibly be any value from $-\infty$ to $+\infty$ depending on the particular process. A constant-volume process is represented by n equals

(A) zero
(B) 2.0
(C) K
(D) 3.0
(E) ∞

50. An ideal gas ($k = 1.4$) is expanded in a nozzle from $P_0 = 1.5$ MPa, $T_0 =$ 150°C, ($V_0 = 0$), to $P_2 = 0.3$ MPa. Assuming reversible adiabatic process the nozzle should be

(A) diverging
(B) converging
(C) convergent-divergent
(D) constant cross-section
(E) divergent-convergent

51. A large class of devices (such as turbines, centrifugal compressors and pumps) involve work. Assuming reversible steady-state, steady-flow process with negligible changes in kinetic and potential energy, the elementary work is

(A) $V\,dP - P.DV$
(B) $P\,dV$
(C) $V\,dP + P\,dV$
(D) $-V\,dP$
(E) $P\,dV - V\,dP$

QUESTIONS 52 – 54 refer to the following diagram.

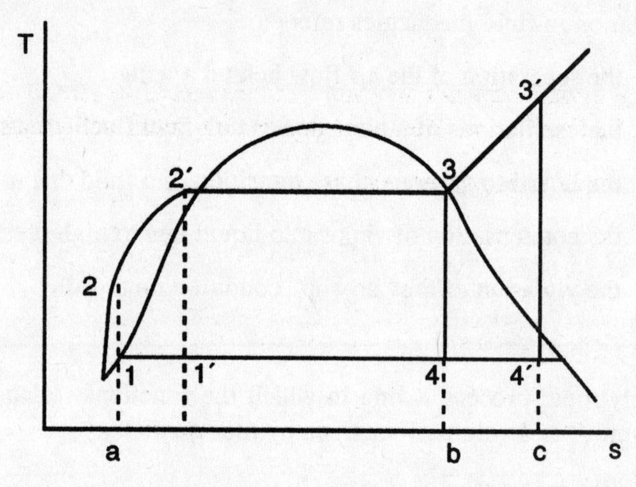

The ideal Rankine cycle and reversible Carnot cycle are shown in the above figure.

52. If changes in kinetic and potential energy are negligible, heat transfer and work may be represented by various areas on the $T - S$ diagram. The thermal efficiency of the Rankine cycle with superheated steam is defined by the relation of areas

(A) $\dfrac{1-2-2'-3-3'-4'-1}{a-2-2'-3-3'-c-a}$

(B) $\dfrac{1'-2'-3-3'-4'-1'}{a-2-2'-3-3'-c-a}$

(C) $\dfrac{1-2-2'-3-3'-4-1}{a-1-4'-c-a}$

(D) $\dfrac{a-2-2'-3-3'-c-a}{1-2-2'-3-3'-4'-1}$

(E) $\dfrac{1-2-2'-3-4-1}{a-2-2'-3-b-a}$

53. Superheating the steam in the reversible Carnot cycle (figure above) will cause

(A) increase in thermal efficiency of the Carnot cycle

(B) decrease in thermal efficiency of the Carnot cycle

(C) no change in thermal efficiency of the Carnot cycle

(D) increase in maximum pressure of the Carnot cycle

(E) decrease in minimum pressure of the Carnot cycle

54. Change in entropy for the reversible Carnot cycle (figure above) is

(A) $dS > 0$

(B) $dS = dQ/T$

(C) $dS = 0$

(D) $dS = dU/T$

(E) $dS < 0$

55. Heat is transferred at constant volume process to the thermodynamic system of a fixed mass. The thermodynamic system will produce

(A) small amount of work

(B) zero work

(C) large amount of work

(D) negative work

(E) positive work

56. In manufacturing process 1000 kw of waste heat at temperature 327° C is available for utilization purposes. If ambient temperature is 27° C, the designed heat engine producing 510 kw of net power output will be

(A) irreversible

(B) reversible

(C) adiabatic

(D) possible

(E) impossible

57. Various thermodynamic processes are shown in the figure below.

$P = \text{Const}$ $\quad PV^r = \text{Const}$ $\quad T = \text{Const}$

$V = \text{Const}$ $\quad PV^k = \text{Const}$

During expansion process (1 – 2) the maximum work done by the thermodynamic system will be obtained in the process:

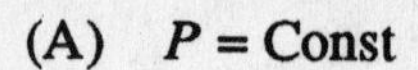
(A) $P = \text{Const}$

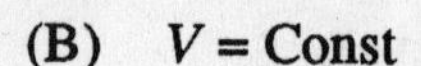
(B) $V = \text{Const}$

(C) $PV^n = \text{Const}$

(D) $Pv^k = \text{Const}$

(E) $T = \text{Const}$

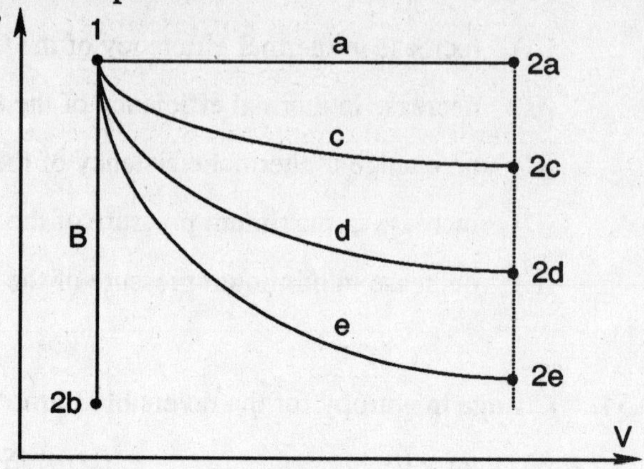

58. The reheat Rankine cycle is proposed to DECREASE:

(A) volumetric flow rate of the working fluid

(B) mass flow rate of cooling water in condenser

(C) back pressure of the turbine

(D) maximum temperature of the cycle restricted by the boiler working conditions

(E) moisture content in the low-pressure stages of the turbine.

59. Compressibility factor Z is a measure of the deviation of the actual gas behavior from ideal gas. An actual gas behavior may be assumed close to the ideal gas when compressibility factor Z approaches:

(A) zero

(B) 0.50

(C) 0.75

(D) 1.0

(E) 1.4

60. Let's assume that gas can be heated either at constant pressure or at constant volume processes. The amounts of heat q_p and q_v are transferred to the gas correspondingly. If temperature increment per unit mass ΔT = Const

(A) $q_p > q_v$

(B) $q_p = q_v - \Delta u$

(C) $q_p = q_v$

(D) $q_v = q_p + h$

(E) $q_p < q_v$

61. Let: W_{st} = isentropic turbine work

W_t = actual turbine work

W_{sc} = isentropic compressor work

W_c = actual compressor work

The turbine and compressor efficiencies 2_T and 2_C are defined correspondingly as

(A) $2_T = \dfrac{W_{ST}}{W_T}$; $2_C = \dfrac{W_C}{W_{SC}}$

(B) $2_T = \dfrac{W_T - W_C}{W_T}$; $2_C = \dfrac{W_{SC} - W_C}{W_{SC}}$

(C) $2_T = \dfrac{W_{ST} - W_{SC}}{W_{ST}}$; $2_C = \dfrac{W_{SC}}{W_{SC} - W_C}$

(D) $2_T = \dfrac{W_T}{W_{ST}}$; $2_C = \dfrac{W_{SC}}{W_C}$

(E) $2_T = \dfrac{W_T}{W_T + W_C}$; $2_C = \dfrac{W_C}{W_T + W_C}$

62. In the boiler, water preheating, vaporization, and superheating processes take place at constant pressure. When water (liquid + vapor) exists at saturation conditions, the state of the water is fixed if all of the following statements are true EXCEPT:

(A) specific volume and quality are known

(B) pressure and temperature are known

(C) pressure and specific volume are known

(D) temperature and specific volume are known

(E) entropy and temperature are known

63. A particle moves from rest at point O in a straight line with a velocity whose square increases linearly with displacement. It reaches point A which is 600 ft from O with velocity 60 **i** ft/sec. The acceleration of the particle is near to

(A) 0.1 **i** ft/sec/sec

(B) 1.0 **i** ft/sec/sec

(C) 2.0 **i** ft/sec/sec

(D) 3.0 **i** ft/sec/sec

(E) 6.0 **i** ft/sec/sec

64. A car moves in a circular road of radius 660 ft with a constant speed of 45 miles/hr. At the shown position, the acceleration of the car is near to

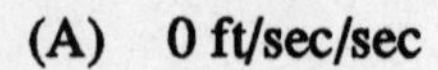

(A) 0 ft/sec/sec

(B) 3.1 **i** ft/sec/sec

(C) – 3.1 **i** ft/sec/sec

(D) 6.6 **i** ft/sec/sec

(E) – 6.6 **i** ft/sec/sec

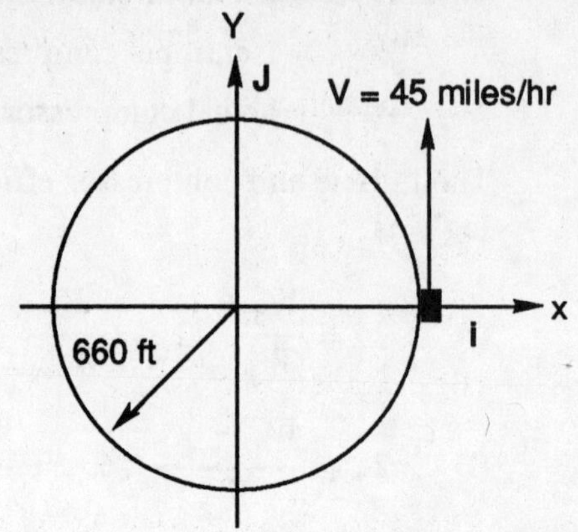

65. The shown disk rotates about a fixed axis *O* with a constant angular velocity of 12**k** rad/sec. The particle *p* is moving with a relative velocity to the straight slot of 2**i** in/sec. The absolute acceleration of the particle as it passes through *O* is near to:

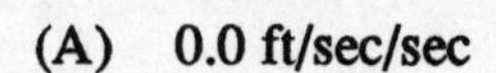

(A) 0.0 ft/sec/sec

(B) 2 **j** ft/sec/sec

(C) – 2 **j** ft/sec/sec

(D) 4 **j** ft/sec/sec

(E) – 4 **j** ft/sec/sec

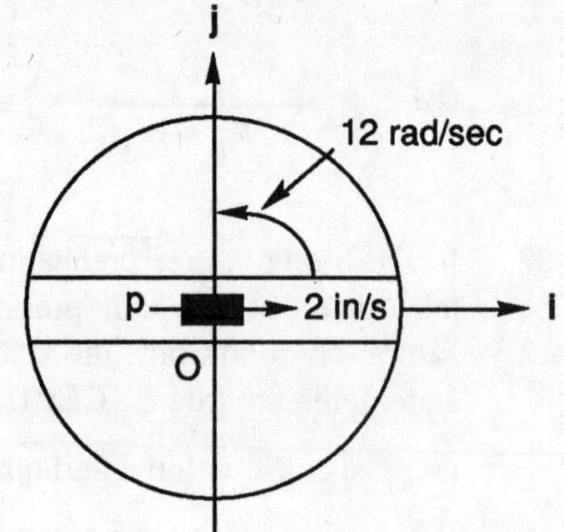

66. The figure shows a uniform thin disk of radius 1-ft and rolls without slipping on a smooth surface. Its geometric center *O* has a constant velocity of 10 **i** ft/sec. In the configuration shown, the line connecting points *O* and *B* is horizontal, the velocity of point *B* is near to

(A) 10 ft/sec →

(B) 10 ft/sec ↓

(C) $10\sqrt{2}$ ft/sec 45° ↙

(D) $10\sqrt{2}$ ft/sec 45° ↘

(E) $10\sqrt{2}$ ft/sec →

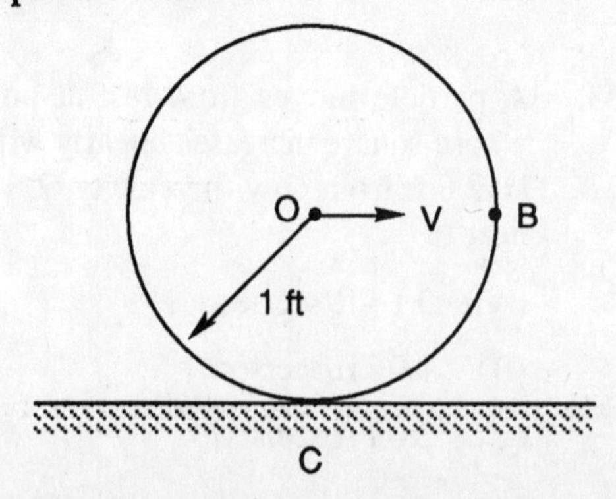

67. The two masses are connected by a massless string which passes over two smooth pulleys such that the friction can be neglected. The acceleration of the 10-kg mass is very close to

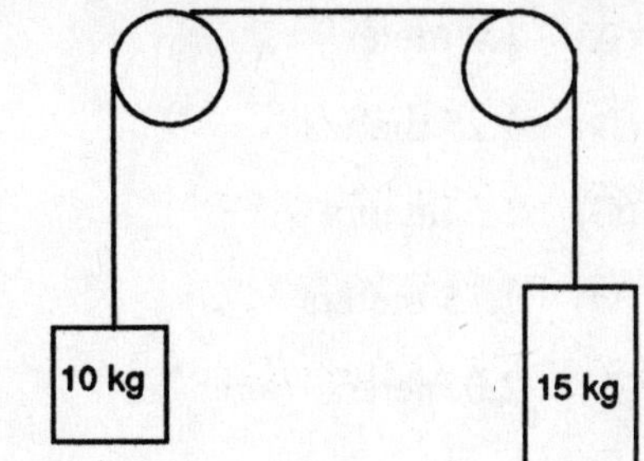

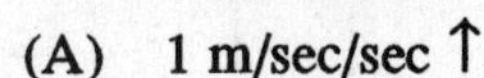

(A) 1 m/sec/sec ↑

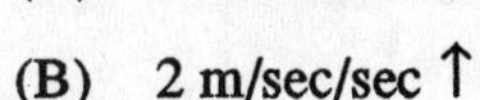

(B) 2 m/sec/sec ↑

(C) 3 m/sec/sec ↑

(D) 4 m/sec/sec ↑

(E) 5 m/sec/sec ↑

68. The mass M is carried by a massless rod of length 2 meters which is allowed to swing as a simple pendulum about the frictionless bearing O. If the velocity of the mass in position A is 2.3 m/sec, then it can move up until it reaches a maximum position B which makes an angle θ close to

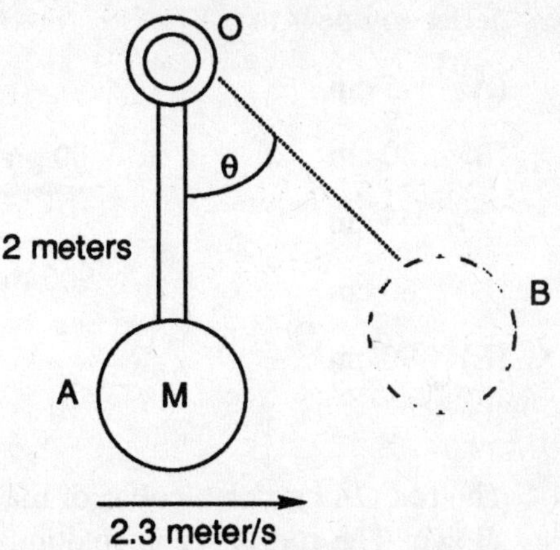

(A) 15°

(B) 30°

(C) 45°

(D) 60°

(E) 70°

69. The static and kinetic coefficients of friction between mass m and the flatbed of the truck are 0.3 and 0.2, respectively. If the speed of the truck is 30 miles/hr, the smallest stopping distance of the truck without allowing the mass to slip is near to

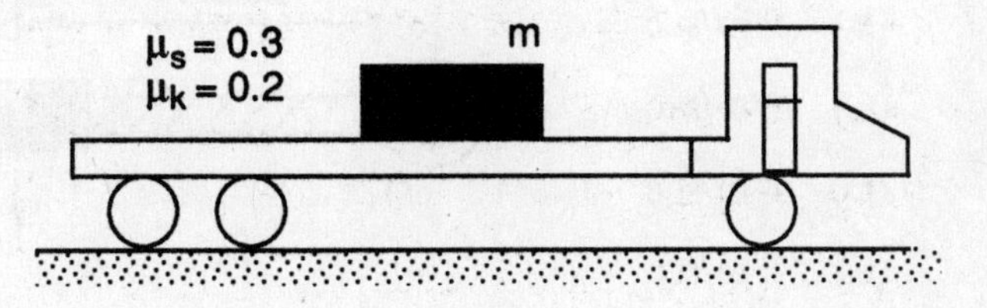

(A) 50 ft

(B) 100 ft

(C) 150 ft

(D) 200 ft

(E) 250 ft

70. A steel ball is dropped from 2 meters height onto a smooth floor. If the

coefficient of restitution between the ball and floor is 0.87, the ball will bound to a height h_1 near to

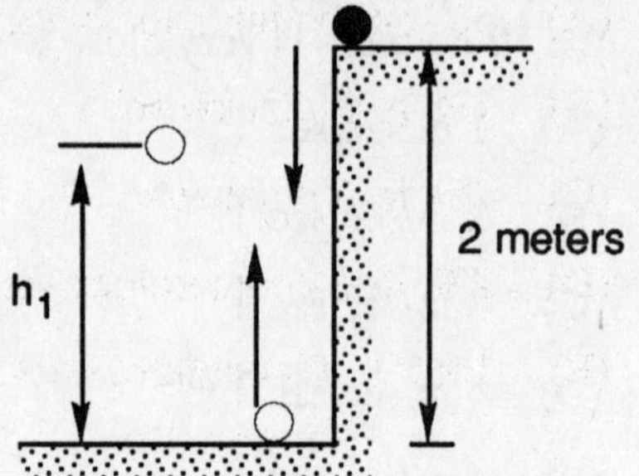

(A) 1.0 meter

(B) 1.25 meters

(C) 1.5 meters

(D) 1.75 meters

(E) 2.0 meters

71. A bullet of mass 50-grams is fired with a velocity of 600-meters/s. The bullet strikes block *A* without rebound. The block has mass of 2-kg and is attached to a spring of stiffness 20-KN/m which is originally undeformed. Just after impact the block and bullet move against the spring until they are brought to a complete stop where the spring has been compressed by a deflection near to

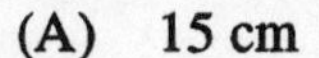

(A) 15 cm

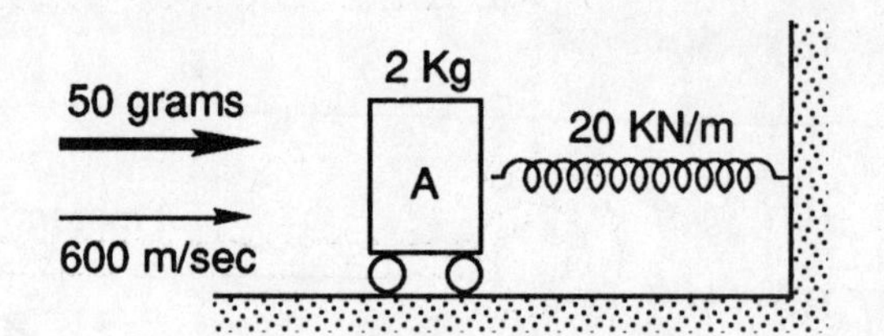

(B) 30 cm

(C) 45 cm

(D) 60 cm

(E) 90 cm

72. The rod *OA* carries a collar of mass *m* which is attached to a coil spring as shown. The rod is set in rotational motion about *O* on a horizontal plane when the length of the spring is 36-cm such that the transverse velocity of the collar is 5-meter/s. When the spring is stretched to a length of 45-cm, the transverse velocity of the collar is near to

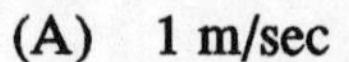

(A) 1 m/sec

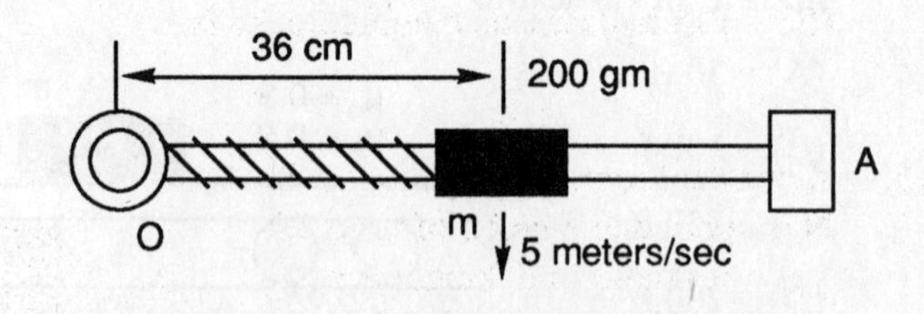

(B) 2 m/sec

(C) 3 m/sec

(D) 4 m/sec

(E) 5 m/sec

73. Block *A* weighs 82-lb and rod *BC* of length 8-ft weighs 46-lb. The rod is welded to block *A* at *B*. The combined system is restricted to move up vertically under the action of an applied force of magnitude 160-lb. The

bending moment exerted by the weld on the rod at B is near to

(A) 46 lb-ft counterclockwise

(B) 138 lb-ft clockwise

(C) 184 lb-ft clockwise

(D) 230 lb-ft counterclockwise

(E) 1656 lb-ft counterclockwise

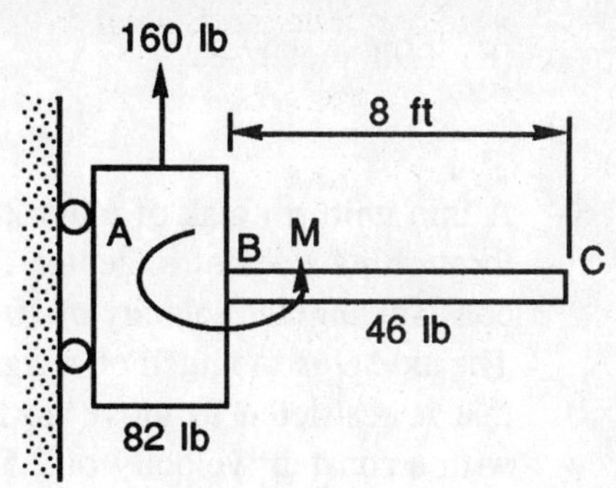

74. The massless rod is welded to a thin disk of weight 20-lb and radius 18-in. The torque $M = 38$ lb-ft is applied on the system at O such that the system is allowed to rotate on a horizontal plane about a vertical axis through O. If the friction in the bearing at O is negligible, the angular acceleration of the system is near to

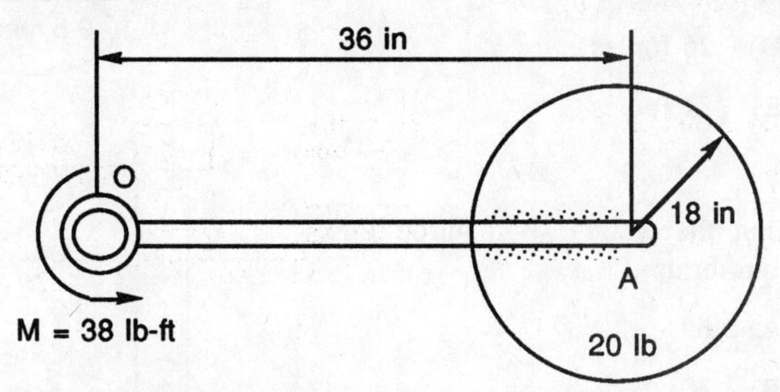

(A) 3.5 rad/sec/sec, clockwise

(B) 6 rad/sec/sec, counterclockwise

(C) 6.4 rad/sec/sec, counterclockwise

(D) 6.8 rad/sec/sec, counterclockwise

(E) 195 rad/sec/sec, counterclockwise

75. A constant force of magnitude 100-N is applied to a block of mass 20-kg. The static and kinetic coefficients of friction are 0.4 and 0.3 respectively. If the force is applied while the block is at rest, after 2-sec the block will reach a velocity near to

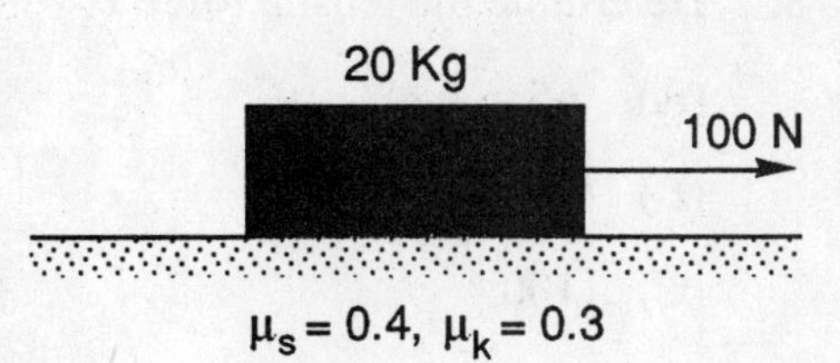

(A) 2.2 meter/sec

(B) 4.1 meter/sec

(C) 5.2 meter/sec (D) 6.3 meter/sec

(E) 9.5 meter/sec

76. A thin uniform disk of mass 2-kg and radius 0.5-meter spins about an axis through its geometric center O with a constant angular velocity of 10-rad/sec. The axle O is mounted on a rigid frame that is restricted to move horizontally with a constant velocity of 2.5 meters/s. The kinetic energy of the disc is near to

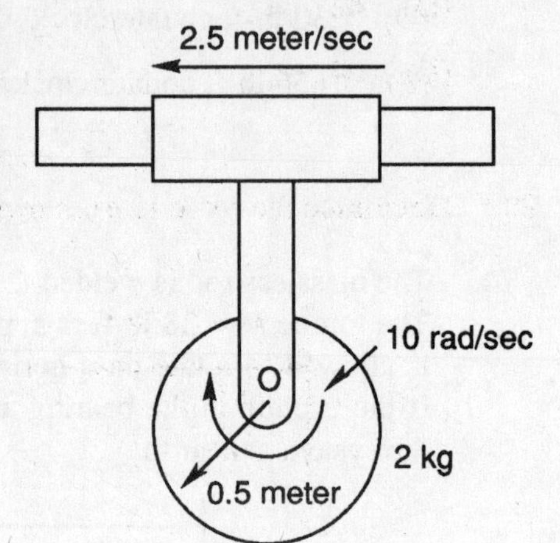

(A) 6 Joules

(B) 12.5 Joules

(C) 19 Joules

(D) 25 Joules

(E) 31 Joules

77. Find the weight W required for equilibrium.

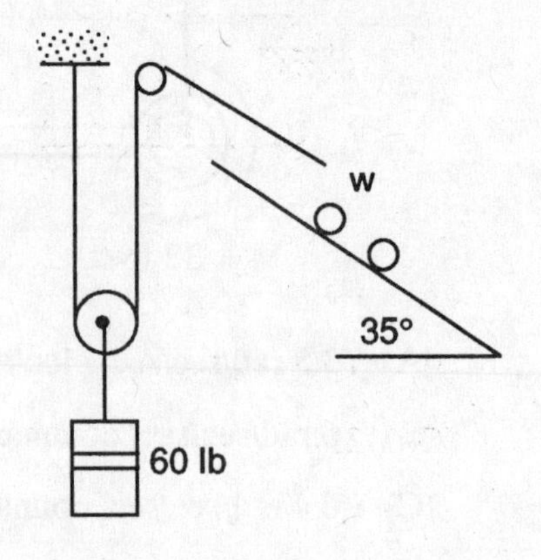

(A) 60

(B) 30

(C) 73.6

(D) 42.8

(E) 52.3

78. Determine the tensile force T_1:

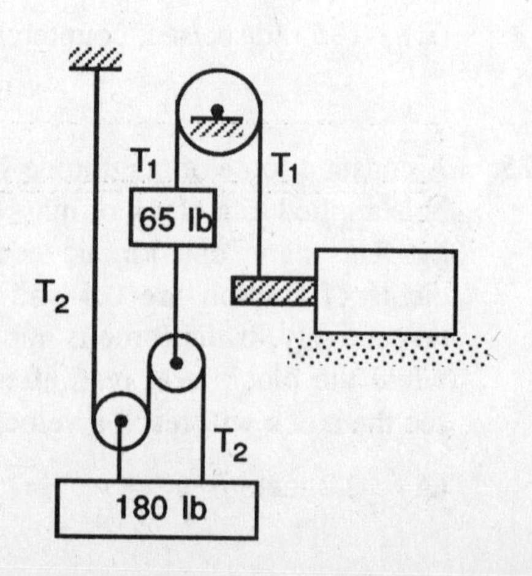

(A) 65

(B) 185

(C) 180

(D) 60

(E) 245

79. The tensile force A is equal to:

(A) 60

(B) 73

(C) 115

(D) 42

(E) 64

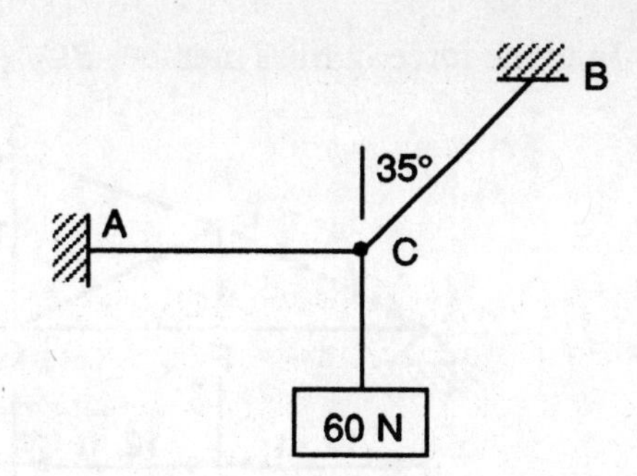

80. Determine the force in truss member AF.

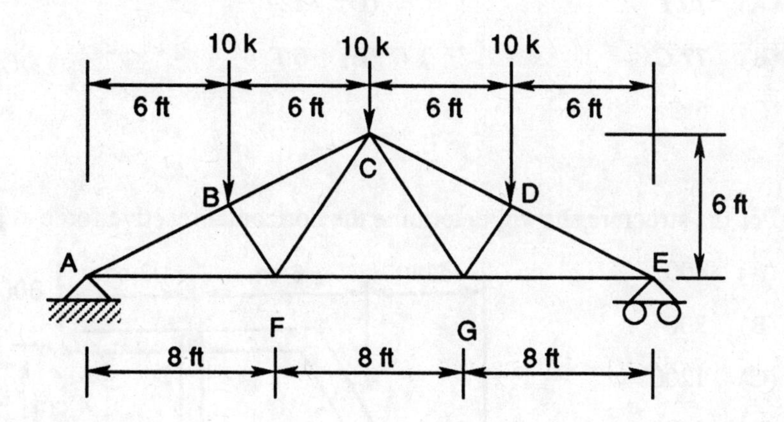

(A) 30 T

(B) 15 C

(C) 33 T

(D) 28 C

(E) 22 T

81. Find the shear force in the hinge at point C.

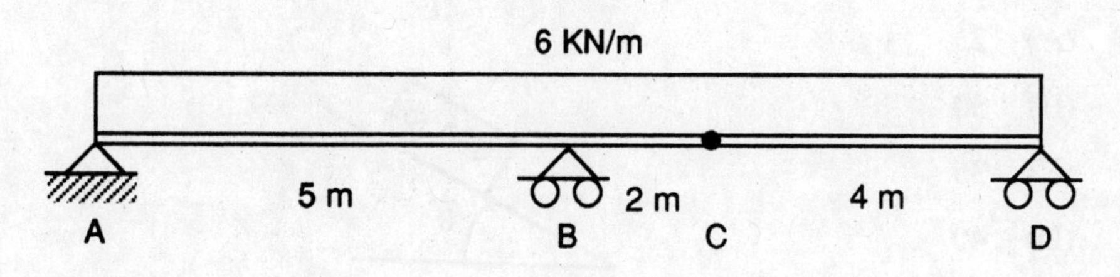

(A) 15

(B) 24

(C) 12

(D) 16

(E) 46

82. Find the force in truss member *BG*.

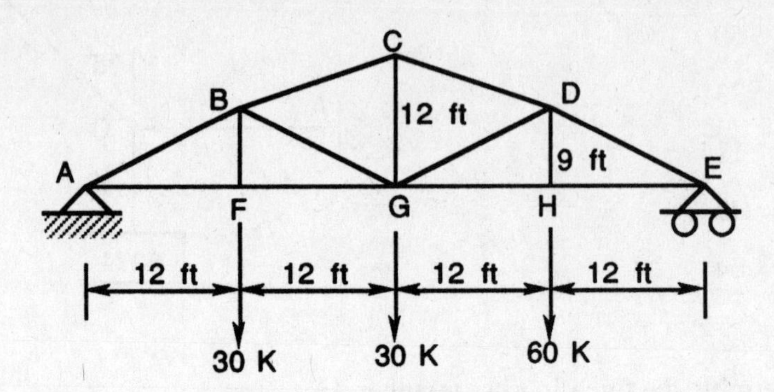

(A) 70 *T*

(B) 77 *C*

(C) 6 *C*

(D) 22 *C*

(E) 6 *T*

83. For the structure shown, determine the horizontal reactive force at joint *E*.

(A) 400

(B) 800

(C) 1200

(D) 2400

(E) 1600

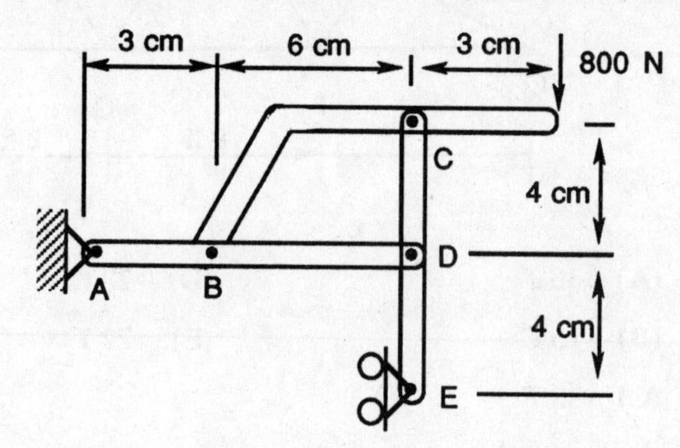

84. A 600 pound box rests on a slope. If the coefficient of friction is 0.70, find the maximum angle, θ n degrees, at which the weight will remain at rest.

(A) 25

(B) 30

(C) 35

(D) 40

(E) 45

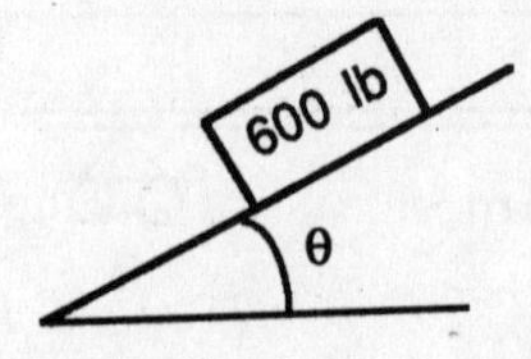

85. Find the reaction at point A for the loaded beam.

(A) 72

(B) 56

(C) 60

(D) 64

(E) 120

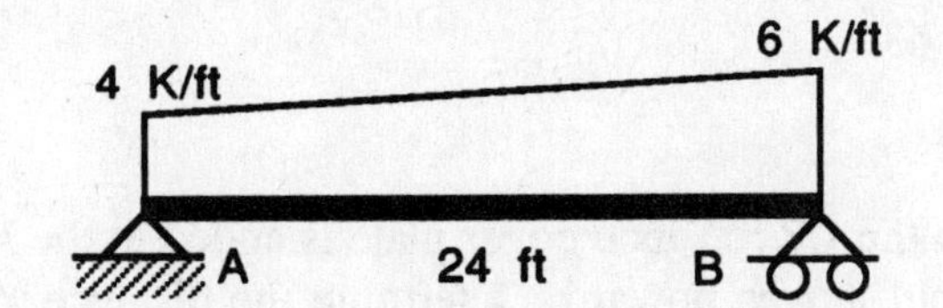

86. Find the total reaction at point B.

(A) 220

(B) 110

(C) 180

(D) 246

(E) 224

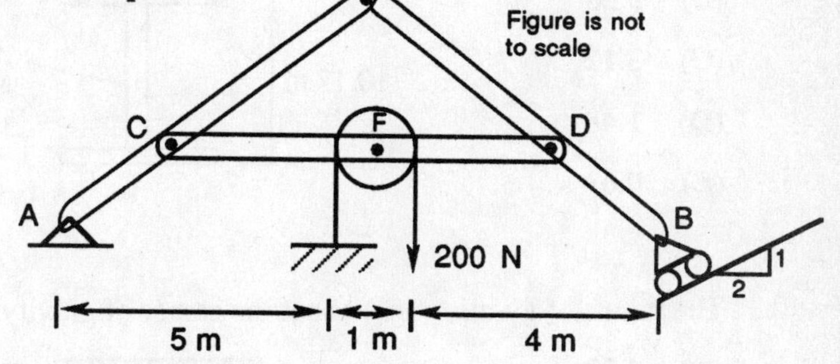

87. Add the three forces by using vector mathematics.

(A) $51i + 61j$

(B) $75i - 130j$

(C) $326i - 69j$

(D) $200i$

(E) $326i + 69j$

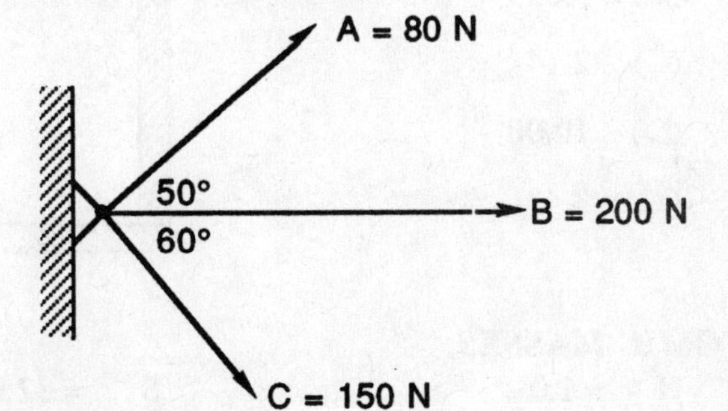

88. Find the vertical reaction at point A.

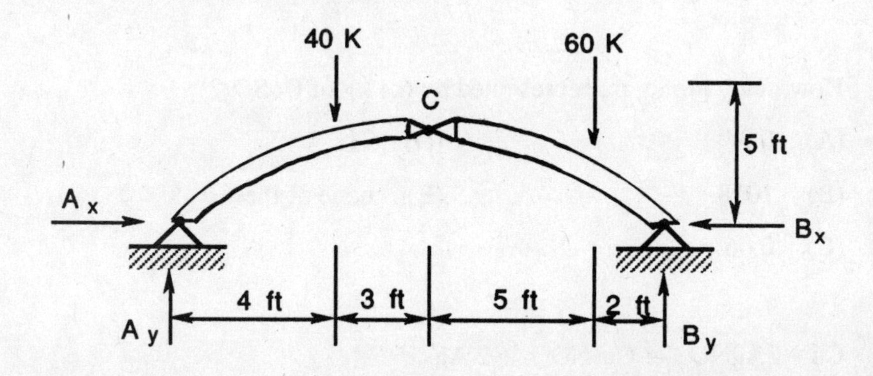

(A) 37
(B) 28
(C) 63
(D) 50
(E) 43

89. When the $1/2 \times 5$ inch cover plate is added to the $W10 \times 22$ steel beam the centroid moves upward. Determine the distance $\overline{Y}$ from the center of the original beam to the centroid of the new beam.

(A) 2.42
(B) 5.08
(C) 3.12
(D) 1.48
(E) 0.63

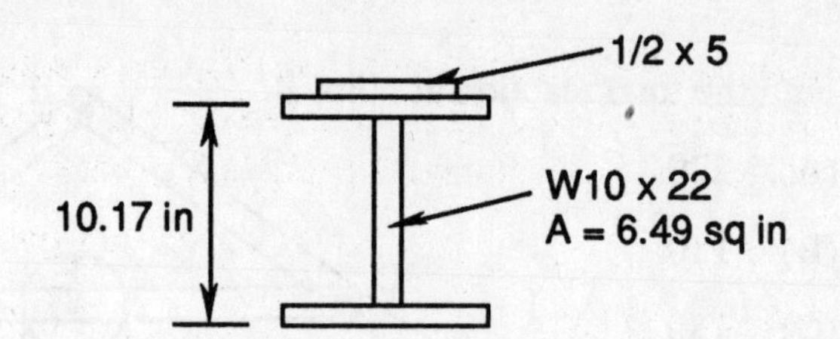

90. The radius of gyration (K_x) about the center of gravity of the beam is:

(A) 4.50
(B) 3.62
(C) 2.14
(D) 10.00
(E) 3.43

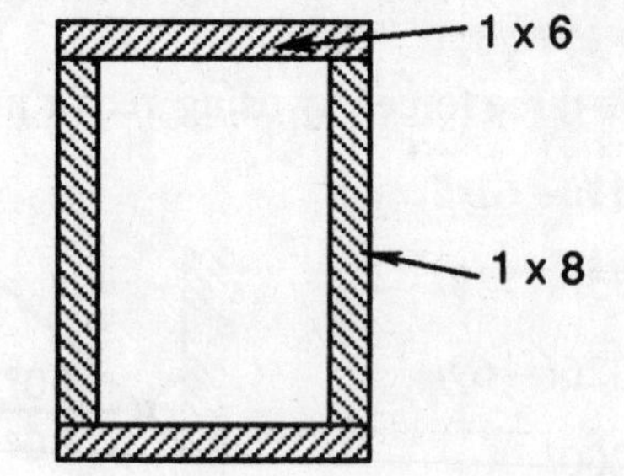

ATOMIC MASSES:

H = 1.0
C = 12.0
Na = 23.0
S = 32.0
Cu = 63.5
O = 16.0

91. How many moles are represented by 6.38 g of $CuSO_4$?

(A) 0.04
(B) 1018
(C) 0.06
(D) 711
(E) none of these

92. $Cu + 2AgNO_3 \rightarrow Cu(NO_3)_2 + 2\,Ag$

The above reaction is an example of a:

(A) decomposition reaction

(B) combination reaction

(C) single displacement reaction

(D) double displacement reaction

(E) oxidation-reduction reaction

93. What is the freezing point (in °C) of a solution containing 93.0 g of ethylene glycol ($C_2H_6O_2$) dissolved in 750.0 g of water? The freezing point constant (K_f) of water = 0.81 deg C Kg/mole.

(A) – 56.5° C

(B) + 1.62° C

(C) + 0.0016° C

(D) – 1.07° C

(E) – 1.62° C

94. 0.180 g of a metal upon reaction with excess hydrochloric acid (HCl) produced 24.6 ml of hydrogen gas (H_2) at 27° C and 1.0 atmospheric pressure. What is the equivalent wt of the metal?

(A) 3.60

(B) 4.50

(C) 0.036

(D) 9.00

(E) 18.00

95. How many electrons are there in the outermost octate of a chlorine (Cl) atom?

(A) 1

(B) 3

(C) 5

(D) 7

(E) 8

96. 15.0 ml of a 0.15 N NaOH solution requires 12.0 ml of sulfuric acid (H_2SO_4) solution for neutralization. What is the molarity of the acid solution?

(A) 0.94

(B) 0.094

(C) 0.38

(D) 3.8

(E) 98

97. Balance the equation:

$$Fe_2O_3 + C \xrightarrow{\Delta} Fe + CO_2$$

The coefficient of CO_2 in the balanced equation is:

(A) 1
(B) 2
(C) 3
(D) 4
(E) 5

98. Balance the red-ox equation:

$$MnO_2 + HCl \xrightarrow{\Delta} MnCl_2 + H_2O + Cl_2$$

The coefficient of HCl in the balanced equation is:

(A) 1
(B) 2
(C) 3
(D) 4
(E) 5

99. The half-life (λ 1/2) of a radioactive isotope is 4.3 days. How long will it take for a soil sample contaminated with the isotope to reduce its radioactivity by 99%?

(A) 6.65 days
(B) 425.7 days
(C) 28.6 days
(D) 23.0 days
(E) 4.26 days

100. What is the missing particle in the following nuclear reaction?

$${}^{9}_{4}Be - ? \quad \rightarrow {}^{9}_{4}B$$

(A) ${}^{0}_{+1}B$
(B) ${}^{0}_{-1}B$
(C) ${}^{1}_{1}H$
(D) ${}^{1}_{0}n$
(E) none of these

101. An isotopic species of lithium hydride (${}^{6}Li\,{}^{2}H$) is a potential nuclear fuel.

$${}^{6}Li^{2}H \rightarrow 2\,{}^{4}He + Energy$$

What is the expected energy production in KJ with the consumption of 1.00 g of the fuel per day at 70% efficiency?

$\overline{c}$, the velocity of light = 3.0×10^8 cm/sec

Atomic Masses: ^6Li = 6.0151 ^2H = 2.0141

^4He = 4.0026

(A) 2.70×10^{11} KJ

(B) 1.89×10^8 KJ

(C) 1.89×10^{11} KJ

(D) 2.70×10^8 KJ

(E) 2.70×10^{18} KJ

102. The number of g of sodium thiosulfate ($Na_2S_2O_3$) needed to prepare 500 ml of 0.4 M solution is:

(A) 790

(B) 1.26

(C) 31.6

(D) 126.4

(E) 158

103. For electronic transition $n_3 \rightarrow n_1$ (i.e., from M shell to K shell), what is the frequency of the photon emitted?

R, the Rydberg Constant = 2.179×10^{-4} J/photon

h, the Planck's Constant = 6.62×10^{-34} J sec/photon

(A) 3.27×10^{15}/sec

(B) 2.93×10^{15}/sec

(C) 1.64×10^{15}/sec

(D) 0.82×10^{15}/sec

(E) 1.94×10^{-4}/sec

104. What is the amount of heat generated (in KJ) by combustion of 1.00 lb of propane?

$$C_3H_8 + 5O_2 \rightarrow 3CO_2 + 4H_20 + \text{Heat}$$

Given the bond dissociation energies:

C – H = 410 KJ/mole   C = O = 731 KJ/mole

C – C = 334 KJ/mole   O – H = 456 KJ/mole

O = O = 497 KJ/mole

(A) 1.60×10^3 KJ

(B) 6.43×10^3 KJ

(C) 8.03×10^3 KJ

(D) 1.65×10^4 KJ

(E) 1.55×10^2 KJ

105. The torque required to produce 2 degrees twist at the end of a hollow rod shown below is given by: (Assume shear modulus $G = 40{,}000$ psi)

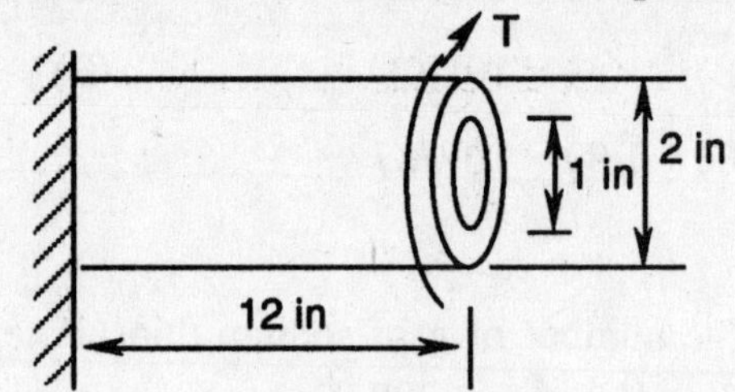

(A) 10,490 lb in

(B) 9,802 lb in

(C) 171 lb in

(D) 183 lb in

(E) 2,052 lb in

106. Which of the following is the correct shear force diagram for the beam?

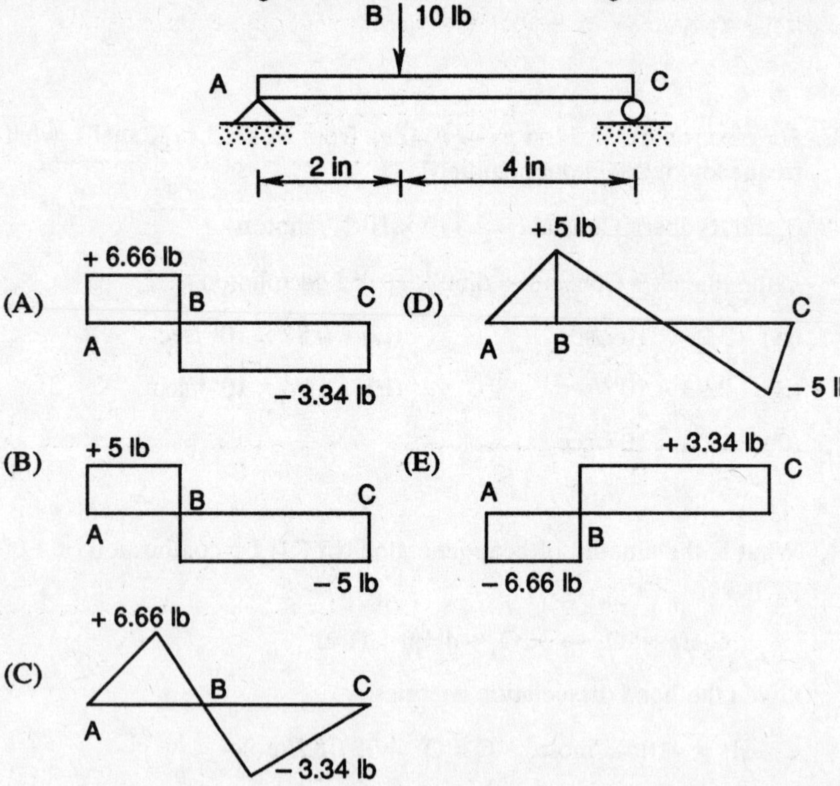

107. For the same beam of Problem 106, which of the following is the correct bending moment diagram?

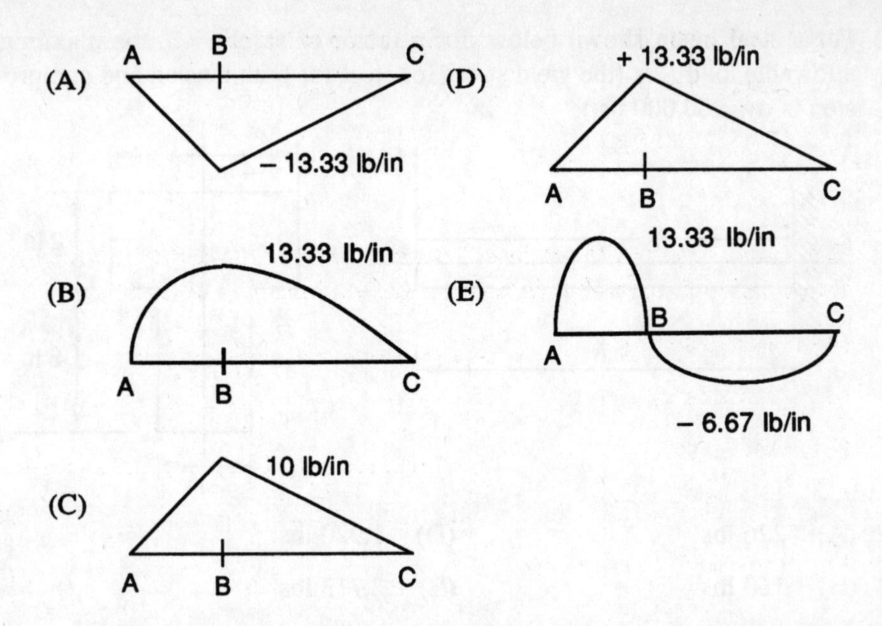

108. The simply supported beam is subjected to uniformly distributed load w lb/ft as shown. The correct bending moment diagram is given by

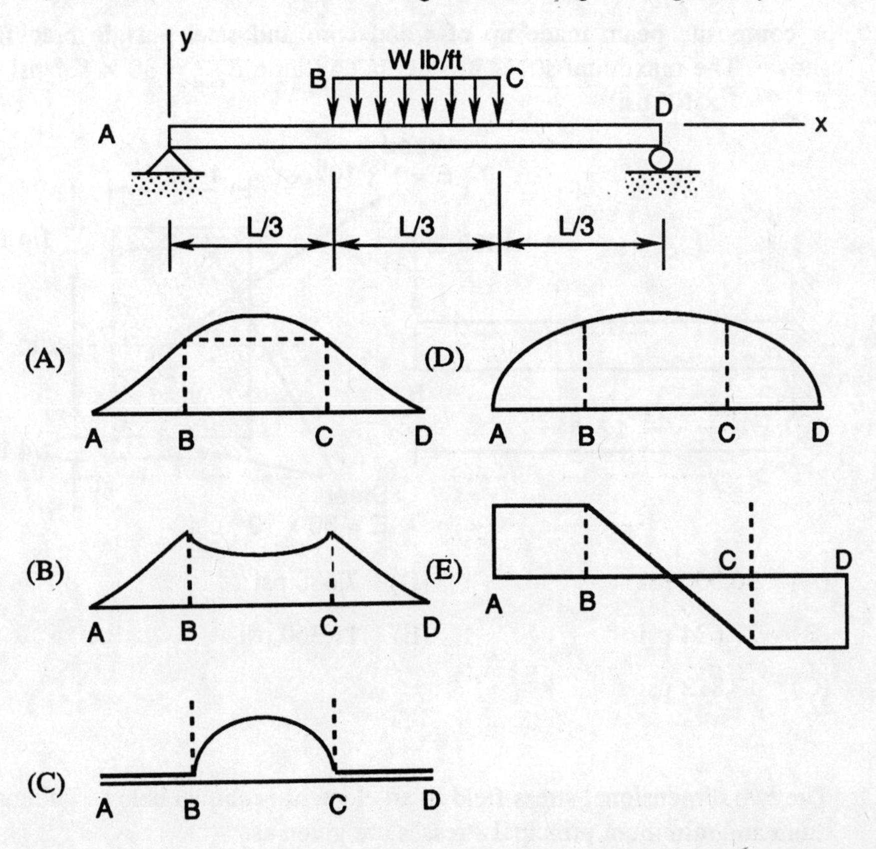

109. For a steel beam shown below, for a factor of safety = 3, the maximum allowable load F is (the yield stress for steel for both tension and compression is $\sigma y = 60{,}000$ psi)

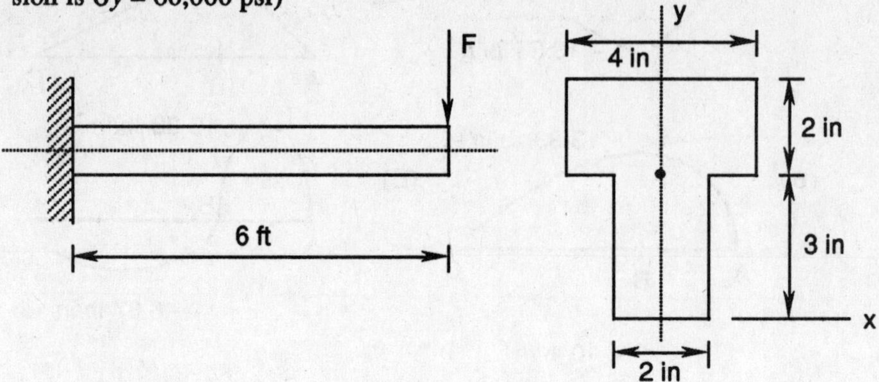

(A) 226 lbs

(B) 160 lbs

(C) 10,000 lbs

(D) 1,920 lbs

(E) 2,713 lbs

110. A composite beam made up of wood core and steel outside bracing is shown. The maximum stress in steel is (Assume $E_{steel} = 30 \times 10^6$ psi and $E_{wood} = 1 \times 10^6$ psi)

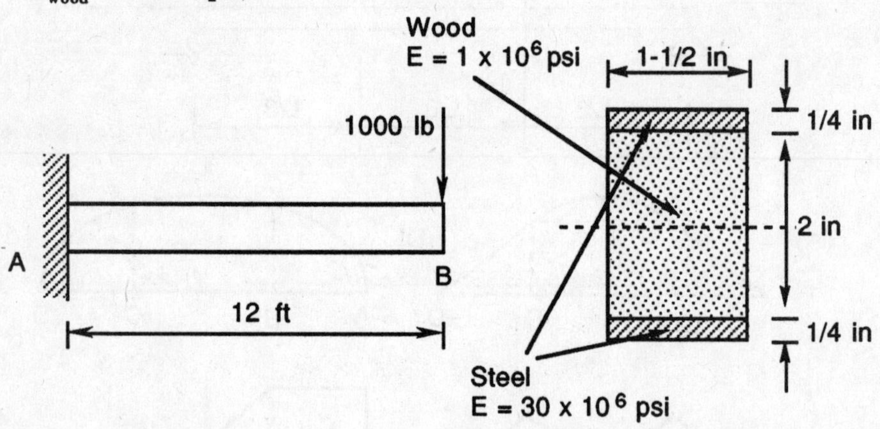

(A) 30,000 psi

(B) 27,624 psi

(C) 13,812 psi

(D) 7,680 psi

(E) 15,360 psi

111. The two dimensional stress field in an element is shown below. The maximum and minimum principal stresses are given as:

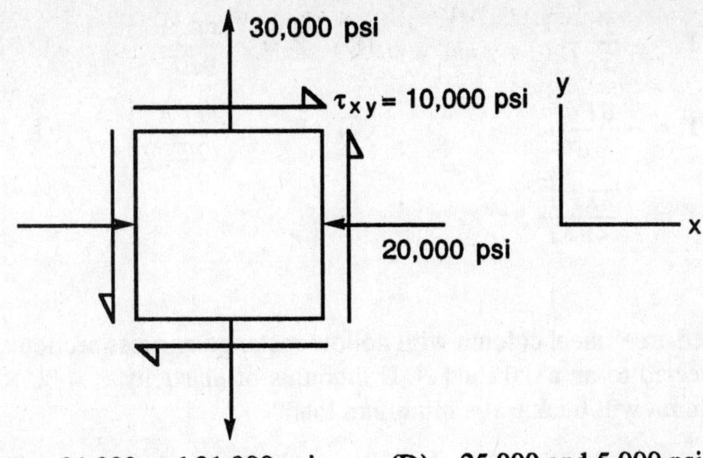

(A) 31,900 and 21,900 psi

(B) 31,900 and 26,900 psi

(C) 31,900 and – 21, 900 psi

(D) 25,000 and 5,000 psi

(E) 50,000 and 10,000 psi

112. In problem 111 the principal planes are at angles of

(A) – 79.5° and – 259.5°

(B) – 47° and – 227°

(C) – 39.75° and – 129.75°

(D) – 60° and – 150°

(E) – 23.5° and – 113.5°

113. The deflection of a simply supported beam at the point of loading is given by:

P = load

a = point of loading measured from left support

E = Young's modulus of beam material

I = moment of inertia of beam cross section about neutral axis

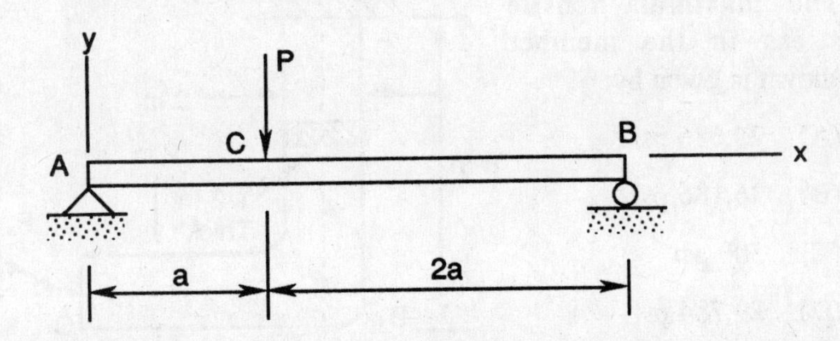

(A) $Y = -\dfrac{Pa^3}{3EI}$

(B) $Y = -\dfrac{8Pa^3}{3EI}$

(C) $Y = -\dfrac{27Pa^3}{48EI}$

(D) $Y = -\dfrac{4Pa^3}{9EI}$

(E) $Y = -\dfrac{27Pa^3}{2EI}$

114. A "fixed-free" steel column with hollow rectangular cross section as shown is subjected to an axial load P. If modulus of elasticity $E = 30 \times 10^6$ psi, the column will buckle at a minimum load

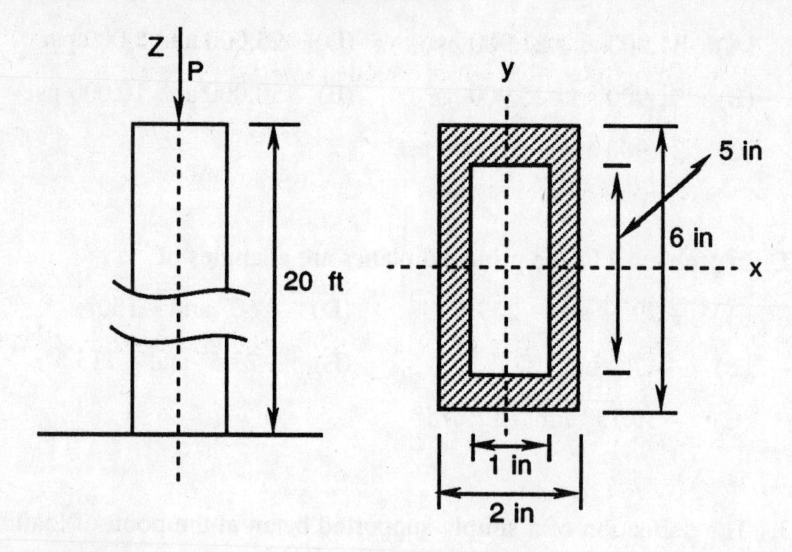

(A) 3,200 lbs

(B) 18,400 lbs

(C) 4,600 lbs

(D) 9,743 lbs

(E) 38,974 lbs

115. The maximum tensile stress in the member shown is given by:

(A) 25,686 psi

(B) 26,186 psi

(C) 500 psi

(D) 29,784 psi

(E) 30,284 psi

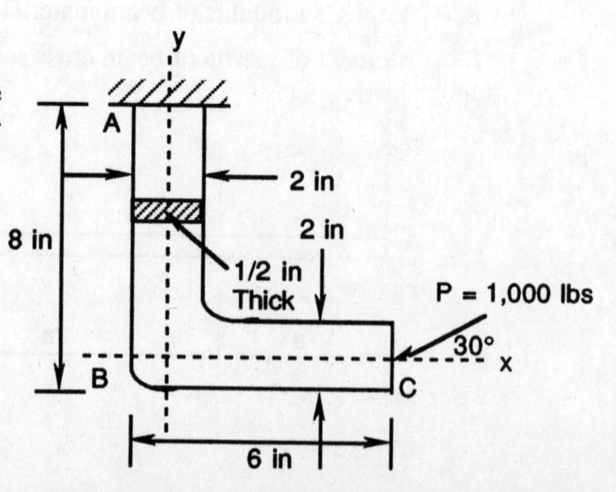

116. Which of the following statements defines best the capitalized cost?

(A) It is the present worth of a specified uniform cash flow for an infinite analysis period.

(B) It is the initial cost of some equipment.

(C) It is the salvage value of an equipment.

(D) It is the amount of money equal to cost minus benefit for an equipment.

(E) None of the above.

117. A credit card company compounds monthly and charges an interest of $1\frac{1}{2}$% per month. What is the effective interest rate per year?

(A) 18%

(B) 19.56%

(C) 4.35%

(D) 1.015%

(E) none of the above

118. If nominal interest rate per year is 12%, and compounding is continuous, what is the effective interest rate per year?

(A) 0.127%

(B) 1.000%

(C) 1.500%

(D) 11.275%

(E) 12.75%

119. Which of the following is not taken into account in making an economic decision between two alternative equipments?

(A) Reliability

(B) Operating cost

(C) Rate of return

(D) Salvage value

(E) Interest rate

120. A large profitable corporation has purchased a jet plane for use by its executives. The cost of the plane is $76 million. It has a useful life of 5 years. The estimated resale value at the end of five years is six million dollars. Using straight-line depreciation, what is the book value of the jet plane at the end of 3 years?

(A) $14 million

(B) $15.2 million

(C) $20 million (D) $34 million

(E) $48 million

121. In a before-tax comparison of alternatives, what effect does the method of depreciation have on which alternative is preferred?

(A) The method of depreciation has no effect on the alternative selected.

(B) The method of depreciation determines the alternative to be selected.

(C) Straight line depreciation leads to the selection of the best alternative.

(D) The ACRS depreciation leads to the selection of the best alternative.

(E) None of the above.

122. The University Foundation has received a gift of $1 million from an alumnus toward the construction and continued upkeep of an engineering laboratory. Annual maintenance cost is estimated at $30,000. In addition, $50,000 will be needed every 10 years for major repairs. How much will be left for initial construction after funds are allocated for perpetual upkeep? The foundation invests the funds and earns a nominal interest rate of 10%.

(A) $331,350 (D) $1 million

(B) $668,650 (E) $1.3373 million

(C) $662,700

123. A center for senior citizens is considering three different plans for providing transportation to its members. Each plan requires a different size of van each with its associated initial cost, annual operating cost and useful economic life. The plans are summarized below.

	Plans		
	A	B	C
Initial Cost	$10,000	$40,000	$50,000
Annual Operating Cost	$16,000	$3,500	$2,500
Useful Life (Years)	6	3	4

For each plan, the salvage value at the end of its useful life is zero. At the end of its useful life, each alternative is replaced with an identical plan for which the above cost data is applicable. Use a 12 year analysis period. The minimum attractive rate of return is 7%. What is the present cost of alternative plan C?

(A) $137,103
(B) $143,751
(C) $148,861
(D) $150,000
(E) $157,566

124. Consider problem 123 above. What is the present cost of alternative plan A? Round off the cost to the nearest hundred dollars.

(A) $106,000
(B) $133,800
(C) $137,100
(D) $143,800
(E) $148,900

125. Consider problem 123 above. What is the present cost of alternative plan B? Round off the cost to the nearest hundred dollars.

(A) $90,500
(B) $141,000
(C) $148,900
(D) $202,000
(E) $242,000

126. Instead of taking a vacation I decide to invest $2,000 in a venture that yields an annual interest rate of 10% compounded continuously. How much would I get approximately at the end of five years from my investment? Round off to the nearest ten dollars.

(A) $2,050
(B) $2,200
(C) $3,220
(D) $3,290
(E) $3,300

127. What type of bonds exists predominantly in a sodium chloride crystal?

(A) metallic bond
(B) ionic bond
(C) covalent bond
(D) van der Waals bond
(E) polar bond

128. The Miller indices of the direction of the arrow in the accompanying figure of unit cell of a cubic lattice is

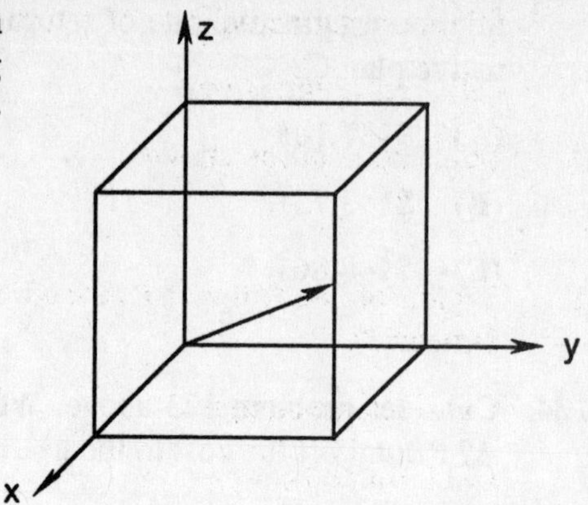

(A) 110

(B) 111

(C) 211

(D) 221

(E) 112

129. Pure metal *A* undergoes an isothermal transformation in which its crystal structure changes from face centered cubic (fcc) to body centered cubic (bcc). As a result, the volume of a piece of metal *A*

(A) increases.

(B) decreases.

(C) remains the same.

(D) decreases up to the midpoint of the transformation and then asymptotically reaches its original value.

(E) increases up to the midpoint of the transformation and then asymptotically reaches its original value.

130. Which of the following leads to a reduction in the electrical resistivity of a pure metal?

(A) cold working

(B) annealing

(C) grain refinement

(D) addition of alloying elements

(E) quenching from near the melting point

131. The ridigity of polymer can be increased by

(A) increasing the degree of polymerization

(B) increasing the extent of cross linking

(C) crystallization

(D) all of the above

(E) none of the above

132. Which of the following properties of a metal is insensitive to the micro-structure?

(A) tensile strength

(B) ductility

(C) modulus of elasticity

(D) hardness

(E) work hardening coefficient

133. The dominant charge carriers in a phosphorus-doped silicon crystal at room temperature are

(A) electron holes

(B) protons

(C) phosphorus ions

(D) silicon ions

(E) electrons

134. Metals are conductive because:

(A) they have a characteristic metallic luster.

(B) they have extra electrons as exhibited by normally positive valence states.

(C) the electrons are loosely bound to the nuclei and, therefore, mobile.

(D) they are on the left side of the Periodic Table.

(E) none of the above

135. An x-ray is:

(A) It is not known what it is; hence the name "x" rays.

(B) electromagnetic radiation.

(C) a ray of mixed subatomic particles containing electrons, protons, and neutrons.

(D) a ray of electrons.

(E) a ray of helium nuclei.

136. Group Ia elements are easily ionized because:

(A) they have a single "s" electron in the outer orbit.

(B) they are metals and, therefore, conductive.

(C) they react violently with water to liberate hydrogen

(D) the reason is not known, but is observed experimentally

(E) they have relatively low melting points for metals.

137. The rare earth metals all have very similar chemical properties because:

(A) the reason is not known, but is observed experimentally.

(B) they are rare, hence little is know about their chemistry.

(C) they melt at extreme temperatures, hence are nearly inert.

(D) successive members of the series are formed by adding 4f electrons which have little effect on reactivity.

(E) they are in a separate row at the bottom of the Periodic Table.

138. An electron volt is

(A) a voltage unit commonly used when measuring the voltage of electrons.

(B) the unit of electrical charge of one electron.

(C) a unit of energy equal to the energy possessed by an electron accelerating through a potential of one volt.

(D) can be any of the above depending on the context.

(E) none of the above.

139. Which of the following group of elements ALL form crystals of the diamond structural lattice?

(A) carbon, boron, and aluminum

(B) carbon, cobalt, and nickel

(C) carbon, silicon, and germanium

(D) lithium, sodium, and potassium

(E) calcium, magnesium, and strontium

140. The stress or load where a crystalline material fails is usually less than that predicted from calculated bond strengths in the crystal lattice because

(A) It is, to date, impossible to calculate something as complicated as the stress at failure.

(B) The stress at failure is governed by crystal defects, not crystal bond strength.

(C) The stress at failure is not reproducible experimentally.

(D) Plastic deformation occurs before failure.

(E) None of the above.

TEST 3 (AM)

ANSWER KEY

1.	(D)	26.	(D)	51.	(D)	76.	(C)
2.	(E)	27.	(B)	52.	(A)	77.	(E)
3.	(D)	28.	(A)	53.	(C)	78.	(B)
4.	(D)	29.	(E)	54.	(C)	79.	(D)
5.	(A)	30.	(B)	55.	(B)	80.	(A)
6.	(A)	31.	(A)	56.	(E)	81.	(C)
7.	(E)	32.	(C)	57.	(A)	82.	(E)
8.	(D)	33.	(D)	58.	(E)	83.	(D)
9.	(D)	34.	(B)	59.	(D)	84.	(C)
10.	(B)	35.	(D)	60.	(A)	85.	(B)
11.	(D)	36.	(C)	61.	(D)	86.	(D)
12.	(B)	37.	(B)	62.	(B)	87.	(C)
13.	(A)	38.	(D)	63.	(D)	88.	(A)
14.	(D)	39.	(A)	64.	(E)	89.	(D)
15.	(D)	40.	(C)	65.	(D)	90.	(E)
16.	(A)	41.	(B)	66.	(D)	91.	(A)
17.	(A)	42.	(E)	67.	(B)	92.	(C)
18.	(D)	43.	(A)	68.	(B)	93.	(E)
19.	(E)	44.	(B)	69.	(B)	94.	(D)
20.	(E)	45.	(B)	70.	(C)	95.	(D)
21.	(A)	46.	(C)	71.	(A)	96.	(B)
22.	(B)	47.	(D)	72.	(D)	97.	(C)
23.	(B)	48.	(C)	73.	(D)	98.	(D)
24.	(B)	49.	(E)	74.	(B)	99.	(C)
25.	(E)	50.	(C)	75.	(B)	100.	(B)

101.	(B)	111.	(C)	121.	(A)	131.	(D)
102.	(C)	112.	(E)	122.	(B)	132.	(C)
103.	(B)	113.	(D)	123.	(A)	133.	(E)
104.	(D)	114.	(C)	124.	(D)	134.	(C)
105.	(C)	115.	(B)	125.	(C)	135.	(B)
106.	(A)	116.	(A)	126.	(E)	136.	(A)
107.	(D)	117.	(B)	127.	(B)	137.	(D)
108.	(A)	118.	(E)	128.	(D)	138.	(C)
109.	(E)	119.	(A)	129.	(A)	139.	(C)
110.	(C)	120.	(D)	130.	(B)	140.	(B)

DETAILED EXPLANATIONS OF ANSWERS
TEST 3
MORNING (AM) SECTION

1. **(D)**

The dot product of perpendicular vectors is zero. **A** is the known vector. **B** represents the unknown vector.

$$\mathbf{A} \cdot \mathbf{B} = (a_1\mathbf{i} + a_2\mathbf{j}) \cdot (b_1\mathbf{i} + b_2\mathbf{j}) = a_1 \cdot b_1 + a_2 \cdot b_2 = 0$$

The values of a_1 and a_2 are specified as

$$a_1 = 1 \quad \text{and} \quad a_2 = 2$$

Thus $\quad 1 \cdot b_1 + 2 \cdot b_2 = 0$

The value of b_1 is given as -6.2

Thus $\quad 2b_2 = 6.2 \quad \text{or} \quad b_2 = 3.1.$

2. **(E)**

In this differential equation, t is the independent variable, and y is the dependent variable. The order of a differential equation is defined by the highest order derivative of the dependent variable. This equation contains the second derivative of y and is thus a second order equation. A differential equation is homogeneous if all terms contain the dependent variable or a derivative of the dependent variable. This equation contains a term containing only a function of t, sin $(2t)$, which makes it non-homogeneous. A differential equation is linear if the coefficients of the dependent variable and its derivatives are not functions of the dependent variable. The coefficients of y and its derivatives in the problem equation are 3, t, and -4, none of which are functions of y. This equation is linear. One of the coefficients is t, the independent variable. In most applications, t represents time, and this equation would describe a "time-varying-parameter system.

3. **(D)**

When the value of 0 is substituted into the function, the result is 0/0, which is indeterminate. L'Hopital's Rule is useful for indeterminate forms.

$$\lim_{x \to a} \frac{F(x)}{G(x)} = \lim_{x \to a} \frac{F'(x)}{G'(x)}$$

where $F'(x) = dF/dx$ and $G'(x) = dG/dx$.

L'Hopital's Rule is valid only if the right-hand side exists.

When the numerator and denominator of the given function are differentiated, the result is

$$\frac{d}{dx}(x + \tan x) = 1 + \sec^2 x$$

$$\frac{d}{dx}(\sin 3x) = 3\cos 3x$$

$$\lim_{x \to 0} \frac{x + \tan x}{\sin 3x} = \lim_{x \to 0} \frac{1 + (\sec x)^2}{3\cos 3x} = \frac{1 + 1^2}{3 \cdot 1} = \frac{2}{3}.$$

4. **(D)**

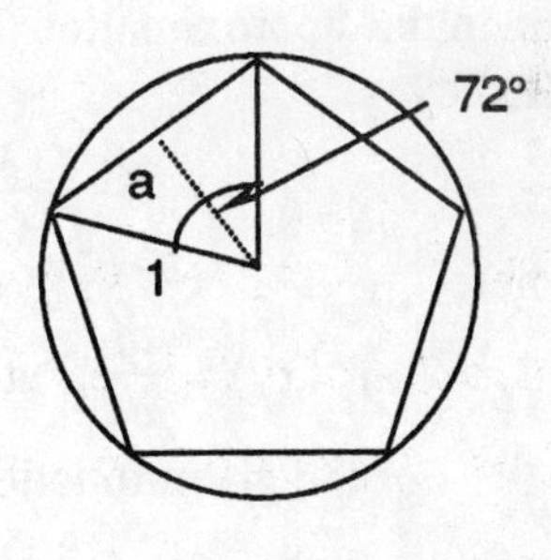

A regular pentagon has five sides of equal lengths. The unit circle has a radius of 1 unit. Each side of the pentagon shown in the figure has a length of $2a$. The chord of length $2a$ subtends an angle of 360/5 = 72°. The radius of the circle is 1.

$$\frac{a}{1} = \sin\left(\frac{72°}{2}\right) = 0.588$$

The length of the side is $2a$ or 1.18.

5. **(A)**

The imaginary number i is defined as $\sqrt{-1}$. Thus $i^2 = -1$. The two factors are multiplied algebraically.

$$\begin{aligned}(2 + 4i) \cdot (1 - 7i) &= 2 - 14i + 4i - 28i^2 \\ &= 2 - 10i - 28 \cdot (-1) \\ &= 30 - 10i.\end{aligned}$$

6. **(A)**

This pair of simultaneous linear equations can be solved by several methods. The easiest method is substitution.

Solve the second equation for X and substitute the result into the first equation.

$$X - 5y = -13$$

$$X = -13 + 5y$$

Substituting into the first equation

$$4(-13 + 5y) + 3y = 17$$

which simplifies to

$$-52 + 20y + 3y = 17$$

$$23y = 69$$

or $\qquad y = 3.$

This value of y, when substituted into the second equation gives

$$X = -13 + 5(3) = 2.$$

7. **(E)**

Newton's method is an iterative procedure to find the root of an equation

$$f(x) = 0$$

An initial approximation is assumed and successive approximations are calculated by

$$x_{n+1} = x_n - \frac{f(x_n)}{f'(x_n)}$$

where

$$f'(x_n) = \frac{df}{dx} \quad \text{at} \quad x = x_n$$

For the given function

$$f(x) = x^3 + 4x^2 + 2x - 2$$

$$f'(x) = 3x^2 + 8x + 2$$

For $x_n = 2$

$$f(x_n) = f(2) = 26$$

$$f'(x_n) = f'(2) = 30$$

Thus,

$$x_{n+1} = 2 - \frac{26}{30} = 1.13$$

which is answer (E).

8. **(D)**

This is an exponential decay problem in which a quantity decreases at a rate proportional to the amount present.

$$\frac{dQ}{dq} = -\frac{1}{\theta}Q \quad Q_{(t=0)} = Q_0$$

where $\quad Q$ = quantity of material

t = time

Θ = time constant of decay.

The solution to this equation is

$$Q = Q_0 e^{-t/\theta}$$

The value of θ is determined from the problem statement

$$Q = {}^1/_2\, Q_0 \quad \text{for} \quad t = 30 \text{ yrs}$$

$${}^1/_2\, Q_0 = Q_0 e^{-30 \text{ yrs}/\theta}$$

$${}^1/_2 = e^{-30 \text{ yrs}/\theta}$$

$$\ln .5 = -0.693 = \ln(e^{-30/\theta}) = -30/\theta$$

Thus, θ = + 30 years (0.693) = 20.8 years.

If 90% had decomposed then

$$\frac{Q}{Q_0} = 0.1 = e^{-t/20.8} \qquad Q = .10 Q_0$$

or $\ln(0.1) = -t/20.8$

from which t = 47.9 years, which corresponds to answer (D).

9. **(D)**

Note that the derivative of the denominator $(6 + x^2)$ is $2xdx$, and there is an xdx in the numerator. When the numerator contains the derivative of the denominator, the antiderivative has the form of the logarithm of the denominator,

$$\int \frac{du}{u} = \ln u$$

Evaluate the antiderivative at the limits to find the result.

$$2\int_1^3 \frac{2x\,dx}{6+x^2} = 2\left[\ln(6+x^2)\right]_1^3$$

$$= 2\left[\ln(6+3)^2 - \ln(6+1^2)\right]$$

$$= 1.52.$$

10. **(B)**

The problem requires the evaluation of

$$dF = \frac{\partial F}{\partial x}dx + \frac{\partial F}{\partial y}dy + \frac{\partial F}{\partial z}dz$$

where $\frac{\partial F}{\partial x}$ is the partial derivative of $F(x, y, z)$ with respect to x, which defines how F changes with x while y and z are held constant. The three partial derivatives are evaluated as follows.

$$\frac{\partial F}{\partial x} = 0 + ze^{xz}\sin 4y$$

$$\frac{\partial F}{\partial y} = (1 + e^{xz})4\cos 4y$$

$$\frac{\partial F}{\partial z} = 0 + xe^{xz} \sin 4y$$

These derivatives are substituted into the definition to give

$$dF = z\, e^{xz} \sin (4y)\, dx + 4 \cos (4y)\, (1 + e^{xz})\, dy$$
$$+ x\, e^{xz} \sin (4y)\, ds,$$

which is answer (B).

11. **(D)**

The simplest way to determine the constants A, B, and C is to multiply the equation by the denominator of the left-hand side.

$$s - 3 = A(s + 2)\,(s + 4) + B(s + 3)\,(s + 4) + C(s + 2)\,(s + 3)$$

By choosing appropriate values of s, the three constants can be determined.

If $s = -3$, the coefficients of B and C are zero and

$$-3 - 3 = A(-3 + 2)\,(-3 + 4) + 0 + 0$$

or $$A = \frac{-6}{(-1)\,(1)} = 6$$

If $s = -2$, the coefficients of A and C are zero and

$$-2 - 3 = 0 + B(-2 + 3)\,(-2 + 4) + 0$$

or $$B = \frac{-5}{(1)\,(2)} = -5/2$$

If $s = -4$, the coefficients of A and B are zero and

$$-4 - 3 = 0 + 0 + C(-4 + 3)\,(-4 + 2)$$

or $$C = \frac{-7}{(-1)\,(-2)} = -7/2\,.$$

12. **(B)**

The probability density function for a normal distribution is

$$f(Z) = \frac{1}{\sqrt{2\pi}}\, e^{-Z^2/2}$$

where $$Z = \frac{X - \text{Mean}}{\text{Standard Deviation}}$$

The probability of measuring values between Z_1 and Z_2 is obtained by integrating the density function between these two limits.

$$\int_{Z_1}^{Z_2} f(Z)\, dZ$$

In problem 12, Z_1 is $(15 - 16)/2$, and Z_2 is $(19 - 16)/2$.

13. **(A)**

The equation of a straight line is

$$y = mx + b$$

where m and b are the slope and intercept, respectively. When the values of the two points are substituted into the above equation, the result is two linear equations in two unknowns.

$$1 = m(3) + b$$

$$15 = m(10) + b$$

These equations are easily solved by subtracting the second from the first so as to eliminate b

$$1 - 15 = m(3 - 10)$$

$$-14 = m(-7)$$

Thus, $m = 2$. When this value of m is substituted into either equation, b is found to be -5.

$$y = 2x - 5.$$

14. **(D)**

Local minimum and maximum values of $y(x)$ will occur at x values where

$$\frac{dy}{dx} = 0$$

$$\frac{dy}{dx} = 6x^2 - 24 = 0$$

or $$x^2 = 4$$

$$x = +2 \quad \text{or} \quad x = -2.$$

To determine which value gives the minimum x, the second derivative is evaluated, when $x = +2$,

$$\frac{d^2y}{dx^2} = 12x$$

When $x = +2$,

$$\frac{d^2y}{dx^2} = 12(2) = 24 > 0$$

A positive value of the second derivative indicates a local minimum. When $x = -2$,

$$\frac{d^2y}{dx^2} = 12(-2) = -24 < 0$$

A negative value of the second derivative indicates a local maximum, thus $x = +2$ is the location of the minimum.

15. **(D)**

There are a total of one hundred engineers at the banquet. The probability that representatives from all disciplines are selected is the product of the probabilities of selecting someone from each discipline. Note that after each representative is selected, the pool from which to select decreases by one.

$$\frac{40}{100} \times \frac{20}{99} \times \frac{30}{98} \times \frac{10}{97} = \frac{240{,}000}{94{,}109{,}400}.$$

16. **(A)**

The volume can be determined by defining a circular slice of thickness dy and radius x.

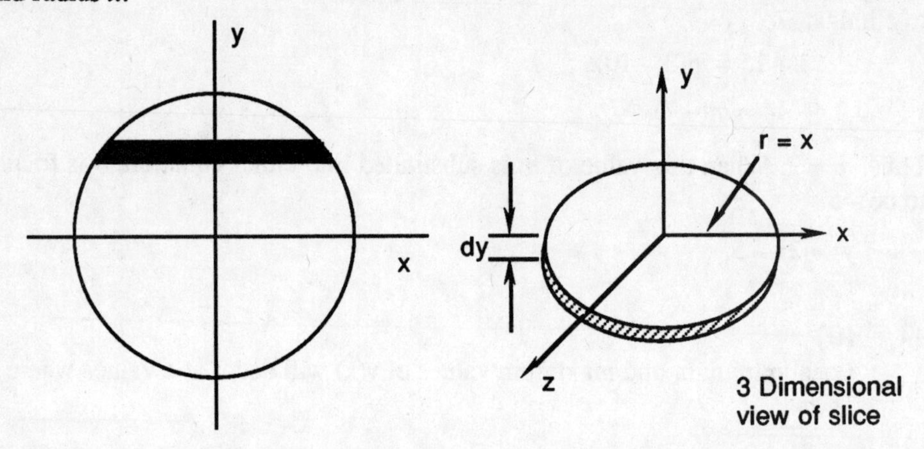

The volume of this slice is $\pi x^2\, dy$. The total volume is found by integration or accumulating the volumes of the slices.

$$\text{Volume} = 2\int_0^1 \pi x^2\, dy$$

$$x^2 + y^2 = 1 \qquad x^2 = 1 - y^2$$

$$= 2\pi \int_0^1 (1 - y^2)\, dy$$

$$= 2\pi \left[y - \frac{y^3}{3} \right]_0^1$$

$$= \frac{4\pi}{3}$$

Note that this is the volume of the unit sphere, which is the volume obtained by rotating the unit circle.

17. **(A)**

Both sides of the equation are multiplied by $\mathbf{A}^{-1}$.

$$\mathbf{A}^{-1}\mathbf{A}x = \mathbf{A}^{-1}b$$

The product of **A** inverse and **A** is the unit matrix, which when multiplied by x gives x.

$$x = \mathbf{A}^{-1}b$$

18. **(D)**

The equation needs to be converted to standard form

$$(x - x_0)^2 + (y - y_0)^2 = r^2$$

x_0, y_0 define the location of the center, r is the radius.

This is done by forming quadratic terms which are squares.

$$x^2 - 4x + y^2 - 2y - 4 = 0$$

$$x^2 - 4x + y^2 - 2y = 4$$

Add the squares of $^1/_2$ the x and y coefficients to both sides of the equation

$$(x^2 - 4x + 4) + (y^2 - 2y + 1) = 4 + 4 + 1$$

$$(x^2 - 2)^2 + (y - 1)^2 = 9$$

This circle is centered at

$$x = 2 \quad y = 1$$

and has a radius of 3.

19. **(E)**

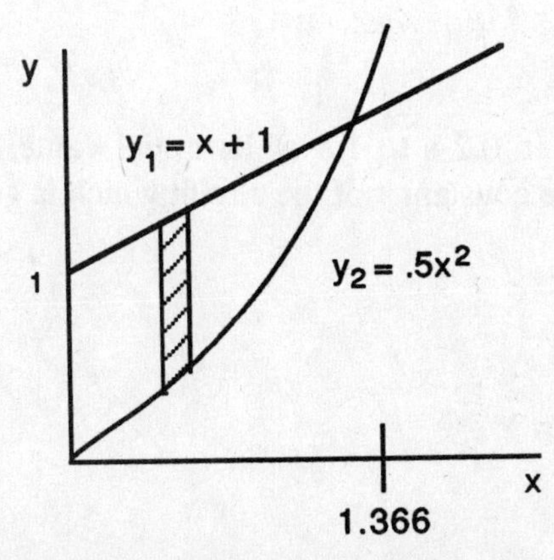

The intersection of the two functions is found by

$$y_1 = x + 1 = .5x^2 = y_2$$

$$.5x^2 - x - 1 = 0$$

This has a positive root at $x = 1.366$.

The area is found by integrating the difference between the functions for x between 0 and the intersection

$$\int_0^{1.366} (y_1 - y_2)\,dx = \int_0^{1.366} (x + 1 - .5x^2)\,dx$$

$$= \left[\frac{x^2}{2} + x - .5\left(\frac{1}{3}\right)x^3\right]_0^{1.366}$$

$$= 0.933 + 1.366 - 0.425$$

$$= 1.874.$$

20. **(E)**

The tangent to the curve is the derivative of the function for y

$$\frac{dy}{dx} = xe^x + e^x$$

when $x = 0.8$, the derivative is

$$0.8e^{0.8} + e^{0.8} = 4.0,$$

which is answer (E).

21. **(A)**

The voltage response of a discharging RC circuit is

$$V(t) = V_0 e^{-t}/RC.$$

The ratio between the initial voltage of 100 V and the voltage after 0.2 sec is

$$\frac{36.8}{100} = 0.368$$

$$= \frac{1}{e}$$

as the voltage decreased in 0.2 s to $1/e$ of its initial value, the elapsed time is therefore equal to the time constant τ of the circuit which is given by

$$\tau = RC$$

$$C = \frac{0.2}{100 \cdot 10^3}$$

$$C = .000002 \text{ F}$$

$$= 2\mu\text{F}$$

22. **(B)**

The voltage across a capacitor cannot change instantaneously, and as the initial voltage was zero, immediately after closing the switch, it still is zero. Therefore the capacitor can be modeled as a short circuit. The current through an inductor cannot change instantaneously, and as its initial value was zero, initially the inductor will be modeled as an open circuit. The model of the circuit immediately after closing the switch is shown below.

By Ohm's law the current is

$$i = \frac{V}{R}$$
$$= 12/3$$
$$= 4.0 \text{ A}$$

3 Ω
12 V
2 Ω

23. **(B)**

For a parallel circuit the admittance is given by:

$$\mathbf{Y} = G + jB_C - jB_L$$

where **Y** is the admittance, in general a complex number, G the conductance, B_C the capacitive suceptance and B_L the inductive suceptance. Under resonance

$$B_C = B_L$$

and therefore

$$\mathbf{Y} = G$$

and the admittance is a minimum and as the current phasor I is given by:

$$I = \mathbf{V} \cdot \mathbf{Y}$$

where **V** is the voltage phasor, the current will be a minimum under resonance conditions.

24. **(B)**

The two 12 Ω inductors are connected in parallel and therefore their equivalent reactance X_{eq} is given by:

$$\frac{1}{X_{eq}} = \frac{1}{X_1} + \frac{1}{X_2}$$

$$\frac{1}{X_{eq}} = \frac{1}{12} + \frac{1}{12}$$

$$X_{eq} = 6\ \Omega$$

This equivalent reactance and the two other elements of the circuit are all connected in parallel and therefore the impedance **Z**, a complex number is given by:

$$\mathbf{Z} = R + jX_L - jX_C$$
$$= 4 + j6 - j3$$
$$= 4 + j3$$

in polar form the impedance **Z** is given by:

$$\mathbf{Z} = \sqrt{R^2 + X^2} \underline{\lfloor a\tan X/R}$$
$$= \sqrt{4^2 + 3^2} \underline{\lfloor a\tan 3/4}$$
$$= 5\underline{\lfloor 37°}\ \Omega$$

the current phasor **I** is therefore given by:

$$\mathbf{I} = \frac{\mathbf{V}}{\mathbf{Z}}$$
$$= \frac{100 \underline{\lfloor 0°}}{5 \underline{\lfloor 37°}}$$
$$= 20 \underline{\lfloor -37°}\ \text{amp}$$

and the voltage phasor $\mathbf{V_R}$ across the 4W resistor is given by:

$$\mathbf{V_R} = \mathbf{I} \cdot R$$
$$= 20 \underline{\lfloor -37°}\ 4\underline{\lfloor 0°}$$
$$= 80 \underline{\lfloor -37°}\ \text{V}$$

25. **(E)**

By definition the impedance is a complex number, its magnitude is the ratio of the amplitude of the voltage to the amplitude of the current and its angle is the lead angle of the voltage with relation to the current. From the graph shown in the problem the voltage amplitude is 10 and the current amplitude 20, therefore the magnitude of the impedance is 10/20 = 0.5 Ω. The voltage lags the current by 0.0021s, as the period of the signal is 0.017s and the ratio between the lag time and the period is 0.0021/0.017 = 1/8, the phase angle is therefore 1/8 · 360° = 45°. This angle is negative, as the voltage lags the current. The impedance is therefore $0.5 \underline{\lfloor -45°}$.

26. **(D)**

The series combination has an input impedance Z_{in} given by:

$$Z_{in} = R + jX$$
$$= 5 + j12$$

or in polar form:

$$= \sqrt{5^2 + 12^2} \underline{\lfloor a\tan X/R}$$
$$= 13 \underline{\lfloor 67.38°}\ \Omega$$

The parallel combination has an input admittance Y_{in} given by:

$$Y_{in} = G_p - jB_p$$

The series and parallel combinations will have the same input impedance if:

$$Y_{in} = \frac{1}{Z_{in}}$$

therefore

$$G_p - jB_p = \frac{1}{13\,\underline{|67.38°}}$$

$$= 0.0296 - j0.0711 \text{ mhos}$$

$$G_p = 0.0296 \text{ mhos}$$

$$B_p = 0.0711 \text{ mhos}$$

and therefore

$$R_p = \frac{1}{G_p}$$

$$= \frac{1}{0.0296}\ \Omega$$

$$= 33.78\ \Omega$$

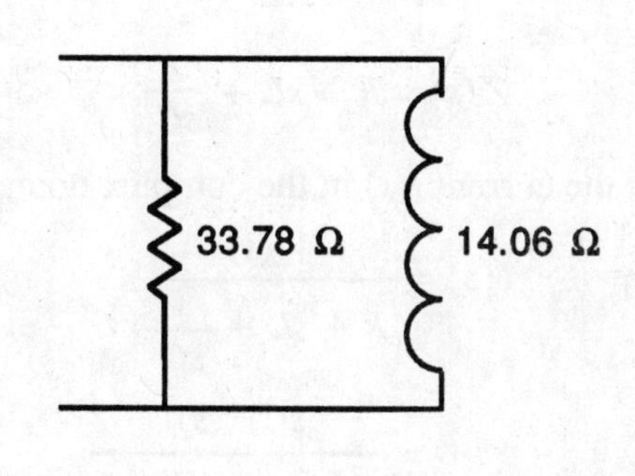

$$X_p = \frac{1}{B_p}$$

$$= \frac{1}{0.0711}\ \Omega$$

$$= 14.06\ \Omega$$

The equivalent parallel circuit is shown here.

27. **(B)**

A circuit is resonant if its input impedance or admittance is a real number.

As the elements are connected in series it is easier to work with the admittance:

$$Y_{in} = \frac{1}{R} + j\omega X - j\frac{1}{\omega L}$$

the circuit would be in resonance if the imaginary part of the input admittance is zero:

$$j\omega C - j\frac{1}{\omega L} = 0$$

$$\frac{\omega^2 LC - 1}{L} = 0$$

$$\omega^2 LC - 1 = 0$$

$$\omega = \sqrt{\frac{1}{LC}}$$

$$= \sqrt{\frac{1}{50 \cdot 10^{-3} \cdot 10 \cdot 10^{-6}}}$$

$$= 1414\,s^{-1}.$$

28. **(A)**

To find the frequency of oscillation of the transient response, the characteristic equation of the circuit has to be found. In the "s" domain the impedances of the three elements in the circuit are given by

$$Z_R = R$$

$$Z_L = SL$$

$$Z_C = \frac{1}{sC}$$

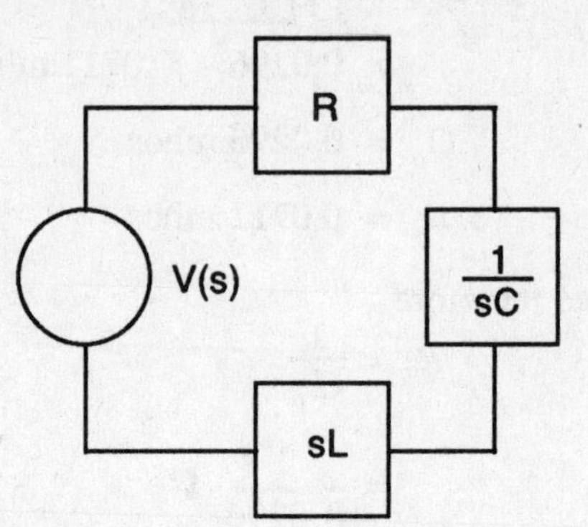

In the "s" domain the circuit diagram would be:

The complex impedance $Z(s)$ is therefore:

$$Z(s) = R + sL + \frac{1}{sC}$$

and the current $I(s)$ in the complex domain will be:

$$I(s) = \frac{V(s)}{R + sL + \frac{1}{sC}}$$

$$= \frac{sCV(s)}{s^2LC + sRC + 1}$$

the characteristic equation is the denominator of the above expression, the roots of this equation determine the nature of the transient behavior. We therefore set:

$$s^2LC + sRC + 1 = 0$$

$$s^2 + s\frac{R}{L} + \frac{1}{LC} = 0$$

and the roots are given by:

$$s_{1,2} = \frac{\frac{-R}{L} \pm \sqrt{\left(\frac{R}{L}\right)^2 - \frac{4}{LC}}}{2}$$

$$s_{1,2} = \frac{\frac{-10}{0.01} \pm \sqrt{\left(\frac{10}{0.01}^2\right) - \frac{4}{0.01 \cdot 10 \cdot 10^{-6}}}}{2}$$

$$s_{1,2} = -500 \pm j3122.5$$

as the roots of the characteristic equation are complex,

$$S_{1,2} = -\alpha \pm j\omega_d,$$

the transient response is under-damped, therefore oscillatory with an angular frequency, $\omega_d = 3122.5s^{-1}$ or a frequency:

$$f = \frac{\omega}{2\pi}$$

$$= \frac{3122.5}{2\pi}$$

$$= 497.5 \text{ Hz}$$

29. **(E)**

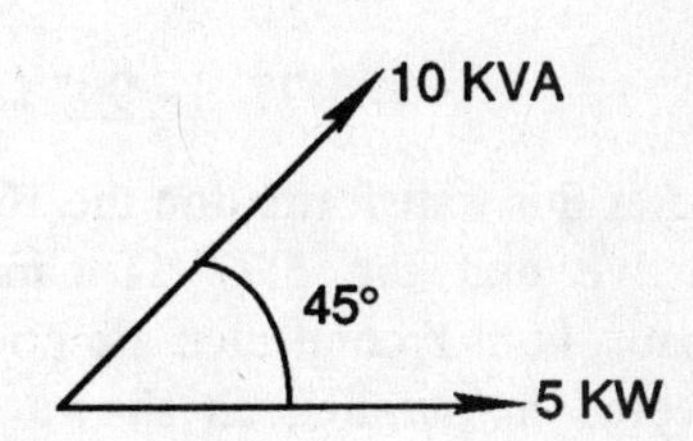

The power diagram shows a real load of 5 KW and lagging load of 10 KVA, as cos 45° = 0.707 the angle between the two loads is 45°.

For the 10 KVA load the real power P_1 and the reactive power Q_1 are given by

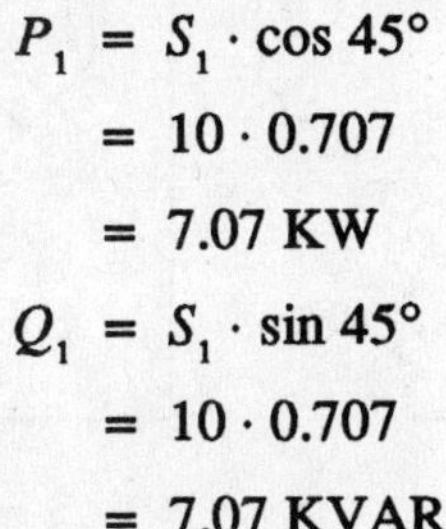

$$P_1 = S_1 \cdot \cos 45°$$

$$= 10 \cdot 0.707$$

$$= 7.07 \text{ KW}$$

$$Q_1 = S_1 \cdot \sin 45°$$

$$= 10 \cdot 0.707$$

$$= 7.07 \text{ KVAR}$$

the 5 KW load has only a real component P_2

$$P_2 = 5 \text{ KW}$$

$$Q_2 = 0 \text{ KVA}$$

the total real P and reactive Q powers are therefore

$$P = P_1 + P_2$$

$$= 5 + 7.07$$

$$= 12.07 \text{ KW}$$

$$Q = Q_1$$

$$= 7.07 \text{ KVAR}$$

and the power factor of the two loads combined will be

$$\text{power factor} = \frac{P}{\sqrt{P^2 + Q^2}}$$

$$= \frac{12.07}{\sqrt{12.07^2 + 7.07^2}}$$

$$= 0.863$$

because the total reactive power of the two loads is positive, the power factor is lagging.

30. **(B)**

The connected load can be transformed in a Y connected load using the relation

$$Z_y = \frac{Z}{3}$$

$$= \frac{17.32 \angle -90^\circ}{3}$$

$$= 5.77 \angle -90^\circ \ \Omega$$

After this transformation the 10 Ω resistive and the 5.77 Ω capacitive loads, both Y connected are now connected in parallel, as shown in the diagram, therefore the equivalent impedance of each branch will be

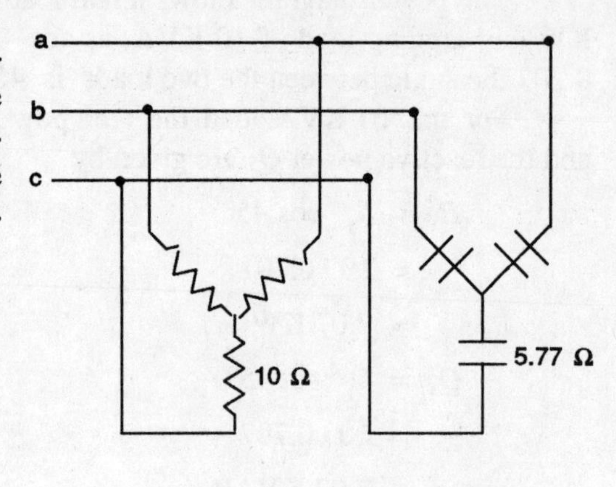

$$\frac{1}{Z_{eq}} = \frac{1}{Z_1} + \frac{1}{Z_2}$$

$$= \frac{1}{10} + \frac{1}{5.77 \angle -90^\circ}$$

$$= 0.1 + j0.173$$

or in polar form

$$= 0.2 \angle 60^\circ \text{ Siemens}$$

or $$Z_{eq} = 5 \angle -60^\circ \ \Omega$$

The equivalent load is Y connected with a $5 \angle -60^\circ \ \Omega$ impedance per branch.

The voltage between line and the neutral n of the Y load is

$$V_{an} = \frac{V_{ab}}{\sqrt{3}} \angle -30^\circ$$

$$= 127 \angle -30^\circ \text{ V}$$

The current is therefore

$$I_a = \frac{V_{an}}{Z_{eq}}$$

$$= \frac{127 \angle -30^\circ}{5 \angle -60^\circ}$$

$$= 25.43 \angle 30^\circ \text{ A}$$

31. **(A)**

The capacitive load is connected to the secondary of a transformer. This load is reflected to the primary side as a capacitive impedance of

$$Z_2^1 = Z_2 (n_2 / n_1)^2$$

$$= 0.12 \angle -90° \, (10/1)^2$$

$$= 12 \angle -90°$$

The total impedance of the circuit seen by the source is therefore

$$Z_1 = 5 \angle 0° + 12 \angle -90°$$

or

$$= 5 - j12$$

or in polar form

$$= \sqrt{5^2 + 12^2} \; a \tan \frac{-12}{5}$$

$$= 13 \angle -67° \; \Omega$$

and the current is

$$\mathbf{I} = \frac{\mathbf{V}}{Z}$$

$$= \frac{220 \angle 90°}{13 \angle -67°}$$

$$= 16.9 \angle 157° \text{ A}$$

32. **(C)**

Given their positive sequence the line voltages are:

$$\mathbf{V}_{ab} = 220 \angle 0° \text{ V},$$

$$\mathbf{V}_{bc} = 220 \angle -120° \text{ V},$$

$$\mathbf{V}_{ca} = 220 \angle +120°$$

The currents through the two resistive loads as shown in the circuit diagram are therefore

$$\mathbf{I}_{ab} = \frac{\mathbf{V}_{ab}}{R_1}$$

$$= \frac{220 \angle 0°}{12 \angle 0°}$$

$$= 18.3 \angle 0° \text{ A}$$

and

$$\mathbf{I}_{bc} = \frac{\mathbf{V}_{bc}}{R_2}$$

$$= \frac{220 \angle -120°}{5 \angle 0°}$$

$$= 44.0 \angle -120° \text{ A}$$

Applying Kirchoff's current law to the node shown in the diagram

$$I_b = I_{bc} - I_{ab}$$
$$= 44.0 \angle -120° - 18.3 \angle 0°$$
$$= -22 - j\,38.10 - 18.3$$
$$= -40.3 - j\,38.10$$

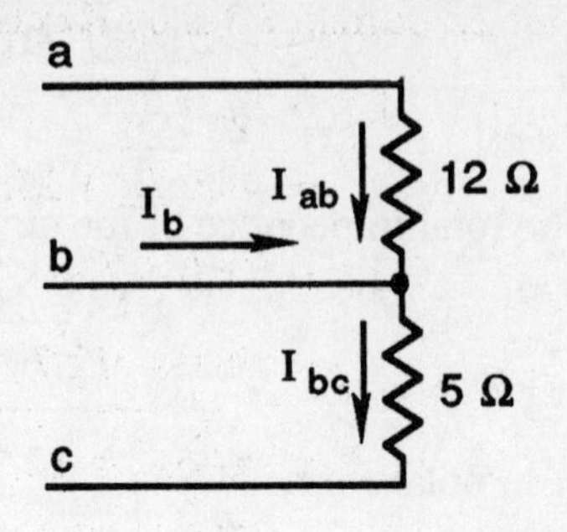

or in polar form

$$= 55.5 \angle -136.7° \text{ A.}$$

33. **(D)**

As the applied voltage has a positive sequence the three line to line voltages are:

$$V_{ab} = 220 \angle 0° \text{ V},$$
$$V_{bc} = 220 \angle -120° \text{ V},$$
$$V_{ca} = 220 \angle +120° \text{ V}$$

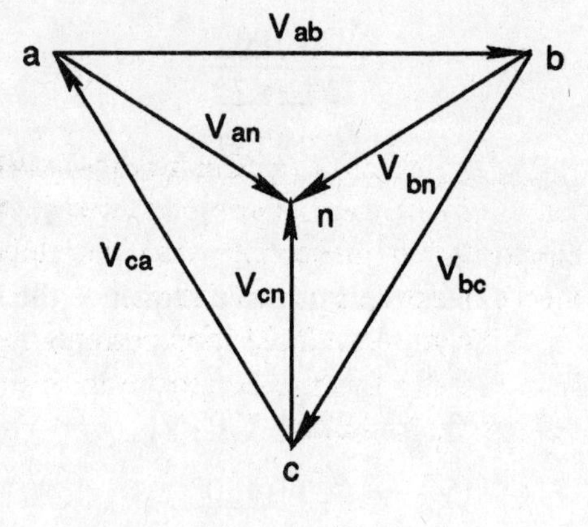

These voltages and the three line to neutral voltages are shown in the phasor diagram and the three line to neutral voltages are:

$$V_{an} = V_{ab}/\sqrt{3} \angle -30° \text{ V}$$
$$= 220/\sqrt{3} \angle -30° \text{ V}$$
$$= 127 \angle -30° \text{ V}$$
$$V_{bn} = 127 \angle -150° \text{ V}$$
$$V_{cn} = 127 \angle +90° \text{ V}$$

The current I_a is therefore:

$$I_a = \frac{V_{an}}{R_1}$$
$$= \frac{127 \angle -30°}{5}$$
$$= 25.4 \angle -30° \text{ A}$$

and the current I_{ab}

$$I_{ab} = \frac{V_{ab}}{R_1}$$
$$= \frac{220 \angle 0°}{5}$$
$$= 44.0 \angle 0° \text{ A}$$

and the current I_a^1 is by Kirchoffs law:

$$I_a^1 = I_{ab} + I_a$$
$$= 44.0\,\underline{/0^\circ} + 25.4\,\underline{/-30^\circ}$$
$$= 44.0 + 22 - j\,12.7$$
$$= 66.0 - j12.7$$
$$= \sqrt{66^2 + 12.7^2}\ \underline{/a\tan(-12.7/66)}$$
$$= 67.2\,\underline{/-10.9^\circ}\ \text{A}$$

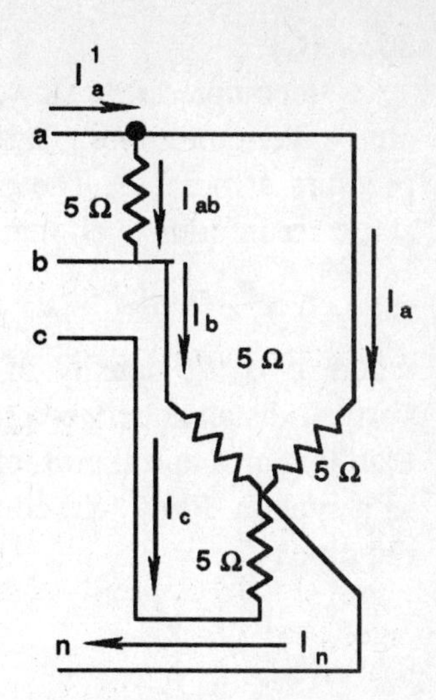

34. **(B)**

The circuit is a planar circuit with three windows and therefore three loop equations are needed to find the three loop circuits I_1, I_2, I_3 as shown.

From these three loop currents the line currents and the current in the neutral can be found:

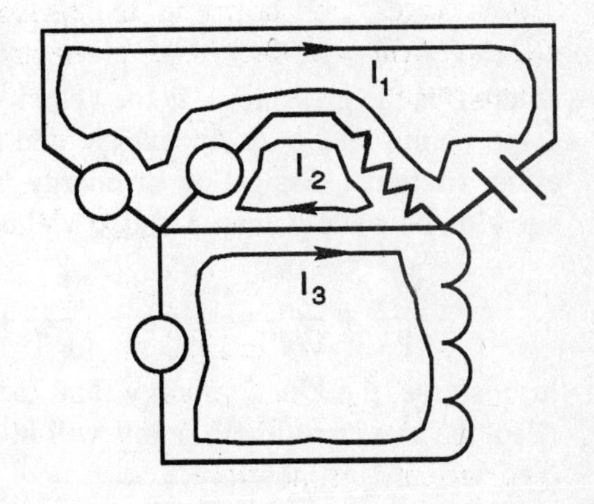

$$I_a = I_1$$
$$I_b = I_2$$
$$I_c = -I_3$$
$$I_n = I_2 - I_3$$

35. **(D)**

A Newtonian fluid behaves according to the following stress/strain rate relationship:

$$\tau = \mu\left(\frac{du}{dy}\right)$$

where τ is the shear stress acting on a fluid body, μ is the absolute viscosity of the fluid and du/dy is a velocity gradient of the fluid (or the shear strain rate). All real fluids show some resistance of motion to an imposed shear force. Fluids showing a non-linear proportionality between shear force and shearing motion are called "non-Newtonian" fluids. If the shear rate decreases with time the fluid would be called "shear thickening."

36. **(C)**

Incompressible flow, like that of water, always leaves an exit of a pipe at atmospheric pressure. Therefore the pressure at point *A* is atmospheric. So is the pressure at point *B*. The pressure at point *C* can be determined using a hydrostatic assumption such that:

$$P_C = \rho g z + P_A$$

where ρ is the density of water, g is the acceleration of gravity, and z is the vertical distance between points *C* and *A*. Therefore the pressure at *C* is greater than that at *A* and therefore flow will result. This device is commonly referred to as a syphon. Flow will continue until the water level (*B*) reaches the entrance of the pipe (*C*).

37. **(B)**

The solution is arrived at by an energy balance, usually between points where something is known. In this case, the pressure and velocity are known at point *A* (P = 0 psig, V = 0 ft/sec) and point *D*, as well as their elevations. The energy following a streamline can be determined by using Bernoulli's equation which can be written:

$$\frac{P}{\rho g} + \frac{V^2}{2g} + z + h_p = \text{constant}$$

where P is the pressure, V is the velocity, z is the vertical position with respect to some datum and h_p is an energy addition or loss term which could represent either frictional dissipation or energy input through a pump. Using Bernoulli's equation between points, *A* and *D* yields:

$$\frac{P_A}{\rho g} + \frac{V_A^2}{2g} + z_A + h_p = \frac{P_D}{\rho g} + \frac{V_D^2}{2g} + z_D$$

In this case, the kinetic energy (and therefore the velocity) can be ignored at *A*. Also, the pressure at both points will be atmospheric (0 psig). Therefore

$$V_D = \sqrt{2g_-^* (z_A - z_D + h_p)} = \sqrt{2(32.2 \text{ ft/sec}^2)(20 \text{ ft})} = 35.9 \text{ ft/sec.}$$

38. **(D)**

The velocity at discharge will be the same as the velocity at the entrance to the pump because of continuity of mass. The energy of a fluid following a streamline can be determined by using Bernoulli's equation which can be written:

$$\frac{P}{\rho g} + \frac{V^2}{2g} + z = \text{constant}$$

where P is the pressure, V is the velocity, and z is the vertical position with respect to some datum. Therefore, information can be gained using Bernoulli's equation between points *A* and *B*. This would be

$$\frac{P_A}{\rho g} + \frac{V_A^2}{2g} + z_A = \frac{P_B}{\rho g} + \frac{V_B^2}{2g} + z_B$$

Again, P_A will be 0 psig, and V_A = 0 ft/sec. Note that the power input by the pump does not enter into the equation because the pump is not between the two points in question. The velocity at the pump is known to be 50 ft/sec. Therefore, solving for the pressure yields:

$$P_B = \rho g * \{(z_A - z_B) - \frac{V_B^2}{2g}\} = (1.94)(32.2)(60 - 5 - 38.8) = 7.0 \text{ psig}$$

Note that the pressure is still above atmospheric.

39. **(A)**

The Pitot-static tube measures the difference between static and dynamic pressure. Using Bernoulli's equation then yields:

$$\frac{V_{duct}^2}{2g_c} = \frac{P_d}{\rho} - \frac{P_s}{\rho}$$

where P_d is the dynamic pressure and P_s is the static pressure. To determine the change in pressure from the Pitot-static tube, it is necessary to consider the weight of the water.

$$P_d - P_s = \rho_{water} \times g/g_c \times 1.6/12 \text{ ft}$$

$$= \left((62.27 \text{ lbm/ft}^3) \times \frac{(32.1739 \text{ ft/sec}^2)}{32.1739 \frac{\text{ft} \cdot \text{lbm}}{\text{lbf} \cdot \text{s}^2}} \right) \times (0.13)$$

$$= 8.3 \text{ lbf/ft}^2$$

and therefore:

$$V_{duct} = \left(2 \times 32.1739 \frac{\text{ft} \cdot \text{lbm}}{\text{lbf} \cdot \text{s}^2} \times \frac{8.3 \text{ lbf/ft}^2}{0.075 \text{ lbm/ft}^3} \right)^{1/2} = 84 \text{ ft/sec.}$$

40. **(C)**

A hydraulic jump occurs in open channel flow when the fluid velocity goes from supercritical to subcritical, resulting in a sharp increase in the depth of the fluid. All flows will eventually become subcritical because viscosity will slow the flow. Therefore, an increase in volumetric flow rate will push the point at which the hydraulic jump occurs downstream. Also, the size of the jump can be derived as:

$$q^2 = \frac{h^3 \cdot g}{2}(r + r^2)$$

where q is the volumetric flow rate, h is the depth of the fluid before the jump and r is the ratio of the depth after the jump to before the jump. It is clear from the equation that if q increases while keeping h constant, then r must increase. Therefore the height difference across the jump will increase with increased volumetric flow rate.

41. **(B)**

Most flow measuring devices need to be calibrated. Both the venturi meter and the orifice meter depend on frictional head losses to produce a pressure drop which is related to velocity so they need to be calibrated. A hot-wire anemometer depends on matching heat transfer from a hot wire to a flow with a predetermined calibration curve. A weir, while accurate after calibration needs to have an experimentally determined calibration factor. Only the Pitot-Static tube, based on Bernoulli's equation, can provide velocity information without having to be calibrated.

42. **(E)**

To maintain dynamic similarity, the Reynolds number of the prototype must match the Reynolds number of the full size vessel. The Reynolds number is defined as:

$$R_e = \frac{VL}{\nu}$$

where V is the velocity of the body, L is a length scale associated with the body and ν is the kinematic viscosity of the fluid. The Reynolds number represents the ratio of inertial to viscous forces. For dynamic similarity, the Reynolds numbers should be the same. Since the viscosity will be the same for the submarine and the model, this results in:

$$V_s L_s = V_m L_m$$

and

$$V_m = \frac{120\text{ ft}}{4\text{ ft}} \times 3.5\text{ ft/sec} = 122.5\text{ ft/sec}$$

But, the drag coefficients on the two bodies must also be the same:

$$C_{fm} = \frac{F_m}{\rho_m V_s^2 L_s^2} = C_{fs} = \frac{F_s}{\rho_s V_s^2 L_s^2}$$

so that the force needed to move the full scale submarine will be (noting that the density of water falls out of the equation):

$$F_s = \frac{3.5^2 \times 12^2}{122.5^2 \times 4^2} \times 200\text{ lbf} = 146.9\text{ lbf.}$$

43. **(A)**

Impulse-momentum equations for incompressible liquids (such as water) indicate that a jet impacting on a plate will transfer its momentum such that:

$$F = MV = \rho VA \times V = \rho AV^2$$

so:

$$V = \sqrt{F/\rho A} = \sqrt{\frac{250 \times g_c}{62.4\text{ lbm/ft}^3 \times (1/24)^2 \times \pi}} = 153.7\text{ ft/sec}$$

44. **(B)**

A normal shock is assumed to be only one dimensional, that is that the shock wave is perpendicular to the direction of fluid flow. If this is not the case, the shock wave is referred to as an oblique shock wave which is the only type of shock wave which can have supersonic flow velocities on both sides of the shock. This is the same whether the fluid is air or any other compressible liquid. All shock waves have dissipation.

45. **(B)**

The surface tension of a liquid is the dominant property in determining the behavior of a fluid in a thin capillary tube. The surface tension is related both to the cohesive forces within the liquid as well as the adhesive forces of the liquid to the tube. Motion is usually slow, so the other parameters listed will have small effects.

46. **(C)**

Weight of the sphere is equal to the weight of water it displaces. The volume of water displaced is:

$$\text{Vol} = 1/2 \times (4/3\ \pi r^3) = 16.7\ \text{in}^3$$

$$\text{Weight} = \rho_{H_2O} \times \frac{g}{g_c} \times \text{Vol}$$

$$= 62.4 \times 16.7 \times (1/12)^3 = 0.6\ \text{lbf}$$

47. **(D)**

One needs to insure a Reynold's number of 2000 where:

$$R_e = \frac{Vd}{\nu}$$

Also, Flow Rate

$$M = 150\ \text{ft}^3/\text{min} = VA = V\pi d^2/4 \ \text{ so}$$

$$V = \frac{4M}{\pi d^2}$$

Solving for d gives:

$$d = \frac{4M}{\nu R_e \pi} = 33.5\ \text{ft}$$

48. **(C)**

Cavitation is the time dependent "boiling" of a liquid due to the local reduction of pressure below the saturation temperature. This normally occurs in high speed machinery such as a propeller in water, where pressures on the low speed side get so low that water changes to vapor, creates bubbles and therefore a "cavity" behind the prop.

49. **(E)**

Equation PV^n = Const can be rearranged

$$P^{1/n} \cdot V = \text{Const}$$

At $n = \infty$ $\quad P^{1/n} = 1$ and V = Const.

50. **(C)**

The critical pressure ratio for our conditions can be determined:

$$\frac{P^*}{P_0} = \left(\frac{2}{K+1}\right)^{\frac{K}{K-1}} = \left(\frac{2}{1.4+1}\right)^{\frac{1.4}{1.4-1}} = 0.5283$$

where $\quad P_0$ = stagnation pressure at the nozzle inlet (1.5 MPa)

$P*$ = critical pressure at the throat of the nozzle

Then $\quad P* = 0.6283P_0 = 0.5283 \times 1.5$ MPa = 0.792 MPa.

At pressure $P*$ = 0.792 MPa critical velocity is reached at the throat of the nozzle (Mach number M = 1) and further expansion takes place outside the converging nozzle.

Therefore combined convergent-divergent nozzle should be used to utilize pressure difference $\Delta P = P* - P_2 = 0.792 - 0.3 = 0.492$ MPa with supercritical velocity achieved at the nozzle exit.

Also note that divergent-convergent nozzles and the nozzles of constant cross section do not exist.

51. **(D)**

Turbines, compressors and pumps are well insulated devices (q = 0). Then from the first law of thermodynamics (2) at $V_1 = V_2$ and $Z_1 = Z_2$

$$W = h_1 - h_2 \quad \text{or} \quad dw = -dh$$

it follows from the property relation

$$T \cdot dS = dh - v \cdot dP \quad \text{at} \quad dq = T \cdot dS = 0 \quad \text{that } dh = V \cdot \text{dp}.$$

Substituting dh we have

$$dW = V \cdot dP.$$

52. **(A)**

The heat transfer q_{in} to the working fluid in process 2 – 2′ – 3 – 3′ is represented by area a – 2 – 2′ – 3 – 3′ – c – a, and the heat transferred from the working fluid q_{out} in process 4′ – 1 is represented by area a – 1 – 4′ – c – a. According to the second law of thermodynamics, the thermal efficiency of any cycle is defined as

$$2_t = \frac{W_{net}}{q} = \frac{q_{in} - q_{out}}{q_{in}}$$

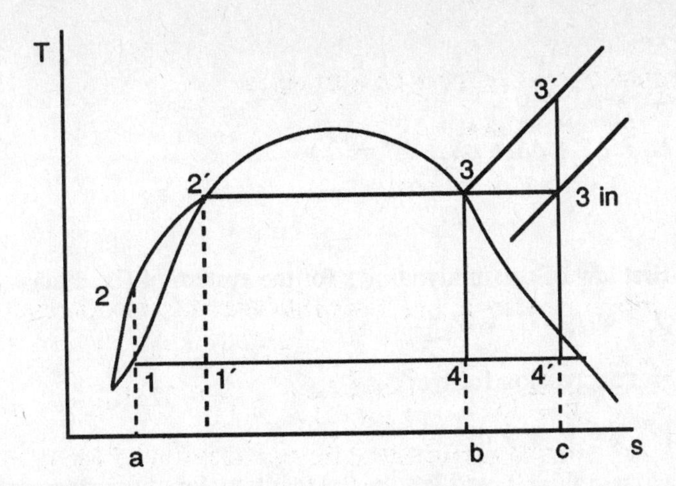

The net work of the cycle $W_{net} = q_{in} - q_{out}$ = AREA 1 - 2 - 2′ – 3 – 3′ – 4′ – 1 and

$$2_t = \frac{1 - 2 - 2' - 3 - 3' - 4' - 1}{a - 2 - 2' - 3 - 3' - c - a}$$

53. **(C)**

The thermal efficiency of the reversible Carnot cycle does not depend on the nature of working fluid and is a function of maximum and minimum temperatures of the hot and cold sources only:

$$2_t = 1 - \frac{T_{min}}{T_{max}}$$

Therefore, (see the figure above) superheating the steam (process 3 – 3′′) in the reversible Carnot cycle 1′ – 2′ – 3′′ – 4′ – 1′ causes no change in thermal efficiency of the Carnot cycle since T_{max} and T_{min} remain constant.

54. **(C)**

Consider reversible Carnot cycle 1′ – 2′ – 3 – 4 – 1′ (see figure above). Change in entropy during heat transfer to the working fluid in isothermal process 2′ – 3 is:

$$\Delta S_1 = S_3 - S_{2'}$$

Change in entropy during heat transfer from the working fluid in isothermal process 4 – 1′ is

$$\Delta S_2 = S_{1'} - S_4$$

$$\Delta S_3 = S_{2'} - S_{1'}$$

$$\Delta S_4 = S_4 - S_3$$

Since in adiabatic processes of expansion (3 – 4) and compression (1′ – 2′) $S_4 = S_3$ and $S_{1'} = S_{2'}$, for the reversible Carnot cycle

$$\Delta S_1 = -\Delta S_2$$
$$\Delta S_3 = \Delta S_4 = 0$$
$$\Delta S = \Delta S_1 + \Delta S_2 + \Delta S_3 + \Delta S_4 = 0.$$

55. **(B)**

The first law of thermodynamics for the system of fixed mass:

$$Q_{1-2} = U_2 - U_1 + W_{1-2}$$

where general expression for work is

$$W_{1-2} = \int_1^2 P \cdot dV.$$

Since V = Const, $dV = 0$, and $W_{1-2} = 0$. Heat transfer at constant volume process will change only the internal energy of the system between states 1 and 2.

56. **(E)**

If available waste heat $\overline{Q}_1$, = 1000 kw and designed heat engine has net power output $\overline{W}$ = 510 kw, thermal efficiency of the proposed engine can be determined:

$$2_t^e = \frac{\overline{W}}{\overline{Q}} = \frac{510}{1000} = 0.51$$

$$2_t^e = 51.0\%$$

Thermal efficiency of the reversible Carnot cycle at a given temperature range can be also calculated:

$$2_t^c = 1 - \frac{T_{min}}{T_{max}} = 1 - \frac{(27 + 273)}{(327 + 273)} = 1 - \frac{300}{600} = 0.50$$

$$2_t^c = 50.0\%$$

2_t^c is maximum possible thermal efficiency of any reversible and irreversible cycle at a given temperature difference. $\Delta T = T_{max} - T_{min}$.

Since $2_t^E > 2_t^c$, such a heat engine is impossible.

57. **(A)** There is a graphical solution to determine work done ON or BY the thermodynamic system during the process. Since

$$W_{1-2} = \int_1^2 P \cdot dV$$

work is represented by area under the process on $P - V$ diagram, as shown on the following page.

Therefore, during expansion 1 – 2 (work is done BY the system and is positive) the most desirable process is constant pressure process 1– $2a$ (maximum area under the process).

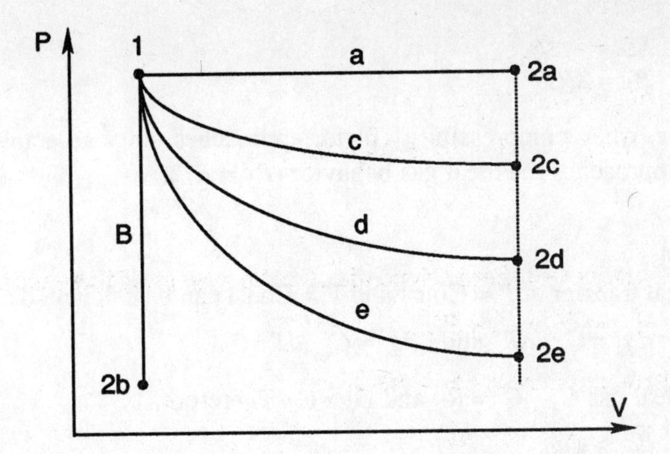

58. **(E)**

The quality of the steam in the low-pressure stage of the turbine affects significantly internal turbine efficiency and, as a result, thermal efficiency of the Rankine cycle. Also, increased moisture content can cause erosion of the last turbine stages. The basic Rankine cycle 1 – 2 – 3 – 4′ – 1 and the reheat cycle 1 – 2 – 3 – 4 – 5 – 6 are represented in the following figure.

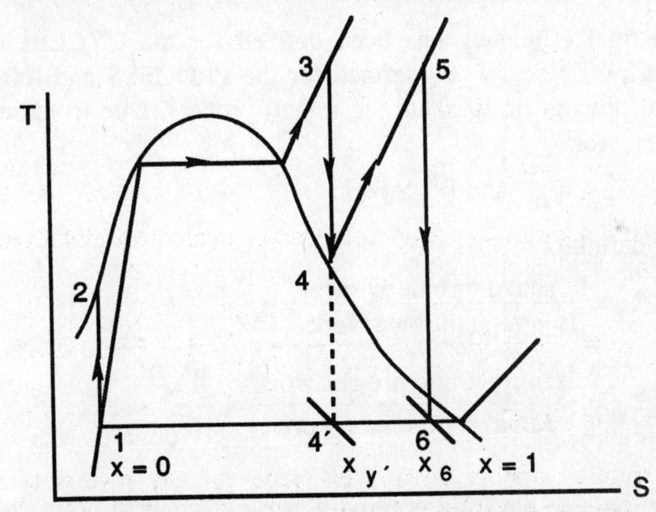

It follows from *T–S* diagram that the main advantage of the reheat cycle is in decreasing the moisture content of the working fluid in the low-pressure stages of the turbine:

$$(1 - X)_6 < (1 - X)_{4'}$$

where *X* is the quality of the steam.

59. **(D)**

Compressibility factor is defined by the relation

$$Z = \frac{P\nu}{RT}$$

and

$$Pv = ZRT$$

Therefore, when compressibility factor approaches unity an actual gas behavior closely approaches the ideal gas behavior ($Pv = RT$).

60. **(A)**

Heat transfer at P = Const and V = Const can be calculated as

$$dq_p = C_p \cdot dT \quad \text{and} \quad dq_v = C_v \cdot dT$$

For an ideal gas $C_p - C_v = R$ and $C_p > C_v$. Therefore:

$$dq_p > dq_v$$

Also, $dq_p = dU + dW$. However,

$dq_v = dU$ (since $dw = 0$). As a result:

$$dq_p > dq_v.$$

61. **(D)**

Thermal efficiency has been defined for the CYCLE. The turbine and compressor efficiencies are defined for the PROCESS and reflect deviation of the actual process from isentropic (ideal) process. Due to losses in the turbine and compressor:

$$W_T < W_{ST} \quad \text{and} \quad W_C > W_{SC}$$

Then, by definition:

$$2_T = \frac{\text{actual turbine work}}{\text{isentropic turbine work}} = \frac{W_T}{W_{ST}} < 1$$

$$2_C = \frac{\text{isentropic compressor work}}{\text{actual compressor work}} = \frac{W_{SC}}{W_C} < 1$$

62. **(B)**

The state of the fluid is fixed if two independent properties are known. It follows from T–S diagram (shown above in problem 58) that in saturation region (process $2' - 3$) constant-temperature and constant-pressure lines coincide along line $2' - 3$. In saturation region temperature and pressure are not independent properties and, therefore, they cannot be selected as a couple of properties to fix the state of the fluid.

63. **(D)**

We know that

$$V^2 = cS.$$

c is the constant of proportionality. Find the relationship between velocity and acceleration:

$$V = \frac{dS}{dt} \cdot \frac{dV}{dt}\frac{dt}{dV}$$

$$V = a \cdot \frac{dS}{dV}$$

$$\int_0^v V\, dV = \int_0^S a\, dS$$

or $V^2 = 2aS$, thus $c = 2a$.

$$a = (60)^2 / 600 = 3\ \mathbf{i}\ \text{ft/sec/sec.}$$

64. **(E)**

Convert velocity to ft/sec using ratio to miles/hr (22/15):

$$V = 45 \times (22/15) = 66\ \text{ft/sec}$$

$$a = V^2 / r = (66)^2/660 = -6.6\ i\ \text{ft/sec/sec.}$$

Because the car's speed is constant, it only has a normal component of acceleration

$$a_n = V^2 / r = (66)^2/660 = -6.6\ \mathbf{i}\ \text{ft/sec/sec.}$$

The negative sign is there because the acceleration is directed toward the center of curvature.

65. **(D)**

Using relative motion analysis applied to rotating axes, the following relationship is attained:

$$\mathbf{a}_p = \mathbf{a}_0 + \dot{\omega} \times (\omega \times \mathbf{r}) + 2\omega \times \mathbf{V}_{rel} + \mathbf{a}_{rel}$$

The terms $\mathbf{a}_0$, $\dot{\omega} \times (\omega \times \mathbf{r})$, and $\mathbf{a}_{rel}$ zeros because the disc is rotating about a fixed axis at a constant angular velocity, the slot is straight and the relative velocity is constant, thus

$$\mathbf{a}_p = 2\omega \times \mathbf{V}_{rel}$$

$$= 2\,(12\mathbf{k}) \times (2\mathbf{i}) / 12 = 4\mathbf{j}\ \text{ft/sec/sec.}$$

66. **(D)**

At the instant we are calculating, assume the wheel is pinned at point C and point B rotates around it.

$$\mathbf{V}_B = \omega \times \mathbf{CB}$$

$$\mathbf{CB} = \sqrt{1^2 + 1^2} = \sqrt{2}\ \text{ft}$$

$$\mathbf{V}_B = (10/1)\sqrt{2}\ \searrow 45^\circ$$

67. **(B)**

Equations of motion of the two masses

$$10a = T - 10g \uparrow$$

$$15a = 15g - T \downarrow$$

adding the two equations gives

$$25a = 5g$$

or $a = g/5 = 1.962$, or 2 m/sec/sec ↑ for the 10 kg mass.

68. **(B)**

Using conservation of energy,

$$K_1 + U_1 = K_2 + U_2$$

$$\tfrac{1}{2} MV^2 = MgH$$

$$0.5\, M(2.3)^2 = Mg\, 2(1 - \cos\theta).$$

Solving gives $\theta = 30°$.

69. **(B)**

$$V = 30(22/15) = 44 \text{ ft/sec}$$

Equate forces on the mass m in the x direction:

$$ma = \mu_s mg$$

$$a = 32.2(0.3) = -9.66 \text{ ft/sec/sec.}$$

As in problem 63, find the relationship between velocity and acceleration:

$$\int_v^0 V\, dV = \int_0^S a\, dS$$

$$-(44)^2/2 = -9.66\, S$$

$$S = 100 \text{ ft.}$$

70. **(C)**

From the definition, the velocity of the particle just before it hits the floor:

$$V_a = \sqrt{2\,\text{gh}}\sqrt{2(9.81)2} = 6.26 \text{ meter/ sec}$$

$$e = \frac{\text{rel. vel. of rebound}}{-\text{rel. vel. of approach}}$$

$$0.87 = -V_r/(-V_a)$$

$$V_r = 0.87(6.26) = 5.45 \text{ meter/ sec}$$

$$= \sqrt{2 gh_1},$$

which gives $h_1 = 1.5$ meter.

71. **(A)**
Conservation of linear momentum gives

momentum before impact = momentum after impact

$0.05(600) = (2 + 0.05)\ V$

$V = 14.634$ meter/sec.

Energy conservation:

$0.5\ mv^2 = 0.5KX^2$

$X^2 = 2.05(14.634)^2/20{,}000$
$= 0.02195$

$X = 0.148$ meter = 15 cm.

72. **(D)**
Conservation of angular momentum:

$r_1 \times m\mathrm{V}_1 = r_2 \times mV_2$

$0.36\ (m5) = 0.45(mV_2)$

$V_2 = 4$ meter/sec.

73. **(D)**

$ma = \Sigma F$

$[(82 + 46) / 32.2)a = 160 - (82 + 46)$

$a = 8.05$ ft/sec/sec

Moments about B:

$mad = \Sigma M_B$

$(46 / 32.2)\ 8.05(4) = M_B - (46)4$

$M_B = 46 + 184 = 230$ lb-ft CCW.

74. **(B)**
Use the relationship between the turning moment and acceleration:

$I_0\alpha = \Sigma M_0$

Find the moment of inertia, I_0, of the disk about point 0 using the parallel axis theorem:

$$I_0 = {}^1/_2\, mr^2 + md^2.$$

$$a = 38 / [0.5(20/32.2)(1.5)^2 + (20/32.2)3^2)]$$

$$= 6 \text{ rad/sec/sec } CCW.$$

75. **(B)**

Applying impulse momentum relationship gives:

$$[F - \mu_s mg]\,\Delta t = m\Delta V$$

$$[100 - 0.3(20)\,9.81]2 = 20(V - 0)$$

$$V = 4.1 \text{ m/sec}$$

This problem can also be solved by using Newton's second law to determine the acceleration, and from $dV/dt = a$, the velocity can be determined.

76. **(C)**

$$K = 0.5I\omega^2 + 0.5mV^2$$

$$= 0.5(2/2)(0.5)^2(10)^2 + 0.5(2)(2.5)^2$$

$$= 12.5 + 6.25 = 18.75 \text{ joules.}$$

77. **(E)**

Tension force in the rope is 30 pounds. Next, determine the weight of the block. Draw a free body diagram and sum forces along the plane.

30 lb
W sin 35°
W cos 35°
35°
n

$$\Sigma F = 0$$

$$W(\sin 35°) - 30 = 0$$

$$W = 30 / \sin 35° = 52.3 \text{ lbs}$$

78. **(B)**

Two free body diagrams are needed to find the forces in all ropes. Draw a free body diagram for the 180 pound weight, and next draw the free body diagram for the 65 pound weight.

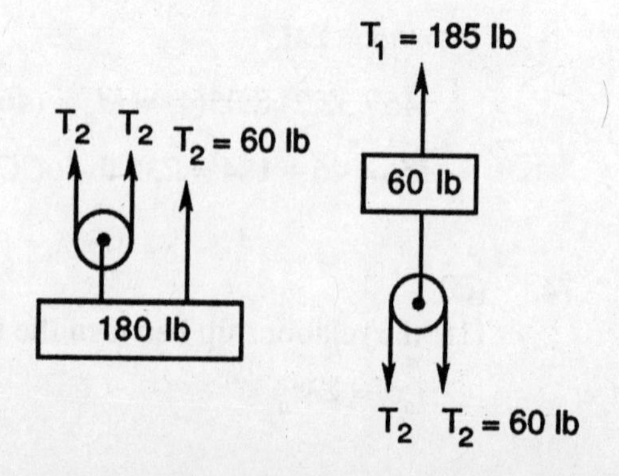

$$T_1 = 2T_2 + 65 \text{ lbs.}$$

79. **(D)**

Draw the free body diagram of point C and sum forces.

$$\Sigma F_y = 0: \; B_y = 60 \text{ N}$$

$$B_x = B_y (\tan 35°) = 42 \text{ N}$$

$$\Sigma F_x = 0: \; 42 - A = 0$$

$$A = 42 \text{ N}$$

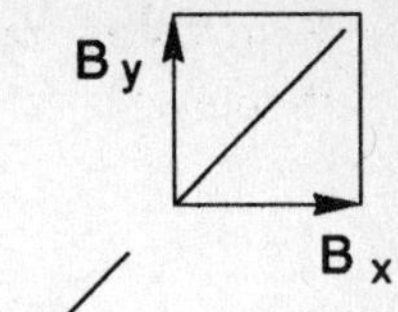

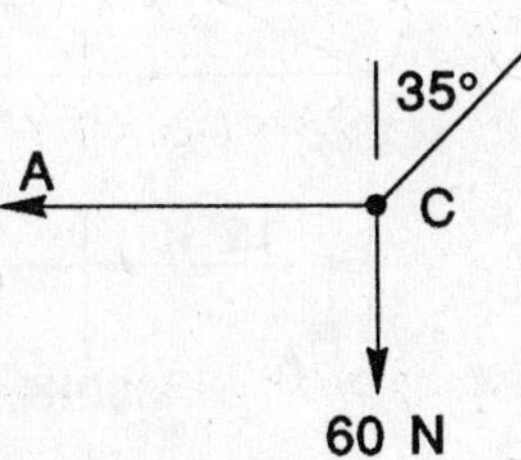

80. **(A)**

Use the method of joints at reaction A

$$\Sigma M_B = 0: \; R_A \times 24 - 10 \times 18 - 10 \times 12 - 10 \times 6 = 0$$

$$R_A = 15$$

$$\Sigma F_y = 0: \; AB_y = 15 \text{ k and } AB_x = 2AB_y = 30 \text{ k}$$

$$\Sigma F_x = 0: \; AF = 30 \text{ k tension}$$

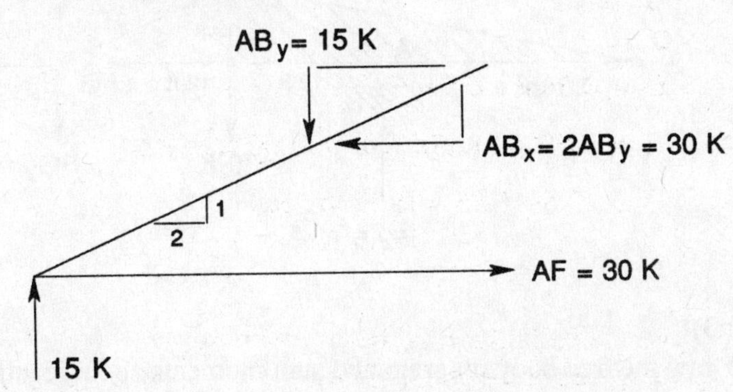

81. **(C)**

Draw two free body diagrams and work with portion CD. Because of symmetry

$$C_y = D_y = \frac{6 \times 4}{2} = 12 \text{ KN}$$

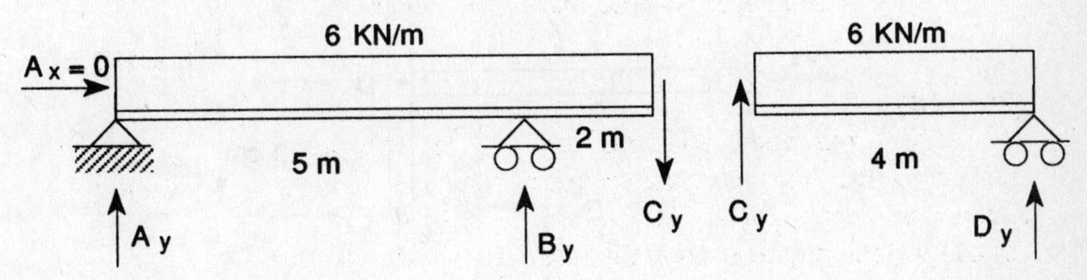

82. **(E)**

Find the reaction at A.

$\Sigma M_E = 0$: $60 \times 12 + 30 \times 24 + 30 \times 36 - R_A(48) = 0$
$R_A = 52.2$ k

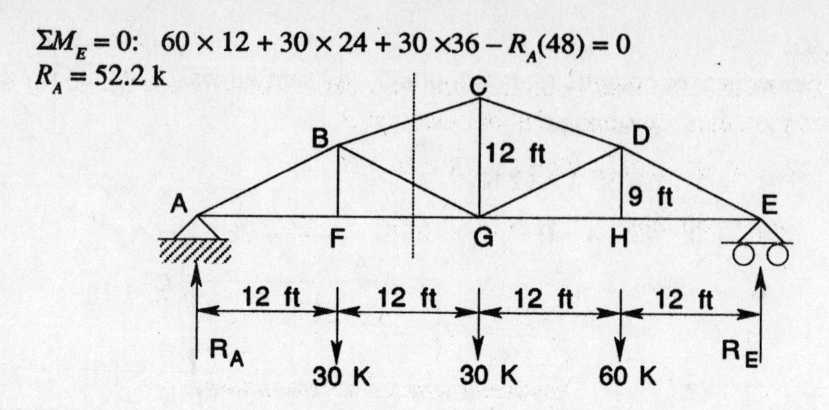

Then solve the force *BG* by the method of sections. Find the intersection at point *O* of the other two unknowns *BC* and *FG*. Sum moments about point O.

$$\Sigma M_o = 0: \quad 36 \times 30 + 48BG_y - 24 \times 52.5 = 0$$

$$BG_y = 3.75 \text{ k and } BG = (5/3)\,3.75 = 6.25 \text{ k tension}$$

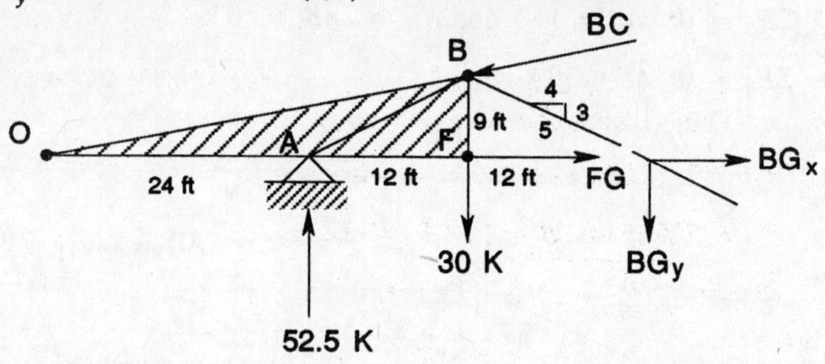

83. **(D)**
Draw the free body diagram and sum moments about reactions *A*.

$$\Sigma M_A = 0: \quad 12 \times 800 - 4E_x = 0$$

$$E_x = 2400 \text{ N}$$

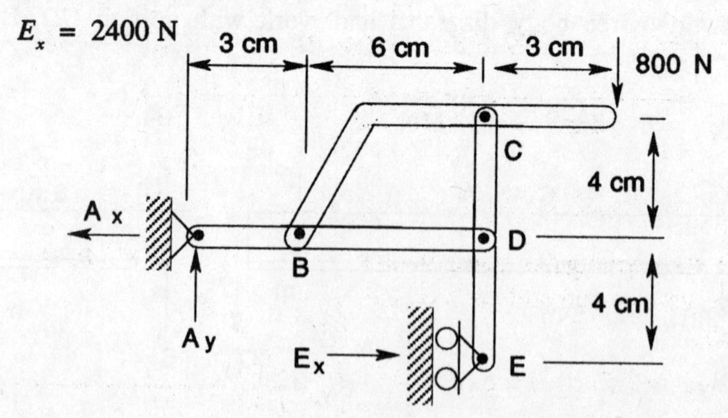

84. **(C)**
The weight will slide when the slope angle exceeds the angle of friction. For this problem the friction angle is 35 degrees where $\tan\phi = 0.70$.

85. **(B)**

The distributed load is broken into two parts. Total load is found for each part, and moments are summed about reaction *B*.

$\Sigma M_B = 0\text{:}\;\; 24R_A - 72 \times 8 - 48 \times 16 = 0$

$R_A = 56$ kips

$\frac{4 \times 24}{2} = 48$ K $\quad \frac{6 \times 24}{2} = 72$ K

4 K/ft 6 K/ft

A B

8 ft 8 ft 8 ft

R_A R_B

86. **(D)**

Draw the free body diagram and sum moments about reaction *A* to find B_y. The horizontal component of reaction *B* is $^1/_2 B_y$.

$\Sigma M_a = 0\text{:}\;\; 200 \times 5 + 200 \times 6 - 10B_y = 0$

$B_y = 220$ N and $B_x = (^1/_2)\, B_y = 110$ N

$B = \sqrt{220^2 + 110^2} = 246$ N

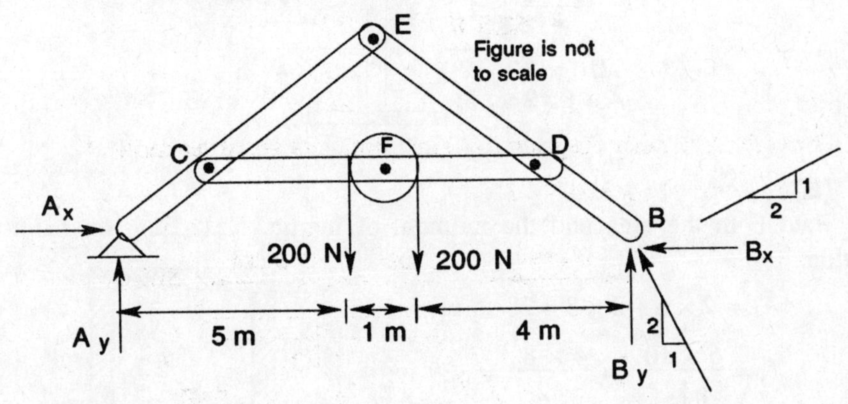

87. **(C)**

Find the rectangular components of the three vectors and add each of the components.

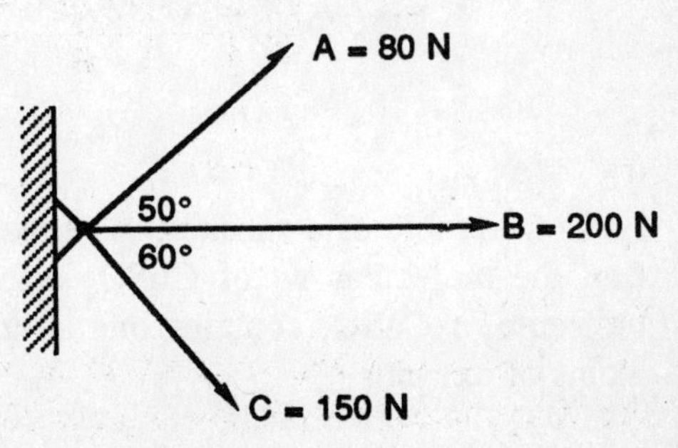

$A = 51.4i + 61.3j$ N

$B = 200\, i$ N

$C = 75Ii - 129.9j$ N

$R = 326.4i - 68.6j$ N

88. **(A)**

Draw the free body diagram and sum moments about reaction B.

$$\Sigma M_B = 0: \; 14A_y - 10 \times 40 - 2 \times 60 = 0$$

$$A_y = 37.1 \text{ kips}$$

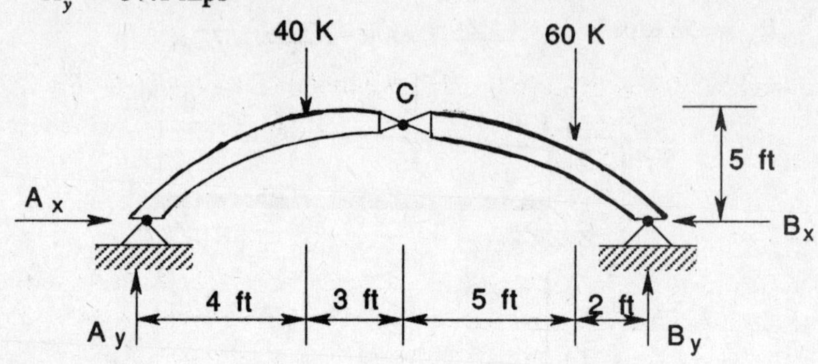

89. **(D)**

Sum moments of area about the center of the original beam area.

$$\overline{Y} = \frac{\Sigma A\overline{Y}}{\Sigma A} = \frac{2.5(10.17/2 + 0.25)}{6.49 + 2.5} = 1.484 \text{ in}$$

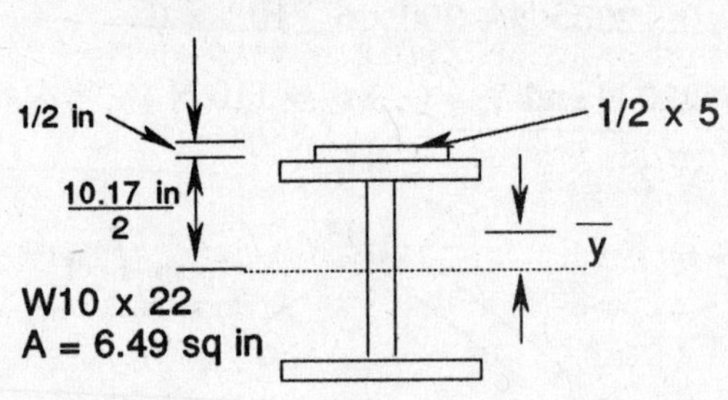

90. **(E)**

Find both the area and the moment of inertia. Next find the radius of gyration.

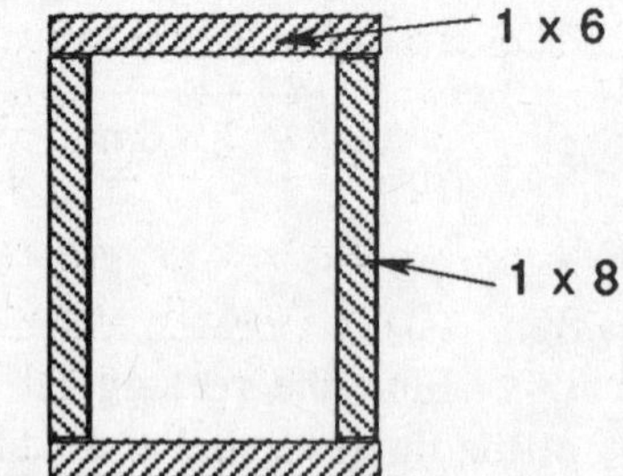

$$A = 2 \times 6 + 2 \times 8 = 28 \text{ sq-in}$$

$$I_x = \frac{6 \times 10^3}{12} = \frac{4 \times 8^3}{12} = 329 \text{ in}^3$$

$$k_x = \sqrt{I_x/A} = \sqrt{329/28} = 3.43 \text{ in.}$$

91. **(A)**

One mole of a substance is equal to its molecular wt expressed in g. First find the molecular wt of $CuSO_4$ using atomic wts of respective elements. A molecule of $CuSO_4$ contains one atom of copper, one atom of sulfur and four atoms of oxygen.

$$1 \times Cu = 1 \times 63.5 = 63.5$$
$$1 \times S = 1 \times 32.0 = 32.0$$
$$4 \times O = 4 \times 16.0 = 64.0$$
$$159.5$$

Thus, the molecular wt of $CuSO_4$ is 159.5 g/mole. Now convert 6.38 g to moles.

$$6.38 \text{ g} \times \frac{1 \text{ mole}}{159.5 \text{ g}} = 0.04 \text{ mole}$$

92. **(C)**

$AgNO_3$ and $Cu(NO_3)_2$ are ionic substances. In $AgNO_3$, silver ion (Ag^+) is the positive ion or cation, and nitrate ion (NO^-_3) is the negative ion or anion. Cu by itself, and Ag by itself, are in metallic form. The reaction shows Cu metal displacing NO^-_3 from $AgNO_3$. Only one ion (nitrate) is displaced. Thus, this reaction is an example of a single displacement reaction.

93. **(E)**

If a non-volatile solute is dissolved in a solvent, the vapor pressure of the solution at any given temperature is lower than the vapor pressure of pure solvent. This property results in elevation (raising) of boiling point and depression (lowering) of freezing point of pure solvent.

Ethylene glycol is a liquid of very low vapor pressure as compared to water. A solution of ethylene glycol in water is commonly used as antifreeze in car radiators. The freezing point depression (Δf) is equal to the freezing point constant of the solvent (K_f) times the molality (m) of the solution.

$$\Delta f = K_f \cdot m \qquad (1)$$

Molality is defined as moles of solute per Kg of solvent. Using atomic weights, the molecular weight of ethylene glycol ($C_2H_6O_2$) is found to be 62.0 g/mole. Therefore, 93.0 g of ethylene glycol represents

$$93.0 \text{ g} \times \frac{1 \text{ mole}}{62.0 \text{ g}} = 1.50$$

moles of ethylene glycol. The molality of the solution is:

$$\frac{1.50 \text{ mole solute}}{0.750 \text{ Kg solvent}} = 2.00 \text{ mole/Kg}$$

(Remember that the mass of solvent is expressed in Kg in calculating molality).

Now, using the (1) equation shown above,

$$\Delta^f = 0.81 \frac{\text{deg C kg}}{\text{mole}} \times 2.00 \frac{\text{mole}}{\text{kg}} = 1.62 \text{ deg C.}$$

Thus, the freezing point depression is 1.62° C. The freezing point of pure water is 0.00° C. Therefore, the freezing point of the solution is – 1.62° C.

94. **(D)**

The equivalent weight of an element is defined as that weight which reacts with, or generates, 1.00 g of hydrogen, 16.00 g of oxygen, 32.00 g of sulfur, 35.5 g of chlorine, etc.

Let us first find the mass of hydrogen evolved, using the General Gas Equation, which s

$$PV = nRT$$

or $$Pv = \frac{mRT}{\text{M.W.}}$$

Where, P = pressure of gas in atm

V = volume of gas in lit

m = mass of gas in g

R = the Gas Constant = 0.082 lit atm/deg mole

M.W. = Molecular wt of gas in g/mole

Rearranging the above equation,

$$m = \frac{(P)\,(V)\,(\text{M.W.})}{(R)\,(T)}$$

$$= \frac{(1.0 \text{ atm})\,(0.0246 \text{ lit})\,(2.00 \text{ g/mole})}{(0.082 \text{ lit atm/deg mole})\,(300 \text{ deg})}$$

$$= 0.002 \text{ g}$$

This amount of hydrogen is produced by 0.180 g of the metal. Find the mass of the metal which would produce 1.00 g of hydrogen.

$$\frac{0.180 \text{ g metal}}{0.002 \text{ g hydrogen}} = 9.00 \text{ g}$$

metal per g of H_2.

Therefore, the equivalent wt of the metal is 9.00.

95. **(D)**

The K shell (the lowest energy shell) octate consists of a capacity for 2 electrons. Octates of other shells represent a capacity for 8 electrons. Elements are placed in various groups according to the number of electrons present in the outermost octates of their atoms. Atoms of Group VIII (also called group O) elements contain complete outermost octates.

Chlorine (Cl) belongs to Group VII A. Therefore, an atom of Cl contains 7 electrons in its outermost octate.

96. **(B)**

Normality (N) of H_2SO_4 can be found using the equation:

$$(N_A)(V_A) = (N_B)(V_B)$$

where, N_A = Normality of acid

V_A = Volume of acid

N_b = Normality of base

V_b = Volume of base

Make sure to use same unit for volume on both sides of the equation.

Substituting,

$$(N_A)(12.0 \text{ ml}) = (0.15 \text{ N})(15.0 \text{ ml})$$

$$N_A = 0.187$$

The normality (N) and molarity (M) are two different ways of expressing concentration. For HCl, HNO_3, NaOH or KOH, the normality is the same as the molarity, because each of these provide one mole of H^+ or one mole of OH^- per mole of the acid or base. Sulfuric acid (H_2SO_4) however provides 2 moles of H^+ per mole of the acid ($H_2SO_4 \rightarrow 2H^+ + SO_4^{-2}$). Therefore, the molarity of H_2SO_4 is half its normality.

Thus, M of

$$H_2SO_4 = \frac{0.187}{2} = 0.094$$

97. **(C)**

A chemical equation is balanced if both sides of the equation show the same number of atoms of each element. During the process of balancing an equation, one can change the coefficient (i.e., the number before the formula) of a given substance in the reaction; but cannot change its formula.

In the present equation, there are two Fe atoms on the left, therefore, we can change the coefficient of Fe on the right side to 2.

On the left side, we have 3 oxygen atoms which go to form CO_2. From 3 oxygen atoms, we get 1.5 molecules of CO_2. So the coefficient of CO_2 becomes 1.5. For 1.5 molecules of CO_2, we require 1.5 atoms of C, thus the coefficient of C becomes 1.5 atoms of C, thus the coefficient of C becomes 1.5.

$$Fe_2O_3 + 1.5\,C \rightarrow 2Fe + 1.5\,CO_2$$

Traditionally, a balanced equation is written using the smallest ratio of molecules in whole numbers. Therefore, the above equation will have to be doubled.

$$2\,Fe_2O_3 + 3C \rightarrow 4Fe + 3CO_2$$

The coefficient of CO_2 in the balanced equation is 3.

98. **(D)**

This is a simple red-ox equation which can be balanced by inspection as described in answer to question 8. However, red-ox equations are easier to balance by using "electron-balance method." The procedure is as follows:

First write the oxidation numbers of each element on both sides of the equation. A knowledge of oxidation numbers of common elements is extremely useful. It is also very helpful to know that the total oxidation number of a molecule or the formula is zero. The oxidation number of an element by itself is also zero.

$$Mn^{+4} O_2^{-4} + H^{+1} Cl^{-1} \rightarrow Mn^{+2} Cl_2 + H_2^{+2} O^{-2} + Cl_2^{0}$$

Inspect the oxidation number of each element *per atom* on both sides of the equation. Which elements have changed their oxidation numbers? Those elements that have changed the oxidation numbers have taken part in the reduction and oxidation.

$$Mn^{+4} \xrightarrow{+2e} Mn^{+2} \quad \text{Half equation for reduction}$$

$$2\,Cl^{-} \xrightarrow{-2e} Cl_2^{0} \quad \text{Half equation for oxidation}$$

(Note: reduction is defined as gain of electron(s), while oxidation is defined as loss of electron(s).)

We can multiply each of the above half equations by any factor so that the total number of electrons lost is equal to the total number of electrons gained.

In the above half equations, we already have 2 electrons gained and 2 electrons lost. Now adding both equations, we get:

$$Mn^{+4} + 2Cl^{-} \rightarrow Mn^{+2} + Cl_2^{0}$$

This completes the balancing of the red-ox part of the equation. The rest of the equation must be balanced by inspection.

We need 2 more Cl^- ions from left to form $MnCl_2$ on the right.

$$MnO_2 + 2Cl^{-} + 2Cl^{-} \rightarrow MnCl_2 + Cl_2^{0}$$

Total of 4 Cl^- come from 4 HCl

$$MnO_2 + 4\,HCl \rightarrow MnCl_2 + Cl_2$$

How many H_2O molecules are formed? We have 2 O atoms and 4 H atoms on the left side. These form 2 H_2O.

Therefore, the complete balanced equation is:

$$MnO_2 + 4\,HCl \rightarrow MnCl_2 + Cl_2 + 2\,H_2O$$

The coefficient of HCl in the balanced equation is 4.

99. **(C)**

The amount of a radioactive isotope can be expressed in terms of its mass (such as mg or ng) or in terms of its activity (such as counts/min). The half-life of a radioactive isotope is defined as the time it takes to reduce the original amount or activity by half. Thus, each half-life cycle reduces the amount by half.

The following equation can be used:

$$\text{Remainint amount or activity} = \frac{\text{Original amount or activity}}{2^n}$$

where n = the number of half-life cycles. To find out how many half-life cycles are needed to reduce original amount or activity to 1%, we substitute in the above equation.

$$1 = \frac{100}{2^n} \quad or \quad 2^n = 100$$

$$n = 6.65$$

That is, it takes 6.65 half-life cycles to reduce the original amount or activity by 99%.

Each half-life cycle is 4.3 days.

$$6.65 \text{ cycles} \times \frac{4.3 \text{ days}}{1 \text{ cycle}} = 28.6 \text{ days}$$

is the required elapsed time.

100. **(B)**

First let us look at some fundamental sub-atomic particles.

	Mass	Charge
Proton (H)	1.00 amu	+ 1
Neutron (N)	1.00 amu	0
Electron ($_{-1}^{0}\beta$)	1/1800 amu	− 1
Positron ($_{+1}^{0}\beta$)	1/1800 amu	+ 1

Now looking at the equation,

$$_{4}^{9}Be - ? \rightarrow {}_{5}^{9}Be$$

We find that the mass of the nucleus has remained the same (i.e., 9). However, the atomic number (which is the number of protons in the nucleus) is changed from 4 to 5. The original nucleus has 4 protons and 5 neutrons. The nucleus produced has 5 protons and 4 neutrons. Obviously, the number of protons has increased by one, while the number of neutrons has decreased by one. What sub-atomic particle much be removed from a neutron to convert it to a proton? An electron. The answer is $_{-1}^{0}\beta$.

101. **(B)**

Nuclear reactions often involve loss in mass as the reactants form products. This loss in mass is called the "mass defect." According to Einstein's theory of relativity, energy is produced at the cost of the mass lost.

Let us first find what the mass defect is. Using atomic wts which are given, the total mass on the left side of the equation is 6.0151 + 2.0141 = 8.0292. On the right side, we have two atoms of ^{4}He. The total mass on the right = 2 ×

4.0026 = 8.0052. The mass defect = 8.0292 – 8.0052 = 0.0240 g per 8.0292 g of the fuel. Therefore, per 1.00 g of the fuel, the mass defect is 0.0030 g.

Now substituting in Einstein's famous equation,

$$E = mc^2$$

where m = mass defect in g

c = the velocity of light

$= 3.0 \times 10^{10}$ cm/sec

$$E = (0.0030 \text{ g})(3.0 \times 10^{10} \text{ cm/sec})^2$$
$$= 2.7 \times 10^{18} \text{ gcm}^2/\text{sec}^2 = 2.7 \times 10^{18} \text{ ergs}$$

(Note that the units gcm^2/sec^2 is the same as the unit erg)

Now convert ergs to kJ

$$(2.7 \times 10^{18} \text{ ergs})\left(\frac{1 \text{ J}}{1.0 \times 10^{7} \text{ ergs}}\right)\left(\frac{1.0 \text{ kJ}}{1.0 \times 10^{3} \text{ J}}\right) = 2.7 \times 10^{8} \text{ kJ}$$

102. **(C)**

Molarity (M) is defined as moles of solute per lit of solution.

$$\text{M} = \frac{\text{moles of solute}}{\text{lit of solution}}$$

Moles of solute, in turn, is equal to the g of solute divided by its molecular wt or formula wt. Therefore, the above equation can be written as:

$$\text{M} = \frac{\text{g solute}}{(\text{M. W. solute})\ (\text{lit of solution})}$$

Using atomic wts, the M.W. of $Na_2S_2O_3$ comes to 158.0 g/mole.

Substituting in the above equation and solving for g of solute

$= (\text{M})(\text{M.W. solute})(\text{lit of solution})$

$= (0.4 \text{ moles/lit})(158.0 \text{ g/moles})(0.500 \text{ lit})$

$= 31.6$ g.

103. **(B)**

The electronic shells, K, L, M, N, O, etc. can also be referred to by using Principal Quantum Numbers 1, 2, 3, 4, 5, etc., respectively. Higher the Principal Quantum Number of an electron, the higher is its energy.

The ionization energy of an electron (i.e., the energy required to remove it from an atom) is given by:

$$E = -R\left(\frac{1}{n^2}\right)$$

where R = the Rydberg Constant

If an electron relaxes (i.e., comes from a higher energy level to a lower energy level), it gives out energy in the form of electro-magnetic radiation (photon or quantum), of energy equal to the energy difference of the electron while in $n_{initial}$ and n_{final}.

In this problem

$$E = 2.179 \times 10^{-18}\ \text{J/photon}\ \left(\frac{1}{n^2_{final}} - \frac{1}{n^2_{initial}}\right)$$

$$= 2.179 \times 10^{-18}\ \text{J/photon}\ \left(\frac{1}{1^2} - \frac{1}{3^2}\right)$$

The frequency of the emitted photon can be found as follows:

$E = hf,$

where h = the Planck's Constant and

f = frequency

1.94×10^{-18} J/photon = $(6.62 \times 10^{-34}$ J. sec/photon) (f)

$f = 2.93 \times 10^{15}$ / sec.

104. **(D)**

Let us first find the heat of this reaction per mole. If the structures are written as follows, it is easier to see which bonds are broken and which bonds are made.

```
     H   H   H                                        H
     |   |   |                                       / \
 H — C — C — C — H + 5 O = O → 3 O = C = O + 4 H      O
     |   |   |
     H   H   H
```

The process of breaking the bonds is endothermic (ΔH = +), while the process of making the bonds exothermic (ΔH = –).

Bonds Broken:

8 C – H	= 8 × 410	= + 3280 kJ
2 C – C	= 2 × 334	= + 668 kJ
5 O = O	= 5 × 497	= + 2485 kJ
Total		= + 6433 kJ (1)

Bonds Formed:

6 C = O	= 6 × 731	= – 4386 kJ
8 H – O	= 10 × 456	= – 3648 kJ
Total		= – 8034 kJ (2)

The overall ΔH of the reaction is the algebraic sum if (1) and (2)

ΔH = – 1601 kJ / mole of propane.

Using atomic wts, we find that the molecular wt of propane is 44 g/mole.

1.00 lb of propane represents 10.32 moles of propane.

Total heat generated by burning 20.32 moles of propane = 10.32 moles × 1601 kJ/mole = 1.65×10^4 kJ.

105. **(C)**

The twist angle is converted to radians:

$$\phi = 2° = 2\left(\frac{2\pi}{360}\right) = 0.03489 \text{ Rad.}$$

$r_1 = 0.5$ in; $r_2 = 1$ in

Polar Moment of Inertia:

$$J = \frac{\pi}{2}\left(r_2^4 - r_1^4\right)$$

$$J = \frac{\pi}{2}\left((1\,\text{in})^4 - (0.5\,\text{in})^4\right) = 1.47\,\text{in}^4$$

$$T = \frac{GJ}{L}\phi = \frac{(40,000\,\text{psi})\,(1.47\,\text{in}^4)}{12\text{in}}\,(0.3489)$$

$$= 171\ \text{lb} \cdot \text{in}.$$

106. **(A)**

$\Sigma M_c = 0$; R_A (6 in) – 10 lb (4 in) = 0 gives $R_A = 6.66$ lb

Hence $R_c = 10$ lb – 6.66 lb = 3.34 lb.

Starting from the left, upward force is positive shear. Hence, between *A* and *B* shear is +6.66 lbs. At *B*, shear drops by 10 lbs; hence from *B* to *C* shear is 6.66 – 10 = – 3.34 lbs.

Also, since there is no external force between *A* and *B*, shear must remain constant at + 6.66 lbs. For similar reason constant shear – 3.34 lb acts between *B* and *C*.

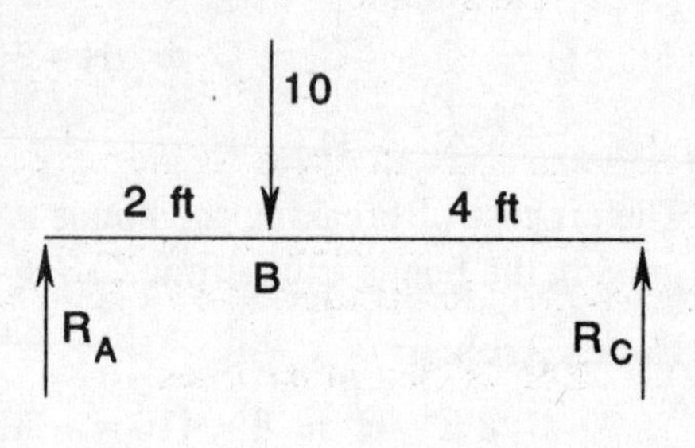

107. **(D)**

$$V = \frac{dM}{dx},$$

where V = shear force and

M = bending moment.

$dM = V\,dx$;

$\phi M = V\,dx$

means increase of moment is equal to V vs x diagram. Also, since V is constant (problem 106), moment would change linearly along x. This rejects (B) and (E).

$M = 0$ at A (due to hinge support); hence moment at B is area of shear diagram (problem 106) up to B = (6.66 lb) (2 in) = 13.33 lb·in. Furthermore, moment at $C = 0$ (roller support). This makes the correct answer (D).

Note:

+ −

The relation of + and – moment with curvature is shown.

Alternate Method: Moment from A to B;

$$M_{AB} = R_A X;$$

a straight line with a value of 0 at $x = 0$ (Point A) and a value of (6.66 lb) (2 in) = 13.33 lb·m at $x = 2$ (Point B).

Moment from B to C;

$$M_{BC} = R_A X - 10(x - 2)$$

$$M_{BC} = -3.34\,x + 20;$$

a straight line with negative slope as shown in (D).

108. **(A)**

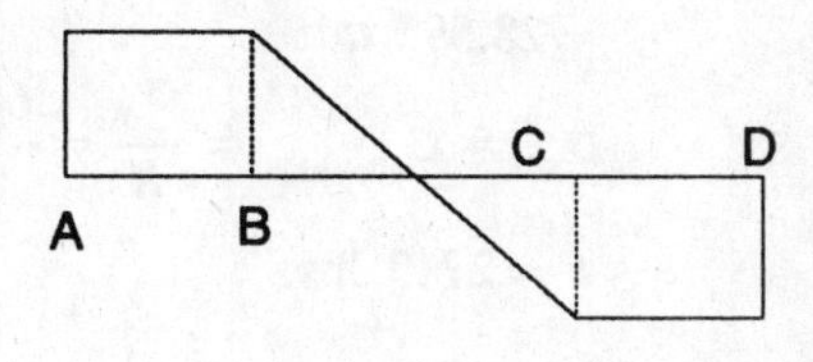

BM (bending moment) diagrams are understood well with a good understanding of SF (shear force) diagrams.

From symmetry; end reaction

$$\uparrow R_A = \uparrow R_B = \frac{1}{2}\left[\omega \cdot \frac{L}{3}\right] = \frac{\omega L}{6} = \text{constant}$$

The shear force diagram on the right suggests

1) Linear variation of moment from A to B *and* C to D (Note: BM is obtained by integration of *SF*). Hence, reject (C), (D) and (E).

2) BM is maximum when

$$\frac{dM}{dx} = 0\ ;\ \text{but}\ \frac{dM}{dx} = V$$

From SF diagram, maximum BM will be halfway between B and C. Hence (A) is the right choice.

Note: BM equation from B to C;

$$M = R_A x - \omega\left(x - \frac{L}{3}\right)\left[\frac{1}{2}\left(x - \frac{L}{3}\right)\right]$$

BM equation from A to B does not contain the second term making the variation linear up to B. The second term (quadratic in x) makes the variation parabolic from B to C.

109. **(E)**

Location of Neutral Axis:

$$\overline{Y} = \frac{\Sigma A_i y_i}{\Sigma A_i} = \frac{(3\text{ in} \times 2\text{ in})\,(1.5\text{ in}) + (2\text{ in} \times 4\text{ in})(3\text{ in} + 1\text{ in})}{(3\text{ in} \times 2\text{ in}) + (2\text{ in} \times 4\text{ in})}$$

$$= 2.928\text{ in}$$

N.A. is located 2.928 in from bottom and (5 in – 2.928 in) = 2.072 in from top.

Stress Equation:

$$\sigma_{mx} = \frac{M_{mx} C_{mx}}{I};$$

$$M_{mx} = (6\text{ ft} \times 12\text{ in/ft})\, F = 72\, F\text{ lb·in}$$

and $C_{mx} = Y_{mx}$

(on compression side, i.e., bottom surface of beam) = 2.928 in.

Moment of Inertia about N.A.

$$= {}^1/_{12}\,(4\text{ in} \times (2\text{ in})^3) + (4\text{ in} \times 2\text{ in})\,(2.072–1\text{ in} – 1\text{ in})^2$$

$$+ {}^1/_{12}\,(2\text{ in} \times (3\text{ in})^3) + (2\text{ in} \times 3\text{ in})\,(2.928\text{ in} – 1.5\text{ in})^2$$

$$= 28.595\text{ in}^4$$

$$\sigma_{mx} = \sigma_{\text{allowable}} = \frac{\sigma_y}{K} = \frac{60{,}000\text{ psi}}{3} = 20000\text{ psi} = \frac{72\,\text{F}(2.928)}{28.595}$$

$$F = 2713\text{ lbs.}$$

110. **(C)**

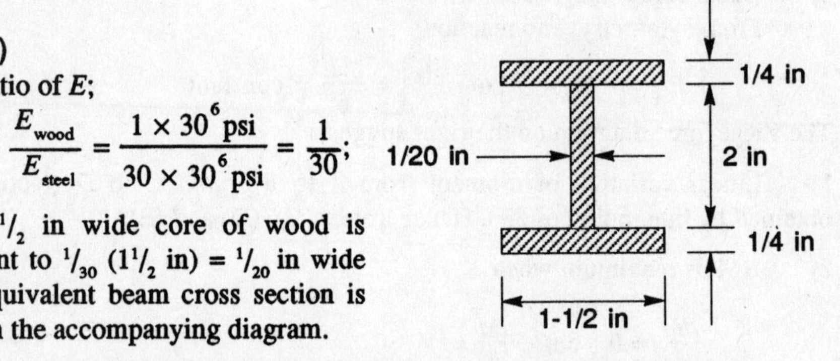

Ratio of E;

$$\frac{E_{\text{wood}}}{E_{\text{steel}}} = \frac{1 \times 30^6\text{ psi}}{30 \times 30^6\text{ psi}} = \frac{1}{30};$$

hence $1^1/_2$ in wide core of wood is equivalent to ${}^1/_{30}$ ($1^1/_2$ in) = ${}^1/_{20}$ in wide steel. Equivalent beam cross section is shown in the accompanying diagram.

$$I_{\text{NA}} = ({}^1/_{12})\,(1.5\text{ in})\,(2.5\text{ in})^3 – 2\,[({}^1/_{12})\,(.75\text{ in} – .1\text{ in})\,(2\text{ in})^3]$$

$$= 1.086\text{ in}^4$$

$$\sigma_{\text{mx}} = \frac{M_{\text{mx}} C}{I} = \frac{(1000\text{ lb} \times 12\text{ in})(1.25\text{ in})}{(1.086\text{ in}^4)} = 13{,}812\text{ psi.}$$

111. **(C)**

The Principal Stresses are given by:

$$\sigma_{mx.min} = \left[\frac{\sigma_x + \sigma_y}{2}\right] \pm \sqrt{\left(\frac{\sigma_x - \sigma_y}{2}\right)^2 + \tau_{xy}^{\ 2}}$$

$$= \left[\frac{-20,000 + 30,000}{2}\right] \pm \sqrt{\left(\frac{-20,000 - 30,000}{2}\right)^2 + (10,000)^2}$$

$$= 5,000 \pm 26,900$$

$$= 31,900 \text{ psi and } -21,900 \text{ psi}$$

Note:

$$\tau_{mx} = \sqrt{\left(\frac{\sigma_x - \sigma_y}{2}\right)^2 + \tau_{xy}^{\ 2}} = 26,900 \text{ psi}$$

112. **(E)**

Principal planes are at angle θ_p where,

$$\tan 2\theta_p = \frac{2\tau_{mx}}{\sigma_x - \sigma_y}.$$

From Problem 111;

$$\tan 2\theta_p = \frac{2(26,900)}{-20,000 - 30,000} = -1.076$$

$$2\theta_p = \tan^{-1}(-1.076) = -47° \text{ and } -227°$$

Hence, $\theta_p = -23.5°$ and $-113.5°$

Note:

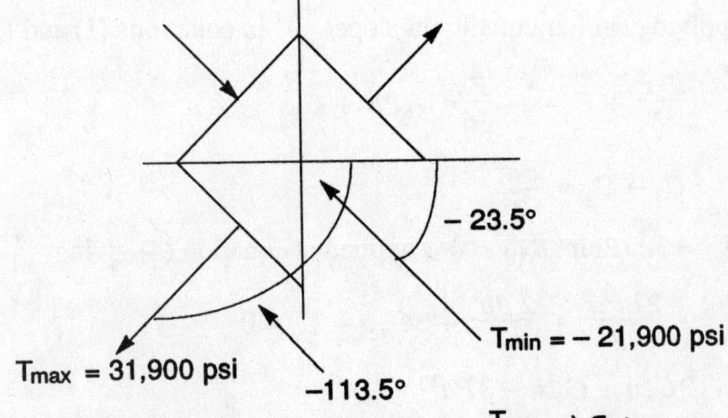

113. **(D)**

Reaction at A;

$$\Sigma M_B = 0; \quad R_A(3a) - P(2a) = 0; \uparrow R_A = {}^2/_3\, P$$

Between 0 x a (A to C) $M = {}^2/_3\, Px$; hence,

$$EI\frac{d^2y}{dx^2} = \frac{2}{3}Px$$

Integrating;

$$EI\,\frac{dy}{dx} = \frac{Px^2}{3} + C_1 \qquad (1)$$

Integrating again

$$EIy = \frac{Px^2}{3} + C_1 x + C_2 \qquad (2)$$

Between $a \quad x \quad 3\,a$ (*C to* B); $M = {}^2/_3\,Px - P(x-a)$; hence

$$EI\,\frac{d^2y}{dx^2} = \frac{-P}{3}x + Pa$$

Integrating;

$$EI\,\frac{dy}{dx} = -\frac{Px^2}{6} + Pax + C_3 \qquad (3)$$

Integrating again

$$EIy = -\frac{Px^3}{18} + \frac{Pax^2}{2} + C_3 x + C_4 \qquad (4)$$

The four constants of integration in equations (1) – (4) are determined from the following boundary conditions:

1. At $x = 0$; $y = 0$ is applied at equation (2) giving $C_2 = 0$

2. At $x = a$ (point C) y is the same if approached from right or left. Using equation (2) and equation (4)

$$\frac{Pa^3}{9} + C_1 a = -\frac{Pa^3}{18} + \frac{Pa^3}{2} + C_3 a + C_4$$

giving $\quad C_1 a - C_3 a - C_4 = \dfrac{Pa^3}{3} \qquad (5)$

3. Applying similar equality in slope$(\frac{dy}{dx})$ in equations (1) and (3)

$$\frac{Pa^2}{3} + C_1 = -\frac{Pa^2}{6} + Pa^2 + C_3$$

giving $\quad C_1 - C_3 = \dfrac{Pa^2}{2} \qquad (6)$

4. At $x = 3a$ (Point B) $y = 0$ is applied at equation (4) giving

$$-\frac{27\,Pa^3}{18} + \frac{2\,Pa^3}{2} + 3C_3 a + C_4 = 0$$

giving $\quad 3C_3 a + C_4 = -3Pa^3 \qquad (7)$

Three unknowns C_1, C_3, C_4 are solved from (5), (6) and (7) as;

$C_1 = -5Pa^2/9$

$C_3 = -19Pa^2/18$

$C_4 = Pa^3/16$

Substituting C_1 in equation (2)

$$EIY = \frac{Px^3}{9} - \frac{5Pa^2}{9}\,x$$

at Load Point $x = a$;

$$y = \frac{1}{EI}\left[\frac{Pa^3}{9} - \frac{5Pa^3}{9}\right] = -\frac{4Pa^3}{9EI}.$$

114. **(C)**

Euler Buckley Formula:

$$P_{CR} = \frac{\pi^2 EI}{L'^2}$$

Therefore, minimum load P_{CR} for column to buckle will be related to the axis about which moment of inertia I is minimum. So,

$$I_{yx} = {}^1/_{12}\,(6\text{ in})\,(2\text{ in})^3 - {}^1/_{12}\,(5\text{ in})\,(1\text{ in})^3 = 3.5833\text{ in}^4$$

For "Fixed-Free" Column; effective length $L' = 2L$. Hence,

$$P_{CR} = \pi^2\,\frac{(30\times 10^6\text{ psi})\,(3.5833\text{ in}^4)}{[2\,(20\text{ ft}\times 12\text{ in/ft})]^2} = 4600\text{ lbs}$$

115. **(B)**

Maximum Moment is produced at fixed support A.

Moment due to x component of

$$P = (1000\cos 30°)\,(7) = 6062\text{ lb·in}$$

Moment due to y component of

$$P = (1000\sin 30°)\,(5) = 2500\text{ lb·in}$$

Total moment = 2500 lb·in + 6062 lb·in = 8562 lb·in

Maximum tensile stress at section A due to bending =

$$\frac{MC}{I} = \frac{(8562\text{ lb}\cdot\text{in})(1\text{ in})}{\frac{1}{12}\,(\frac{1}{2}\text{ in})(2\text{ in}^3)} = 25,686\text{ psi}$$

Tensile stress at section A due to y component of

$$P = (1000\sin 30°)\,/\,(2\text{ in})\,({}^1/_1\text{ in}) = 500\text{ psi}$$

Total maximum tensile stress

$$= 25,686\text{ psi} + 500\text{ psi} = 26,186\text{ psi}$$

116. (A)

Capitalized cost is the present sum of money that would be set aside at some interest rate to yield the funds to provide the desired task or service forever.

117. (B)

The effective interest rate per year

$$= (1 + i)^m - 1$$

$$= (1 + 0.015)^{12} - 1 = 0.1956 \text{ or } 19.56\ \%.$$

118. (E)

If r is the nominal interest rate per year, the effective interest rate per year with continuous compounding (i) is given by

$$i = e^r - 1$$

Here, $r = 0.12$. Therefore

$$i = e^{0.12} - 1$$

$$= 0.1275 \text{ or } 12.75\%$$

119. (A)

Reliability of the alternative equipments is not included in the economic analysis.

120. (D)

Year	Book Value Before Depreciation Charge	Straight line Depreciation For The Year	Book Value After Depreciation Charge
1	76.00	$\frac{76-6}{5} = 14$	62.00
2	62.00	14	48.00
3	48.00	14	34.00
4	34.00	14	20.00
5	20.00	14	6.00

Note: The book values and depreciation in the above table are in millions of dollars.

121. (A)

In a before-tax comparison of alternatives, depreciation is not taken into account.

122. **(B)**

First consider the following cash flow and find the equivalent uniform annual cash flow A.

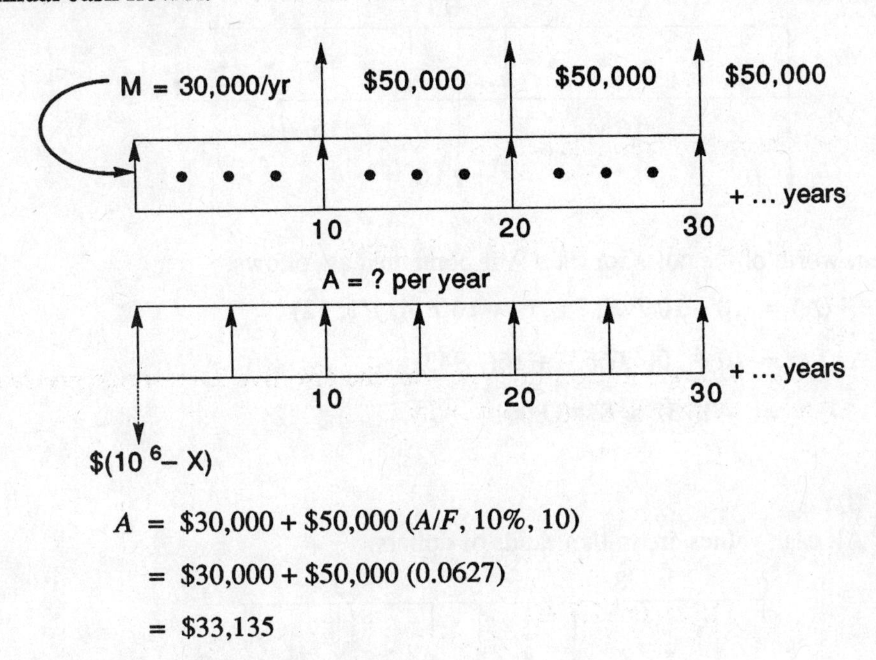

$$A = \$30{,}000 + \$50{,}000\,(A/F, 10\%, 10)$$
$$= \$30{,}000 + \$50{,}000\,(0.0627)$$
$$= \$33{,}135$$

Let X be the cost of construction.

$$\$(10^6 - X) = \frac{A}{i} = \frac{\$33,135}{0.1} = \$\,331,350$$

$$X = \$668{,}650$$

123. **(A)**

All cash values are in thousands of dollars.

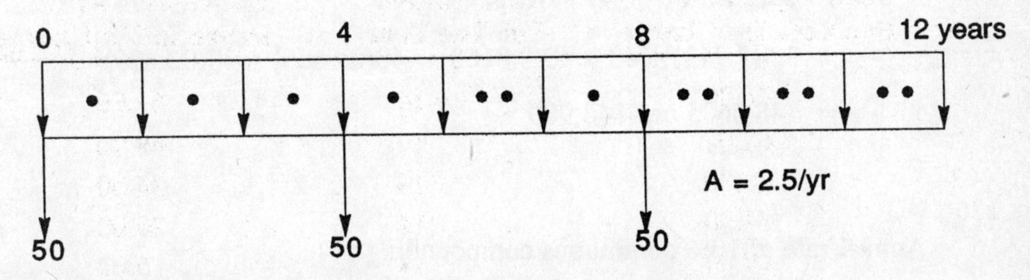

Present worth of the costs for Plan C is computed as follows:

$$PW(C) = 2.5\,(P/A, 7\%, 12) + 50 + 50(P/F, 7\%, 4) + 50(P/F, 7\%, 8)$$
$$= 2.5(7.943) + 50 + 50(0.7629) + 50(0.582)$$
$$= 137.1025$$
$$= \$137{,}103.00$$

124. **(D)**

All cash values are in thousands of dollars.

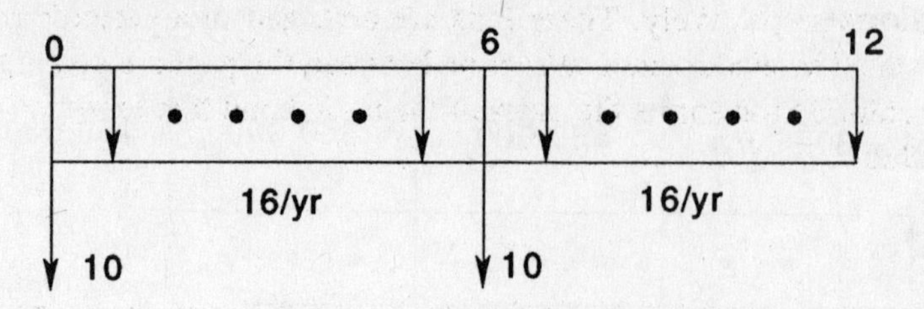

Present worth of the costs for Plan A is computed as follows:

$$PW(\text{A}) = 10 + 10(P/F, 7\%, 6) + 16(P/A, 7\%, 12)$$
$$= 10 + 10(0.6663) + 16(7.943)$$
$$= 143.751 \text{ or } \$143,800$$

125. **(C)**

All cash values are in thousands of dollars.

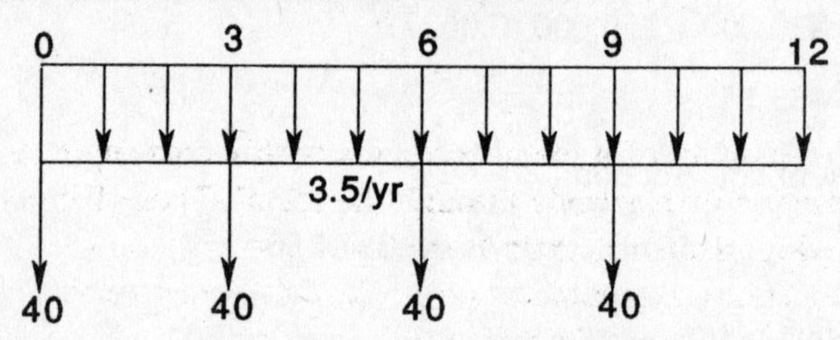

Present worth of the costs for Plan B is computed as follows:

$$PW(\text{B}) = 3.5\,(P/A, 7\%, 12) + 40 + 40(P/F, 7\%, 3) + 40(P/F, 7\%, 6)$$
$$+ 49(P/F, 7\%, 9)$$
$$= 3.5(7.943) + 40 + 40(0.8163) + 40(0.6663) + 40(0.5439)$$
$$= 148.8605 \text{ or } \$148,900$$

126 **(E)**

Annual rate of 10% continuous compounding

$$F = Pe^{rn} = \$2,000.00\, e^{(0.1\times 5)}$$
$$= \$3,297.44 \text{ or } \$3,300.00$$

or

$$F = P(\text{F/P}, 10\%, 5)$$
$$= \$2,000\ (1.649)$$
$$= \$3,298.00 \text{ or } \$3,300.00$$

F = ?
0 1 2 3 4 5
time in years
$2,000

127. **(B)**

In a sodium chloride crystal, sodium and chlorine are present as sodium and chloride ions respectively. These ions are arranged in a periodic pattern to form the crystal. The electrostatic attraction between the positive sodium ion and the negative chloride ion forms the basis of the ionic bond that exists in a sodium chloride crystal.

128. **(D)**

The direction vector begins at the origin and the coordinates of the position where it ends are 1, 1, 1/2. Thus the direction represented by the arrow is 221.

129. **(A)**

The change in crystal structure occurs isothermally and hence there are no thermally induced volume changes occurring. The volume change is entirely due to the transformation. The bcc structure is less densely packed than the fcc structure and hence the transformation from fcc to bcc structure is accompanied by an increase in the volume.

130. **(B)**

Electrical resistivity of a metal increases with increasing extent of factors which impede motion of electrons through the metal. These factors include the presence of alloying elements, grain boundaries and crystal defects such as vacancies and dislocations. Except annealing, all of the other treatments lead to an increase in the amount of resistivity due to increasing one or more of the factors listed above. Annealing helps to remove structural defects and hence results in reduced resistivity.

131. **(D)**

The rigidity of a polymer increases with increasing length of the molecules, with the formation of network structure and with the formation of crystalline regions in the polymer. Thus all of the three factors mentioned lead to an increase in the rigidity of the polymer.

132. **(C)**

Modulus of elasticity is determined by the binding forces between atoms, which is only dependent on the type of atoms that are present. All of the other properties depend very strongly on microstructural features such as grain size, dislocation density and texture.

133. **(E)**

A phosphorus doped silicon crystal is a *n*-type semiconductor. Phosphorus has five outer shell electrons of which four are used up in bonding. The extra electron is normally present in the donor level which is very close to the conduction band of the semiconductor. The thermal activation present at room temperature can elevate these electrons into the conduction band and these electrons are the major charge carriers in this material.

134. **(C)**

This question tests knowledge of the mechanism of conductivity in metals. The correct answer is (C)

Answer (A) is incorrect. Luster does not effect conductivity even though some feel that the characteristic luster and high conductivity both result from free electrons. Answer (B) is incorrect. "Extra" electrons must be carefully defined. Electrical neutrality requirements dictate that all atoms will have equal numbers of protons and electrons. Positive valence does not mean "extra" electrons. Answer (D) is incorrect. Position in the Periodic Table is quite arbitrary and not the cause of any physical or chemical property; but rather the converse — properties are organized in the table to reflect their periodically recurring nature.

135. **(B)**

This question tests knowledge of the nature of x-rays and sub-atomic particles. The correct answer is (B).

Answer (A) is incorrect. When first discovered, nature of the rays was unknown, hence the name "x" rays. However, the nature of x-rays has been known for many decades. Answer (C) is incorrect. A ray of mixed sub-atomic particles is not given a unique name. Answer (D) is incorrect. A ray of electrons is called a β ray (or, more commonly, β particles) when it comes from the nucleus of an atom. If it is from an accelerator, it is called an electron beam. Answer (E) is incorrect, A ray of helium nuclei is called an α ray (or, more commonly, α particles).

136. **(A)**

This question tests knowledge of Group Ia elements and their low values of electronegativity. The correct answer is (A). The Group Ia elements have a single "s" electron which is easily lost to form an alkali metal ion.

Answer (B) is incorrect. Conductivity is not related to the tendency of a metal to become ionized. The most conductive metal, silver, is not ionized nearly as easily as many less conductive metals. Answer (C) is incorrect. The violent reaction of the alkali metals with water to liberate hydrogen results from the position of the alkali metals on the emf series as does their ability to be ionized. But the reaction with water does not affect their ability to be ionized. Answer (D) is incorrect. The reason is well known. Answer (E) is incorrect. The tendency of a metal to be ionized, and the melting point are not related.

137. **(D)**

This question tests knowledge of the electronic configuration of the rare earth metals and how it relates to their chemical properties. The correct answer is (D). The valence, or reactive, electrons are the same for each of the rare earth metals. They are $6s^2$ electrons. Each successive member of the series is made by adding a 4f electron which is two orbits down from the reactive $6s^2$ electrons and has a minimal effect on the chemical properties.

Answer (A) is incorrect. The reason is well known. Answer (B) is incorrect. The materials are sufficiently plentiful that their properties have been well studied. Answer (C) is incorrect. The melting points of the rate earth metals is not particularly high or low, but typical of that of other metals in the Periodic Table. Answer (E) is incorrect. The position of the rare earth metals in the Periodic Table is arbitrary and cannot effect properties.

138. **(C)**

This question tests knowledge of the electron volt — an important unit of energy that is commonly used in the quantum mechanical descriptions of the structure of matter and energy. The correct answer is (C). An electron volt (abbreviated eV) is an energy unit. It is the energy possessed by an electron after it has accelerated through an electrical potential of one volt. The common conversion factors are:

$$1 \text{ Joule} = 6.24 \times 10^{18} \text{ eV}$$

$$1 \text{ calorie} = 2.61 \times 10^{19} \text{ eV}$$

$$1 \text{ erg} = 6.24 \times 10^{11} \text{ eV}$$

Answer (A) is incorrect, an electron volt is not a voltage unit. Answer (B) is incorrect, an electron volt is not a unit of charge. Answer (D) is incorrect, an electron volt is a precisely defined unit of energy, and its definition does not depend on the context. Answer (E) is incorrect since answer (C) is correct.

139. **(C)**

This question tests knowledge of how to find the crystalline structure of the elements from the Periodic Table which, presumably, will be available to most examinees during the test. The correct answer is (C). Complete forms of the Periodic Table show the most common crystalline structure of the elements. Although the most common structure for carbon is the hexagonal (graphite) structure, it, of course, also forms diamond crystals. Silicon and germanium are the only elements whose most common crystalline structure is that of diamond.

Answer (A) is incorrect. Neither boron nor aluminum form diamond-type crystals. Boron crystals are hexagonal, aluminum is face centered cubic. Answer (B) is incorrect. Cobalt forms hexagonal crystals, nickel face centered cubic. Answer (D) is incorrect. All group Ia elements, of which lithium, sodium, and potassium are examples, form body centered cubic crystals. Answer (E) is incorrect. Magnesium forms hexagonal crystals; calcium and strontium form face centered cubic crystals.

140. **(B)**

This question tests knowledge of defects in crystal lattice structures which form weak points in the crystal. The correct answer is (B). Failure, of course, occurs at the weakest points in the crystal lattice — i.e., at the crystal defects.

Answer (A) is incorrect. Very small crystals without defects, such as "whiskers" fail very near the calculated crystal bond strength. Answer (C) is incorrect. The stress at failure is quite reproducible for most materials if samples are prepared properly. Answer (D) is incorrect. Plastic deformation, indeed, occurs before failure for many materials, but is not the reason that the stress at failure does not correspond to the crystalline bond strength.

Fundamentals of Engineering

P.M. SECTION

Test 3

Fundamentals of Engineering

P.M. SECTION

TEST 3 – ANSWER SHEET

1. (A) (B) (C) (D) (E)
2. (A) (B) (C) (D) (E)
3. (A) (B) (C) (D) (E)
4. (A) (B) (C) (D) (E)
5. (A) (B) (C) (D) (E)
6. (A) (B) (C) (D) (E)
7. (A) (B) (C) (D) (E)
8. (A) (B) (C) (D) (E)
9. (A) (B) (C) (D) (E)
10. (A) (B) (C) (D) (E)
11. (A) (B) (C) (D) (E)
12. (A) (B) (C) (D) (E)
13. (A) (B) (C) (D) (E)
14. (A) (B) (C) (D) (E)
15. (A) (B) (C) (D) (E)
16. (A) (B) (C) (D) (E)
17. (A) (B) (C) (D) (E)
18. (A) (B) (C) (D) (E)
19. (A) (B) (C) (D) (E)
20. (A) (B) (C) (D) (E)
21. (A) (B) (C) (D) (E)
22. (A) (B) (C) (D) (E)
23. (A) (B) (C) (D) (E)
24. (A) (B) (C) (D) (E)
25. (A) (B) (C) (D) (E)
26. (A) (B) (C) (D) (E)
27. (A) (B) (C) (D) (E)
28. (A) (B) (C) (D) (E)
29. (A) (B) (C) (D) (E)
30. (A) (B) (C) (D) (E)
31. (A) (B) (C) (D) (E)
32. (A) (B) (C) (D) (E)
33. (A) (B) (C) (D) (E)
34. (A) (B) (C) (D) (E)
35. (A) (B) (C) (D) (E)
36. (A) (B) (C) (D) (E)
37. (A) (B) (C) (D) (E)
38. (A) (B) (C) (D) (E)
39. (A) (B) (C) (D) (E)
40. (A) (B) (C) (D) (E)
41. (A) (B) (C) (D) (E)
42. (A) (B) (C) (D) (E)
43. (A) (B) (C) (D) (E)
44. (A) (B) (C) (D) (E)
45. (A) (B) (C) (D) (E)
46. (A) (B) (C) (D) (E)
47. (A) (B) (C) (D) (E)
48. (A) (B) (C) (D) (E)
49. (A) (B) (C) (D) (E)
50. (A) (B) (C) (D) (E)
51. (A) (B) (C) (D) (E)
52. (A) (B) (C) (D) (E)
53. (A) (B) (C) (D) (E)
54. (A) (B) (C) (D) (E)
55. (A) (B) (C) (D) (E)
56. (A) (B) (C) (D) (E)
57. (A) (B) (C) (D) (E)
58. (A) (B) (C) (D) (E)
59. (A) (B) (C) (D) (E)
60. (A) (B) (C) (D) (E)
61. (A) (B) (C) (D) (E)
62. (A) (B) (C) (D) (E)
63. (A) (B) (C) (D) (E)
64. (A) (B) (C) (D) (E)
65. (A) (B) (C) (D) (E)
66. (A) (B) (C) (D) (E)
67. (A) (B) (C) (D) (E)
68. (A) (B) (C) (D) (E)
69. (A) (B) (C) (D) (E)
70. (A) (B) (C) (D) (E)

FUNDAMENTALS OF ENGINEERING EXAMINATION

TEST 3

AFTERNOON (PM) SECTION

TIME: 4 Hours
70 Questions

DIRECTIONS: For each of the following questions and incomplete statements, choose the best answer from the five answer choices. You must answer all questions.

QUESTIONS 1–10 refer to the following matrices.

$$A = \begin{bmatrix} 2 & 0 & 4 \\ 1 & 4 & -1 \\ 3 & 1 & 2 \end{bmatrix} \qquad B = \begin{bmatrix} 3 & 5 & 2 \\ 0 & 8 & 4 \\ 1 & 0 & 9 \end{bmatrix}$$

$$C = \begin{bmatrix} 1 \\ 2 \\ 3 \end{bmatrix} \qquad D = \begin{bmatrix} 1 & 4 \\ -2 & 3 \end{bmatrix}$$

1. The transpose of A is

(A) $\begin{bmatrix} 2 & 0 & 4 \\ 3 & 1 & 2 \\ 1 & 4 & -1 \end{bmatrix}$

(B) $\begin{bmatrix} 2 & 1 & 3 \\ 0 & 4 & 1 \\ 4 & -1 & 2 \end{bmatrix}$

(C) $\begin{bmatrix} 3 & 1 & 2 \\ 3 & 4 & 1 \\ 4 & -1 & 2 \end{bmatrix}$

(D) $\begin{bmatrix} 2 & 3 & 1 \\ 1 & 4 & 0 \\ 2 & -1 & 4 \end{bmatrix}$

(E) $\begin{bmatrix} 2 & 0 & 4 \\ 0 & 4 & 1 \\ 3 & 2 & 1 \end{bmatrix}$

2. The determinant of A is

(A) 4

(B) – 12

(C) 16

(D) – 26

(E) 48

3. The inverse of D is

(A) $\frac{1}{11}\begin{bmatrix} 2 & -4 \\ 3 & 1 \end{bmatrix}$

(B) $\frac{1}{11}\begin{bmatrix} 3 & -4 \\ 2 & 1 \end{bmatrix}$

(C) $\frac{1}{12}\begin{bmatrix} 5 & -3 \\ 2 & 1 \end{bmatrix}$

(D) $\frac{1}{4}\begin{bmatrix} -2 & 10 \\ 4 & 1 \end{bmatrix}$

(E) $\frac{1}{16}\begin{bmatrix} 6 & -4 \\ -2 & 3 \end{bmatrix}$

4. The product of $A \times C$ is

(A) $\begin{bmatrix} 1 & 4 & 2 \\ 2 & 6 & 1 \\ -2 & 0 & 5 \end{bmatrix}$

(B) $[3\ 2\ 1]$

(C) $\begin{bmatrix} -2 & 2 \\ 8 & 0 \\ 1 & 5 \end{bmatrix}$

(D) $\begin{bmatrix} 2 & 5 & 8 \\ -1 & -4 & 6 \end{bmatrix}$

(E) $\begin{bmatrix} 14 \\ 6 \\ 11 \end{bmatrix}$

5. The rank of B is

(A) 0

(B) 1

(C) 2

(D) 3

(E) undefined

6. The sum of $A + B$ is

(A) $\begin{bmatrix} 5 & 1 & 11 \\ 1 & -2 & 1 \\ 11 & 1 & 0 \end{bmatrix}$

(B) $\begin{bmatrix} 5 & 5 & 6 \\ 1 & 12 & 3 \\ 4 & 1 & 11 \end{bmatrix}$

(C) $\begin{bmatrix} 5 & 1 & 4 \\ 5 & 12 & 1 \\ 6 & 1 & 11 \end{bmatrix}$

(D) $\begin{bmatrix} 0 & 0 & -3 \\ -2 & 2 & 3 \\ 3 & 1 & 4 \end{bmatrix}$

(E) $\begin{bmatrix} 4 & 0 & 2 \\ -1 & 4 & 1 \\ 2 & 1 & 3 \end{bmatrix}$

7. The product of $3 \times D$ is

(A) $\begin{bmatrix} 1 & 12 \\ -2 & 9 \end{bmatrix}$

(B) $\begin{bmatrix} 3 & 12 \\ -2 & 3 \end{bmatrix}$

(C) $\begin{bmatrix} 3 & 12 \\ -6 & 9 \end{bmatrix}$

(D) $\begin{bmatrix} 3 & -12 \\ 14 & 9 \end{bmatrix}$

(E) $\begin{bmatrix} 3 & -6 \\ 6 & 9 \end{bmatrix}$

8. The product of $C \times D$ is

(A) $\begin{bmatrix} 1 & 8 \\ -2 & 6 \end{bmatrix}$

(B) $\begin{bmatrix} 2 & 8 \\ -2 & 6 \\ -6 & 9 \end{bmatrix}$

(C) $\begin{bmatrix} 2 & -2 & -6 \\ 8 & 6 & 9 \end{bmatrix}$

(D) $\begin{bmatrix} 1 \\ 8 \\ 12 \end{bmatrix}$

(E) undefined

9. The matrix equation

$$Ax = C$$

represents the following set of equations:

(A) $2x_1 + x_2 + 4x_3 = 3$
$2x_1 + 4x_2 - x_3 = 2$
$3x_1 + x_2 + 3x_3 = 1$

(B) $2x_1 + 2x_2 + 4x_3 = 1$
$x_1 + 4x_2 - x_3 = 2$
$3x_1 + x_2 + 2x_3 = 3$

(C) $2x_1 + 4x_3 = 1$
$x_1 + 4x_2 - x_3 = 2$
$3x_1 + x_2 + 2x_3 = 3$

(D) $2x_1 + x_2 + 3x_3 = 1$
$4x_2 + x_3 = 2$
$4x_1 - x_2 + 2x_3 = 3$

(E) $2x_1 + 2x_2 + 4x_3 = 1$
$3x_1 + x_2 + 2x_3 = 2$
$x_1 + 4x_2 - x_3 = 3$

10. All of the following statements are true EXCEPT

(A) $A \times B \neq B \times A$

(B) $A \times B \times C$ is an array with 3 rows and 1 column.

(C) $A + B \neq B + A$

(D) If a square matrix has two identical rows, its determinant is zero.

(E) Matrix division is performed by multiplying by the inverse of the matrix in the denominator.

QUESTIONS 11–15 refer to the following differential equation

$$\frac{dy^2}{dt^2} + 5\frac{dy}{dt} + 6y = 4 \quad y(0) = \frac{dy}{dt}(0) = 0$$

11. The roots of the characteristic equation are

(A) 1, 5

(B) −2, −3

(C) 2, 3

(D) 3, 5

(E) 4, 6

12. The complementary solution is

(A) $C_1e^{2t} + C_2e^{-3t}$

(B) $C_1e^{2t} + C_2e^{3t}$

(C) $C_1e^{t/2} + C_2e^{t/3}$

(D) $C_1e^{-4t} + C_2e^{-t}$

(E) $C_1e^{-2t} + C_2e^{-3t}$

13. The particular solution is

(A) 2/3

(B) 4

(C) 6

(D) 4/5

(E) 1/6

14. The Laplace Transform of y is

(A) $\dfrac{24}{s^2(s^2+5s+6)}$

(B) $\dfrac{12s}{(s^2+5s+6)}$

(C) $\dfrac{4}{s(s^2+5s+6)}$

(D) $\dfrac{6}{s(s^2+5s+6)}$

(E) $\dfrac{s+2}{s(s^2+5s+6)}$

15. The value of y as $t \to$ infinity is

(A) 4

(B) 6

(C) 0

(D) 3/2

(E) 2/3

QUESTIONS 16–20 refer to the following set of numbers.

–3, 1, 2, 2, 4, 4, 4, 8, 23

16. The arithmetic mean of the set is

(A) 2

(B) 3

(C) 4

(D) 5

(E) 9

17. The mode is

(A) 2

(B) 3

(C) 4

(D) 5

(E) 9

18. The median is

(A) 2

(B) 3

(C) 4

(D) 5

(E) 9

19. The range is

(A) 2
(B) 4
(C) 9
(D) 26
(E) 48

20. The variance is

(A) 2
(B) 4
(C) 9
(D) 26
(E) 48

QUESTIONS 21–30 refer to the following problem.

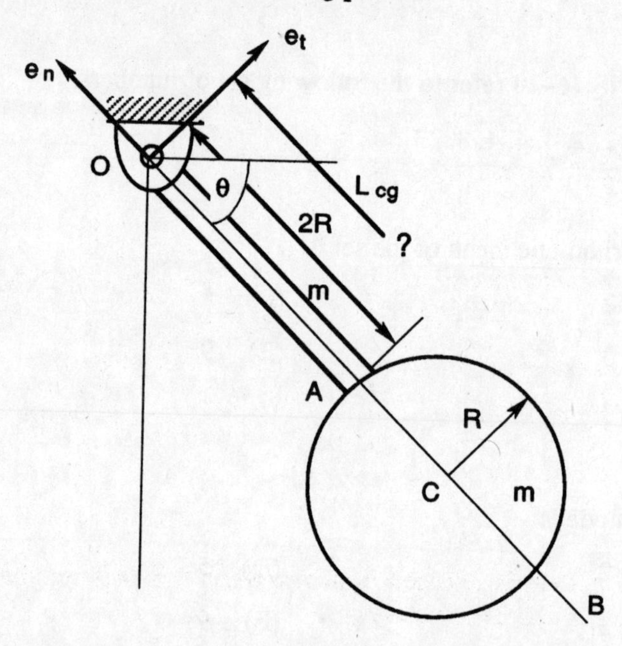

The system shown above consists of a uniform rod of mass *m* and length 2*R*. The rod is hinged at *O* and is welded to a uniform thin disk at *A*. The mass of the disk is *m* and its radius is *R*.

21. The center of mass of the system is at a distance L_{cg} from *O* whose value is

(A) 1.5*R*
(B) 2*R*
(C) 2.5*R*
(D) 2.75*R*
(E) 3*R*

22. The mass moment of inertia of the system about O is

(A) $(56/6)mR^2$
(B) $(57/6)mR^2$
(C) $(63/6)mR^2$
(D) $(127/12)mR^2$
(E) $(65/6)mR^2$

QUESTIONS 23–30. Consider the following data

$m + m = 20$ kg,

$L_{cg} = 2$ meters,

radius of gyration of the system from $O = 2.33$ meters.

23. The distance of the center of percussion from O *is near to*

(A) 2.25 meters
(B) 2.5 meters
(C) 2.71 meters
(D) 3 meters
(E) 3.25 meters

24. If the system is released from rest when $\theta = 60°$, the angular acceleration of the system just after release is near to

(A) 0 rad/sec/sec
(B) 0.33 rad/sec/sec
(C) 1.81 rad/sec/sec
(D) 3.27 rad/sec/sec
(E) 10.73 rad/sec/sec

25. If the system is released from rest when $\theta = 0°$, then the tangential force along e_t when $\theta = 60°$ is near to

(A) 163 e_t N
(B) 98 e_t N
(C) 65 e_t N
(D) 26 e_t N
(E) 43 e_t N

26. If the system is released from rest when $\theta = 0°$, then the normal force along e_n when $\theta = 60°$ is near to

(A) 420 e_n N
(B) – 420 e_n N
(C) 227 e_n N
(D) – 227 e_n N
(E) 170 e_n N

27. If the system is released from rest when θ = 0, the kinetic energy of the system when θ = 60° is near to

(A) 113 joules
(B) 226 joules
(C) 286 joules
(D) 340 joules
(E) 680 joules

28. If the system is released from rest when θ = 0, the angular momentum of the system about *O* when θ = 60° is near to

(A) 113 Kg m²/sec Clockwise
(B) 226 Kg m²/sec Clockwise
(C) 271 Kg m²/sec Clockwise
(D) 340 Kg m²/sec Clockwise
(E) 680 Kg m²/sec Clockwise

29. If the system is released from rest when θ = 0, the angular velocity of the system about *O* when θ = 90° is near to

(A) 6.5 rad/sec
(B) 2.7 rad/sec
(C) 2.3 rad/sec
(D) 1.9 rad/sec
(E) 1.6 rad/sec

30. If the system starts from rest at an angle slightly smaller than 90°, the system will oscillate with a frequency close to

(A) 0.3 Hz
(B) 0.6 Hz
(C) 0.9 Hz
(D) 1.2 Hz
(E) 1.8 Hz

QUESTIONS 31–40 relate to the following phase diagram.

Assume equilibrium conditions unless specified otherwise. α, σ, β are solid solutions.

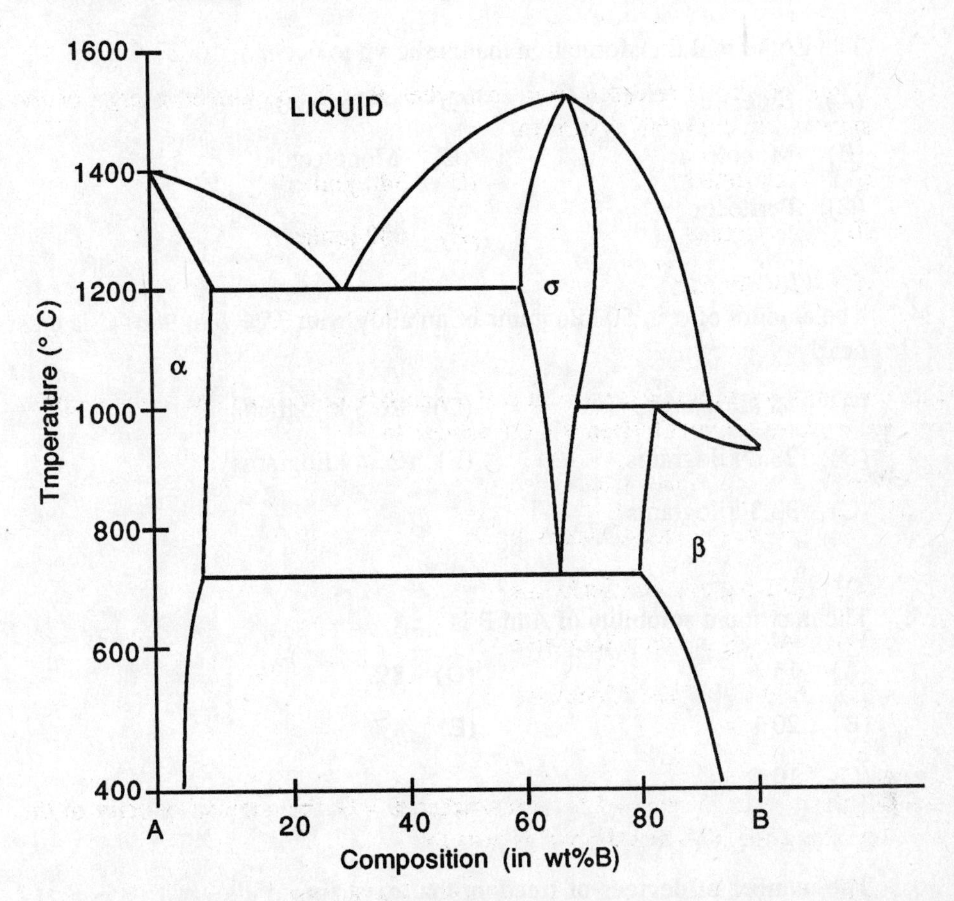

31. The highest temperature at which an alloy in this system can be completely solid is

    (A) 1400° C
    (B) 1000° C
    (C) 750° C
    (D) 1500° C
    (E) 800° C

32. The phases present and their compositions in an alloy with 50 weight percent *B* at 1300° C are

    (A) Liquid with 41% *B* and σ with 61% *B*
    (B) Liquid I with 41% *B* and Liquid II with 84% *B*
    (C) Liquid with 20% *B* and α with 5% *B*

(D) α with 5% *B* and Liquid with 84% *B*

(E) σ with 69% *B* and Liquid with 84% *B*

33. The isothermal transformation that is shown to occur at 1000° *C is*

(A) Eutectic
(B) Monotectic
(C) Peritectic
(D) Peritectoid
(E) Monotectoid

34. The amount of σ in 50 kilograms of an alloy with 35% *B* at 900° C is most nearly

(A) 15 kilograms
(B) 23.6 kilograms
(C) 33.3 kilograms
(D) 16.5 kilograms
(E) 26.4 kilograms

35. The maximum solubility of *A* in *B* is

(A) 15%
(B) 20%
(C) 10%
(D) 8%
(E) 4%

36. The number of degrees of freedom that exist for an alloy with 3% *B* at a constant pressure of 1 atmosphere at temperatures below 1200° C is

(A) 0
(B) 1
(C) 2
(D) 3
(E) 4

37. If the presence of 10% σ in the alloy makes the *A*-rich alloys brittle at 950° *C*, then the upper limit on the composition of an alloy that can be forged at 950° C is most nearly

(A) 4% *B*
(B) 25.3% *B*
(C) 40.2% *B*
(D) 14.5% *B*
(E) 35% *B*

38. Assuming α and β have the same density, the volume percent of a precipitate at 400° C in an alloy with 82% *B* is most nearly

(A) 12.5%
(B) 87.5%
(C) 9.3%
(D) 90.7%
(E) 18%

39. When an alloy with 20% *B* (by weight) is cooled slowly from 1600° C to 900° C, the cooling curve will most closely resemble

(A)
Tmperature (° C)
1200° C
Time

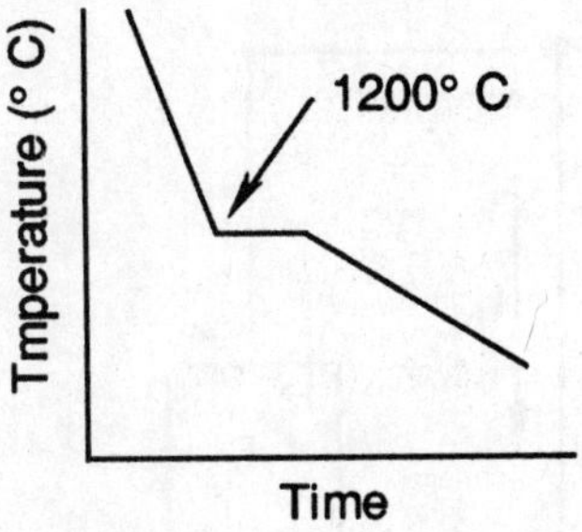

(B)
Tmperature (° C)
1300° C
Time

(C)
Tmperature (° C)
1300° C
1200° C
Time

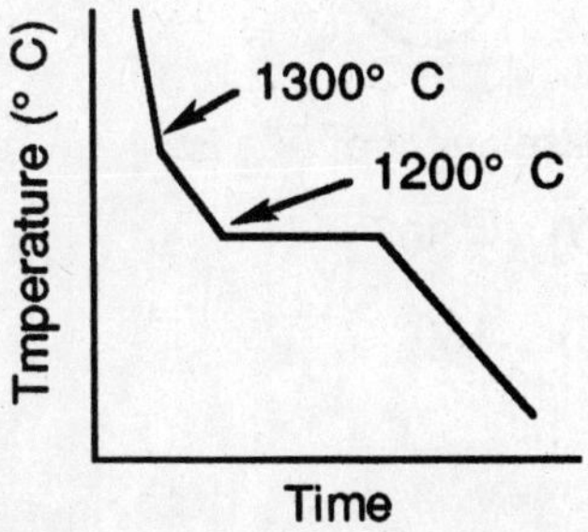

(D)
Tmperature (° C)
1360° C
1000° C
Time

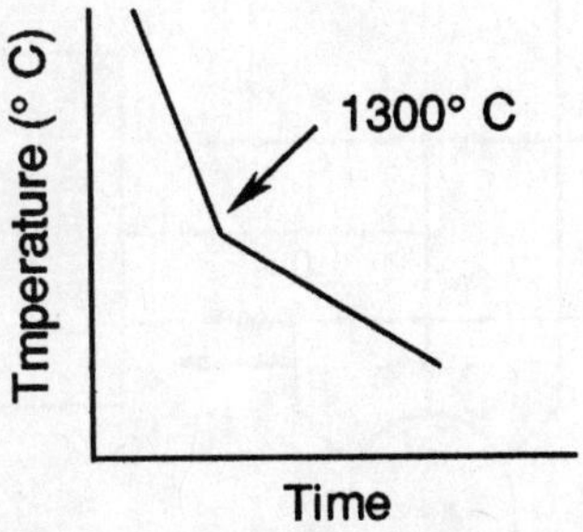

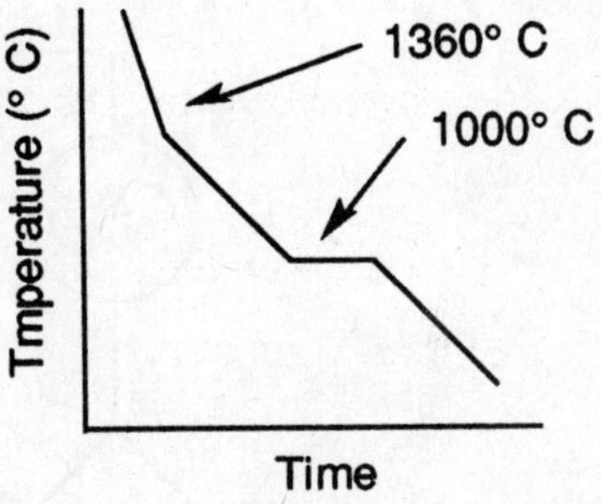

(E)
Tmperature (° C)
1300° C
Time

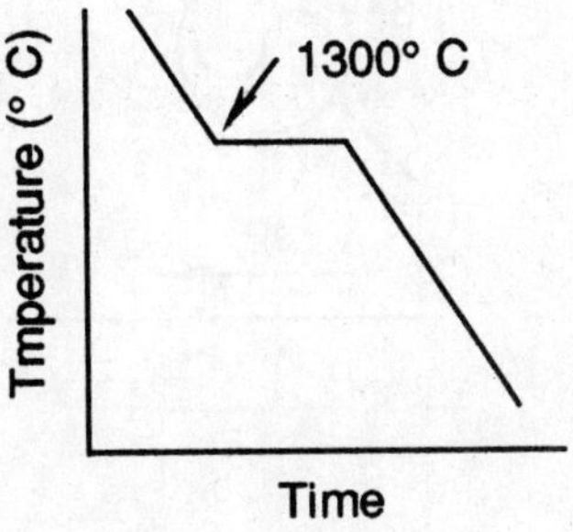

40. The freezing range of an alloy with 80% *B* is nost nearly

(A) 660° C
(B) 1360° C
(C) 1000° C
(D) 140° C
(E) 360° C

QUESTIONS 41–50 relate to the circuit shown below.

For questions 41–48 the capacitor bank is not connected to the circuit. *L* is a lighting load of 5 KW and a power factor of 1 and M is a three phase 5 HP induction motor with an efficiency of 0.8 and a lagging power factor of .85. The line to line voltages are $|V_{12}| = |V_{23}| = |V_{31}| = 220$ volts and have phase rotation of 1–2–3. All magnitudes are given in root mean square values. The magnitude of a phasor is equal to the root mean square value of the amplitude of the periodic function represented by the phasor. Bold letters are used for phasors and plain letters for their magnitude. **V** would indicate a phasor while V its magnitude.

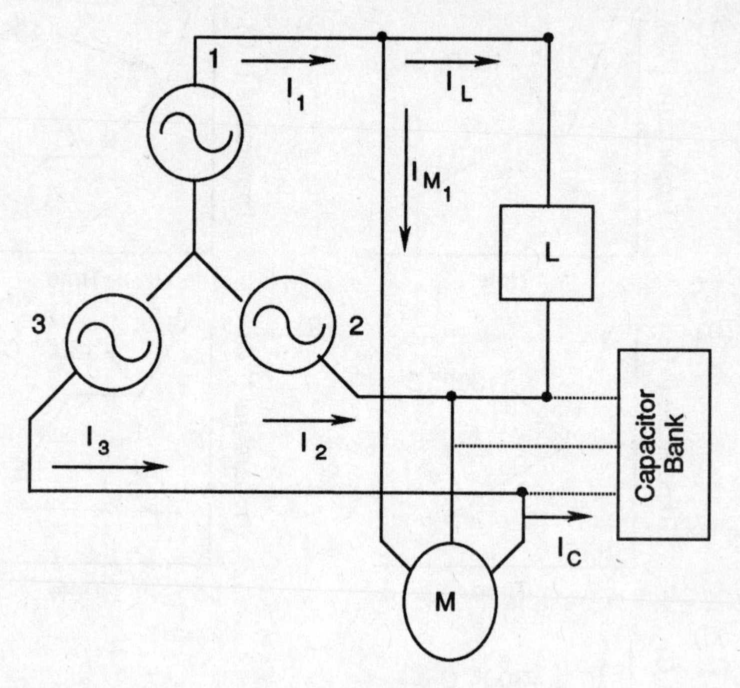

41. If the phase angle of $\mathbf{V}_{12}$ is 0° the phase angle of $\mathbf{V}_{31}$ is:

    (A) 120°
    (B) – 120°
    (C) 90°
    (D) – 90°
    (E) 180°

42. The current through the load *L* is

    (A) 44.54 A
    (B) 11.36 A
    (C) 22.72 A
    (D) 19.7 A
    (E) 39.27 A

43. The phase angle of the load current I_L is:

(A) – 30° (D) + 60°

(B) + 30° (E) – 60°

(C) 0°

44. The electric power delivered to the motor M is

(A) 6.7 KW (D) 4.66 KW

(B) 3.73 KW (E) 2.98 KW

(C) 5.0 KW

45. The current I_{M1} flowing to the motor is given by

(A) 22.92 A (D) 12.23 A

(B) 8.31 A (E) 14.39 A

(C) 24.91 A

46. The voltage phasor V_{1n} is given by:

(A) 127 $\angle +60°$ V (D) 381 $\angle -30°$ V

(B) 381 $\angle +30°$ V (E) 127 $\angle -30°$ V

(C) 127 $\angle +30°$ V

47. The motor current phasor I_{M1} for the motor current is given by:

(A) 12.23 $\angle -30°$ A (D) 8.31 $\angle -181.78°$ A

(B) 14.39 $\angle -61.78°$ A (E) 8.31 $\angle +61.78°$ A

(C) 14.39 $\angle -31.78°$ A

48. The line current phasor I_1 is given by

(A) 37.11 $\angle -31.78°$ A (D) 37.11 $\angle +61.78°$ A

(B) 32.13 $\angle -23.23°$ A (E) 37.11 $\angle -181.78°$ A

(C) 32.13 $\angle +31.78°$ A

49. The rating of a three phase capacitor bank, connected to the motor terminals, needed to improve the power factor at the terminals to 1 is given by:

(A) 2.88 KVAR
(B) 5.66 KVAR
(C) 9.59 KVAR
(D) 4.75 KVAR
(E) 1.44 KVAR

50. The currents I_c in the connection to the capacitor bank is:

(A) 4.35 A
(B) 7.57 A
(C) 9.6 A
(D) 4.8 A
(E) 14.6 A

QUESTIONS 51–60 refer to the following problem.

The Space Cast Foundry makes small castings. The company has been acquired by a large corporation. The new management has decided to make changes in several areas of operation. The following problems have been sent to you for analysis and recommendation.

51. The foundry has been buying cores for a casting at $90 per unit. The foundry can make its own cores and the associated costs and investment are as follows:

(i)	direct cost of each core	$ 26.00
(ii)	overhead costs per year	$ 9,000.00
(iii)	core making machine	$150,000.00
(iv)	useful life of core making machine	10 yrs.
(v)	salvage value of the core making machine after 10 years	$ 10,000.00

A Minimum Attractive Rate of Return (MARR) of 15% is desired for this investment. What is the break-even value for number of cores per year beyond which producing the cores is better than buying them?

(A) 200
(B) 400
(C) 600
(D) 800
(E) 1,000

52. The shipping area has to be expanded to meet the production schedule for the next 10 years. The initial cost of investment and annual operating

benefits and costs are very different due to various degrees of automation. These can be summarized as follows:

	Alternative Y	Alternative Z
Initial Investment Cost	$180,000	$100,000
Annual Operating Costs	$ 10,000	$ 9,000
Annual Benefits	$ 63,000	$ 30,000
Salvage Value	$ 0	$ 5,000
Useful Life (yrs)	10	10

What is the benefit/cost ratio for alternative Y? Use MARR of 5%.

(A) 1.066
(B) 1.374
(C) 1.478
(D) 2.944
(E) 3.396

53. Consider alternative Z presented in problem 52 above. What is the benefit to cost ratio for this alternative? Use MARR of 15%.

(A) 1.066
(B) 1.037
(C) 1.478
(D) 2.150
(E) 3.846

54. Consider alternatives Y and Z in problem 52 above. What is the incremental benefit-cost ratio (ΔB / ΔC) on the increment of investment between the alternatives? Use MARR of 15%.

(A) 1.066
(B) 1.478
(C) 1.948
(D) 2.500
(E) 2.963

55. Consider the alternative Y in problem 52 above. What is the present worth of this alternative if bank loan can be secured at a nominal annual interest rate of 10%? Round off your answer to the nearest thousand.

(A) $146,000
(B) $180,000
(C) $325,000
(D) $350,000
(E) None of the above

56. Consider the alternative Y in problem 52 above. What is the approximate rate of return for this alternative? Round off to the nearest integer value.

(A) 15% (D) 27%

(B) 17% (E) 30%

(C) 20%

57. Consider alternative Z in Problem 52 above. Compute the rate of return for this alternative. Round off to the nearest integer.

(A) 8% (D) 14%

(B) 10% (E) 16%

(C) 12%

58. Consider the cash flow for (Y – Z) in problem 52 above. Compute the incremental rate of return (Δ *ROR*). Round off to the nearest integer.

(A) 10% (D) 25%

(B) 15% (E) 30%

(C) 20%

59. The foundry is in the 34% income tax bracket. By purchasing a robot at $30,000, it saves $5,000 during year 1, $4,000 during year 2, $3,000 during year 3, $2,000 during year 4 and $1,000 during year 5. The company depreciated the robot using the sum-of-years-digits (SOYD) depreciation method over its four-year depreciable life, while assuming a zero salvage value for depreciation purposes. The company wants to sell the robot at the end of year 5. Assume that the capital gains are taxed as ordinary income and there is no investment tax credit.

What resale value will yield a 5% after-tax rate of return for the company? Round off the answer to the nearest hundred.

(A) $11,900 (D) $22,000

(B) $15,900 (E) $23,000

(C) $18,100

60. Consider problem 57 above. If the depreciation is changed to the straight

line method, what resale value for the robot will yield a 5% after-tax rate of return for the company? Round off the answer to the nearest hundred.

(A) \$18,100
(B) \$22,000
(C) \$23,000
(D) \$23,400
(E) None of the above

QUESTIONS 61–70 refer to the following problem and pertains to the frictionless piston cylinder and pressure-volume diagram shown below.

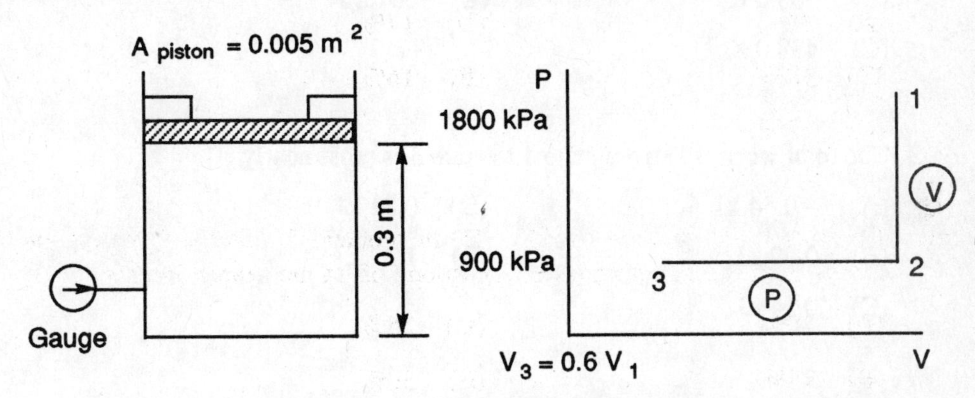

Gas	Molecular Weight	Specific heats, kJ/kg-K Cp	C_v	Gas Constant kJ/kg-K
O_2	32.00	0.9216	0.6618	0.2598
N_2	28.013	1.0416	0.7748	0.2968

QUESTIONS 61–66 pertain to the piston cylinder device filled with oxygen alone. The device is exposed to the atmosphere with a pressure of one atmosphere. Initially, the pressure in the device is sufficient to push the piston up against the stops and the temperature at state 1 is 730°C. The device is cooled from state 1 to state 3 as depicted in the *P-V* diagram. The pressure required to support the piston is 900 kPa.

61. The mass of the oxygen is most nearly

(A) 0.00907 kg
(B) 0.01036 kg
(C) 0.01246 kg
(D) 0.01423 kg
(E) 0.01657 kg

62. The pressure on the gauge at state 1 is most nearly

(A) – 1799 kPa
(B) – 1698.7 kPa
(C) 1698.7 kPa
(D) 1799 kPa
(E) 99525 kPa

63. The temperature at state 2 is most nearly

(A) 300.9 K
(B) 365.0 K
(C) 438.0 K
(D) 501.5 K
(E) 601.8 K

64. The total work (kJ) from state 1 to state 3 is most nearly

(A) – 0.54 kJ
(B) 0.00 kJ
(C) 0.54 kJ
(D) 0.81 kJ
(E) 1.89 kJ

65. The mass of the piston is most nearly

(A) 0.4071 kg
(B) 0.5103 kg
(C) 3.9896 kg
(D) 407.10 kg
(E) 510.30 kg

66. The change in the entropy (kJ/kg-K) from state 1 to state 3 is most nearly

(A) – 0.7761 kJ/kg-K
(B) – 0.9295 kJ/kg-K
(C) – 0.9769 kJ/kg-K
(D) – 1.1277 kJ/kg-K
(E) – 1.2897 kJ/kg-K

QUESTIONS 67–70 pertain to the piston cylinder device as shown in the diagrams. Nitrogen and Oxygen are placed into the device in the following proportions.

Gas	Mole Fraction
O_2	0.60
N_2	0.40

67. The mass of the mixture is most nearly

(A) 0.008761 kg
(B) 0.009311 kg
(C) 0.009844 kg
(D) 0.010331 kg
(E) 0.013525 kg

68. The temperature of the mixture at state 2 is most nearly

(A) 300.9 K
(B) 365.0 K
(C) 438.0 K
(D) 501.5 K
(E) 601.8 K

69. The constant volume specific heat of the mixture is most nearly

(A) 0.682 kJ/kg-K
(B) 0.707 kJ/kg-K
(C) 0.718 kJ/kg-K
(D) 0.729 kJ/kg-K
(E) 0.752 kJ/kg-K

70. The partial pressure of the oxygen at state 1 is most nearly

(A) 540 kPa
(B) 720 kPa
(C) 900 kPa
(D) 1080 kPa
(E) 1500 kPa

TEST 3 (PM)

ANSWER KEY

1.	(B)	21.	(B)	41.	(A)	61.	(B)
2.	(D)	22.	(E)	42.	(C)	62.	(C)
3.	(B)	23.	(C)	43.	(C)	63.	(D)
4.	(E)	24.	(C)	44.	(D)	64.	(A)
5.	(D)	25.	(D)	45.	(E)	65.	(D)
6.	(B)	26.	(A)	46.	(E)	66.	(B)
7.	(C)	27.	(D)	47.	(B)	67.	(C)
8.	(E)	28.	(C)	48.	(B)	68.	(D)
9.	(C)	29.	(B)	49.	(A)	69.	(B)
10.	(C)	30.	(A)	50.	(B)	70.	(D)
11.	(B)	31.	(D)	51.	(C)		
12.	(E)	32.	(A)	52.	(B)		
13.	(A)	33.	(C)	53.	(B)		
14.	(C)	34.	(B)	54.	(C)		
15.	(A)	35.	(B)	55.	(A)		
16.	(D)	36.	(C)	56.	(D)		
17.	(C)	37.	(D)	57.	(E)		
18.	(C)	38.	(C)	58.	(D)		
19.	(D)	39.	(C)	59.	(E)		
20.	(E)	40.	(E)	60.	(D)		

DETAILED EXPLANATIONS OF ANSWERS

TEST 3

AFTERNOON (PM) SECTION

QUESTIONS 1–10, it is convenient to use the following subscript notation for vectors and matrices. a_{ij} is the value in row i and column j of the matrix **A**. The first subscript defines the row position. The second subscript defines the column position. A vector is defined as having one column and more than one row.

1. **(B)**

The transform of a matrix is formed by interchanging rows and columns.

If the transform of **A** is $\mathbf{A}^T$, then $a^T_{ij} = a_{ji}$

Answer (B) is correct.

2. **(D)**

The determinant of a 3 by 3 matrix is a single number (scalar) calculated by

$$
\begin{aligned}
|A| &= a_{11}(a_{22}\,a_{33} - a_{32}\,a_{23}) - a_{12}(a_{21}\,a_{33} - a_{31}\,a_{23}) \\
&\quad + a_{13}\,(a_{21}\,a_{32} - a_{31}\,a_{22}) \\
&= 2(4 \cdot 2 - 1 \cdot (-1)) - 0(1 \cdot 2 - 3 \cdot (-1)) + 4(1 \cdot 1 - 3 \cdot 4) \\
&= 18 - 0 - 144 \\
&= -26
\end{aligned}
$$

This is answer (D).

3. **(B)**

The inverse of a square matrix is a square matrix of the same size. The product of a square matrix and its inverse is the identity matrix. The inverse of a two-by-two matrix D

$$\mathbf{D} = \begin{bmatrix} d_{11} & d_{12} \\ d_{21} & d_{22} \end{bmatrix} \text{ is}$$

$$D^{-1} = \frac{1}{|D|}\begin{bmatrix} d_{22} & -d_{12} \\ -d_{21} & d_{11} \end{bmatrix}$$

where $|D|$ is the determinant of D, which is found by simplifying the formula in problem 2.

$$\begin{aligned} |D| &= d_{11}(d_{22}) - d_{21}(d_{12}) \\ &= 1(3) - (-2)4 \\ &= 3 + 8 \\ &= 11 \end{aligned}$$

Thus the inverse of D is

$$\frac{1}{11}\begin{bmatrix} 3 & -4 \\ 2 & 1 \end{bmatrix}$$

which is answer (B).

4. **(E)**

The elements of a product $P = A \times C$ are defined as

$$P_{ij} = \sum_{k=1}^{N} a_{ik} c_{kj}$$

where N is the number of columns in A and the number of rows in C.

For

$$\mathbf{A} = \begin{bmatrix} 2 & 0 & 4 \\ 1 & 4 & -1 \\ 3 & 1 & 2 \end{bmatrix} \quad \mathbf{C} = \begin{bmatrix} 1 \\ 2 \\ 3 \end{bmatrix}$$

$$\begin{aligned} P_{11} &= (2 \times 1) + (0 \times 2) + (4 \times 3) = 14 \\ P_{21} &= (1 \times 1) + (4 \times 2) - (1 \times 3) = 6 \\ P_{31} &= (3 \times 1) + (1 \times 2) + (2 \times 3) = 11 \end{aligned}$$

$$P = \begin{bmatrix} 14 \\ 6 \\ 11 \end{bmatrix}$$

which is answer (E)

Note that multiplication of matrices cannot be done unless number of columns in the first matrix equals the number of rows in the second matrix.

5. **(D)**

The rank of a matrix is the number of independent rows and can be determined by calculating the determinant of the given matrix or determinants of smaller matrices composed of rows and columns of the original matrix. The determinant of the matrix **B** is 220. A non-zero determinant means that all rows are independent. The rank of **B** is 3, which is answer (D).

6. **(B)**

If two matrices have the same number of rows and the same number of columns, the sum is determined by adding corresponding elements.

$$S = \mathbf{A} + \mathbf{B}$$

$$s_{ij} = a_{ij} + b_{ij}$$

$$S = \begin{bmatrix} 2+3 & 0+5 & 4+2 \\ 1+0 & 4+8 & -1+9 \\ 3+1 & 1+0 & 2+9 \end{bmatrix} = \begin{bmatrix} 5 & 5 & 6 \\ 1 & 12 & 3 \\ 4 & 1 & 11 \end{bmatrix}$$

The answer is (B).

7. **(C)**

The product of a single number (scalar) and a matrix is another matrix obtained by multiplying every element of the original matrix by the scalar.

$$\mathbf{P} = 3 \times \mathbf{D}$$

$$P_{ij} = 3d_{ij}$$

$$= \begin{bmatrix} 3 & 12 \\ -6 & 9 \end{bmatrix}$$

The answer is (C).

8. **(E)**

Multiplication of two matrices $\mathbf{M}_1 \times \mathbf{M}_2$ is defined only if the number of columns in $\mathbf{M}_1$ is the same as the number of rows in $\mathbf{M}_2$. Recall the definition given in problem 4. Matrices C and D cannot be multiplied. The answer is (E).

9. **(C)**

The definition of the product **Ax** is given in the solution to problem 4. The first element of **Ax** is

$$A_{11}x_1 + A_{12}x_2 + A_{13}x_3 = 2x_1 + 0x_2 + 4x_3$$

When this is equated to the first element of **C**, the result is the first equation of the set.

$$2x_1 + 0 + 4x_3 = 1$$

The other two equations are found in a similar manner. The correct answer is (C).

10. **(C)**

Statement (C) is incorrect. Addition is commutative, but multiplication is not.

11. **(B)**

The characteristic equation is

$$r^2 + 5r + 6 = 0$$

$$(r + 2)(r + 3) = 0$$

Where the exponent of r corresponds to the degree of differentiation at y.

i.e., $y'' \Rightarrow r^2 \quad y' \Rightarrow r$

The roots are -2 and -3, which correspond to answer (B).

12. **(E)**

The complementary solution is the solution to the homogeneous equation

$$\frac{d^2y}{dt^2} + 5\frac{dy}{dt} + 6y = 0$$

The complementary solution has the form

$$\sum C_i e^{r_i t}$$

where r_i = a root of the characteristic equation.

C_i = a coefficient to be determined from the initial conditions.

The roots of the characteristic equation were determined in problem 11. The complementary solution is

$$C_1 e^{-2t} + C_2 e^{-3t}$$

which is answer (E).

13. **(A)**

The particular solution is found by assuming that the particular solution has the same form as the right hand side of the equation. Because the right hand side is a constant, 4, the particular solution is assumed to be a constant, K. This assumed form is substituted into the equation to obtain

$$0 + 0 + 6K = 4$$

Thus, $K = 4/6$ or $2/3$, which is answer (A).

14. **(C)**

The definition of the Laplace Transform is

$$Y(s) = \int_0^\infty y(t)e^{-st}\,dt$$

This integral has been evaluated for most functions of interest. A Table of Transforms is used, an example of which is the following.

$y(t)$	$Y(s)$
1	$\frac{1}{s}$
e^{at}	$\frac{1}{s-a}$
$\frac{dy}{dt}$	$SY(S) - y(0)$
$\frac{d^2y}{dt^2}$	$s^2Y(s) - s\frac{dy}{dt}(0) - y(0)$

Each term of the problem equation is transformed using these transform pairs.

$$\left(s^2Y(s) - s\frac{dy}{dt}(0) - y(0)\right) + 5(sY(s) - y(0)) + 6Y(s) = \frac{4}{s}$$

The initial conditions are zero. $Y(s)$ can be factored in the left-hand side.

$$(s^2 + 5s + 6)\,Y(s) = \frac{4}{s}$$

or

$$Y(s) = \frac{4}{s(s^2 + 5s + 6)}$$

which is answer (C).

This transform can easily be expanded by partial fractions.

$$Y(s) = \frac{4}{s(s^2 + 5s + 6)} = \frac{4}{s(s+2)(s+3)}$$

$$= \frac{A}{s} + \frac{B}{s+2} + \frac{C}{s+3}$$

Solve for constants.

$$= \frac{2}{3s} - \frac{2}{s+2} + \frac{4}{3(s+3)}$$

Each term can be inverted to the time domain using the transform table.

$$y = \frac{2}{3} - 2e^{-2t} + \frac{4}{3}e^{-3t}$$

15. **(A)**

The total solution is the sum of the complementary solution and the particular solution.

$$Y = C_1 e^{-2t} + C_2 e^{-3t} + {}^2/_3$$

The constants C_1 and C_2 are evaluated with initial conditions.

$$y = -2e^{-2t} + {}^4/_3 e^{-3t} + {}^2/_3$$

$$y(0) = C_1 + C_2 + {}^2/_3 = 0 \quad (1)$$

$$\frac{dy}{dt}(0) = -2C_1 - 3C_2 = 0 \quad (2)$$

From equation (1)

$$C_1 = -{}^2/_3 - C_2$$

Substituting into equation (2)

$$-2(-{}^2/_3 - C_2) - 3C_2 = 0$$

$${}^4/_3 + 2C_2 - 3C_2 = 0$$

$$C_2 = {}^4/_3$$

Then

$$C_1 = -{}^2/_3 - {}^4/_3 = -2$$

or $\quad y = -2e^{-2t} + {}^4/_3 e^{-3t} + {}^2/_3$

at $t = \infty$, the first two terms approach zero. $y \Rightarrow {}^2/_3$.

This is the same solution as obtained by inverting the Laplace Transform in problem 14.

The limit of y as t becomes very large is ${}^2/_3$, which is answer (A).

Note that the roots of the characteristic equation are negative. This means that all exponentials in the solution will decay to zero, and that all time derivatives will approach zero. The differential equation at large values of time approaches

$$0 + 0 + 6y = 3$$

At large time

$$y = {}^4/_6 = {}^2/_3.$$

16. **(D)**

The arithmetic mean is the sum of the numbers in the set divided by how many numbers are in the set. There are nine values.

$$-3 + 1 + 2 + 2 + 4 + 4 + 4 + 8 + 23 = 45$$

$${}^{45}/_9 = 5$$

which is answer (D).

17. **(C)**

The mode is the value that occurs most frequently. The value "4" occurs three times and, therefore, is the mode. This is answer (C).

18. **(C)**

The median is the middle value when all data are arranged in increasing or decreasing order. The set of 9 numbers is arranged in increasing order. The fifth value, which is 4, is the middle one. This is answer (C). (When there is an even number of values, the median is defined as the mean of the middle two values).

19. **(D)**

The range is the difference between the maximum value and the minimum value.

$$\text{Range} = 23 - (-3) = 26$$

This is answer (D).

20. **(E)**

The variance of a set of values is defined by

$$\sigma^2 = \frac{\Sigma(X_i - \text{Mean})^2}{n}$$

It is convenient to prepare a table. The mean is 5.

X_i	$(X_i - \text{Mean})$	$(X_i - \text{Mean})^2$
−3	−8	64
1	−4	16
2	−3	9
2	−3	9
4	−1	1
4	−1	1
4	−1	1
8	3	9
23	18	324
	Sum	434

The variance is

$$\frac{434}{9} = 48.2$$

Which corresponds to answer (E).

21. **(B)**

Equate individual centers of mass to the total system:

$$2mL_{cg} = mR + m(3R)$$
$$L_{cg} = 2R.$$

22. **(E)**

Determine the individual moments from the general formulas. Then use the parallel axis theorem to find the moments of inertia about point O.

$$I_{OO} = (1/12)m\,(2R)^2 + mR^2 + (1/2)mR^2 + m(3R)^2$$
$$= (65/6)mR^2.$$

23. **(C)**

The center of percussion is

$$q = K_0^2/L_{cg}$$
$$q = (2.33)^2/2 = 2.71 \text{ meters.}$$

24. **(C)**

$$I_{OO}\alpha = \Sigma M_O$$
$$I_{OO} = k^2 m$$
$$[20(2.33)^2]\alpha = 20(9.81)\,2\cos 60$$
$$\alpha = 1.81 \text{ rad/sec/sec.}$$

25. **(D)**

The angular acceleration will be the same as in problem 24. Applying Newton's second law along the tangential direction gives

$$ma_t = \Sigma F_t = mg\cos\theta - O_t$$
$$20(2)\,1.81 = 20(9.81)\cos 60 - O_t$$
$$O_t = 25.7\, e_t \text{ Newton}$$

26. **(A)**

The angular acceleration in terms of the angle θ is obtained from the equation

$$I_{OO}\alpha = \Sigma M_O = mgL_{cg}\cos\theta, \text{ or}$$
$$\alpha = (gL_{cg} / K_0^2)\cos\theta.$$

Applying the well-known relationship

$$\omega\, d\omega = \alpha\, d\theta$$

after integration gives

$$\omega^2 = (2gL_{cg} / K_0^2) \sin \theta.$$

Now applying Newton's second law along e_n gives

$$m\omega^2 L_{cg} = O_n - mg \sin 60.$$

Thus $\quad O_n = 20[2(9.81)\ 2 \times 2 \sin 60 / (2.33)^2 + 9.81 \sin 60]$

$$= 20\ [12.52 + 8.4957] = 420.3\ e_n \text{ N}.$$

27. **(D)**

Use the formula for rotational kinetic energy and substitute in the formula from problem 26 for angular velocity:

$$K = 0.5\, I_{OO}\, \omega^2$$

$$\omega^2 = 2gL_{cg} \sin 60 / K_0^2$$

$$= 2(9.81)\ 2(0.866) / (2.33)^2$$

$$= 6.26.$$

$$\omega = 2.5 \text{ rad/sec.}$$

$$K = 0.5\ (20)\ (2.33)^2\ (2.5)^2$$

$$= 340 \text{ joules.}$$

28. **(C)**

Using the definition of angular momentum,

$$H_o = I_{OO}\omega = 20(2.33)^2\ 2.5$$

$$= 271 \text{ Kg m}^2\text{/sec CW.}$$

29. **(B)**

Using the angular velocity relation found in problem 26,

$$\omega^2 = 2gL_{cg} \sin 90 / K_0^2$$

$$= 2(9.81)2 / (2.33)^2$$

$$= 7.23$$

$$\omega = 2.68 \text{ rad/sec.}$$

30. **(A)**

Let the angle that exceeds 90° be ϕ. The moment equation of motion of the system about O in terms of ϕ is:

$$I_{oo}\ddot{\phi} = -mgL_{cg} \sin \phi \approx -mgL_{cg} \phi$$

or

$$\ddot{\phi} + \frac{mgL_{cg}}{I_{oo}} \phi = 0$$

Thus the frequency of oscillation is

$$f = \sqrt{\frac{mgL_{cg}}{I_{oo}}} / 2\pi = \sqrt{\frac{20(9.81)2}{20(2.33)^2}} / 2\pi$$

$$= 1.9010 / 2\pi = 0.303 \approx 0.3 \text{ Hz.}$$

31. **(D)**

The 65% B alloy exists as the solid solution σ in the temperature range of 800° C to 1500° C which is the melting point of this alloy.

32. **(A)**

The 50% B alloy at 1300° C is in the $L + \sigma$ two phase region. The tie line for this alloy at this temperature intersects the phase boundaries at 41% B on the liquid side and at 61% B on the σ side.

33. **(C)**

The transformation occurring at 1000° C is one in which liquid and s combine to form β. This type of transformation is the peritectic transformation.

34. **(B)**

Application of lever rule to the 35% B alloy at 900° C yields for the fraction of σ in the alloy

$$f_\sigma = \frac{35 - 9}{64 - 9} = 0.473$$

α
900° C
σ
9%
35%
64%

The amount of σ in 50 kg of the alloy is given as 50 × 0.473 = 23.6 kg.

35. **(B)**

The maximum solubility of A in B is given by the maximum horizontal extent of the terminal solid solution β in the phase diagram. This is seen from the phase diagram as 20% at 700° C.

36. **(C)**

The 3% B alloy exists in the single phase α region at temperatures below 1200° C. The phase rule states that, at constant pressure, the number of degrees of freedom, F, is given by

$$F = C - P + 1$$

where C is the number of components and P is the number of phases. In this case, there are two components, (A and B), one phase (α) and hence there are two degrees of freedom.

37. **(D)**

The tie line through the $\alpha + \sigma$ two phase region at 950° C is shown below. The composition of the alloy that will contain 10% σ can be calculated as

$$\frac{x-9}{64-9} = .10$$

$$x = 14.5$$

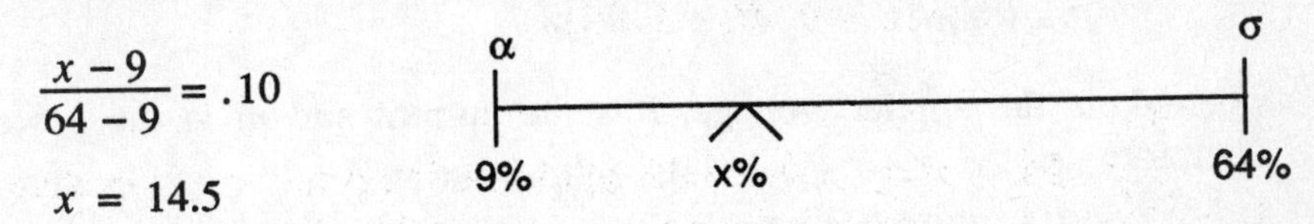

$\therefore$ alloy is 14.5% B.

38. **(C)**

Since the densities of α and β are the same, the volume percentage of the precipitate is the same as the weight percentage of the precipitates. Application of lever rule for this alloy at 400° C yields

$$\% \alpha = \frac{90-82}{90-4} \times 100$$

$$= 9.3\%$$

α
400° C
β
4%
82%
90%

39. **(C)**

From the phase diagram, it can be seen that the 20% B alloy begins to solidify as a single phase solid (α) at 1300° C. This is accompanied by a reduction in the slope of the cooling curve due to the liberation of latent heat of fusion. The alloy continues to solidify as α until the temperature of 1200° C is reached at which point the remaining liquid (which is of eutectic composition) solidifies as a eutectic. The latter reaction occurs isothermally at 1200° C. This is seen as an arrest in the cooling curve until solidification is complete. The solid alloy continues to cool beyond this point. The cooling curve which shows all of these features is the one shown in (C).

40. **(E)**

The 80% B alloy is completely liquid at temperatures above 1360° C and is completely solid at temperatures below 1000° C. When this alloy is cooled from the liquid region, solidification begins at 1360° C and ends at 1000° C resulting in a freezing range of 360° C.

41. (A)

For a phase rotation 1 – 2 – 3 the phasor diagram for the line voltages is shown here.

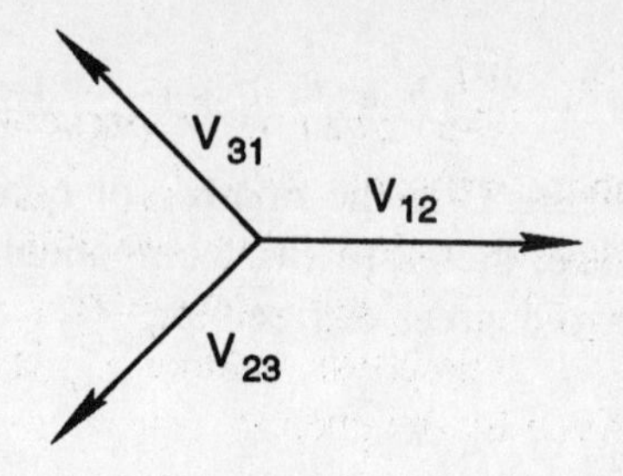

The phase angle of V_{31} is therefore 120°.

42. (C)

The power P for a one phase load is given by the relation

$$P = V\,I\,pf$$

where V is the applied voltage, I is the current and pf is the power factor. Therefore

$$5{,}000 = 220 \cdot I \cdot 1.00$$

$$I = 22.72 \text{ A}$$

43. (C)

As the power factor of the load **L** is 1, the current $\mathbf{I}_L$ must be in phase with the applied voltage $\mathbf{V}_{12} = 220\,\underline{|0°}$ V therefore the current $\mathbf{I}_L$ has a phase angle of 0°, and

$$\mathbf{I}_L = 22.72\,\underline{|0°} \text{ A}$$

44. (D)

The electrical power P_e in a motor is related to the mechanical shaft power given in HP and the efficiency η by:

$$P_e = \frac{\text{HP} \cdot 0.746}{\eta}$$

$$= \frac{5 \cdot 0.746}{0.8}$$

$$= 4.66 \text{ KW}$$

45. (E)

The real power P_e in a three phase balanced network is related to the line voltage V_1, the current I_{M1} and the power factor pf by the relation

$$P_e = \sqrt{3}\,V_1 I_1 pf$$

$$4660 = \sqrt{3} \cdot 220 \cdot I_{M1} \cdot 0.85$$

and therefore

$$I_{M1} = 14.39 \text{ A}$$

46. **(E)**

The phasors representing the phase voltages V_{1n}, V_{2n}, V_{3n} and the phasors representing the line voltages V_{12}, V_{23}, V_{31} are shown here.

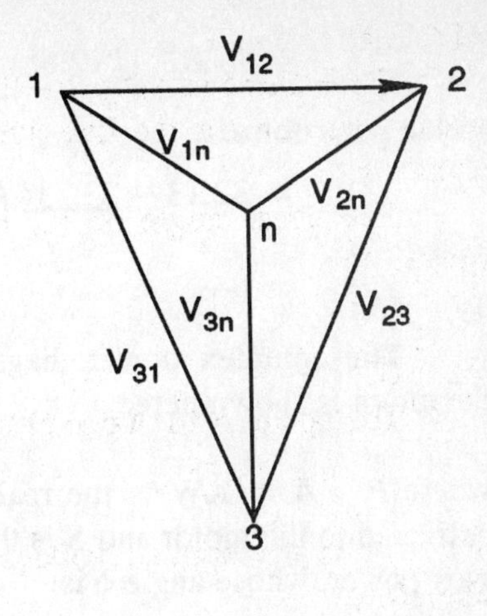

The phase voltage V_{1n} is therefore given by the phasor:

$$\mathbf{V}_{1n} = \frac{V_{12}}{\sqrt{3}} \cdot \underline{|-30°} \, V$$

$$= \frac{220}{\sqrt{3}} \cdot \underline{|-30°} \, V$$

$$= 127 \, \underline{|-30°} \, V$$

47. **(B)**

The phase angle between the phase voltage V_{12} and the phase current I_{M1} in the motor is given by:

$$\phi = -a \cos (pf)$$

$$= -a \cos 0.85°$$

$$= -31.78°$$

The negative angle indicates that the current lags the voltage because load M is an induction motor. As the phase voltage V_{12} has an angle of – 30°, and I_{M1} lags – 31.78° behind, the phase angle for I_{M1} is – 61.78°. As the motor is a balanced load the currents in the other two lines to the motor will be

– 181.78° and – 301.78°.

From problem 45, it was obtained that the magnitude I_{M1} = 14.39 A. The three motor currents are given by:

$$I_{M1} = 14.39 \, \underline{|-61.78°} \text{ A}$$

$$I_{M2} = 14.39 \, \underline{|-181.78°} \text{ A}$$

$$I_{M3} = 14.39 \, \underline{|-301.78°} \text{ A}$$

48. **(B)**

By Kirchoff's current law the line current I_1 is related to the motor current I_{M1} and to the load current I_L by the relation:

$$I_1 = I_{M1} + I_L$$

$$= 14.39 \, \underline{|-61.78°} + 22.72 \, \underline{|0°}$$

transforming from polar to rectangular form:

$$= 6.80 - j\,12.68 + 22.72 + j0$$

$$= 29.52 - j\,12.68$$

and in polar form:

$$= 32.13 \angle -23.23° \text{ A}$$

49. **(A)**

The complex power diagram for the motor is shown here.

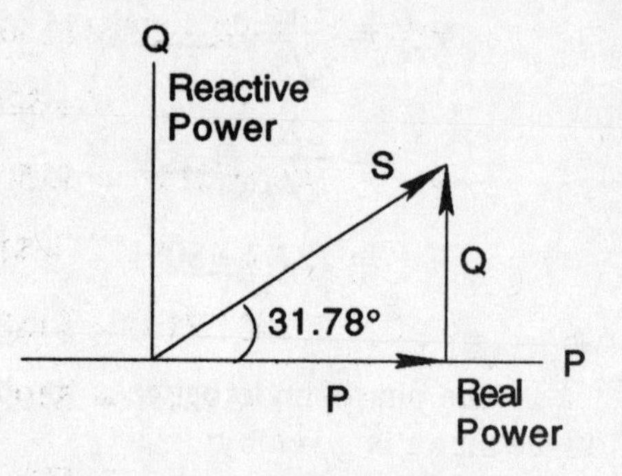

where P = 4.66 KW is the real power delivered to the motor and S is the complex power, whose angle ϕ is:

$$\phi = a\cos(pf)$$

$$= a\cos 0.85$$

$$= 31.78°$$

The reactive power Q delivered by the network to the motor is therefore:

$$Q = P \cdot \tan 31.78°$$

$$= 4.66 \cdot \tan 31.78°$$

$$= 2.88 \text{ KVAR}$$

To change the power factor to 1 the capacitor bank has to supply this reactive power, the capacitor bank will be rated 2.88 KVAR.

50. **(B)**

The current I_c can e found form the relation:

$$Q = \sqrt{3} V I_c$$

from which

$$I_c = \frac{2880}{\sqrt{3} \cdot 220}$$

$$= 7.57 \text{ A}$$

51. **(C)**

Let X be the number of cores to be produced or bought per year.

Total cost of buying core/yr = \$90 X.

Total cost of producing X units per yr

$$= \$26X + \$9{,}000 + \$150{,}000(A/P, 15\%, 10)$$

$$- \$10{,}000(A/F, 15\%, 10)$$

$$= \$26X + \$9{,}000 + \$150{,}000(0.1993) - \$10{,}000(0.0493)$$

$$= \$26X + \$38{,}402$$

The break-even number of cores per year is given by

$$\$90X = \$26X + \$38{,}402$$

or $$X = 600.313 \text{ or } 600$$

Alternative Solution

$$\$90X(P/A., 15\%, 10) = \$150{,}000 + \$9{,}000\ (P/A, 15\%, 10) + \$26X(P/A, 15\%, 10) - \$10{,}000(P/F, 15\%, 10)$$

$$\$90X(5.019) = \$150{,}000 + \$9{,}000(5.019) + \$26X(5.019) - \$10{,}000(0.2472)$$

$$\$451.71X = \$150{,}000 + \$45{,}171 + \$130{,}494X - \$2{,}472$$

$$\$321X = \$192{,}699$$

$$X = 600.31 \text{ or } 600$$

52. **(B)**

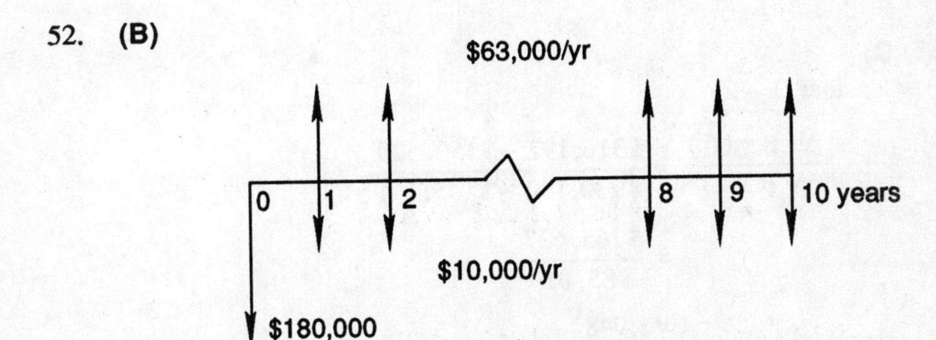

$$\text{Present Worth of benefits} = \$63{,}000(P/A, 15\%, 10)$$
$$= \$63{,}000(5.019)$$
$$= \$316{,}197$$

$$PW(\text{Cost}) = \$180{,}000 + \$10{,}000(1A, 15\%, 10)$$
$$= \$180{,}000 + \$10{,}000(5.019)$$
$$= \$230{,}190$$

$$\frac{PW\ (\text{Benefit})}{PW\ (\text{Cost})} = \frac{\$316{,}197}{\$230{,}190} = 1.374$$

53. **(B)**

$$\text{Present Worth of Costs} = \$100{,}000 + \$9{,}000(P/A, 15\%, 10)$$

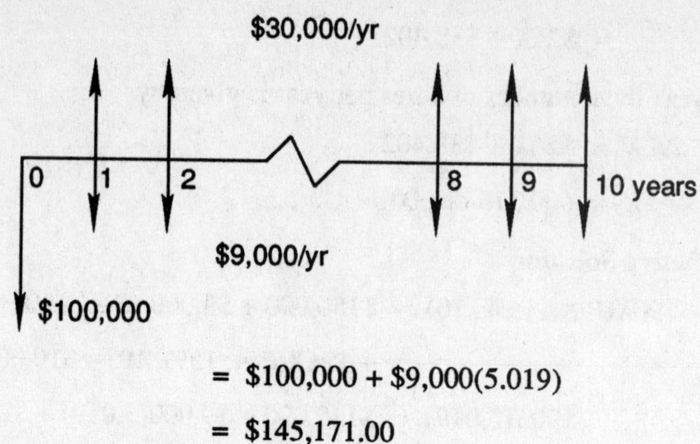

$$= \$100{,}000 + \$9{,}000(5.019)$$

$$= \$145{,}171.00$$

$$\text{Present Worth of Benefits} = \$30{,}000(P/A,\ 15\%,\ 10)$$

$$= \$30{,}000(5.019)$$

$$= \$150{,}570$$

$$\frac{PW\ (\text{Benefits})}{PW\ (\text{Cost})} = \frac{\$\,150{,}570}{\$\,145{,}172} = 1.037$$

54. **(C)**

Consider $(Y - Z)$

$$\frac{\Delta\ (\text{Benefit})}{\Delta\ (\text{Cost})} = \frac{\$316{,}197 - \$150{,}570}{\$230{,}190 - \$145{,}171}$$

$$= \frac{\$165{,}627}{\$85{,}019}$$

$$= 1.948$$

55. **(A)**

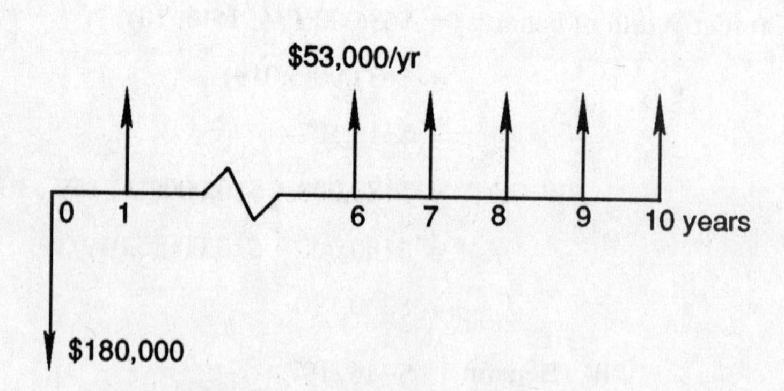

$$PW = -\$180{,}000 + \$53{,}000(P/A,\ 10\%,\ 10)$$

$$= -\$180{,}000 + \$53{,}000(6.145)$$

$$= \$145{,}685 \text{ or } \$146{,}000$$

56. **(D)**

$$\$180{,}000 = \$53{,}000(P/A, i\%, 10)$$

$$(P/A, i\%, 10) = \frac{\$\,180{,}000}{\$\,53{,}000} = 3.396$$

From the tables for compound interest factors,

$i\%$	$(P/A, i\%, 10)$
25%	3.571
30%	3.092

So the rate of return is about 27%.

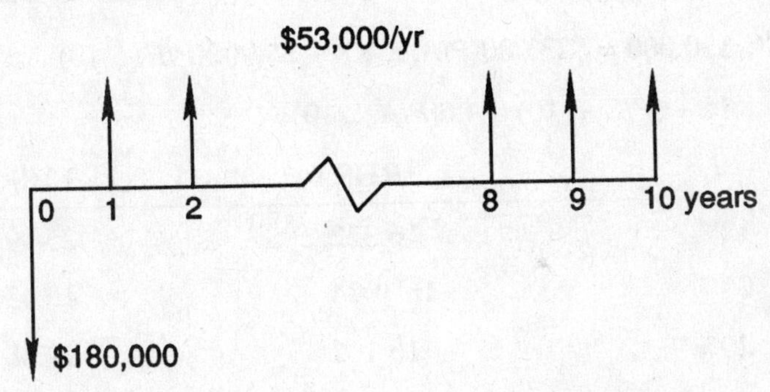

57. **(E)**

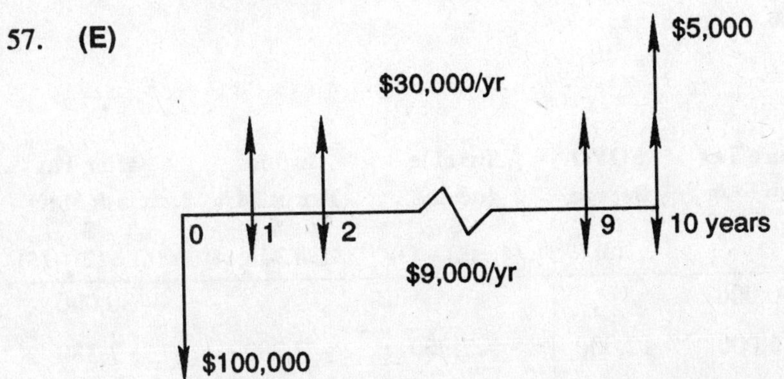

$$\$100{,}000 = \$21{,}000(P/A, i, 10) - \$5{,}000(P/F, i, 10)$$

$$20 + (P/F, i\%, 10) = 4.2\,(P/A, i, 10)$$

i	RHS	LHS	RHS – LHS
10%	20.385	25.809	
12%	20.322	23.73	
15%	20.2472	21.08	– 0.83
18%	20.1911	18.87	+ 1.32

$$i = 15 + \frac{.83}{2.15} \cdot 3$$

$$\approx 16\%$$

58. (D)

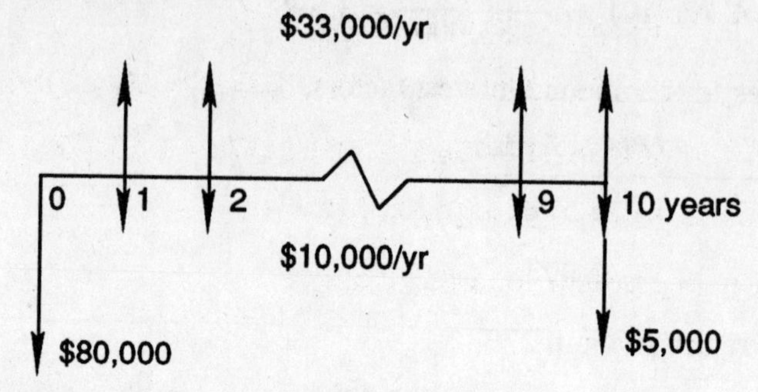

$$\$80{,}000 = \$23{,}000(P/A, i, 10) - \$5{,}000(P/F, i, 10)$$

$$16 + (P/F, i, 10) = 4.6(P/A, i, 10)$$

i	RHS	LHS
12%	16.322	25.99
25%	16.1074	16.43
20%	16.165	19.28

Δ ROR $\approx$ 25%.

59. (E)

Year (1)	Before Tax Cash Flow $ (2)	SOYD Deprec. $ (3)	Taxable Income $ (4)=(2) – (3)	Income Tax at 34% $ (5)=0.34 × (4)	After Tax Cash Flow $ (6) = (2) – (5)
0	– 30,000				– 30,000
1	5,000	12,000	– 7,000	– 2,380	7,380
2	4,000	9,000	– 5,000	– 1,700	5,700
3	3,000	6,000	– 3,000	– 1,020	4,020
4	2,000	3,000	– 1,000	– 340	2,340
5	{ 1,000 S	0	1,000	{ 340 0.34S	{ 660 0.66S

$$\$30{,}000 = \$7{,}380(P/F, 5\%, 1) + \$5{,}700(P/F, 5\%, 2)$$
$$+ \$4{,}020(P/F, 5\%, 3) + \$2{,}340(P/F, 5\%, 4)$$
$$+ \$660(P/F, 5\%, 5) + 0.66S(P/F, 5\%, 5)$$

$$S \approx \$23{,}000$$

Alternative Solution

$$\begin{aligned} \$30{,}000 &= \$7{,}380(P/A, 5\%, 4) - \$1{,}680(P/G, 5\%, 4) \\ &\quad + (\$660 + \$0.66S)\,(P/F, 5\%, 5) \\ &= \$7{,}380(3.546) - \$1{,}680(5.103) + (\$660 + \$0.66S)\,(0.7835) \\ &= \$26{,}169.48 - \$8{,}573.04 + \$517.11 + \$0.51711S \\ &= \$18{,}113.55 + \$0.51711S \\ &= \$22{,}986.30 \text{ or } \$23{,}000 \end{aligned}$$

60. **(D)**

Year (1)	Before Tax Cash Flow $ (2)	Straight Line Deprec. $ (3)	Taxable Income $ (4)=(2) – (3)	Income Tax at 34% $ (5)=0.34 × (4)	After Tax Cash Flow $ (6) = (2) – (5)
0	– 30,000				– 30,000
1	5,000	7,500	– 2,500	– 850	5,850
2	4,000	7,500	– 3,500	– 1,190	5,190
3	3,000	7,500	– 4,500	– 1,530	4,530
4	2,000	7,500	– 5,500	– 1,870	3,870
5	{ 1,000	0	1,000	{ 340	{ 660
	S			0.34S	0.66S

$$\begin{aligned} \$30{,}000 &= \$5{,}580(0.9524) + \$5{,}190(0.9070) \\ &\quad + \$4{,}530(0.8638) + \$3{,}870(0.8227) \\ &\quad + \$660(0.7835) + \$6.66(0.7835)S \\ S &\approx \$23{,}400 \end{aligned}$$

Alternative Solution

$$\begin{aligned} \$30{,}000 &= \$5{,}850(P/A, 5\%, 4) - \$660(P/G, 5\%, 4) \\ &\quad + (\$660 + \$0.66S)\,(P/F, 5\%, 5) \\ &= \$5{,}850(3.546) - \$660(5.103) + (\$660 + \$0.66S)\,(0.7835) \\ &= \$17{,}893.23 + \$0.51711S \\ S &= \$23{,}412.37 \quad \text{or} \quad \$23{,}400.00 \end{aligned}$$

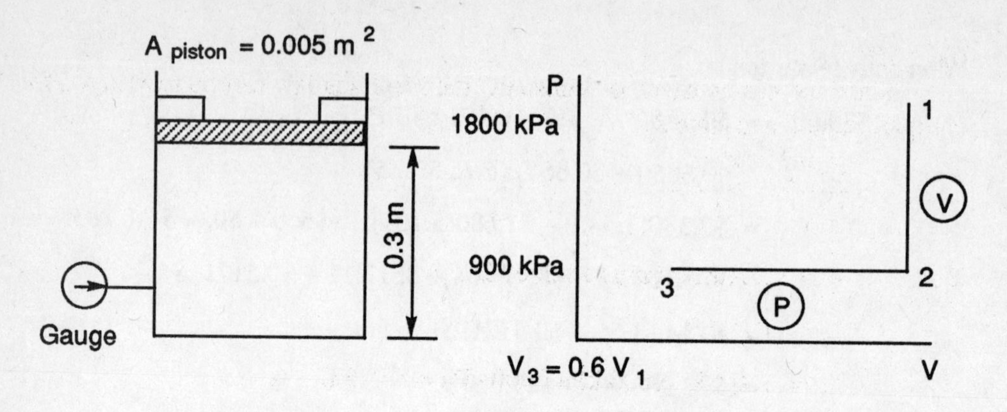

Gas	Molecular Weight	Specific heats, kJ/kg-K C_p	Specific heats, kJ/kg-K C_v	Gas Constant kJ/kg–K
O_2	32.00	0.9216	0.6618	0.2598
N_2	28.013	1.0416	0.7748	0.2968

61. **(B)**

The mass can be obtained from the application of the ideal gas law at state point 1

$$PV = \mathrm{m}RT$$

The volume is calculated from the geometry

$$V = A_{piston} \cdot \text{height} = 0.005\ m^2 \cdot 0.3\ m = 0.0015\ m^3$$

solving for m and substituting values produces

$$m = PV/RT = (1800 \text{ kPa} \cdot 0.0015 m^2) / (0.2598 \text{ kJ/kg–K} \cdot 1003 \text{ K})$$

$$m = 0.01036 \text{ kg}$$

62. **(C)**

The gauge pressure can be calculated from the application of the pressure equation; the absolute pressure is equal to the gauge pressure plus the reference pressure (often the atmospheric pressure). $P_{atm} = 101.3$ kPa.

$$P_{abs} = P_{gage} + P_{ref}$$

$$P_{gage} = P_{abs} - P_{ref}$$

$$P_{gage} = 1800 \text{ kPa} - 101.3 \text{ kPa}$$

$$P_{gage} = 1698.7 \text{ kPa}$$

Remember that P_{abs} is always the pressure that the gauge feels.

63. **(D)**

Since the gas constant is "constant" the ideal gas law can be rearranged to compare state 1 and state 2

$$\frac{P_1V_1}{T_1} = \frac{P_2V_2}{T_2}$$

But $V_1 = V_2$, rearranging to solve for T_2

$$T_2 = T_1P_2 / P_1$$

$$T_2 = 1003 \text{ K} \cdot 1800 \text{ kPa} / 900 \text{ kPa} = 501.5 \text{ K}$$

64. **(A)**

The total work is equal to the work from state 1 to state 2 plus the work from state 2 to state 3. Since the volume is constant from state 1 to state 2 the work during this process is zero. This is true since the only work mode is boundary work.

$$W_{13} = W_{12} + W_{23}$$

$$W_{13} = W_{23} = \int_2^3 PdV$$

$$W_{13} = P_{2 \text{ or } 3}(V_3 - V_2)$$

Now, $V_2 = V_1 = 0.0015m^3$

and $V_3 = 0.6\ V_2 = 0.009m^3$

$$W_{13} = 900 \text{ kPa } (0.009 - 0.0015)m^3$$

$$W_{13} = -0.54 \text{ kPa–}m^3 = -0.54 \text{ (kN/}m^2)m^3 = -0.54 \text{ kJ}$$

The negative sign indicates that the work is being performed on the gas. Since process 2 – 3 is a compression the work is on the gas and must be negative. All answers except (A) are impossible.

65. **(D)**

Realizing that pressure is force per unit area, a free body diagram about the piston produces the following pressure relations

$$P_2 = P_{piston} + P_{atm}$$

P_2 is used since 900 kPa are required to support the piston. Solving for P_{piston}

$$P_{piston} = P_2 - P_{atm} = 900 \text{ kPa} - 101.3 \text{ kPa} = 798.7 \text{ kPa}$$

The piston pressure is merely the force of the piston divided by the piston area. Extreme care must be taken when considering units.

$$P_{piston} = [(m_{piston} \cdot g) / A_{piston}] \cdot 1 \text{ kPa} / 1000 \text{ Pa}$$

$m_{piston} = [(P_{piston} \cdot A_{piston}) / g] \cdot 1000$ Pa/kPa

$m_{piston} = (798.3 \text{ kPa}) (0.005 m^2) (1000 \text{ Pa/kPa}) / 9.8 \, m/s^2$

$m_{piston} = 407.1$ kg

66. **(B)**

The change in entropy can be accomplished in parts, 1 – 2 plus 2 – 3, or in one step from state 1 to state 3. The latter method will be used here.

$$\Delta s = C_p \ln (T_3 / T_1) - R \ln (P_3 / P_1)$$

Need to calculate T_3. With R equal to a constant and $P_3 = P_2$ the ideal gas law produces

$$T_3 / V_3 = T_2 \ / V_2$$

$$T_3 = T_2(V_3 / V_2) \text{ but } V_3 = 0.6 \, V_2$$

$$T_3 = 0.6 \, T_2 = 0.6 \cdot 501.5 \text{ K} = 300.9 \text{ K}$$

$$\Delta s = 0.9216 \text{ kJ/kg–k} \ln (300.9 \text{ K} / 1003 \text{ K})$$

$$- 0.2598 \text{ kJ/kg–K} \ln (900 \text{ kPa} / 1800 \text{ kPa})$$

$$\Delta s = - 0.9295 \text{ kJ/kg–K}$$

67. **(C)**

The ideal gas law can be used to calculate the mass of the mixture of two ideal gases. First the molecular weight of the mixture must be calculated, then the gas constant, and finally the mass of the mixture. The analysis is best done at state 1 since most of the data is known.

Gas	Fraction	Mole Wt	mass/kmole mix
O_2	0.6	32.00	19.2
N_2	0.4	28.013	11.2052
		Total	30.4053

$R_{mix} = R/M$

$R_{mix} = (8.31434 \text{ kJ/kmole–K})/(30.4052 \text{ kg/kmole})$

$R_{mix} = 0.27345$ kJ/kg–K

$PV = mRT$

$m = PV/RT$

$m = (1800 \text{ kPa} \cdot 0.005 m^2 \cdot 0.3 m)/(0.27345 \text{ kJ/kg–K} \cdot 1003 \text{ K})$

$m = 0.009844$ Kg

68. **(D)**

The answer and solution to this question is exactly the same as problem 63. Since the only difference with the mixture is the gas constant, and since the gas constant does not enter into the calculations, the temperature is the same as long as an ideal gas is the medium.

69. **(B)**

The constant volume specific heat is calculated from the sum of the mole fractions times the individual specific heats for the ideal gas.

$$C_{v\ mix} = y_{oxygen} C_{v\ oxygen} + y_{nitrogen} C_{v\ nitrogen}$$

$$C_{v\ mix} = 0.6 \cdot 0.6618\ \text{kJ/kg–K} + 0.4 \cdot 0.7748\ \text{kJ/kg–K}$$

$$C_{v\ mix} = 0.707\ \text{kJ/kg–K}$$

70. **(D)**

The partial pressure is equal to the mole fraction times the total pressure of the mixture.

$$P_{oxygen} = y_{oxygen} \cdot P_1$$

$$P_{oxygen} = 0.6 \cdot 1800\ \text{kPa}$$

$$P_{oxygen} = 1080\ \text{kPa}$$

Fundamentals of Engineering

EXTRA PRACTICE PROBLEMS

Fundamentals of Engineering

EXTRA PRACTICE PROBLEMS

ANSWER SHEET

1. Ⓐ Ⓑ Ⓒ Ⓓ Ⓔ
2. Ⓐ Ⓑ Ⓒ Ⓓ Ⓔ
3. Ⓐ Ⓑ Ⓒ Ⓓ Ⓔ
4. Ⓐ Ⓑ Ⓒ Ⓓ Ⓔ
5. Ⓐ Ⓑ Ⓒ Ⓓ Ⓔ
6. Ⓐ Ⓑ Ⓒ Ⓓ Ⓔ
7. Ⓐ Ⓑ Ⓒ Ⓓ Ⓔ
8. Ⓐ Ⓑ Ⓒ Ⓓ Ⓔ
9. Ⓐ Ⓑ Ⓒ Ⓓ Ⓔ
10. Ⓐ Ⓑ Ⓒ Ⓓ Ⓔ
11. Ⓐ Ⓑ Ⓒ Ⓓ Ⓔ
12. Ⓐ Ⓑ Ⓒ Ⓓ Ⓔ
13. Ⓐ Ⓑ Ⓒ Ⓓ Ⓔ
14. Ⓐ Ⓑ Ⓒ Ⓓ Ⓔ
15. Ⓐ Ⓑ Ⓒ Ⓓ Ⓔ
16. Ⓐ Ⓑ Ⓒ Ⓓ Ⓔ
17. Ⓐ Ⓑ Ⓒ Ⓓ Ⓔ
18. Ⓐ Ⓑ Ⓒ Ⓓ Ⓔ
19. Ⓐ Ⓑ Ⓒ Ⓓ Ⓔ
20. Ⓐ Ⓑ Ⓒ Ⓓ Ⓔ
21. Ⓐ Ⓑ Ⓒ Ⓓ Ⓔ
22. Ⓐ Ⓑ Ⓒ Ⓓ Ⓔ
23. Ⓐ Ⓑ Ⓒ Ⓓ Ⓔ
24. Ⓐ Ⓑ Ⓒ Ⓓ Ⓔ
25. Ⓐ Ⓑ Ⓒ Ⓓ Ⓔ
26. Ⓐ Ⓑ Ⓒ Ⓓ Ⓔ
27. Ⓐ Ⓑ Ⓒ Ⓓ Ⓔ
28. Ⓐ Ⓑ Ⓒ Ⓓ Ⓔ
29. Ⓐ Ⓑ Ⓒ Ⓓ Ⓔ
30. Ⓐ Ⓑ Ⓒ Ⓓ Ⓔ
31. Ⓐ Ⓑ Ⓒ Ⓓ Ⓔ
32. Ⓐ Ⓑ Ⓒ Ⓓ Ⓔ
33. Ⓐ Ⓑ Ⓒ Ⓓ Ⓔ
34. Ⓐ Ⓑ Ⓒ Ⓓ Ⓔ
35. Ⓐ Ⓑ Ⓒ Ⓓ Ⓔ
36. Ⓐ Ⓑ Ⓒ Ⓓ Ⓔ
37. Ⓐ Ⓑ Ⓒ Ⓓ Ⓔ
38. Ⓐ Ⓑ Ⓒ Ⓓ Ⓔ
39. Ⓐ Ⓑ Ⓒ Ⓓ Ⓔ
40. Ⓐ Ⓑ Ⓒ Ⓓ Ⓔ
41. Ⓐ Ⓑ Ⓒ Ⓓ Ⓔ
42. Ⓐ Ⓑ Ⓒ Ⓓ Ⓔ
43. Ⓐ Ⓑ Ⓒ Ⓓ Ⓔ
44. Ⓐ Ⓑ Ⓒ Ⓓ Ⓔ
45. Ⓐ Ⓑ Ⓒ Ⓓ Ⓔ
46. Ⓐ Ⓑ Ⓒ Ⓓ Ⓔ
47. Ⓐ Ⓑ Ⓒ Ⓓ Ⓔ
48. Ⓐ Ⓑ Ⓒ Ⓓ Ⓔ
49. Ⓐ Ⓑ Ⓒ Ⓓ Ⓔ
50. Ⓐ Ⓑ Ⓒ Ⓓ Ⓔ

FUNDAMENTALS OF ENGINEERING EXAMINATION

EXTRA PRACTICE PROBLEMS

DIRECTIONS: The following problem sets are for your practice use. They are not connected to any of the tests in this book. They are simply for the practice of various engineering theories which appear throughout the tests.

QUESTIONS 1–10 refer to the following problem.

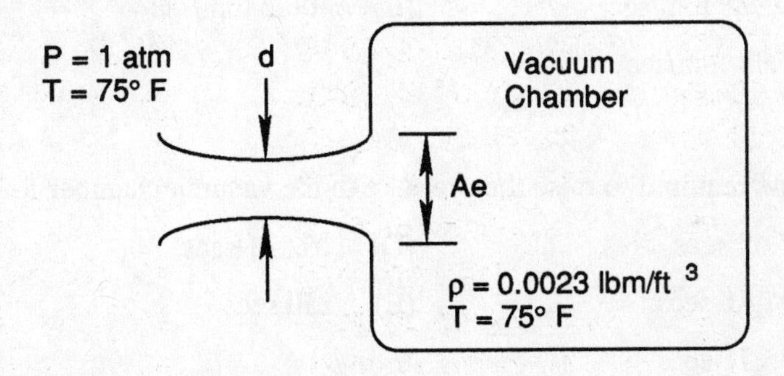

A vacuum chamber, shown in the figure above, is isolated by a thin gate located at the throat of the nozzle (*d*). Suddenly, the gate is opened and the chamber becomes connected to the atmosphere through a converging/diverging nozzle. The volume of the chamber is 450 ft^3, the width of the throat (*d*) is 0.04 ft^2, and the area of the nozzle at the chamber (*Ae*) is 0.14 ft^2.

1. The pressure in the chamber before the gate is opened is:

   (A) – 14.7 psia

   (B) – 1.6 psia

   (C) 0.455 psia

   (D) 7.0 psia

   (E) 102.2 psia

2. The speed of sound in the chamber before the gate is opened is:

(A) 204 ft/sec
(B) 309.2 ft/sec
(C) 24.1 ft/sec
(D) 1133 ft/sec
(E) 3307 ft/sec

3. The net force exerted on the gate prior to opening is: (positive value means force is acting left to right as shown in the figure, negative value means the force is acting right to left)

(A) – 4 lbf
(B) 82 lbf
(C) 1040 lbf
(D) – 820 lbf
(E) 1200 lbf

4. The mass flow rate of air into the chamber, while the flow is supersonic is:

(A) 0.0124 lbm/sec
(B) 0.821 lbm/sec
(C) 1.95 lbm/sec
(D) 21.4 lbm/sec
(E) 306.2 lbm/sec

5. The time required to raise the pressure in the vacuum chamber to 7 psia is:

(A) 7.62 secs
(B) 21.41 secs
(C) 35.21 secs
(D) 102.31 secs
(E) 250 secs

6. The initial Mach number at the chamber entrance (*Ae*) is:

(A) 0.8
(B) 1.0
(C) 1.8
(D) 2.4
(E) 2.8

7. What is the temperature of the air at the chamber entrance when air flow initially commences?

(A) – 252° F
(B) 321° F
(C) 75° F
(D) – 100° F
(E) 2.8° F

8. What is the stagnation pressure in the chamber when a shock wave sits at the chamber entrance?

   (A) – 14.7 psia (D) 7.0 psia

   (B) – 1.6 psia (E) 5.72 psia

   (C) 0.455 psia

9. What is the change in stagnation temperature across the shock wave as discussed in the previous question?

   (A) – 252° F (D) – 100° F

   (B) 321° F (E) 2.8° F

   (C) 0° F

10. What is the velocity of the air at the entrance to the chamber when the gate is initially opened?

   (A) 1979 ft/sec (D) 1133 ft/sec

   (B) 309.2 ft/sec (E) 3307 ft/sec

   (C) 24.1 ft/sec

QUESTIONS 11–15 refer to the following problem.

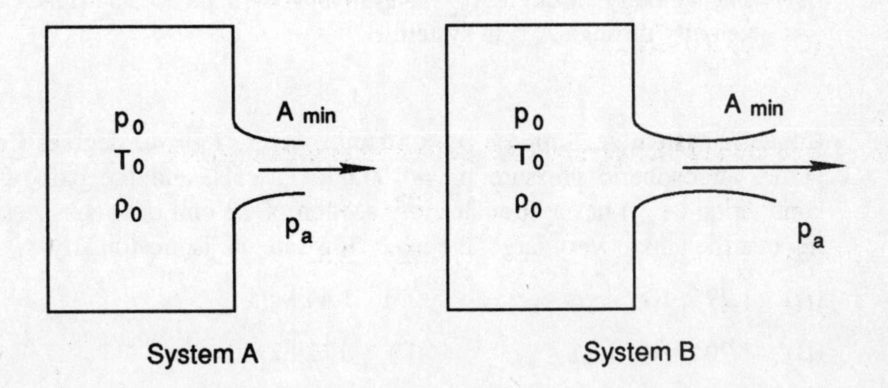

Systems A and B are identical except that system B has a conical section attached to its exit. Air is exhausting from the pressure vessels, and p_0, T_0, ρ_0, and p_a are known. Assume inviscid, isentropic flow (unless a shock is present).

The converging section in system A and the converging-diverging section in system B are axisymmetric (i.e. they have circular cross-sections everywhere).

11. If the flow in BOTH systems is *subsonic* everywhere, choose the correct statement:

    (A) The mass flow rate in system B will be GREATER THAN that in system A.

    (B) The mass flow rate in system B will be LESS THAN that in system A.

    (C) The velocity at the exit plane of system B will be LARGER THAN the velocity at the exit plane of system A.

    (D) A shock wave may be present in system B, but no shock wave can exist in system A.

    (E) In both systems, the velocity at A_{min} (i.e. at the minimum area location) will be IDENTICAL .

12. If the flow at A_{min} is *sonic* for both systems, choose the correct statement:

    (A) The mass flow rate in system B will be GREATER THAN that in system A.

    (B) The mass flow rate in system B will be EQUAL TO that in system A.

    (C) A shock wave may be present in system A, but no shock wave can exist in system B.

    (D) A shock wave cannot exist in either system.

    (E) The velocity through A_{min} in system A will be LESS THAN the velocity through A_{min} in system B.

13. Consider system A. Suppose p_0 = 5 atmospheres, T_0 = 40 degrees Centigrade, atmospheric pressure p_a = 100,000 Pascals, and the exit of the contraction (A_{min}) has a circular cross-section of 12 mm diameter. Assuming that the tank is very large, the mass flow rate, $\dot{m}$, is most nearly

    (A) 1.29×10^{-6} kg/s

    (B) 1.90×10^{-5} kg/s

    (C) 0.361 kg/s

    (D) 1.44 kg/s

    (E) 0.129 kg/s

14. Consider system B. Suppose p_0 = 500,000 Pascals, T_0 = 40 degrees Centigrade, the diameter of the throat is 12 mm, and the diameter at the exit plane is 24 mm. The maximum atmospheric pressure, p_a, which will en-

sure that the flow at the exit plane is supersonic is closest to

(A) 2,890 Pascals
(B) 14,900 Pascals
(C) 147,800 Pascals
(D) 256,200 Pascals
(E) 492,500 Pascals

15. Consider system B. Suppose p_0 = 500,000 Pascals, T_0 = 40 degrees Centigrade, the diameter of the throat is 12 mm, and the diameter of the exit plane is 24 mm. Suppose a normal shock wave forms right at the exit plane. The velocity *just upstream* of this shock is closest to

(A) 53 m/s
(B) 225 m/s
(C) 438 m/s
(D) 556 m/s
(E) 631 m/s

QUESTIONS 16–20 refer to the following problem.

The manometer below is riding in an elevator.

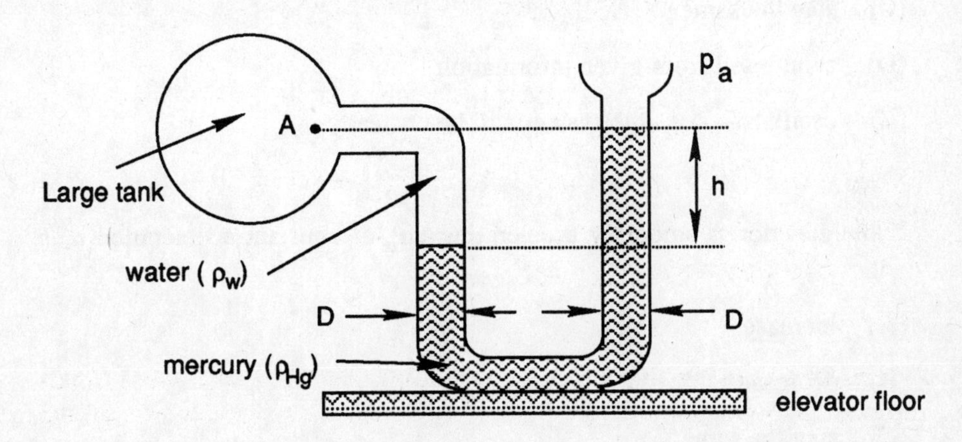

In the diagram shown above, the mercury in the right side of the U-tube rises a height h above the mercury in the left side of the U-tube. The height of point A and the height of the mercury on the right side of the U-tube are identical when the elevator is *stationary*.

Assuming constant atmospheric pressure p_a in the elevator, and the fact that the spherical tank is very large while not completely full of water, answer each of the following questions.

16. For a *stationary* elevator, the gage pressure at point A is

(A) $(\rho_{Hg} - \rho_w)\, gh$

(B) $\rho_{Hg}\, gh$

(C) $\rho_w\, gh$

(D) $p_a + \rho_{Hg}\, gh$

(E) $p_a + (\rho_{Hg} - \rho_w)\, gh$

17. For a stationary elevator the absolute pressure at point A is

(A) $(\rho_{Hg} - \rho_w)\, gh$

(B) $\rho_{Hg}\, gh$

(C) $\rho_w\, gh$

(D) $p_a + \rho_{Hg}\, gh$

(E) $p_a + (\rho_{Hg} - \rho_w)\, gh$

18. If the elevator is smoothly moving UP at constant speed V, h will

(A) increase

(B) decrease

(C) stay the same

(D) cannot tell from given information

(E) oscillate — i.e., the system will be unsteady

19. If the elevator is smoothly accelerating UP, at constant acceleration a, h will

(A) increase

(B) decrease

(C) stay the same

(D) cannot tell from given information

(E) oscillate — i.e., the system will be unsteady

20. If the elevator now starts accelerating uniformly upward with an acceleration of one g (i.e., $a = gk$), what will the new height H be, assuming that p_a and p_A remain constant? Note: H is the height for the accelerating case, while h is the original stationary height as indicated in the first figure.

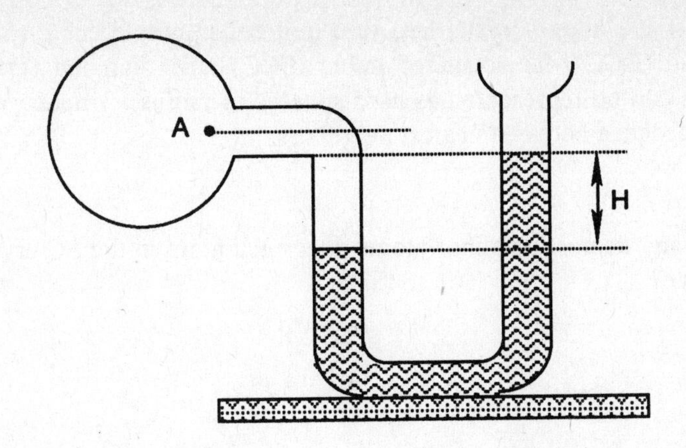

(A) $H = h$ (i.e., no change)

(B) $H = h/2$

(C) $H = 2h$

(D) $H = \dfrac{\rho_{Hg} h}{2\rho_{Hg} - \rho_w}$

(E) $H = \dfrac{1}{2\rho_{Hg} - \rho_w}(\rho_{Hg} h - p_a/g)$

QUESTIONS 21–30 refer to the following problem.

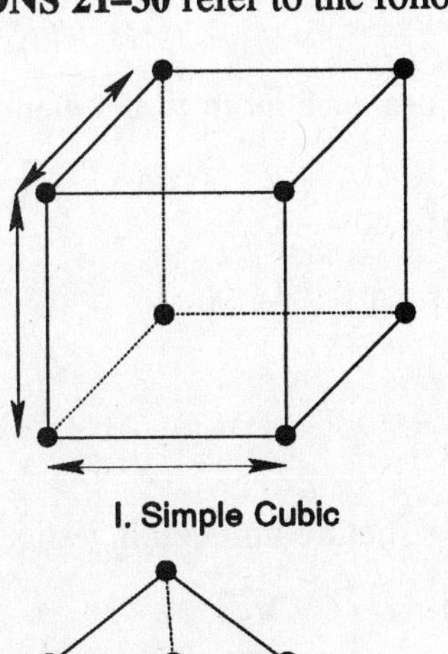

I. Simple Cubic

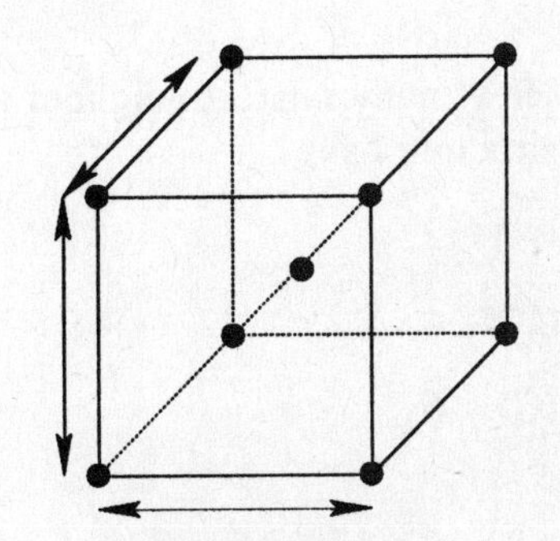

II. Body-Centered Cubic

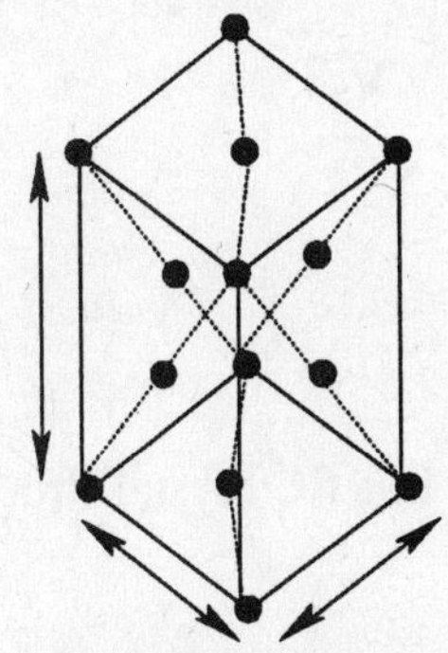

III. Face-Centered Cubic

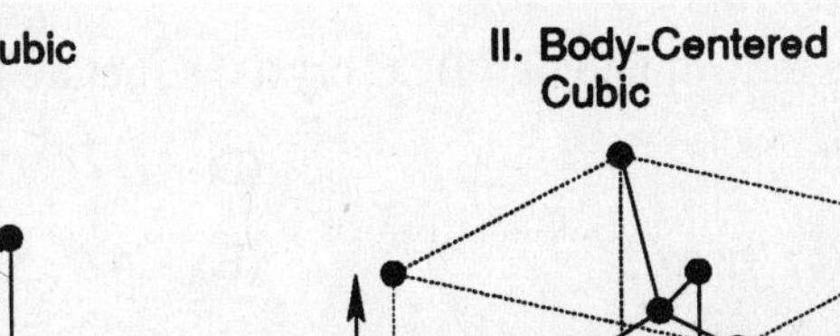

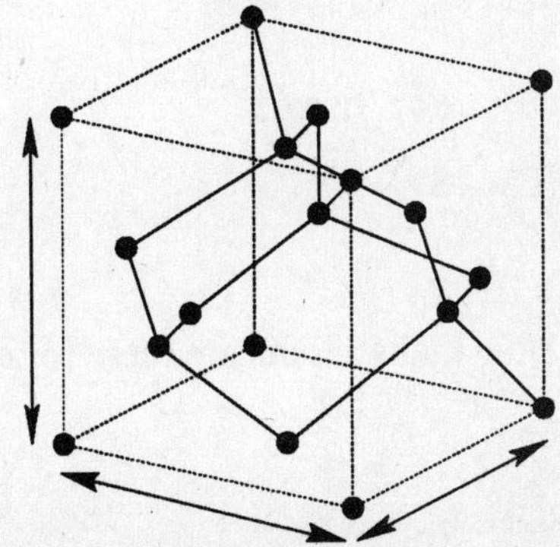

IV. Diamond

Consider the above crystal structure unit cells: simple cubic (SC), body-centered cubic (BCC), face-centered cubic (FCC), and diamond (D). Assume that all atoms can be represented as hard spheres of radius r where r is equal to one-half the distance between nearest neighbors.

21. How many nearest neighbor atoms does each atom in the SC crystal structure have?

(A) 2
(B) 4
(C) 6
(D) 8
(E) 12

22. How many nearest neighbor atoms does each atom in the FCC crystal structure have?

(A) 2
(B) 4
(C) 6
(D) 8
(E) 12

23. How many nearest neighbor atoms does each atom in the diamond crystal structure have?

(A) 2
(B) 4
(C) 6
(D) 8
(E) 12

24. What is the volume of a BCC crystal structure unit cell in terms of r?

(A) r^3
(B) $64r^3/3\sqrt{3}$
(C) $8r^3$
(D) $32r^3/\sqrt{2}$
(E) $4r^3/\sqrt{2}$

25. How many atoms are there for each unit cell in a SC crystal structure?

(A) 1
(B) 2
(C) 3
(D) 4
(E) 5

26. How many atoms are there for each unit cell in a BCC crystal structure?

(A) 1

(B) 2

(C) 5

(D) 7

(E) 9

27. How many atoms are there for each unit cell in a FCC crystal structure?

(A) 2

(B) 4

(C) 7

(D) 9

(E) 14

28. How many atoms are there for each unit cell in a diamond crystal structure?

(A) 4

(B) 8

(C) 12

(D) 14

(E) 16

29. In terms of the lattice constant a, what is the distance between nearest-neighbor atoms in the BCC crystal structure?

(A) $a/2$

(B) $\sqrt{3}a/2$

(C) $\sqrt{2}a/2$

(D) a

(E) $\sqrt{a}$

30. In terms of the lattice constant a, what is the distance between nearest-neighbor atoms in the FCC crystal structure?

(A) $a/2$

(B) $\sqrt{3}a/2$

(C) $\sqrt{2}a/2$

(D) a

(E) $\sqrt{2}a$

QUESTIONS 31–40 refer to the following problem.

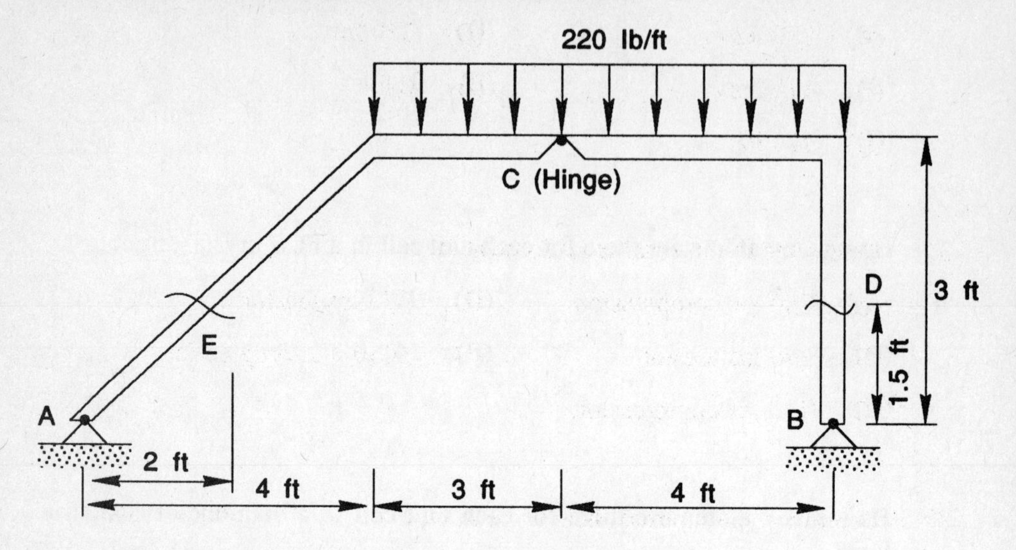

31. Determine the horizontal force in the hinge at point C.

    (A) 810 lbs
    (B) 880 lbs
    (C) 170 lbs
    (D) 0 lbs
    (E) 490 lbs

32. Determine the vertical reaction A_y.

    (A) 1540 lbs
    (B) 770 lbs
    (C) 490 lbs
    (D) 370 lbs
    (E) 1050 lbs

33. Determine the vertical reaction B_y.

    (A) 1540 lbs
    (B) 770 lbs
    (C) 490 lbs
    (D) 370 lbs
    (E) 1050 lbs

34. Determine the horizontal reaction B_x.

    (A) 880 lbs
    (B) 170 lbs
    (C) 770 lbs
    (D) 810 lbs
    (E) 0 lbs

35. Determine the vertical shear force in the hinge at point *C*.

(A) 880 lbs
(B) 170 lbs
(C) 770 lbs
(D) 370 lbs
(E) 0 lbs

36. Determine the internal axial force at *D*.

(A) 880 lbs compression
(B) 170 lbs tension
(C) 770 lbs compression
(D) 370 lbs tension
(E) 1050 lbs compression

37. Determine the internal shear force at *D*.

(A) 170 lbs
(B) 810 lbs
(C) 370 lbs
(D) 770 lbs
(E) 880 lbs

38. Determine the internal moment at *D*.

(A) 880 ft-lbs
(B) 0 ft-lbs
(C) 1220 ft-lbs
(D) 770 ft-lbs
(E) 1050 ft-lbs

39. Determine the internal axial force at point *E*.

(A) 940 lbs compression
(B) 650 lbs compression
(C) 100 lbs compression
(D) 430 lbs compression
(E) 370 lbs tension

40. Determine the internal moment at point *E*.

(A) 240 ft-lbs
(B) 980 ft-lbs
(C) 1540 ft-lbs
(D) 1220 ft-lbs
(E) 430 ft-lbs

QUESTIONS 41–50 refer to the following problem.

The Rankine cycle with reheat is shown below. Properties of the working fluid at fixed points of the cycle are represented in the following table.

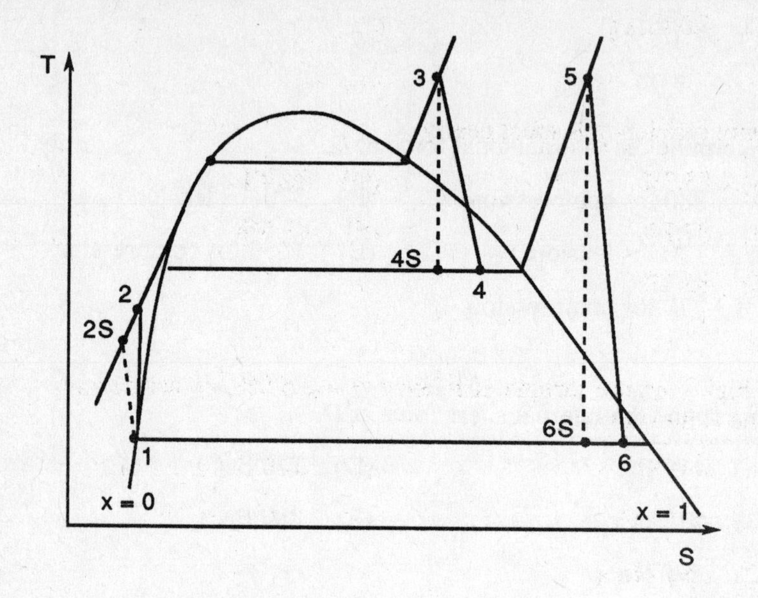

	P [kPa]	*v* [m³ / kg]	*h* [kJ / kg]
1	7.5	0.001	169
2*S*	5000	—	—
2	5000	—	175
3	—	—	3316
4*S*	—	—	2746
4	—	—	—
5	—	—	3378
6*S*	7.5	—	2477
6	7.5	—	2570

Also Note:

Process 1 – 2*S* is isentropic compression
Process 1 – 2 is actual compression
Processes 3 – 4*S* and 5 – 6*S* are isentropic expansion
Processes 3 – 4 and 5 – 6 are actual expansion

41. The isentropic pump work is most nearly

(A) 6.5 kJ/kg
(B) 6.0 kJ/kg
(C) 5.0 kJ/kg
(D) 4.5 kJ/kg
(E) 4.0 kJ/kg

42. Pump efficiency is most nearly

(A) 83.3%
(B) 87.1%
(C) 85.7%
(D) 90.0%
(E) 81.5%

43. If high pressure turbine efficiency η_T = 85.0%, an actual work of this turbine is most nearly

(A) 480 kJ/kg
(B) 475 kJ/kg
(C) 490 kJ/kg
(D) 495 kJ/kg
(E) 485 kJ/kg

44. Low pressure turbine efficiency is most nearly

(A) 89.1%
(B) 89.7%
(C) 88.6%
(D) 88.2%
(E) 79.9%

45. Heat transfer to the working fluid during the process of preheating, vaporization, and superheating is most nearly

(A) 3156 kJ/kg
(B) 3151 kJ/kg
(C) 3146 kJ/kg
(D) 3141 kJ/kg
(E) 3136 kJ/kg

46. Heat transfer to the working fluid during reheating process is most nearly

(A) 547 kJ/kg
(B) 549 kJ/kg
(C) 551 kJ/kg
(D) 553 kJ/kg
(E) 555 kJ/kg

47. An actual work of the low pressure turbine is most nearly

(A) 800 kJ/kg
(B) 804 kJ/kg
(C) 816 kJ/kg
(D) 812 kJ/kg
(E) 808 kJ/kg

48. Thermal efficiency of the Rankine cycle is most nearly

(A) 34.5%
(B) 34.7%
(C) 34.9%
(D) 35.1%
(E) 35.3%

49. If net turbine power output $\overline{W_T}$ = 100,000 kw, the required mass flow rate of working fluid is most nearly

(A) 78.0 kg/sec
(B) 77.7 kg/sec
(C) 77.4 kg/sec
(D) 77.1 kg/sec
(E) 78.3 kg/sec

50. If enthalpies of the cooling water at condenser inlet and outlet are h_{in} = 84 kj/kg and h_{out} = 124 kj/kg, the mass flow rate of cooling water is most nearly

(A) 4685 kg/sec
(B) 4678 kg/sec
(C) 4671 kg/sec
(D) 4664 kg/sec
(E) 4657 kg/sec

EXTRA QUESTIONS

ANSWER KEY

1.	(C)	14.	(C)	27.	(B)	40.	(A)
2.	(D)	15.	(E)	28.	(B)	41.	(C)
3.	(B)	16.	(A)	29.	(B)	42.	(A)
4.	(C)	17.	(E)	30.	(C)	43.	(E)
5.	(A)	18.	(C)	31.	(A)	44.	(B)
6.	(E)	19.	(B)	32.	(C)	45.	(D)
7.	(A)	20.	(D)	33.	(E)	46.	(A)
8.	(E)	21.	(B)	34.	(D)	47.	(E)
9.	(C)	22.	(D)	35.	(B)	48.	(C)
10.	(A)	23.	(A)	36.	(E)	49.	(B)
11.	(A)	24.	(B)	37.	(B)	50.	(D)
12.	(B)	25.	(A)	38.	(C)		
13.	(E)	26.	(B)	39.	(A)		

DETAILED EXPLANATIONS OF ANSWERS

Extra Practice Problems

1. **(C)**

Air can be treated as a perfect gas. Therefore the pressure, temperature and density are related by:

$$P = \rho RT$$

where P is in lbf/ft², ρ is in lbm/ft³, R is the "engineering" gas constant which changes for different species, and T is the *absolute* temperature in Rankines. For air,

$$R = 53.34 \frac{\text{ft} \cdot \text{lbf}}{\text{lbm} \cdot {}^{\circ}\text{R}}$$

Inserting the values given for temperature and pressure and noting that:

$$1 \text{ lbf/ft}^2 = 144 \text{ psia}$$

and $\quad T(°R) = T(°F) + 459.9$

yields $\quad P = 0.455$ psia.

2. **(D)**

The speed of sound in a perfect gas is determined to be a function of the temperature only. In this problem, even though the vacuum chamber has very low density and pressure, its temperature is the same as that outside of the chamber. The equation for sound speed in a perfect gas is

$$c = \sqrt{g_c kRT}$$

where c is in ft/sec, g_c is the unit conversion between lbm and lbf, k is the ratio of specific heats and R and T are as defined above. More specifically,

$$g_c = 32.1739 \frac{\text{ft} \cdot \text{lbm}}{\text{lbf} \cdot s^2}$$

Using a temperature of 75° F and $k = 1.4$ for air yields:

$$c = 1133 \text{ ft/sec}$$

Remember to use absolute temperature in your calculations.

3. **(B)**

Pressure is force per unit area. The net force will be determined by calculating the net pressure difference across the gate and multiplying by the surface area of the gate.

Therefore:

$$F_{net} = (P_{atm} - P_{chamber}) * A_{gate}$$
$$= (14.7 \text{ lbf/in}^2 - 0.455 \text{ lbf/in}^2) * 144 \text{ in}^2/\text{ft}^2 * 0.04 \text{ ft}^2$$
$$= 82.05 \text{ lbf}$$

Note that the direction of the force is from left to right because the value is positive.

4. **(C)**

To determine the mass flow rate into the chamber, it is first necessary to determine if the flow is supersonic. The pressure at a point in compressible flow is related to the stagnation pressure by the following equation which is based on isentropic flow:

$$\frac{P_0}{P} = \left(1 + \frac{k-1}{2} M^2\right)^{k/(k-1)}$$

where M is the Mach number and k is the ratio of specific heats (1.4 for air). For small pressure differences, the Mach number of the flow at a point will be small. However, when the Mach number is 1 (sonic flow) the pressure difference between a point and stagnation is:

$$\frac{P_0}{P} = (1 + 0.2)^{3.5} = 1.89$$

So (inverting the above fraction) if

$$\frac{P_{low}}{P_{high}} < 0.528,$$

then the flow through the nozzle becomes supersonic at some point. Note, the pressure difference can be less and still have supersonic flow, so this is a sufficient, but not necessary condition for supersonic flow. For this problem,

$$\frac{P_{low}}{P_{high}} = \frac{0.455 \text{ lbf/in}^2}{14.7 \text{ lbf/in}^2} << 0.528,$$

so supersonic flow will result. When this is the case, the flow is "choked" and the mass flow rate through the throat is determined by Fleigner's formula, namely:

$$M = 0.532 \times A^* \times \frac{P_0}{T_0}$$

Where P_0 is the upstream stagnation pressure, T_0 is the upstream stagnation temperature and $A*$ is the area of the throat. Again, the temperature must be absolute! Therefore:

$$M = 0.532\ \frac{\text{lbm}\cdot{}^\circ\text{R}}{\text{lbf}\cdot\text{s}} \times 0.04\,\text{ft}^2 \times \frac{14.7\ \text{lbf/in}^2 * 144\ \text{in}^2/\text{ft}^2}{(75+459.9)^\circ\text{R}}$$

$$= 1.95\ \text{lbm/sec}$$

5. **(A)**

Enough information is available to determine the amount of mass inside the chamber both at the start and when the pressure rises to 7 psia. Given the initial density of air in the chamber and the chamber volume, yields immediately the amount of mass in the chamber:

$$\text{Mass}_{\text{initial}} = \rho \times \text{Vol} = 0.0023\ \text{lbm/ft}^3 \times 450\ \text{ft}^3 = 1.035\ \text{lbm}$$

When the pressure has risen to 7 psia, the new density can be determined from the perfect gas law such that:

$$\rho_{\text{final}} = \frac{P}{RT} = \frac{7\ \text{lbf/in}^2 \times 144\ \text{in}^2\,\text{ft}^2}{53.34\ \frac{\text{ft}\cdot\text{lbf}}{\text{lbm}\cdot{}^\circ\text{R}} \times (75+459.9)^\circ\text{R}} = 0.0353\ \text{lbm/ft}^3$$

so:

$$\text{Mass}_{\text{final}} = \rho \times \text{Vol} = 0.0353\ \text{lbm/ft}^3 \times 450\ \text{ft}^3 = 15.89\ \text{lbm}$$

Therefore 15.89 – 1.035 lbm of air enter the chamber. The time required for this is:

$$\text{Time} = \frac{(15.89 - 1.035\ \text{lbm})}{1.95\ \text{lbm/sec}} = 7.62\ \text{sec}$$

Note: Please see the explanation to problem 4 for additional information.

6. **(E)**

Assuming an isentropic acceleration, the relationship between Mach number and area is determined through the isentropic relations for compressible flow. Most text books on the subject will have the relations tabulated and one needs only to look up the value from the table. Here, the appropriate relation is:

$$\frac{Ae}{d} = \frac{1}{Me}\left(\left(\frac{2}{k+1}\right)\left(1 + \frac{k-1}{2}Me^2\right)\right)^{(k+1)/2(k-1)}$$

where Me is the Mach number at the chamber entrance and k is the ratio of specific heats (1.4). Solving here for Me yields:

$$\frac{Ae}{d} = 3.5 \Rightarrow Me = 2.8$$

7. **(A)**

The temperature will be determined using isentropic relations, as was the cause in the previous problem. Here, the relation is:

$$\frac{T_0}{Te} = 1 + \frac{k-1}{2} Me^2$$

where the stagnation temperature is in absolute degrees. For $Me = 2.8$ and $T_0 = 534.7°R$, the temperature becomes:

$$Te = 208°R$$

$$= 208°R - 459.9 = -252° \text{ F}$$

8. **(E)**

A shock wave changes the stagnation pressure of a fluid. Normal shock relations for air are normally tabulated. However, expressions are also available for the change in various parameters across the shock. Here, the expression is:

$$\frac{P_{ox}}{P_{oy}} = \left(\left[\frac{2k}{k+1} M_x^2 - \frac{k-1}{k+1}\right]^{1/(k-1)}\right) / \left(\frac{(k+1)M_x^2/2}{1+(k-1)M_x^2/2}\right)$$

which could be solved to provide the answer. However, using normal shock tables will normally be a much easier way to determine the change in stagnation pressure.

If the shock occurs at the entrance to the chamber, the upstream Mach number at that point will be 2.8 (M_x), since $k = 1.4$. Then solving the above equation gives:

$$\frac{P_{ox}}{P_{oy}} = 2.568$$

and therefore $P_{oy} = 14.7$ psia / 2.568 = 5.72 psia.

9. **(C)**

Since no energy is added to or taken from the flow as it is accelerated through the nozzle and across the shock, there is no change in stagnation temperature which is a measure of the total energy content of the gas. Note, shock waves are highly dissipative and energy is transferred between kinetic and internal, but the total energy does not change.

10. **(A)**

The mach number and temperature at the entrance to the chamber when the gate is initially opened have been determined in earlier problems. In problem 6, Me was found to be 2.8. In problem 7, the temperature was found to be $Te = 208°R$. Therefore the speed of sound is:

$$c = \sqrt{g_c kRt} = \sqrt{32.1739 \frac{\text{lbm} \cdot \text{ft}}{\text{lbf} \cdot \text{s}^2} \times 1.4 \times 53.34 \frac{\text{lbf} \cdot \text{ft}}{\text{lbm} \cdot {}^\circ\text{R}} \times 208\,{}^\circ\text{R}}$$

$$= 706.9 \text{ ft/sec}$$

for $Me = 2.8$, this makes $Ve = 2.8 \times 7.06.9$ ft/sec = 1979 ft/sec.

11. **(A)**

For subsonic flow, the conical expanding duct in system B will result in an increase in the mass flow rate. This is due to the fact that the pressure at the exit plane of a subsonic jet is equal to the surrounding ambient pressure, p_a. Thus, in system B, the pressure at the throat is LESS THAN p_a, while in system A, the pressure at the throat is EQUAL TO p_a. For subsonic flow this implies that the velocity at the throat in system B is greater than the velocity at the throat at system A. In follows that since the area a_{min} is the same in both cases, system B will have a larger mass flux.

12. **(B)**

Once the flow is sonic, the system is called "choked." Upstream of the sonic throat, everything must be identical in the two systems. Hence, the mass flow rate will be the same.

Another way of analyzing the problem is that once the flow at the throat is sonic, further changes in downstream conditions will have no effect on the mass flow rate. Thus, it does not matter if there is a conical section in system B.

13. **(E)**

For $p_0/p_a = 5.0$, the flow is most definitely choked. Hence the mass flow rate is a maximum, given by

$$\dot{m} = \dot{m}_{max} = \frac{0.6847\, p_0 A^*}{(RT_0)^{1/2}}$$

for air. Here,

$$p_0 = 5.0\,(100{,}000) = 500{,}000 \text{ P}_a = 500{,}000 \text{ N/m}^2$$

$$A^* = A_{min} = {}^{\pi}/_4\,(0.012 \text{ m})^2 = 1.131 \times 10^{-4} \text{ m}^2$$

$$R = 287 \text{ m}^2/(\text{s}^2{\cdot}\text{K}) \text{ for air}$$

$$T_0 = 40° \text{ C} + 273.15 = 313.15 \text{ K}$$

Thus,

$$\dot{m} = \frac{(0.6847)(500{,}000 \text{ N/m}^2)(1.131 \times 10^{-4} \text{ m}^2)}{[(287 \text{ m}^2/\text{s}^2 \cdot \text{K})(313.15 \text{ K})]^{1/2}} \frac{\text{kg m}}{\text{s}^2\text{N}}$$

$$\dot{m} = 0.129 \text{ kg/s}$$

14. **(C)**

For supersonic isentropic flow with no shock waves inside the diverging part of the duct, atmospheric pressure must be less than or equal to the pressure which will cause a shock to form right at the exit plane.

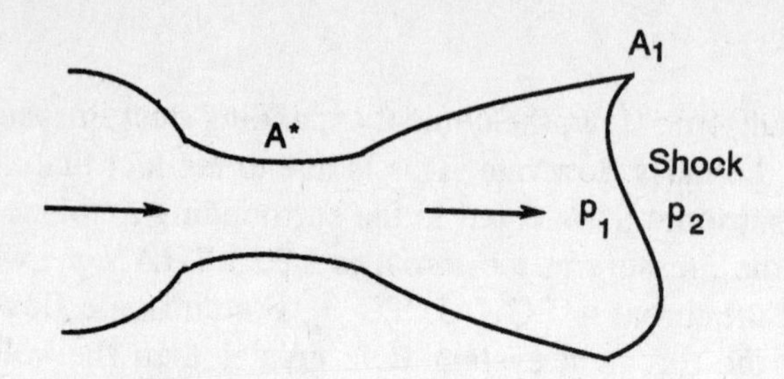

For $A_1 / A^* = D_1^2 / D^{*2} = 24^2/12^2 = 4.0,$

$p_1 / p_0 = 0.0298,$

and $M_{a_1} = 2.94$

(This can be found either from the isentropic equations or from the isentropic flow tables for air).

Across the shock, @

$$M_{a_1} = 2.94, \quad p_2/p_1 = 9.9175$$

(This can be found either from the normal shock equations or the shock tables for air). Thus,

$$p_a = p_2 = \left(\frac{p_2}{p_1}\right)\left(\frac{p_1}{p_0}\right)p_0 = (9.9175)\,(0.0298)\,(500{,}000P_a)$$

or $p_a = 147{,}800\ P_a$

Note: For any p_a less than this value, the shock will move downstream, i.e., out of the nozzle.

15. **(E)**

The conditions here are identical to those of the previous question. Hence,

$$M_{a_1} = 2.94, \quad p_1/p_0 = 0.0298$$

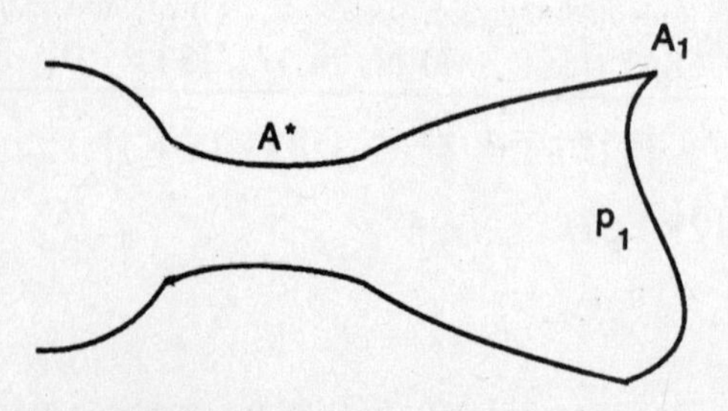

Also, from the isentropic tables,

$$T / T_0 = 0.3665.$$

Then, since

a = speed of sound $\sqrt{\alpha RT}$ for a perfect gas,

$$\frac{a_1}{a_0} = \sqrt{0.3665} = 0.6054$$

Finally, $V_1 = M_{a_1} \cdot a_1$

Hence,

$$V_1 = M_{a_1} \frac{a_1}{a_0} a_0 = M_{a_1} \left(\frac{a_1}{a_0}\right) \sqrt{\alpha RT_0}$$

$$V_1 = (2.94)(0.6054)\left[(1.4)(287 \text{ m}^2/\text{s}^2 \cdot \text{K})(313.15 \text{ K})\right]^{1/2}$$

$$V_1 = 631.4 \text{ m/s}$$

16. **(A)**
For a stationary manometer, the fluid statics equations are valid.

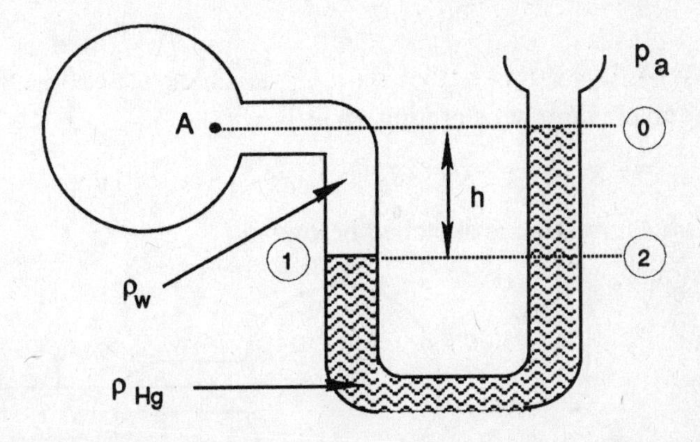

Assuming the fact that the water level is higher than the point A, we state that:

At point 0, the pressure is equal to p_a.

At points 1 and 2, the pressures must match since there is a continuous path from 1 to 2 in stationary mercury, and 1 and 2 are at the same elevation. Thus,

$$p_t = p_A \rho_w gh = p_2 = p_a + \rho_{Hg} gh$$

Gage Pressure at

$$A = p_A - p_a = (\rho_{Hg} - \rho_w)\, gh.$$

17. **(E)**
Absolute pressure is equal to gage pressure plus atmospheric pressure.

Thus

$$p_A = p_a + (\rho_{Hg} - \rho_w)\, gh.$$

18. **(C)**

When the elevator is moving up *at constant speed*, there is no acceleration other than gravity. Thus, there can be no change in manometer fluid height.

19. **(B)**

When the elevator is *accelerating* up, there is an acceleration force which adds to gravity and makes the system "feel" heavier. A good analogy, is when a person is accelerating upward in an elevator he feels heavier. Mathematically, for rigid body linear acceleration, g can be replaced by $G = g - a$ where a is the acceleration vector. Then, the fluid statics equations can be used except substitute G in place of g.

Here $G = g - a$ and a is up while g is down. Thus, G, the magnitude of G will be greater than g, the magnitude of g. Thus, a shorter column height of mercury will be able to measure the same pressure difference. Hence h will decrease.

20. **(D)**

For an elevator accelerating up at

$$a = -g, G = g - a = 2g.$$

This vector summation is sketched below

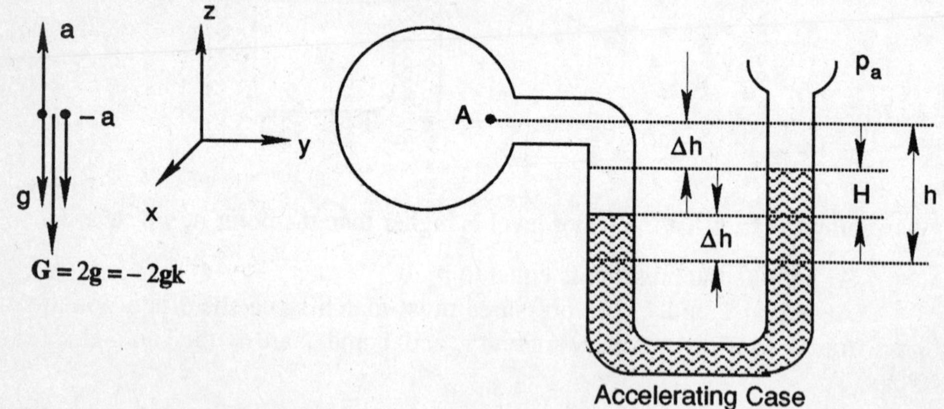

Accelerating Case

Note that since both manometer branches are of equal diameter, a downward shift Δh on the right side must cause an upward shift of the mercury by the same amount (Δh) to conserve mass.

Statics equations with G substitutes for g

$$p_A + (H + \Delta h)\, \rho_w G = p_a + H\rho_{Hg}\, G$$

Since the tank is assumed to be very large, we can neglect the pressure increase of point A due to the slight change in water level inside the tank. Therefore, as we know from the above question that, in terms of h,

$$p_A - p_a = (\rho_{Hg} - \rho_w)\, gh,$$

Also, $G = 2g$ and $\Delta h = {}^1/_2(h - H)$

Hence

$$2H\rho_{Hg}g - 2H\rho_w g - 2\,({}^1/_2)\,(h - H)\,\rho_w g = h\rho_{Hg}\,g - h\rho_w\,g$$

or
$$H = \frac{\rho_{Hg}\,h}{2\rho_{Hg} - \rho_w}$$

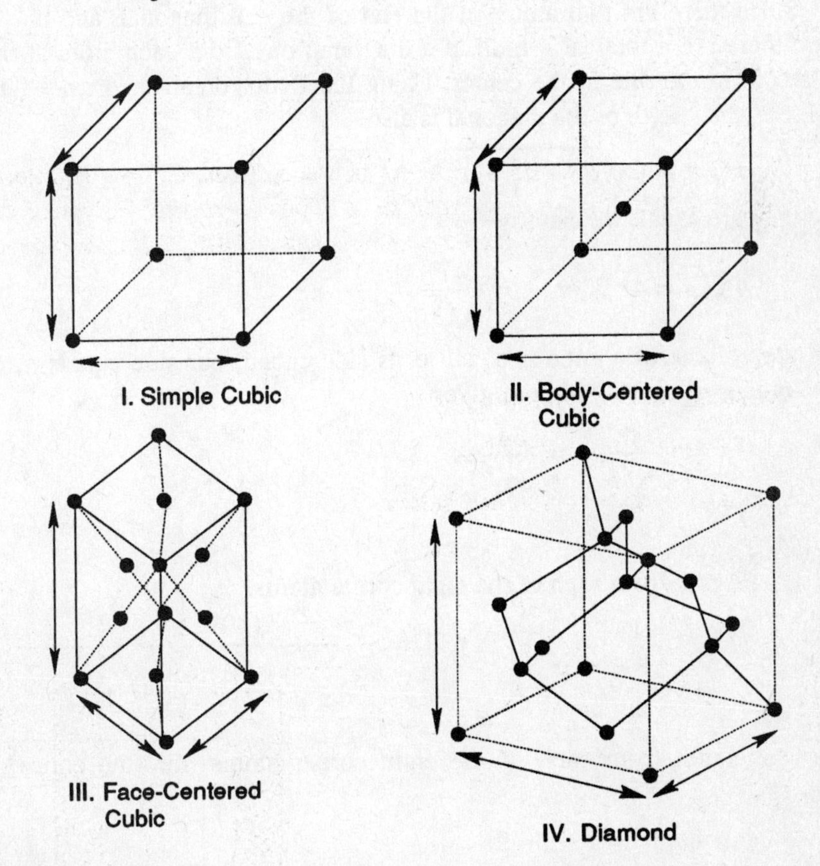

21. **(B)**
Nearest neighbor atoms are located above, below, in front of, in back of, to the right of, and to the left of any single atom.

22. **(D)**
Nearest neighbor atoms of the atom on the front face of a unit cell are located on the corners of the front face (4), the faces of the plane located $a/2$

behind the front face (4), and the faces of the plane located $a/2$ in front of the front face (4).

23. **(A)**

(A) is correct because the nearest neighbor atoms are located in directions indicated by the drawing illustrating the diamond crystal structure unit cell. Remember a carbon atom has to form 4 bonds. For these reasons, all other choices are incorrect.

24. **(B)**

Since there are two atoms at the end of the cell diagonal, and one in the center, there are a total of 4 radii in a diagonal; one from each atom at the end and two from the one in the center. Using the Pythagorean theorem with sides equal to a, the length of the diagonal is also

$$a\sqrt{3} = (\sqrt{a^2 + a^2 + a^2} = \sqrt{3a^2} = \sqrt{3}\,a).$$

Since they are equal, we can say

$$a\sqrt{3} = 4r\,, \quad \text{or} \quad a = \frac{4r}{\sqrt{3}}.$$

Since the volume of a cube is equal to its side cubed, our side equals a, hence, the volume is a^3. Substituting in gives

$$a^3 = \left(\frac{4r}{\sqrt{3}}\right)^3 = \frac{64\,r^3}{3\sqrt{3}}.$$

25. **(A)**

One-eighth from each of the eight corner atoms.

$(8 \times {}^1/_8 = 1)$

26. **(B)**

One-eighth from each of the eight corner atoms plus the entire center atom.

$(1 + 8 \times {}^1/_8 = 1 + 1 = 2)$

27. **(B)**

One-eighth from each of the eight corner atoms plus one-half from each of six face atoms.

$(8 \times {}^1/_8 + 6 \times {}^1/_2 = 1 + 3 = 4)$

28. **(B)**

One-eighth from each of the eight corner atoms plus one-half from each of six face atoms plus four entire atoms contained within the unit cell.

$$(8 \times {}^1/_8 + 6 \times {}^1/_2 + 4 \times 1 = 1 + 3 + 4 = 8)$$

29. **(B)**
As was determined in problem 24, the length of the diagonal is

$$\sqrt{3}\,a,$$

hence half of that distance is $\sqrt{3}\,\frac{a}{2}$.

30. **(C)**
Here we need the length of a face diagonal. Using the Pythagorean theorem we get a value of

$$\sqrt{a^2 + a^2} = \sqrt{2a^2} = \sqrt{2}\,a.$$

Half of this value would be $\sqrt{2}\,\frac{a}{2}$.

31. **(A)**
To determine C_x draw a free body diagram of part BC and sum forces in the horizontal direction.

$$\Sigma F_x = 0$$

$$C_x - 813 \text{ lb} = 0$$

$$C_x = 813 \text{ lbs}$$

220 lb/ft

C_x

C_y

3 ft

B

$B_x = 813$ lb

4 ft

$B_y = 1050$ lb

The frame becomes statically indeterminate when the reaction B is fixed. When reaction B is fixed there are more than six unknown reactive components. This requires more than the six possible statics equations.

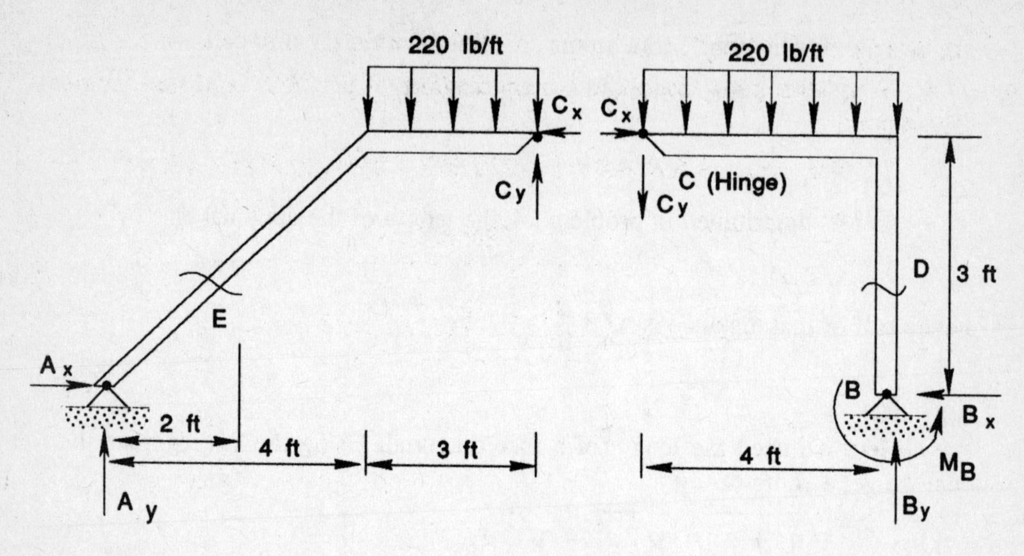

32. **(C)**
Draw a free body diagram of the entire structure and sum moments about point B.

$$\Sigma M_B = 0; \quad 11 \text{ ft } A_y - 7 \text{ ft} \times \ 220 \text{ lb/ft} \times 3.5 \text{ ft} = 0$$

$$A_y = 490 \text{ lbs}$$

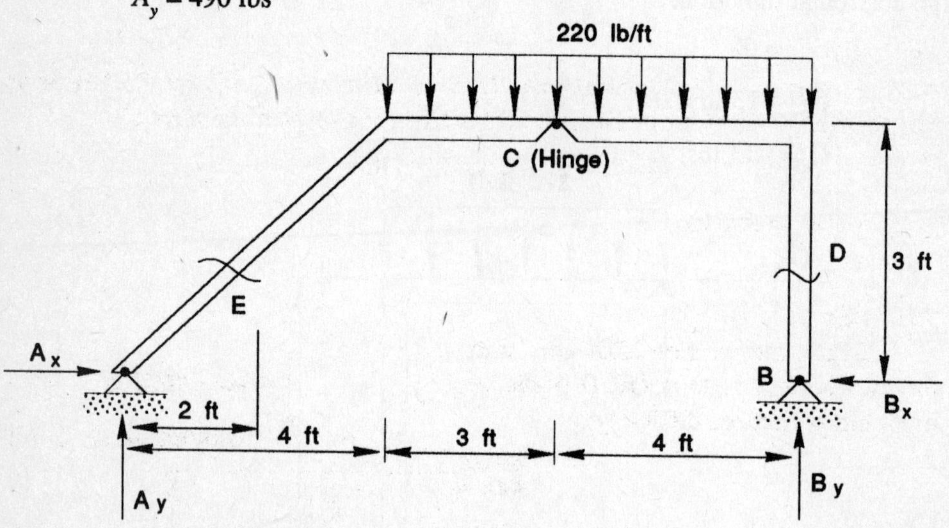

33. **(E)**
Sum forces in the vertical direction using the free body diagram of the entire structure.

$$\Sigma F_y = 0; \quad 490 \text{ lb} + B_y - 7 \text{ ft} \times 220 \text{ lb/ft} = 0$$

$$B_y = 1050 \text{ lbs}$$

34. **(D)**

To determine B_x draw a free body diagram of part BC and sum moments about point C.

$\Sigma M_C = 0; \quad 3\text{ ft } B_x + 4\text{ ft} \times \ 220\text{ lb/ft} \times 2\text{ ft} - 1050\text{ lb} \times 4\text{ ft} = 0$

$B_x = 813\text{ lbs}$

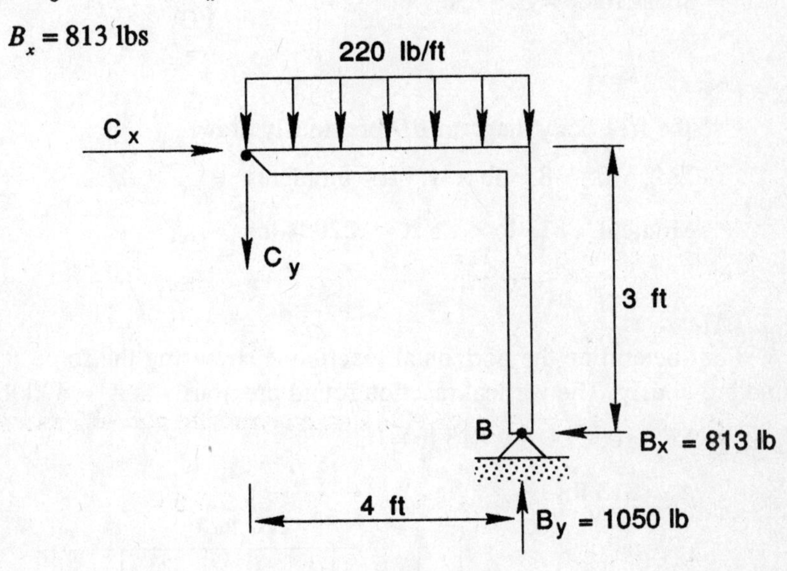

35. (B)

To determine the vertical shear force in the hinge C, we use the free body diagram of the previous problem and sum forces in the vertical direction.

$\Sigma F_y = 0; \quad 1050\text{ lb} - C_y - 4\text{ ft } \times 220\text{ lb/ft} = 0$

$C_y = 170\text{ lbs}$

36. **(E)**

Draw the portion BD and sum forces. The left portion $AECD$ could be used, but with more difficulty.

Axial = 1050 lb
Moment = 1220 ft-lbs
Shear 813 lb
D
1.5 ft

$\Sigma F_y = 0; \quad 1050\text{ lb} - \text{Axial} = 0$

Axial force = 1050 lbs compression

B
Bx = 813 lb
By = 1050 lb

37. **(B)**

Use the free body diagram *BD* previously drawn.

$\Sigma F_x = 0;$ shear − 813 lb = 0

Shear force = 813 lbs

38. **(C)**

Use the free body diagram *BD* previously drawn.

$\Sigma M_D = 0;$ 813 lb × 1.5 ft − moment = 0

Moment = 813 lb × 1.5 ft = 1220 ft-lbs

39. **(A)**

First determine the horizontal reaction A_x by using the force B_x = 813 lbs found previously. The vertical reaction found previously is A_y = 490 lbs.

$\Sigma F_x = 0;$ $A_x -$ 813 lb = 0

A_x = 813 lbs

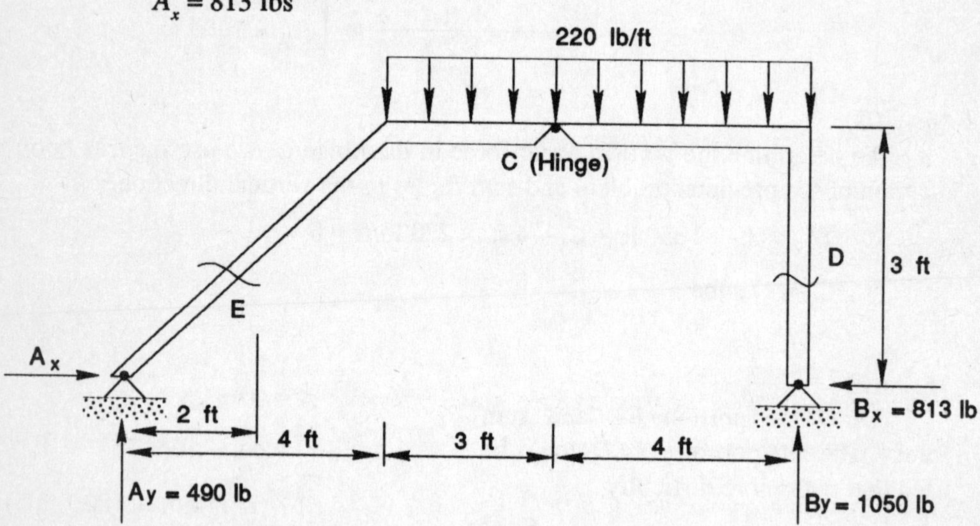

To determine the internal axial force at point *E*, draw a free body diagram of part *AE*. Resolve the reactions (A_x = 813 lbs, A_y = 490 lbs) along and transverse to the part and sum forces.

$\Sigma F = 0 \nearrow;$ $^4/_5$ (813 lb) + $^3/_5$ (490 lb) − Axial = 0

Axial force is = 944 lbs compression

40. **(A)**

Use the free body diagram of part *AE* of the previous problem and sum moments about point *E* using the reactions A_x and A_y.

$\Sigma M_E = 0;$ Moment $- 813$ lb $\times$ 1.5 ft $+$ 490 lb $\times$ 2 ft $= 0$

Moment $= 813$ lb $\times$ 1.5 ft $-$ 490 lb $\times$ 2 ft $= 240$ ft-lbs

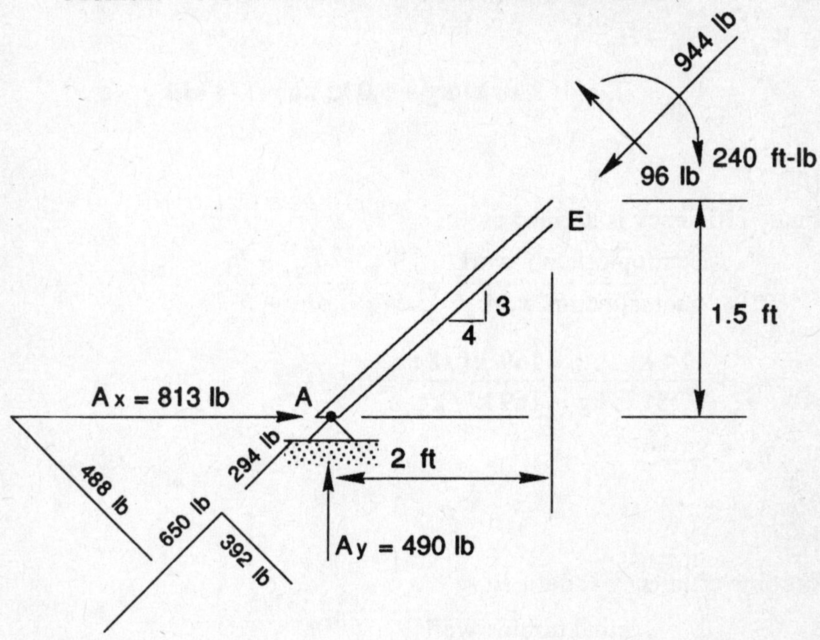

41. **(C)**

Starting with the elementary relation

$$Tds = dh - v\,dp \tag{1}$$

Realizing that for an isentropic process ds = 0

$$0 = dh - v\,dp$$

$$dh = v\,dp \tag{2}$$

From an energy balance (ignoring kinetic and potential effects)

$$W = dh \tag{3}$$

Comparing equations (2) and (3)

$$W = v\,dp$$

The isentropic pump work, W_{sp} can now be determined from the available information

$$W = v\,dp$$

$$W_{sp} = v_1\,(P_2 - D_1)$$

$$W_{sp} = (.001\text{m}^3/\text{kg})(5000\text{ kPa}-7.5\text{ kPa})$$

$$\left(\frac{1000\text{ N/m}^2}{\text{kPa}}\right)\left(\frac{\text{J}}{\text{N}\cdot\text{m}}\right)\left(\frac{\text{KJ}}{1000\text{ J}}\right)$$

$$W_{sp} = 5.0 \text{ kJ/kg}$$

Additionally

$$W_{sp} = H_{2s} - H_1$$

$$H_{2s} = W_{sp} + H_1 = 169.10 \text{ kJ/kg} + 5.0 \text{ kJ/kg} = 174 \text{ kJ/kg}$$

42. **(A)**

Pump efficiency is defined as

$$\eta_P = \frac{\text{isentropic pump work}}{\text{actual pump work}} = \frac{W_{sp}}{W_P} = \frac{h_{2s} - h_1}{h_2 - h_1}$$

Then:

$$\eta_P = \frac{174 \text{ kJ/kg} - 169 \text{ kJ/kg}}{175 \text{ kJ/kg} - 169 \text{ kJ/kg}} = 0.833$$

$$\eta_P = 83.3\%$$

43. **(E)**

Turbine efficiency is defined as

$$\eta_T = \frac{\text{actual turbine work}}{\text{isentropic turbine work}} = \frac{W_T}{W_{ST}}$$

The actual work of the high pressure turbine

$$W_{T1} = W_{ST} \cdot \eta_T = (h_3 - h_{4s}) \cdot \eta_T \text{ kJ/kg}$$

where: W_{ST} = isentropic work of the high pressure turbine.

η_T = turbine efficiency

$$W_{T1} = (3316 \text{ kJ/kg} - 2746 \text{ kJ/kg}) \times 0.85 = 485 \text{ kJ/kg}$$

Then, at point 4:

$$h_4 = h_3 - W_{T1} = 3316 \text{ kJ/kg} - 485 \text{ kJ/kg} = 2831 \text{ kJ/kg}$$

44. **(B)**

Turbine efficiency is defined as

$$\eta_T = \frac{\text{actual turbine work}}{\text{isentropic turbine work}} = \frac{W_{T2}}{W_{ST}} = \frac{h_5 - h_6}{h_5 - h_{6S}}$$

Then:

$$\eta_T = \frac{3378 \text{ kJ/kg} - 2570 \text{ kJ/kg}}{3378 \text{ kJ/kg} - 2477 \text{ kJ/kg}} = 0.897$$

$$\eta_T = 89.7\%$$

45. **(D)**

Heat transfer to the working fluid during the process of preheating, vaporization, and superheating (process 2 – 3):

$$Q'_1 = h_3 - h_2 = 3316 \text{ kJ/kg} - 175 \text{ kJ/kg} = 3141 \text{ kJ/kg}$$

46. **(A)**

Heat transfer to the working fluid during reheating process 4 – 5:

$$Q''_1 = h_5 - h_4 = 3378 \text{ kJ/kg} - 2831 \text{ kJ/kg} = 547 \text{ kJ/kg}$$

47. **(E)**

Actual work of the low pressure turbine (process 5 – 6):

$$W_{T1} = h_5 - h_6 = 3378 \text{ kJ/kg} - 2570 \text{ kJ/kg} = 808 \text{ kJ/kg}$$

48. **(C)**

Thermal efficiency of the Rankine cycle:

$$\eta_t = \frac{W_{NET}}{Q_1} = \frac{(W_{T1} + W_{T2}) - W_P}{Q'_1 + Q''_1} = \frac{(W_{T1} + W_{T2}) - (h_2 - h_1)}{Q'_1 + Q''_1}$$

$$\eta_t = \frac{(485 \text{ kJ/kg} + 808 \text{ kJ/kg}) - (175 \text{ kJ/kg} - 169 \text{ kJ/kg})}{3141 \text{ kJ/kg} + 547 \text{ kJ/kg}} = 0.349$$

$$\eta_t = 34.9\%$$

49. **(B)**

The required mass flow rate of working fluid to develop net turbine power output:

$$\overline{m} = \frac{\text{Net Turbine Power Output}}{\text{Net work of 1 kg of the working fluid}} \text{ [kg/sec]}$$

Net turbine power output:

$$\overline{W_T} = 100{,}000 \text{ kw}$$

Net work developed by 1 kg of the working fluid:

$$W_{NET} = (W_{T1} + W_{T2}) - W_p \text{ kJ/kg}$$

$$W_{NET} = (485 \text{ kJ/kg} + 808 \text{ kJ/kg}) - (175 \text{ kJ/kg} - 169 \text{ kJ/kg}) = 1287 \text{kJ/kg}$$

$$\overline{m} = \frac{W_T}{W_{NET}} = \frac{(100{,}000 \text{ kw})\left(\frac{1 \text{ kJ/s}}{1 \text{ kw}}\right)}{1287 \text{ kJ/kg}} = 77.7 \text{ kg/sec}$$

50. **(D)**

Heat rejected form the working fluid to the surroundings in condensor (process 6 – 1):

$$\overline{Q}_2 = \overline{m}(h_6 - h_1) = 77.7 \text{ kg/s } (2570 \text{ kJ/kg} - 169 \text{ kJ/kg})$$

$$= 186{,}558 \text{ kJ/sec}$$

The same amount of heat is absorbed by the cooling water in condenser. Therefore, mass flow rate of cooling water can be calculated using the expression:

$$\overline{m}_w = \frac{\overline{Q}_2}{h_{out} - h_{in}} == \frac{186{,}558 \text{ kJ/s}}{124 \text{ kJ/kg} - 84 \text{ kJ/kg}} = 4664 \text{ kg/sec}$$